Acta Numerica 2007

Acta

Numerica

Volume 16   2007

CAMBRIDGE UNIVERSITY PRESS
Cambridge, New York, Melbourne, Madrid, Cape Town,
Singapore, São Paulo, Delhi, Tokyo, Mexico City

Cambridge University Press
The Edinburgh Building, Cambridge CB2 8RU, UK

Published in the United States of America by Cambridge University Press, New York

www.cambridge.org
Information on this title: www.cambridge.org/9780521174343

First published 2007
First paperback edition 2011

A catalogue record for this publication is available from the British Library

ISBN 978-0-521-87743-5 Hardback
ISBN 978-0-521-17434-3 Paperback

Contents

Contents

Acta Numerica (2007), pp. 1–65
doi: 10.1017/S0962492906280012

Molecular dynamics and the accuracy of numerically computed averages

Stephen D. Bond
Department of Computer Science,
University of Illinois,
Urbana, IL 61801-2302, USA

Benedict J. Leimkuhler
School of Mathematics,
University of Edinburgh,
Edinburgh EH9 3JZ, UK

Molecular dynamics is discussed from a mathematical perspective. The recent history of method development is briefly surveyed with an emphasis on the use of geometric integration as a guiding principle. The recovery of statistical mechanical averages from molecular dynamics is then introduced, and the use of backward error analysis as a technique for analysing the accuracy of numerical averages is described. This article gives the first rigorous estimates for the error in statistical averages computed from molecular dynamics simulation based on backward error analysis. It is shown that molecular dynamics introduces an appreciable bias at stepsizes which are below the stability threshold. Simulations performed in such a regime can be corrected by use of a stepsize-dependent reweighting factor. Numerical experiments illustrate the efficacy of this approach. In the final section, several open problems in dynamics-based molecular sampling are considered.

CONTENTS

1. Molecular dynamics

Molecular dynamics is a central component of modern simulation in the fields of chemistry, physics, materials science, and medicine (Allen and Tildesley 1987, McCammon and Harvey 1987, Whittle and Blundell 1994, Verlinde and Hol 1994, Frenkel and Smit 2002, Schlick 2002). It is a powerful, general-purpose technique that allows treatment of a wide variety of problems such as optimization of molecular structures, computing probabilities of events or averages of functions of molecular configurations. In some cases, it even allows tracking dynamical processes, such as transitions from one molecular conformation to another. In non-equilibrium modelling, molecular dynamics is the high-accuracy tool that enables simulation of transient behaviour.

In this introductory section, we concentrate on the formulation of molecular dynamics problems, introducing a commonly used classical molecular model, discussing the role of molecular dynamics simulations and several historical and mathematical issues related to the numerical methods used for the purpose. We also introduce the backward error analysis, which is the foundation for MD simulation. In the following section, we focus on the use of dynamics to recover averages for molecular systems, introducing two formulations for Nosé dynamics along with numerical methods. In Section 3, expansions are constructed for Nosé dynamics methods, allowing interpretation of numerical trajectories as exact solutions of perturbed systems. Statistical mechanical implications of these perturbations are described. In particular, we give a correction term for averages computed using the Nosé–Poincaré method. Finally, Section 4 contains descriptions of several open problems and current research topics.

1.1. Molecular models

Molecular dynamics (MD) refers to the simulation of the physical motion of the atoms of some substance. In this article, we focus on N-body (classical) models with configurational interactions modelled by a potential energy function

$$U(q) = U(q_1, q_2, \ldots, q_N),$$

where $q_i \in \mathbb{R}^3$, is the position of the nucleus of the ith atom. The potential energy function typically involves 2-body, 3-body or 4-body terms,

$$U_{ij}(q_i, q_j), \qquad U_{ijk}(q_i, q_j, q_k), \qquad U_{ijkl}(q_i, q_j, q_k, q_l).$$

The precise form of these few-body potential energy functions will be related to the material under study (sometimes this is referred to as the *chemistry* in a molecular dynamics model). The use of a single potential energy function to describe the nuclear interactions makes a simplifying assumption, the Born–Oppenheimer ansatz, that the electronic structure relaxes instantaneously relative to the nuclear motion.

Each interaction term gives rise to certain forces which are obtained as the gradient of potential energy. The typical form of the short-range pairwise interaction used in MD is due to van der Waals and is most often modelled by a Lennard–Jones potential:

$$\phi_{ij}^{LJ}(q_i, q_j) = 4\epsilon_{ij}\left(\left(\frac{\sigma_{ij}}{r_{ij}}\right)^{12} - \left(\frac{\sigma_{ij}}{r_{ij}}\right)^6\right), \qquad r_{ij} = \|q_i - q_j\|.$$

Such a term is incorporated between each pair of atoms, so the coefficients ϵ_{ij} and σ_{ij} depend on the particular types of atoms involved.

Because of the strong repulsion for $r \to 0$, the atoms stay well separated. As the Lennard–Jones potential tends relatively rapidly to zero as $r \to \infty$, we describe this as a short-ranged term. In practice the potential is cut off to zero outside some fixed radius. (How this is done may have important ramifications for the quality of numerical results, as discussed in Section 2.)

In addition to Lennard–Jones potentials, we may have *length bonds* modelled as linear springs with rest-length

$$\phi_{ij}^{l.b.}(q_i, q_j) = \frac{1}{2}k_{ij}(r_{ij} - r_{ij}^0)^2.$$

In many cases these springs are very stiff compared to other potential terms, so the associated time-scales of vibration play an important role in determining the usable simulation time-step.

Another type of two-body interaction is the electrostatic term

$$\phi_{ij}^C(q_i, q_j) = \frac{Q_i Q_j}{D r_{ij}},$$

where D is a constant. Because the Coulomb potential falls off slowly with distance compared to the Lennard–Jones potential, we say that it is *long-ranged*. In practice, all pairs of charged particles will have a nontrivial Coulomb interaction and each of these terms must be computed with some level of precision. Many efforts have been made in recent years to reduce the cost of long-ranged force computation, with some success (Barnes and Hut 1986, Darden, York and Pedersen 1993, Greengard 1994, Krasny and Duan 2002), although we do not consider this issue here.

Three-body potentials often arise as penalty terms to control the angles made by chemically bonded triples of atoms (angle bonds):

$$\phi_{ijk}^{a.b.}(q_i, q_j, q_k) = \frac{1}{2}k_{ijk}(\theta_{ijk} - \theta_{ijk}^0)^2.$$

The angle θ_{ijk} is defined in terms of the positions as

$$\theta_{ijk} = \arcsin\frac{(q_i - q_j) \cdot (q_j - q_k)}{r_{ij} r_{jk}}.$$

Empirical three-body potentials are also used to simulate materials such as graphite. In this way, bond breakage and formation can be simulated in a limited way (Takai, Lee, Halicioglu and Tiller 1990). The cost of simulating with such a potential is large, however, because of the vast number of interaction terms and potentials that need to be computed ($O(N^3)$).

In the most straightforward case, the potential interactions are homogeneous distance potentials,

$$U = \sum_{i=1}^{N-1} \sum_{j=i+1}^{N} \phi(\|q_i - q_j\|),$$

but even in this case, the structure of the potential energy landscape and resulting motion are both very complicated, owing to the combined effects of the many terms.

The subject of molecular dynamics may be deemed unappealing as an area for mathematical research, in part because of the complexity of the molecular description. Fortunately it is generally possible and useful to work with elementary examples and model problems to understand basic principles and to test numerical methods. The simplest of these is the harmonic oscillator

$$\mathrm{d}^2 q / \mathrm{d}t^2 = -\omega^2 q,$$

with energy function $U(q) = \omega^2 q^2/2$. A slightly more interesting model is the 'springy pendulum' in two or three dimensions:

$$U(q) = k(r - L)^2/2, \qquad r = \|q\|.$$

A more sophisticated model problem, which we will consider later, is a chain of seven atoms connected sequentially, the Hamiltonian for this system being

$$H(q,p) = \frac{1}{2} \sum_{i=1}^{N} m_i^{-1} \|p_i\|^2 + \frac{\kappa}{2} \sum_{i=1}^{N-1} (r_{i,i+1} - r^0)^2 + \sum_{i=1}^{N-1} \sum_{j=i+1}^{N} \phi^{\mathrm{LJ}}(q_i, q_j). \quad (1.1)$$

We took $\epsilon = 1$, $\kappa = 1000$, and $r^0 = 1$. Each atom interacts with the other six atoms through the Lennard–Jones potential. With a large value of the constant κ, near the minimum of energy, this system behaves as though the springs were essentially rigid rods.

A classic model which is extremely useful for understanding MD issues for larger systems is the Lennard–Jones system with periodic boundary conditions, described by a Hamiltonian of the form

$$H(q,p) = \frac{1}{2} \sum_{i=1}^{N} m_i^{-1} \|p_i\|^2 + \sum_{\hat{L}} \sum_{i=1}^{N-1} \sum_{j=i+1}^{N} \hat{\phi}^{\mathrm{LJ}}(q_i, q_j + \hat{L}). \qquad (1.2)$$

Here $\hat{L}$ is a multi-index which ranges over vectors $(a, b, c)^T$, where a, b and c are $-L$, 0 or L; L defines the box width; $\hat{\phi}$ is usually a smoothly cut-off version of the Lennard–Jones potential.

1.2. The role of molecular dynamics

This paper is about obtaining trajectories that explore the potential energy landscape or thermodynamics of a molecular system. The most straight-forward approach is to simulate the dynamics of Newtonian equations of motion:

$$m_i \ddot{q}_i = F_i,$$

where $F_i = -\nabla_{q_i} U$ is the force acting on the ith atom, and m_i is the positive mass of the ith atom. The alternative way of writing this system is as follows:

$$\frac{dq}{dt} = \mathbf{M}^{-1}p, \tag{1.3}$$

$$\frac{dp}{dt} = -\nabla U(q), \tag{1.4}$$

where now $q = (q_1, q_2, \ldots, q_N)$ is a $3N$-dimensional vector of all positions, $p = (p_1, p_2, \ldots, p_N)$ is the vector of associated momenta, and $M = \text{diag}(m_1, m_1, m_1, m_2, m_2, m_2, \ldots, m_N, m_N, m_N)$ is a matrix of masses. The system is associated with a Hamiltonian (energy function) of the form

$$H(q_1, q_2, \ldots, q_N, p_1, p_2, \ldots, p_N) = \frac{1}{2} \sum_{i=1}^{N} m_i^{-1} \|p_i\|^2 + U(q_1, q_2, \ldots, q_N)$$

$$= \frac{p^T \mathbf{M}^{-1} p}{2} + U(q), \tag{1.5}$$

which is a first integral of (1.3)–(1.4). There is always a severe limitation in MD simulation because the use of a detailed potential energy function U means that there will be very rapid oscillatory components in the dynamics; this limits the time-step and hence the maximum time interval on which simulation is possible. In current practice, typical simulations are performed on nanosecond time intervals, with unusual examples stretching to a microsecond.

To understand the context in which molecular dynamics is used, it is necessary to understand that molecular simulation comprises a wide range of tasks aimed at understanding different aspects of molecular structure and dynamical behaviour. One important question commonly asked about a given system is this: What is the global minimum of U for all $q = (q_1, q_2, \ldots, q_N)$?

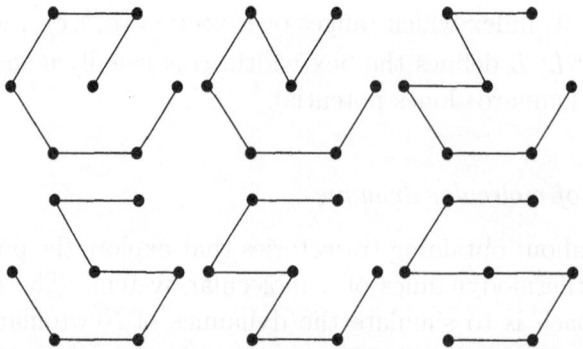

Figure 1.1. The six minimizing configurations of the atoms of a 7-atom chain.

The idealized minimizing structures for the seven-atom chain (in the constraint-chain limit) all consist of the placement of six atoms at equally spaced points on a circle of radius 1, with the seventh at the centre. Up to symmetry, there are six different minimizing configurations, as shown in Figure 1.1. Note that all these structures have zero harmonic energy and the same Lennard–Jones energy.

The determination of global minima in molecular systems becomes rapidly more difficult as the dimension of the system increases, owing to the proliferation of local minima and the general corrugation of the energy landscape. Another problem we sometimes hear spoken of in connection with molecular systems is the *configurational sampling*, or the averaging of a function of q with respect to a suitable density function. A third challenge is to resolve the long-term chaotic behaviour in the potential field U, ascribing to the ith atom some suitable mass. From the point of view of numerical analysis, these are all exceptionally difficult problems

In fact, though, the problems mentioned are all accessible to the powerful ideas of statistical mechanics. In a molecular system, the atomic motion is highly constrained by the laws of probability. The trajectories of the molecular dynamics model may be chaotic, but they fill out a large region of space, mapping the energy surface and allowing us to perform integration (sampling) using the dynamics. The states of lowest energy are visited most frequently, roughly in accordance with Boltzmann's hypothesis, so that minima of the potential are frequented by the dynamics; this allows dynamics to play a role in determining minima. Molecular dynamics, if properly implemented, provides one of the few general-purpose tools for study of the molecular landscape. Indeed, the presentation of molecular dynamics as the numerical solution of a conservative systems of ordinary differential equations is a bit misleading, because it disguises the real use which is made of MD simulation: typically the mapping of the energy surface or, from an alternative perspective, understanding of certain features of a probability distribution function in a $6N$-dimensional phase space. Molecular dynamics should be viewed as one technique of several available for analysing the accessible phase space. In this sense it is an alternative to Metropolis Monte Carlo simulation that samples phase space by taking a sequence of discrete, randomized steps. In some cases MD is a more efficient tool than Monte Carlo, since, in essence, steps taken with MD are automatically 'accepted'. MD is typically also more precise than Monte Carlo in that it offers the potential of recovering *dynamical* information as well. It can answer questions such as: How long does it take on average to make the transition from basin A to basin B? This is not possible with standard Monte Carlo simulation without simplifying assumptions from transition state theory (Voter, Montalenti and Germann 2002).

Because of its flexibility and ease of implementation, molecular dynamics has become the 'high-accuracy' tool of choice in simulation of materials for engineering and biological sciences applications. Despite its simplicity (the *ideal* model is quantum-mechanical, after all), molecular dynamics can afford insight into a surprisingly wide variety of relevant issues, from characterizing the progressive formation of a crack in a crystalline material to protein docking and the study of folding pathways, essential procedures in the field of rational drug design.

1.3. A brief history of MD integrators

In this subsection, we look at the development of molecular dynamics methods from the point of view of the numerical analysis. Let us summarize a few features that we expect to characterize good methods. Because low-accuracy trajectories are generally needed (partly because of the many errors already introduced at an early stage of the modelling, only phenomenological questions are usually asked), numerical analysts would propose to use *low-order methods*. In order to get good long-term stability while approximating the system in the basins where it spends most of its time, one would expect to use numerical methods with *neutral stability*. Because the forces are complicated and dominate the cost of computation, one would expect to find *explicit* methods to be of the greatest value. Although many different methods could be applied for the purpose, the need to be able to take a very large number of steps with stable long-term behaviour places severe constraints on the integration technique.

The first MD simulation is attributed to Alder and Wainwright (1957), a simulation of hard spheres which already demonstrated both the simulation technique and some of the prospects of such simulations for revealing the properties of liquids. Their method used periodic boundary conditions (which had already been introduced much earlier in Monte Carlo simulations) and included computation of radial density functions.

Rahman (1964) showed that a molecular dynamics simulation could be performed using a cut-off continuous potential (based on Lennard–Jones), the system described by (1.3)–(1.4) for 864 argon atoms in a periodic box, and used a predictor–corrector method. Simulations involved a time-step of 10 femtoseconds, each of which took about 40 seconds to compute using the most advanced computers of the day. Total simulation time was limited to about 10 picoseconds. Verlet (1967) described his performance of a similar calculation using enhanced methodology. Like Rahman, he used 864 atoms in cut-off Lennard–Jones interaction, together with a book-keeping device to limit calculation of pairwise forces to only those atoms which are likely to yield a nontrivial interaction.

In Rahman's paper the integration method used consisted of two iterations of a predictor–corrector method based on the trapezoidal rule; this method already shows substantial artificial behaviour in even relatively short simulations due to numerical instability. Verlet suggested instead using a scheme that is equivalent to the leapfrog/Störmer's rule, and this has withstood the test of time. The Verlet method can be written as (superscripts here indicating time-step number):

$$q^{n+1} = q^n + h\mathbf{M}^{-1}p^{n+1/2}, \tag{1.6}$$

$$p^{n+1/2} = p^n - \frac{h}{2}\nabla U(q^n), \tag{1.7}$$

$$p^{n+1} = p^{n+1/2} - \frac{h}{2}\nabla U(q^{n+1}). \tag{1.8}$$

This method is second-order. Verlet was able to compute simulations at about the rate of a second per time-step, benefitting from a combination of his better numerical methods and rapid advances in computer hardware.

During the 1970s, MD methodology continued to mature. While the original simulations of simple liquids such as argon could rely on the pairwise additive nature of the force field, more complicated liquids seemed to require a sophisticated multibody potential. Rahman and Stillinger (1971) represented the many body terms by an effective pair potential and were thus able to perform a simulation of 216 water molecules, using Ewald summation to compute long-ranged forces of interaction, and a fourth-order predictor–corrector method to integrate the rigid body equations of motion.

In addition to the various potential energy terms, molecular dynamics models often incorporate additional modelling complications. For example, it is common in biological molecular modelling to freeze some of the chemical bonds at their minimizers, in which case we introduce constraints such as

$$\|q_i - q_j\| = l_{ij}.$$

Even some of the angle bonds may be so constrained. The m constraints introduced in this way may be described as a system of equations $g(q) = 0$, with a vector function $g : \mathbb{R}^{3N} \to \mathbb{R}^m$. These then modify the form of the dynamical system, resulting in equations of the form

$$\frac{dq}{dt} = \mathbf{M}^{-1}p,$$

$$\frac{dp}{dt} = -\nabla U(q) - g'(q)^T \lambda,$$

$$g(q) = 0.$$

Here λ is a vector of Lagrange multipliers which must be solved for in tandem with the physical variables q and p. One approach that has resurfaced several times over the years is to attempt to reduce this system to ODEs, *e.g.*, by constraint differentiation, but this approach has not been found to be effective in general MD simulation, and the constrained equations are typically solved directly. Ryckaert, Ciccotti and Berendsen (1977) introduced the SHAKE method for this purpose, and used it to model a system with chemical bonds. The method can be written as

$$q^{n+1} = q^n + h\mathbf{M}^{-1}p^{n+1/2},$$

$$p^{n+1/2} = p^{n-1/2} - h\nabla U(q^n) - hg'(q^n)^T \lambda^n,$$

$$g(q^{n+1}) = 0,$$

where one views the momenta and positions as evolving on grids which are staggered by a time offset of $h/2$. It is also possible to rewrite this method in a one-step form and to view it as a mapping of the co-tangent bundle of the constraint manifold $\{q \mid g(q) = 0\}$.

Advances in computer technology, together with improved availability of parametrized force fields, triggered efforts by several groups to apply molecular dynamics machinery for the direct simulation of proteins. McCammon, Gelin and Karplus (1977) reported the results of their study of bovine pancreatic trypsin inhibitor (BPTI), selected due to 'its small size ... high stability ... and accurately determined X-ray structure'. The simulation was facilitated by the incorporation of hydrogen atoms into heavier atoms through 'suitable adjustment of atomic parameters', reducing the number of degrees of freedom and allowing larger time-steps through the elimination of the high-frequency hydrogen bonding interactions. A femtosecond time-step was used and the simulation performed on an interval of 8.8 picoseconds.

At that time, the understanding of the optimal numerical scheme for molecular simulation was skewed by the availability of computer power. The method used by McCammon, Gelin and Karplus was the Gear predictor–corrector scheme. Van Gunsteren and Berendsen (1977) repeated the BPTI study, systematically comparing different numerical schemes in a very short (100-step) simulation, and concluding that a relatively high-order Gear

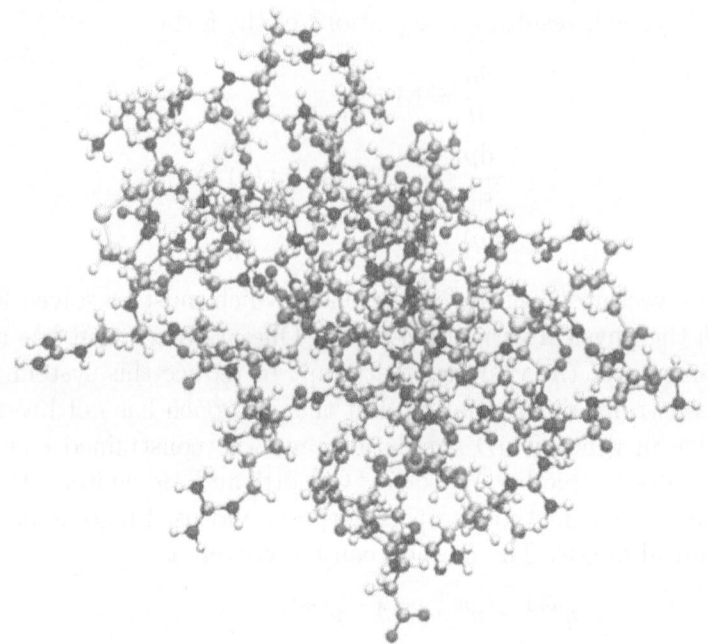

Figure 1.2. BPTI (bovine pancreatic trypsin inhibitor) was one of the first proteins to be investigated by use of molecular dynamics. In this representation, there are 882 atoms, including hydrogens, nitrogens, oxygens, carbon and sulphur. The chemical bonds are shown as sticks joining the atoms. Not shown are the thousands of water molecules which would typically added to the simulation in order to simulate the molecule in its normal environment. (Image created using VMD, courtesy Paul Brenner (Notre Dame).)

predictor–corrector scheme was the optimal choice for general-purpose unconstrained simulation. In molecular dynamics simulations, the evaluation of numerical methods (and determination of the quality of a numerical trajectory) must be based on the magnitude of the observed energy drift. From one time-step to the next, the energy can fluctuate quite considerably in an MD simulation, regardless of the method, and these local fluctuations are generally larger in a low-order method than in a higher-order one, at a given stepsize. A method such as the predictor–corrector scheme does not conserve anything, and it will show greater secular drift if the time interval is long, but it may exhibit a higher efficiency than a low-order method in a short simulation. It was only when longer time simulations became possible that the broad superiority of Verlet's method became obvious.

The 1980s and 1990s saw many advances in force fields, including various models of water (crucial in protein dynamics) based on dipole and quadrupole approximations. At the same time, powerful new methods were being introduced for quantum simulation, such as the technique of Car and Parinello, a scheme for incorporating quantum mechanics, which uses a fictitious dynamics to develop an approximation to the wave function. In this approach orthonormality of the orbitals is maintained using the SHAKE discretization (Car and Parinello 1985). This method was proposed by its authors as both (1) a method for ground state electronic properties, and (2) a way to enhance molecular dynamics by allowing simulations that build the Born–Oppenheimer surface 'on the fly', rather than relying entirely on a parametrized approximation. Although substantially more demanding than the traditional full classical method, Car–Parinello ideas gradually became popular as part of MD simulation and are used for a variety of special purposes. These sorts of techniques can be simplified, as by Sprik and Klein (1988), and used to design a polarizable water model (see also Rick, Stuart and Berne (1994)), as well as for more general purposes (Rappe and Goddard 1991).

Another innovation introduced in the 1980s was Nosé's dynamical method (Nosé 1984a), roughly simultaneous with the use of stochastic-dynamical methods such as Brownian dynamics and Langevin dynamics in molecular simulation (van Gunsteren and Berendsen 1982, Brunger, Brooks and Karplus 1984). These techniques opened the door for molecular dynamics simulations to be used for equilibrium sampling in a constant temperature (or constant temperature and pressure) ensemble and the computation of thermodynamical properties such as diffusion coefficients. A focus of this article is the Nosé-type schemes, and we take up the discussion of their implementation and use for computational statistical mechanics in the next section.

1.4. Geometric integrators

As we have seen, during the first decades after molecular dynamics was introduced, increasing attention was placed on understanding the basic properties of integrators. In addition, many open questions were floating around regarding the suitability of various schemes for the uses to which they were being put. In the 1980s, a new development took hold in the physics and mathematics communities: geometric integration. A geometric integrator is a numerical method which preserves certain geometric structures of the exact flow of a differential equation. From at least the late 1960s, the computational physics community had clearly believed that it was important for a numerical method to mimic the time-reversal symmetry present in physical N-body systems. If we write the numerical method as a map,

$$\begin{bmatrix} q^{n+1} \\ p^{n+1} \end{bmatrix} = \Psi_h \left(\begin{bmatrix} q^n \\ p^n \end{bmatrix} \right),$$

and we define the map R by

$$R\left(\begin{bmatrix} q \\ p \end{bmatrix}\right) = \begin{bmatrix} q \\ -p \end{bmatrix},$$

then we find

$$\Psi_h \circ R \circ \Psi_h \circ R = \text{Id},$$

where Id is the identity map. A map with this property is said to be time-reversible. It is straightforward to show that the Verlet method is time-reversible, but many other methods such as the Gear predictor–corrector method are not. Another important property, particularly for astronomical N-body simulation, is the conservation of the angular momentum. If the potential energy function involves central forces, *i.e.*,

$$\sum_{i=1}^{N} q_i \times \nabla_{q_i} U = 0,$$

so that the angular momentum is conserved,

$$\sum_{i=1}^{N} q_i(t) \times p_i(t) = \text{constant},$$

then the same quantity is conserved from step to step of discretization using Verlet (1.6)–(1.8):

$$\sum_{i=1}^{N} q_i^{n+1} \times p_i^{n+1} = \sum_{i=1}^{N} q_i^n \times p_i^n.$$

Although remarkable, this property is, itself, of limited value, since most MD simulations involve the use of periodic boundary conditions which destroy angular momentum conservation.

Ruth (1983) argued that numerical methods could and should be constructed to preserve the symplectic 2-form:[1]

$$dq \wedge dp = \sum_{i=1}^{N} dx_i \wedge dp_{x_i} + dy_i \wedge dp_{y_i} + dz_i \wedge dp_{z_i}.$$

The property of the invariance of the 2-form can be restated as

$$d\Psi_h^T J \, d\Psi_h = J,$$

where

$$J = \begin{bmatrix} 0 & I_{3N} \\ -I_{3N} & 0 \end{bmatrix}$$

---

[1] The mathematician de Vogelaere had in fact considered the basic idea much earlier (de Vogelaere (1956)).

and $d\Psi_h$ represents the Jacobian matrix of the map Ψ_h. The proof that the Verlet method is symplectic is straightforward. Taking differentials of (1.6)–(1.8) yields

$$dq^{n+1} = dq^n + h\mathbf{M}^{-1} dp^{n+1/2}, \qquad (1.9)$$

$$dp^{n+1/2} = dp^n - \frac{h}{2} d\nabla U(q^n), \qquad (1.10)$$

$$dp^{n+1} = dp^{n+1/2} - \frac{h}{2} d\nabla U(q^{n+1}). \qquad (1.11)$$

A term of the form $d\nabla U(q)$ can be replaced by $U''(q)\,dq$, where U'' is the symmetric Hessian matrix of U. Moreover, it is straightforward to show that $dq \wedge A\,dq = 0$ if A is symmetric. Hence, from (1.11) we have

$$dq^{n+1} \wedge dp^{n+1} = dq^{n+1} \wedge dp^{n+1/2},$$

then, using (1.9) and a similar argument, we have

$$dq^{n+1} \wedge dp^{n+1/2} = dq^n \wedge dp^{n+1/2},$$

and, finally, using (1.10),

$$dq^n \wedge dp^{n+1/2} = dq^n \wedge dp^n.$$

A lot of subsequent research has been performed by mathematicians and physicists both on method construction and on explaining properties of symplectic methods. This work has confirmed that symplectic methods such as Verlet are generally superior choices for computing very long trajectories. However, it is important to state clearly that the discovery that the Verlet method is a good method for atomistic molecular dynamics does not originate in mathematical observations about symplectic structure, but in the numerical experience documented in various papers appearing after 1967 which showed that Verlet's method was more stable and efficient than alternatives for molecular simulation.

Leimkuhler and Skeel (1994) proved that the SHAKE method is equivalent to a symplectic integrator. This paper looked at both SHAKE and Anderson's 'self-starting' alternative to SHAKE, known as RATTLE:

$$q^{n+1} = q^n + h\mathbf{M}^{-1}p^{n+1/2},$$

$$p^{n+1/2} = p^n - \frac{h}{2}\nabla U(q^n) - \frac{h}{2}g'(q^n)^T\lambda^n,$$

$$g(q^{n+1}) = 0,$$

$$p^{n+1} = p^{n+1/2} - \frac{h}{2}\nabla U(q^{n+1}) - \frac{h}{2}g'(q^{n+1})^T\mu^{n+1},$$

$$g'(q^{n+1})\mathbf{M}^{-1}p^{n+1} = 0.$$

Leimkuhler and Skeel showed that RATTLE is a symplectic map of the co-tangent bundle of the constraint manifold associated to this problem, and they also incidentally demonstrated that SHAKE and RATTLE are formally conjugate (*i.e.*, they generate equivalent numerical trajectories up to a modification of the initial condition).

During the 1990s, largely due to the success in providing a theoretical justification for the improved performance of Verlet, SHAKE and RATTLE, symplecticness took hold as a litmus test for integrators for molecular dynamics applications. A particular area where new symplectic integrators were developed during the mid-1990s, and found to enhance simulation efficiency, was in rigid body molecular dynamics. Until this time, the common techniques were based on parametrization of rigid body motion by means of quaternions, resulting in a system which could not be discretized by the symplectic Verlet method; instead, the usual approach was based on extrapolation or non-symplectic Runge–Kutta methods. The results were fine in short-term simulations, but energy drift was evident as integration times increased. A generalized symplectic treatment of the Euler equations for rigid body motion was proposed independently by McLachlan (1993) and Reich (1994), and implemented for systems of rigid bodies, as occur in molecular dynamics, by Dullweber, Leimkuhler and McLachlan (1997). This method is now a standard scheme in molecular simulation, where it is sometimes referred to as the MRDL method. Alternative methods have also been proposed based on rotation matrices (Kol, Laird and Leimkuhler 1997). While new types of rigid body methods have been introduced more recently (McLachlan and Zanna 2005, van Zon and Schofield 2007), with improved accuracy, these methods are not an improvement on MRDL for standard molecular dynamics simulations, owing to the errors introduced because of intermolecular forces.

Multiple time-scale methods (García-Archilla, Sanz-Serna and Skeel 1998, Hochbruck and Lubich 1999, Hairer and Lubich 2000, Izaguirre, Reich and Skeel 1999) have been introduced which preserve the symplectic structure, and have typically demonstrated improved resolution of averaged behaviour compared to non-symplectic alternatives, particularly when the time interval for simulation is very long.

Besides the symplectic structure, one may wonder about the role of first integrals and time-reversal symmetry. Although it seems, from practical experience, that maintaining first integrals alone is not sufficient to allow long-term simulations to be performed with sufficient accuracy for sampling, it is far less obvious to which extent time-reversal symmetry is an appropriate foundation for method building for highly chaotic molecular systems. This property can be mimicked by a numerical discretizaton. There is numerical evidence that time-reversal symmetry does allow, at some sufficiently small stepsize and for some problems, long-term simulations to

proceed stably compared to non-time-reversible and non-symplectic methods, just as a similar statement can be made regarding symplectic methods. The situation is complicated by the fact that time-reversible methods
are, in many cases, easier to construct than their symplectic counterparts
and sometimes more efficient. An example is the reversible multiple time-
scale methods of Leimkuhler and Reich (2001), which allow resonance-free
multiple scale integration. Another example often cited for the practical
value of time-reversible methods is in the context of Nosé–Hoover simulations for constant-temperature molecular dynamics, which are considered
in the following section, although in this case we favour symplectic alternatives (Bond, Laird and Leimkuhler 1999).

The theoretical ground for the study of symplectic (and time-reversible)
integrators is the concept of backward error analysis, which we next describe.

1.5. Backward error analysis

Traditional 'forward error analysis' for ODEs describes the difference between the exact trajectory and numerical trajectory. For an sth-order numerical method, this is typically expressed in terms of bounds of the form

$$\|z(t_n) - \hat{z}_n\| \leq C_1 h^s \exp(C_2 t_n),$$

where $\{z(t) \mid t \geq 0\}$ and $\{\hat{z}_n \mid n = 0, 1 \ldots\}$ are the exact and numerical
trajectories respectively and h is the time-step size. The constants C_1 and
C_2 depend on the vector field and the numerical method. Unfortunately,
in the context of molecular simulation, bounds of this form are not very
useful, since C_2 is often large, and computing averages requires very large
time intervals. The chaotic nature of molecular dynamics means that any
small numerical errors must result in large 'forward error' in the trajectory.

A much better concept of error for molecular dynamics comes from 'backward error analysis'. Here the difference between the numerical and exact
solution is expressed in terms of a perturbation of the problem or vector
field. These perturbations are derived using the method of modified equations (Benettin and Giorgilli 1994, Sanz-Serna 1997, Reich 1999, Neishtadt
1984, Hairer 1994, Sanz-Serna and Calvo 1994, Skeel and Hardy 2001, Hairer
and Lubich 1997). The idea is that the numerical trajectory can be made an
arbitrarily accurate approximation if one modifies the equations or problem.
For an sth-order numerical method applied to $\dot{z} = f(z)$, one says that $\bar{f}_r$ is
an rth-order modified vector field if the numerical trajectory is an rth-order
approximation to the solution of $\dot{z} = \bar{f}_r(z)$. This modified vector field will
be a function of the stepsize, and is typically expressed in a series expansion
in powers of h:

$$\bar{f}_r(z) = f(z) + h^s f_{[s]}(z) + \cdots + h^r f_{[r]}(z).$$

Assuming that f is sufficiently smooth, it can be shown (Hairer 1994) that the terms of the series $f_{[i]}$ can be systematically derived in terms of derivatives of f. The process of deriving these terms involves comparing a series expansion of one step of the numerical method with a similar expansion for the solution of $\bar{f}_i$.

It is tempting to take the limit as $r \to \infty$ to obtain an exact trajectory which interpolates the numerical trajectory. Unfortunately the series does not converge in the typical case. Despite this technical difficulty the modified vector field is still useful as a truncated series. It can be shown that there is an optimal truncation index for which the difference between the numerical method and the modified trajectory is exponentially small (Benettin and Giorgilli 1994, Reich 1999, Neishtadt 1984), and this optimal index increases as the stepsize decreases.

The power of the method of backward error analysis and the method of modified equations can be demonstrated in the context of geometric integrators. It can be shown that if the numerical method preserves a particular geometric structure (*e.g.*, symplectic, time-reversibility), the modified equations must preserve this same structure. Hence the modified equations are in the same geometric class as the original problem. For example, if a symplectic integrator is applied to a Hamiltonian system, the truncated modified equations to order r must be Hamiltonian:

$$\bar{H}_r(q, p) = H(q, p) + h^s H_{[s]}(q, p) + \cdots + h^r H_{[r]}(q, p).$$

It follows that one may view the numerical solution as the nearly exact solution of a slightly different Hamiltonian system. Let us see how to construct the first terms of such an expansion for the Verlet method applied to the Hamiltonian in (1.5). Assuming it exists, it can be shown that (since the Verlet method is time-reversible) $\bar{H}_r$ contains only even order terms in h. Since Verlet is a second-order method, we have $s = 2$, and

$$q^n = \bar{q}_h(nh),$$

$$p^n = \bar{p}_h(nh),$$

where $\bar{q}_h(t), \bar{p}_h(t)$ solves the differential equations

$$\frac{\mathrm{d}\bar{q}_{h,i}}{\mathrm{d}t} = \frac{\partial \bar{H}_r}{\partial p_i} = \frac{p_i}{m_i} + h^2 \frac{\partial H_{[2]}}{\partial p_i} + \cdots + h^r \frac{\partial H_{[r]}}{\partial p_i}, \tag{1.12}$$

$$\frac{\mathrm{d}\bar{p}_{h,i}}{\mathrm{d}t} = -\frac{\partial \bar{H}_r}{\partial q_i} = -\frac{\partial U}{\partial q_i} - h^2 \frac{\partial H_{[2]}}{\partial q_i} - \cdots - h^r \frac{\partial H_{[r]}}{\partial q_i}. \tag{1.13}$$

Expanding the solution of these equations in a Taylor series, we get

$$\bar{q}_h(t + \tau) = \bar{q}_h(t) + \tau \frac{\mathrm{d}\bar{q}_h}{\mathrm{d}t}(t) + \frac{1}{2}\tau^2 \frac{\mathrm{d}^2\bar{q}_h}{\mathrm{d}t^2}(t) + \cdots,$$

$$\bar{p}_h(t + \tau) = \bar{p}_h(t) + \tau \frac{\mathrm{d}\bar{p}_h}{\mathrm{d}t}(t) + \frac{1}{2}\tau^2 \frac{\mathrm{d}^2\bar{p}_h}{\mathrm{d}t^2}(t) + \cdots,$$

which we may evaluate at $\tau = h$. The higher derivatives of $\bar{q}_h$ and $\bar{p}_h$ can be obtained differentiating the differential equations (1.12)–(1.13):

$$\frac{d^2 \bar{q}_{h,i}}{dt^2} = \sum_j \left(\frac{\partial^2 \bar{H}_r}{\partial p_i \partial q_j} \frac{\partial \bar{H}_r}{\partial p_j} - \frac{\partial^2 \bar{H}_r}{\partial p_i \partial p_j} \frac{\partial \bar{H}_r}{\partial q_j} \right),$$

$$\frac{d^2 \bar{p}_{h,i}}{dt^2} = \sum_j \left(\frac{\partial^2 \bar{H}_r}{\partial q_i \partial p_j} \frac{\partial \bar{H}_r}{\partial q_j} - \frac{\partial^2 \bar{H}_r}{\partial q_i \partial q_j} \frac{\partial \bar{H}_r}{\partial p_j} \right).$$

At the same time, we can view the Verlet method as defining q^{n+1}, p^{n+1} in powers of h,

$$q_i^{n+1} = q_i^n + h \frac{p_i^n}{m_i} - \frac{h^2}{2} \frac{1}{m_i} \frac{\partial U}{\partial q_i},$$

$$p_i^{n+1} = p_i^n - h \frac{\partial U}{\partial q_i} - \frac{h^2}{2} \sum_j \frac{\partial^2 U}{\partial q_i \partial q_j} \frac{p_j}{m_j}$$

$$- \frac{h^3}{4} \left(\sum_j \sum_k \frac{\partial^3 U}{\partial q_i \partial q_j \partial q_k} \frac{p_j p_k}{m_j m_k} - \sum_j \frac{\partial^2 U}{\partial q_i \partial q_j} \frac{1}{m_j} \frac{\partial U}{\partial q_j} \right) + \cdots.$$

Matching the terms of these asymptotic expansions yields a partial differential equations for the unknown $H_{[k]}$ terms, e.g.,

$$\frac{\partial H_{[2]}}{\partial p_i} = \frac{1}{6} \sum_j \frac{\partial^2 U}{\partial q_i q_j} \frac{p_j}{m_i m_j},$$

$$\frac{\partial H_{[2]}}{\partial q_i} = \frac{1}{12} \sum_j \sum_k \frac{\partial^3 U}{\partial q_i q_j q_k} \frac{p_j p_k}{m_j m_k} - \frac{1}{12} \sum_j \frac{\partial^2 U}{\partial q_i q_j} \frac{1}{m_j} \frac{\partial U}{\partial q_j}.$$

Fortunately, these equations can be successively solved to yield the desired expressions, e.g.,

$$H_{[2]}(q,p) = \frac{1}{12} \sum_j \sum_k \frac{\partial U}{\partial q_j \partial q_k} \frac{p_j}{m_j} \frac{p_k}{m_k} - \frac{1}{24} \sum_j \frac{\partial U}{\partial q_j} \frac{1}{m_j} \frac{\partial U}{\partial q_j}.$$

For sufficiently small stepsize, the backward error expansion explains the remarkable energy conservation of symplectic integrators (e.g., Störmer/ leapfrog/Verlet) over very long (exponentially long) time intervals. A recent article of Hairer, Lubich, and Wanner in this journal develops the theory of backward error analysis in relation to the Verlet method and explains some of its applications in detail (Hairer, Lubich and Wanner 2003). Very notable in the context of molecular dynamics is the construction of Skeel and Hardy (2001) of the 'interpolated shadow Hamiltonian', by means of which very high-order approximations of the terms in the backward error expansion can be computed using a numerical method. For general discussion of geometric integration and backward error analysis, the reader is referred to the recent books on the subject by Leimkuhler and Reich (2005), and Hairer, Lubich and Wanner (2002).

As an illustration of the usefulness of backward error analysis, a recent paper used it to examine the cut-off smoothness in MD simulation. As we have mentioned previously, in a typical MD simulation a box of atoms is simulated using periodic boundary conditions, *e.g.*, (1.2). The potential $\hat{\phi}$ is a cut-off Lennard–Jones potential with the cut-off distance determined so that interactions are limited to less than the box width. Engle, Skeel and Drees (2005) demonstrated that a highly accurate (24th-order) implementation of the modified equations computed using an improved implementation of the interpolation technique of Skeel and Hardy (2001) helps to clarify the effect of inappropriate cut-off of non-bonded forces. Whereas large fluctuations seen in the energy itself tend to disguise such subtleties until they have accumulated sufficiently to become obvious, the much smoother interpolated shadow Hamiltonian reveals clear jumps at isolated points corresponding to a C^1-smooth restraint function or cut-off. At a given stepsize, for higher-order cut-offs, the effect of truncation is greatly reduced.

Recently, backward error analysis has been applied with great success to hybrid Monte Carlo algorithms (Duane, Kennedy, Pendleton and Roweth 1987). In hybrid Monte Carlo, configurations from a constant-temperature distribution are generated by steps combining randomly sampled momenta and classical constant-energy molecular dynamics. The acceptance or rejection of steps is based on the probabilistic Metropolis–Hastings criterion, which is a function of the change of energy over the step. If the molecular dynamics trajectory is exact, acceptance or rejection can be determined before the trajectory is computed, which significantly reduces the computational cost of the method. Unfortunately, one cannot expect exact conservation of energy for numerically generated molecular dynamics trajectories and computational effort is wasted computing trajectories for rejected steps. To mitigate this problem, the targeted shadowing and shadow hybrid Monte Carlo methods generate a constant-temperature distribution for the modified or shadow Hamiltonian obtained from backward error analysis. The results are then reweighted in post-processing to obtain the desired distribution for the unmodified Hamiltonian (Izaguirre and Hampton 2004, Akhmatskaya and Reich 2005).

2. Stochastic modelling with chaotic dynamics

Consider a trajectory of some Hamiltonian system started from some initial configuration and with certain initial momenta. The energy $E = H(q, p)$ will be conserved. If this energy is low, then the trajectory will be confined to some basin associated to some minimizing configurations or, more likely, some higher-energy metastable state. If the energy is very large, then the trajectory will wander erratically in phase space, spending much of the time far from low-energy configurations. Thus we see that it will be necessary somehow to control the 'energeticness' of the trajectory in order to control the region of phase space that is sampled by the trajectory.

It is difficult to relate the value of energy *per se* directly to the physical environment, since it depends on detailed properties of the system under study. On the other hand, the temperature, typically defined as the *average* kinetic energy per degree of freedom in the system, can be taken to be independent of the specific characteristics of the system under study, and serves as an invariant macroscopic parameter which can be measured by placing the given system in contact with a known quantity of some particular substance (*e.g.*, water, mercury, *etc.*), *i.e.*, a thermometer.

In molecular modelling, we use temperature in much the same way as we use it in the laboratory: to calibrate our simulations to a physical environment. Because temperature (unlike energy) is scale-independent, it is key to development of multiscale approaches.

Now let us suppose that we wish to compute a trajectory of a system consistent with a given temperature, say corresponding to room temperature in a related physical setting. We generally associate the temperature with the average over time and number of particles of the kinetic energy, thus it is easy to find a set of initial conditions which are consistent with this average value – we can just choose the initial momenta appropriately. However, it must be remembered that temperature is a macroscopic parameter, whereas the specific positions and momenta of the system at a given time specify a *microstate*. The detailed assignment of kinetic energy for a given microstate, to be consistent with the target temperature, will have to take into consideration the entropy of the system. It is very unlikely that the simple Maxwellian distribution, or some other arbitrarily chosen distribution of momenta, will be associated to a macrostate at the correct thermal level, at least in a complicated system. This is the reason why a temperature control mechanism is usually needed in molecular dynamics.

The means by which molecular dynamics allows us to perform dynamical sampling has been elucidated in the dynamical systems and ergodic theory communities. The basic idea is that molecular dynamics generates space-filling trajectories which gradually fill the accessible region of phase space. These *sampling trajectories* then become a tool for computing averages of functions with respect to a given density of states.

The fundamental assumption in statistical mechanics – that an ensemble average is equivalent to a trajectory average – is known as ergodicity. Roughly speaking, ergodicity means that almost all trajectories are 'statistically the same' so long as the initial conditions are consistent with the thermodynamic state. Although this assumption seems quite plausible for

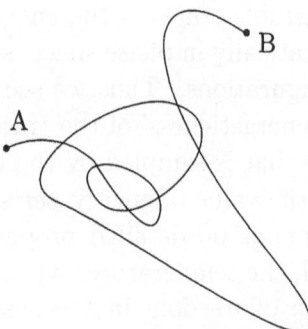

Figure 2.1. Is every pair of phase points on an energy
surface linked by a dynamical path? In an ergodic
Hamiltonian system, almost every trajectory will visit
a neighbourhood of almost every point of the
associated constant energy surface.

many large systems with nonlinear interactions, one should note that in
many important situations this does not hold (*e.g.*, harmonic solids, proteins with strong chemical bonds, *etc.*) Furthermore, even if this assumption
is true in the limit as time approaches infinity, it may not be true for any
practical finite time interval. For example, large energy barriers may be
present which prevent the system from thoroughly sampling configuration
space. Despite this difficulty, it is generally believed that ergodicity holds
for some relevant systems in molecular dynamics, such as liquid argon simulated via Lennard–Jones potential (Hansen and McDonald 1986, Allen and
Tildesley 1987, Frenkel and Smit 2002). For the purposes of our own presentation, we will follow the approach taken in the chemical physics literature
and typically assume ergodicity of a certain system in order to prove a con-
sequent result, or we will discuss the construction of methods which may
'enhance the ergodicity' in a practical sense, that is, accelerate the computation of averages. In all cases, then, the final arbiter of success has to be
numerical experiment, which verifies the claims made for a given method
or formulation, demonstrating that they are valid at least to some approximate degree.

2.1. Statistical mechanics

The macroscopic (thermodynamic) states of a system can be related to the
microscopic (molecular) motion, by averaging with respect to an ensemble
(or distribution) which is invariant under the flow (McQuarrie 1976, Hansen
and McDonald 1986, Chandler 1987, Toda, Kubo and Saitô 1991). An
ensemble is the collection of all the microscopic states which are consistent with the macroscopic description. For example, the microcanonical

ensemble consists of all microscopic states with the same number of particles (N), volume (V), and energy (E). This is the thermodynamic model for an isolated system, with its state completely specified by these three variables (NVE). It is assumed that at equilibrium any thermodynamic measurement should be repeatable, depending only on these variables, *i.e.*, independent of the exact evolution of the microscopic motion. If this assumption holds, the macroscopic properties of a system must be described entirely by the distribution of states.

The time or trajectory average of a function of phase space, B, is simply the normalized integral of its value over the trajectory:

$$\langle B \rangle_{\text{time}} := \lim_{t \to \infty} \frac{1}{t} \int_0^t B(z(\tau))\, d\tau.$$

Here, $z(t)$ is a trajectory or solution to the ordinary differential equation (ODE) $\dot{z} = f(z)$ with initial conditions $z(0) = z_0$. For a typical classical molecular dynamics simulation, $z = (q, p)$, where $q, p \in \mathbb{R}^{3N}$ are the configurations and momenta of N particles in three-dimensional space. For the microcanonical ensemble, the ODE is derived from a Hamiltonian with energy function or Hamiltonian, $H(q, p)$:

$$\dot{q}_i = \frac{\partial H}{\partial p_i},$$

$$\dot{p}_i = -\frac{\partial H}{\partial q_i}, \quad i = 1, \ldots, 3N.$$

In comparison, the ensemble average of a function, B, is calculated using a weighted integral (or sum) over all admissible microscopic states:

$$\langle B \rangle_{\text{ens}} = \int_\Gamma B(z) \rho_{\text{ens}}(z)\, dz.$$

In this formulation, ρ_{ens} is a probability density function which describes the distribution of configurations, Γ. The exact form of this density function will depend on the thermodynamic constraints imposed on the system. In the microcanonical ensemble, the distribution of states is uniform on a surface of constant energy:

$$\rho_{\text{mc}}(z) = \frac{1}{C_{\text{mc}}} \delta[H(z) - E].$$

Here, H is a classical Hamiltonian which provides the energy of each configuration, $z = (q, p)$. The constant of proportionality, C_{mc}, is chosen to normalize ρ_{mc} (*i.e.*, $\int \rho_{\text{mc}}\, d\Gamma = 1$). If the system has other first integrals, *e.g.*, conservation of linear or angular momentum, this must be accounted for in the ensemble, and we find

$$\rho_{\mathrm{mc}}(\Gamma) = \frac{1}{C_{\mathrm{mc}}} \delta\big[H(z) - E\big] \delta\big[I_1(z) - C_1\big] \cdots \big[I_k(z) - C_k\big],$$

where each $I_i : \mathbb{R}^{3N} \to \mathbb{R}$ is constant (with a value of C_i) along the flow.

The connection between the time and ensemble average comes from the ergodic hypothesis. Clearly the time average is a function of the initial conditions. However, if the system of equations given by the ODE has a single invariant measure, ρ, the time average will be independent of the initial conditions for almost all initial conditions. Furthermore, the time average can be transformed to a configuration or ensemble average using this invariant measure to define the ensemble average (Toda *et al.* 1991). Hence the ergodic hypothesis can be stated mathematically as

$$\langle B \rangle_{\mathrm{ens}} = \langle B \rangle_{\mathrm{time}}$$

for all initial conditions outside a set of measure zero.

Computing an invariant distribution for a system of ODEs involves solving a partial differential equation known as the Liouville equation (Toda *et al.* 1991) for the evolution of a distribution of configurations in phase space,

$$\frac{\partial \rho}{\partial t} + \nabla_z \cdot (\rho\, f) = 0,$$

where f is the vector field of the system of ODEs. Rewriting the Liouville equation in terms of a material derivative, one finds

$$\frac{\mathrm{D}\rho}{\mathrm{D}t} + \rho \nabla_z \cdot f = 0. \tag{2.1}$$

In the case of autonomous Hamiltonian systems, the vector field is divergence-free and one finds that ρ is constant (Toda *et al.* 1991). For non-Hamiltonian systems the situation is more complicated. For example, the constant temperature or 'canonical ensemble' prescribes that configurations are distributed according to the Gibbs distribution:

$$\rho_{\mathrm{c}}(z) \propto \exp\left[-\frac{1}{k_B T} H(z)\right], \tag{2.2}$$

where k_B is Boltzmann's constant and T is temperature.

2.2. Nosé dynamics

Let a Hamiltonian system be given with energy function

$$H(q, p) = \frac{1}{2} p^T \mathbf{M}^{-1} p + U(q),$$

with q and p $3N$-dimensional vectors of positions and momenta, respectively. Let $\mathbf{M}$ be a positive definite symmetric mass matrix, and let U be the potential energy function.

If the system is ergodic – and many large-scale molecular dynamics problems are assumed to be (nearly) so – one hopes that all degrees of freedom are quickly brought into equilibration through natural energetic exchange, resulting in a well-defined average kinetic energy which can be identified with the temperature of the system. In practice, however, this process may take an extremely long time or never be observed on the time-scale of simulation. Moreover, in some cases it is desirable to adjust the temperature or other parameters which would require re-equilibration to a specified target temperature, or the system may progress through intermediate, metastable states; during the transitions, thermal equilibrium may be difficult to maintain. In practice, some device, which can be viewed as an artificial thermal bath, is almost always incorporated to maintain the desired target temperature. Nosé dynamics offers the promise of thermal regulation via a simple dynamical device, based on substituting a modestly extended dynamics for the simple constant energy model.

Nosé dynamics is derived from the extended phase space Hamiltonian,

$$H_N = H(q, \tilde{p}/s) + \frac{p_s^2}{2\mu} + gk_BT \ln s,$$

where g is the total number of degrees of freedom (including the thermostatting degrees of freedom), k_B is the Boltzmann constant, and T is the target temperature at which sampling is desired. μ is a parameter that effectively allows the strength of dynamic coupling to be adjusted. The momentum appearing in H_N should be treated as canonical to q, whereas the physical momentum is related to $\tilde{p}$ by the change of variables

$$p = \frac{\tilde{p}}{s},$$

which suggests an intrinsic rescaling of time. It was shown by Nosé that canonical sampling can be obtained along (assumed ergodic) trajectories of H_N via the relation

$$\iint \delta[H_N - H_N^0] \, d\tilde{p}_1 \cdots d\tilde{p}_{3N} \, ds \, dp_s = C \exp\left(-\frac{1}{k_BT} H(q, p)\right) dp_1 \cdots dp_{3N},$$

where the integration is performed over the physically accessible phase space of the thermostatting variables, $(s, p_s) \in (0, \infty) \times \mathbb{R}$.

2.3. Experiment: 256-atom Lennard–Jones system

Without, for the moment, going into the details of how the Nosé dynamics approach is implemented, we will describe the behaviour of the technique when applied to a 256-atom Lennard–Jones system with periodic boundary conditions. The density and temperature in reduced units (Frenkel and Smit 2002) were set to 0.95 and 1.5 respectively. This system is widely assumed to be nearly ergodic. The results, for thermal mass parameter μ ranging from $\mu = 0.01$ to $\mu = 10^6$ and for numbers of time-steps (samples) ranging from 10^3 to 10^6, are shown in Figure 2.2. These figures chart the distribution of kinetic energies obtained with the indicated parameters. What do we observe? For very small μ, we do not sample from the correct distribution. For very large μ, it takes a much longer time to achieve the correct distribution. Notice that it starts off shifted. For the largest values of μ we get the wrong variance. In fact, it starts converging to the microcanonical distribution at a shifted temperature before slowly sliding over to the correct temperature (but wrong distribution).

In Figures 2.3 and 2.4 we show the mean and variance, respectively, in the various simulations performed with given variation of parameters. We expect the distribution to approach a chi-square distribution (sum of squares of normal variables), but with 765 degrees of freedom; this distribution

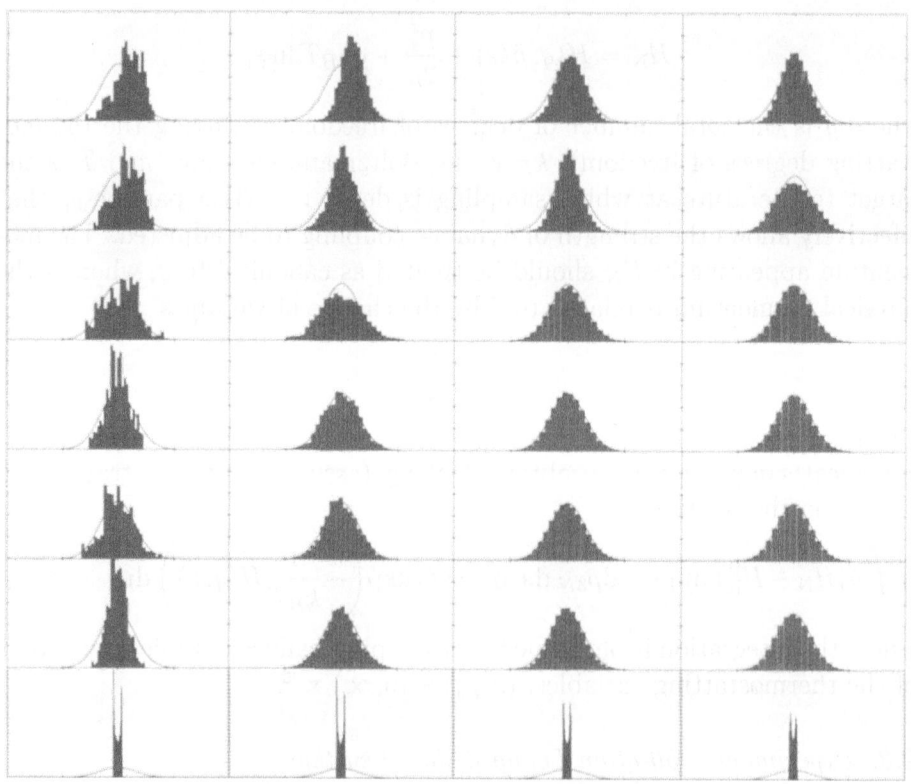

Figure 2.2. The figures show the convergence of the kinetic energy distribution with trajectory length for different choices of the thermal mass parameter μ. *Top to bottom*: $\mu = 10^6, 10^4, 10^2, 10, 1, 0.1, 0.01$. *Left to right*: Number of samples $= 10^3, 10^4, 10^5, 10^6$.

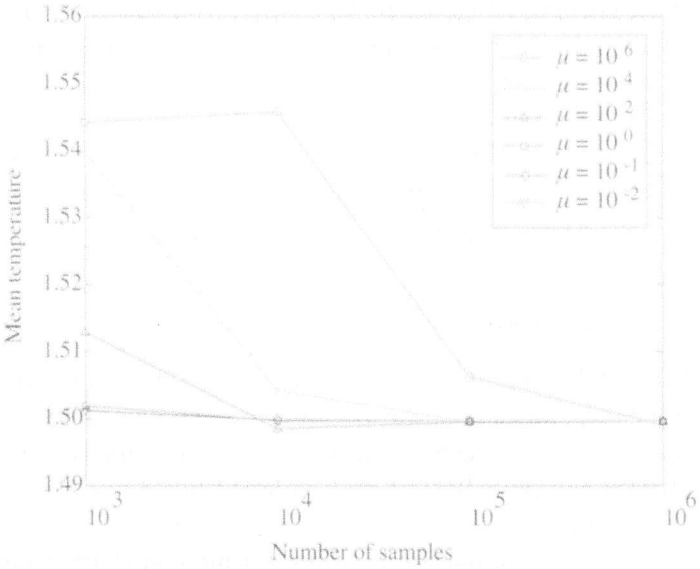

Figure 2.3. Convergence of mean temperature with number of samples, for different choices of thermal mass parameter.

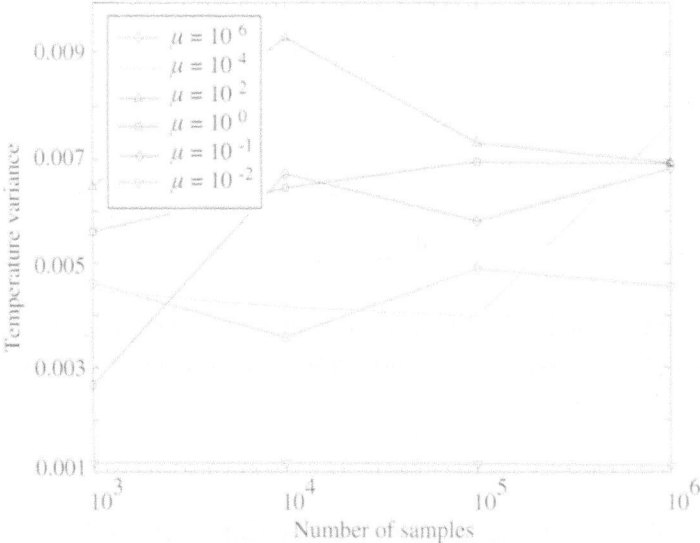

Figure 2.4. Convergence of variance of temperature distribution with number of samples, for different choices of thermal mass parameter.

is very close to a normal distribution. The results suggest that for a broad range of thermal mass, a reasonable sampling is achieved if the interval of simulation is long enough. Note that the mean temperature does not converge to the target temperature of 1.5 but to a slightly different value. This is a consequence of the numerical error which is associated to the numerical method used; both mean and variance could be brought into agreement with the expected values.

2.4. Nosé–Hoover and Nosé–Poincaré

While useful for understanding the concept of Nosé dynamics, H_N is not usually recommended for simulation because, on the one hand, computation of certain types of averages (*e.g.*, autocorrelation functions) requires data at equally spaced points in time, and, more importantly, the equations of motion corresponding to H_N are poorly scaled for $s \to 0$. The key to improving the numerical simulation is to use a time transformation to regularize the equations of motion. Such transformations are widely used in celestial mechanics simulations to improve numerical stability in the vicinity of close approaches of bodies. The idea is to replace the differential equation

$$\frac{dz}{dt} = f(z),$$

by a modified differential equation running in a new time,

$$\frac{dz}{d\tau} = \gamma(z)f(z),$$

where the two time variables are obviously related by

$$\frac{dt}{d\tau} = \gamma(z).$$

This is sometimes called a Sundman time transformation. The Nosé–Hoover reformulation of Nosé dynamics incorporates such a time transformation of the form

$$\frac{d\tilde{t}}{dt} = s, \tag{2.3}$$

where the notation is to suggest that it is the time variable $\tilde{t}$ associated to H_N that one should view as a modification of t.

The equations of motion for H_N are

$$\frac{dq}{d\tilde{t}} = \mathbf{M}^{-1}\tilde{p}/s^2, \tag{2.4}$$

$$\frac{d\tilde{p}}{d\tilde{t}} = F, \tag{2.5}$$

$$\frac{ds}{d\tilde{t}} = p_s/\mu, \tag{2.6}$$

$$\frac{dp_s}{d\tilde{t}} = \frac{\tilde{p}^T \mathbf{M}^{-1} \tilde{p}}{s^3} - \frac{gk_B T}{s}, \tag{2.7}$$

where $F = F(q) = -\nabla_q U$ is the vector of forces acting on the bodies. Replacing $\tilde{p}/s$ by p in (2.4)–(2.7), we find

$$\frac{dq}{d\tilde{t}} = \mathbf{M}^{-1} p/s,$$

$$\frac{dp}{d\tilde{t}} = \frac{1}{s} F - \frac{1}{s} \frac{ds}{d\tilde{t}} p,$$

$$\frac{ds}{d\tilde{t}} = p_s/\mu,$$

$$\frac{dp_s}{d\tilde{t}} = \left[p^T \mathbf{M}^{-1} p - gk_B T \right]/s.$$

Using the time transformation (2.3) and identifying p_s/μ with ξ gives the familiar Nosé–Hoover system:[2]

$$\frac{dq}{dt} = \mathbf{M}^{-1} p, \tag{2.8}$$

$$\frac{dp}{dt} = F - \xi p, \tag{2.9}$$

$$\frac{d\xi}{dt} = (p^T \mathbf{M}^{-1} p - gk_B T)/\mu. \tag{2.10}$$

This time transformation destroys the Hamiltonian structure of the equations of motion. There is, however, a remnant of the original time-reversal symmetry: (2.8)–(2.10) are invariant under the change of variables

$$(q, p, \xi, t) \rightarrow (q, -p, -\xi, -t).$$

The first integral corresponding to the Nosé Hamiltonian still exists, with the 'extended energy'

$$E(q, p, \xi, \eta) = \frac{1}{2} p^T \mathbf{M}^{-1} p + U(q) + \frac{1}{2} \mu \xi^2 + gk_B T \eta,$$

conserved along exact trajectories with

$$\frac{d\eta}{dt} = \xi. \tag{2.11}$$

---

[2] Note that the constant g is the number of degrees of freedom in the 'physical system', $H(q, p)$. This is in contrast to Nosé's Hamiltonian which used the number of degrees of freedom of the 'extended system'.

An alternative, due to Bond *et al.* (1999) (independently derived by Dettmann (1999)) is to use a Poincaré transformation. Simulation with the Nosé–Poincaré method is based on the Hamiltonian

$$H_{NP} = s(H_N - H_N^0), \qquad (2.12)$$

where the constant H_N^0 must be chosen so that H_{NP} vanishes at the initial value, and hence for all time along Hamiltonian dynamics in the extended phase space.

The equations of motion associated to (2.12) are

$$\frac{dq}{dt} = M^{-1}\tilde{p}/s, \qquad (2.13)$$

$$\frac{d\tilde{p}}{dt} = sF, \qquad (2.14)$$

$$\frac{ds}{dt} = sp_s/\mu, \qquad (2.15)$$

$$\frac{dp_s}{dt} = \frac{\tilde{p}^T M^{-1} \tilde{p}}{s^2} - gk_B T - \Delta H_N, \qquad (2.16)$$

where $\Delta H_N = H_N - H_N^0$.

The relation between (2.13)–(2.16) and (2.8)–(2.10) is made clear by the substitution $p = \tilde{p}/s$ and $\xi = p_s/\mu$. Then (2.13)–(2.14) are easily seen to be identical to (2.8)–(2.9), while the last two equations become

$$\frac{ds}{dt} = s\xi,$$

$$\frac{d\xi}{dt} = \frac{1}{\mu}(p^T M^{-1} p - gk_B T - \Delta H_N).$$

The last equation is easily seen to be identical to (2.10) up to the perturbation $\mu^{-1}\Delta H_N$, which vanishes along the exact solution. However, under discretization, we do not expect ΔH_N to vanish, and trajectories obtained from the two systems (2.13)–(2.16) and (2.8)–(2.10) will differ considerably over long time simulations.

To find the ensemble associated with the Nosé–Hoover vector field, we insert (2.8)–(2.10) in the Liouville equation, (2.1), which results in

$$\frac{D\rho}{Dt} = \rho \sum_i^{N_f} \xi = \rho g\xi,$$

since $g = N_f$ is the number of degrees of freedom in the physical system.

Furthermore, one finds that

$$\frac{\mathrm{d}}{\mathrm{d}t}\left(\frac{1}{2}p^T\mathbf{M}^{-1}p + U(q) + \frac{1}{2}\mu\xi^2\right) = -\xi g k_B T,$$

and hence

$$\frac{\mathrm{D}}{\mathrm{D}t}\rho = \frac{-1}{k_B T}\rho\frac{\mathrm{d}}{\mathrm{d}t}\left(\frac{1}{2}p^T\mathbf{M}^{-1}p + U(q) + \frac{1}{2}\mu\xi^2\right).$$

Solving for ρ, we find an invariant distribution in extended phase space:

$$\rho_{\mathrm{NH}}(q,p,\xi) = \frac{1}{C}\exp\left[\frac{-1}{k_B T}\left(\frac{p^T\mathbf{M}^{-1}p}{2} + U(q) + \frac{\mu\xi^2}{2}\right)\right], \qquad (2.17)$$

where C is a normalizing constant. Note that the extended variable, ξ, is not coupled to the variables, (q,p), in the distribution above. Hence, integrating over ξ yields a canonical distribution in (q,p):

$$\int \rho_{\mathrm{NH}}(q,p,\xi)\,\mathrm{d}\xi \propto \exp\left[\frac{-1}{k_B T}H(q,p)\right].$$

A modified version of this argument can be used to show the same result for Nosé–Poincaré.

2.5. Separated form

Nosé dynamics has an interesting alternative formulation based on a simple change of variables. Let $s = e^\theta$, $p_s = e^{-\theta}p_\theta$; then H_N can be seen to be equivalent to the system described by

$$\hat{H}_N = \frac{1}{2}e^{-2\theta}\tilde{p}^T\mathbf{M}^{-1}\tilde{p} + \frac{1}{2\mu}e^{-2\theta}p_\theta^2 + U(q) + g k_B T\theta.$$

We can next introduce a Poincaré-type time transformation equivalent to $H_N \rightarrow s^2(H_N - H_N^0)$, yielding the separated form

$$\hat{H}_N^* = \frac{1}{2}\tilde{p}^T\mathbf{M}^{-1}\tilde{p} + \frac{1}{2\mu}p_\theta^2 + e^{2\theta}(U(q) + g k_B T\theta - H_N^0).$$

This separation of variables was first suggested by Dettmann and Morriss (1997); it was further studied and expanded by Leimkuhler (2002). The method allows a simple visualization of the behaviour of Nosé dynamics. Consider, for example, a double-well potential described by the potential $U(q) = (q^2 - 1)^2$. The effective potential

$$\hat{U}(q,\theta) = e^{2\theta}(U(q) + g k_B T\theta - H_N^0)$$

is graphed in Figure 2.5 for several different values of the parameter $\gamma = g k_B T$ and a fixed energy H_N^0. The figure shows that the introduction of the temperature parameter T via Nosé dynamics can be directly related to a

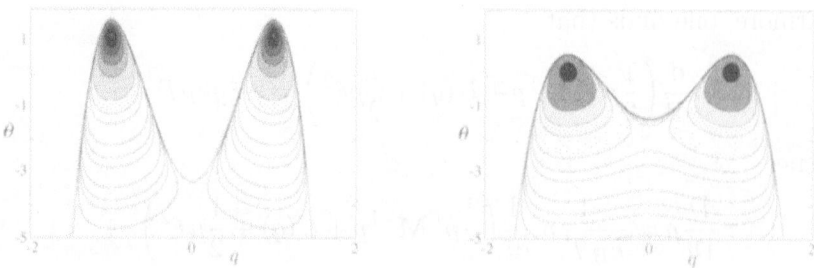

Figure 2.5. Contours of the Nosé effective potential:
$\gamma = 0.2$ (*left*), $\gamma = 0.5$ (*right*).

smoothing of the potential barrier (in a larger-dimensional space). As a consequence, higher temperatures enhance the transition between well basins in the double-well system.

2.6. Algorithms for canonical ensemble sampling

At this juncture, we mention an important theoretical issue. Even if one assumes that the exact flow is ergodic, there is no guarantee that the numerical dynamics will be ergodic as well. If trajectory averages converge in the limit $t \to \infty$, ergodicity may not be 'observed' in practice since one is restricted to finite time intervals. For the class of systems considered in molecular dynamics, most of the key questions cannot be rigorously addressed. Preliminary results obtained by Tupper (2007) are of great interest in this regard, but do not yet give quantifiable predictions which are relevant for general MD simulation. Despite this unresolved concern, we next describe algorithms for sampling a molecular system from the canonical ensemble, based on Nosé dynamics. We give a comparison of several alternative numerical procedures for Nosé–Hoover and Nosé–Poincaré simulation, including one Nosé–Poincaré method proposed by Nosé himself, clarifying the relative advantages of the two types of scheme.

The constraint under which we work in typical molecular dynamics applications is that the potential energy U and vector of force F are relatively expensive to compute. In fact the number of evaluations of potential and/or the force may be taken as the measure of computational work in MD simulation. (The computational cost is not substantially greater to compute both potential *and* force at a given point than would be incurred in computing one or the other of these terms.) In practice it is observed that the only useful algorithms for MD are partially explicit in the sense that no iteration need be performed at each time-step which would lead to multiple force evaluations. We term such partially explicit methods *force-explicit*. All methods considered in this section are force-explicit with a single U and/or F evaluation required at each time-step.

Time-reversible integrators for Nosé–Hoover

The Nosé–Hoover equations are time-reversible. A large number of reversible schemes are available. We mention here only two: a scheme due to Holian, Groot, Hoover and Hoover (1990) which we term *Nosé–Hoover explicit*, abbreviated NHE, and an *implicit* scheme, NHI, from Frenkel and Smit (1996). The formulas for each of these familiar methods are given below. It is seen that NHI requires the solution of a cubic nonlinear equation, but this presents no difficulty in practice and in fact both of these methods can be viewed as force-explicit methods. Both of these methods are also time-reversible.

Nosé–Hoover explicit (NHE)
_____

$$p^{n+1/2} = p^n + \frac{h}{2}(F^n - \xi^n p^{n+1/2}),$$

$$q^{n+1} = q^n + h\mathbf{M}^{-1}p^{n+1/2},$$

$$\xi^{n+1} = \xi^n + h((p^{n+1/2})^T\mathbf{M}^{-1}p^{n+1/2} - gk_BT)/\mu,$$

$$p^{n+1} = p^{n+1/2} + \frac{h}{2}(F^{n+1} - \xi^{n+1}p^{n+1/2}).$$

Nosé–Hoover implicit (NHI)
_____

$$p^{n+1/2} = p^n + \frac{h}{2}(F^n - \xi^n p^n), \tag{2.18}$$

$$\xi^{n+1/2} = \xi^n + \frac{h}{2}((p^n)^T\mathbf{M}^{-1}p^n - gk_BT)/\mu, \tag{2.19}$$

$$q^{n+1} = q^n + h\mathbf{M}^{-1}p^{n+1/2}, \tag{2.20}$$

$$p^{n+1} = p^{n+1/2} + \frac{h}{2}(F^{n+1} - \xi^{n+1}p^{n+1}), \tag{2.21}$$

$$\xi^{n+1} = \xi^{n+1/2} + \frac{h}{2}((p^{n+1})^T\mathbf{M}^{-1}p^{n+1} - gk_BT)/\mu. \tag{2.22}$$

The last two equations here must be solved together for ξ^{n+1} and p^{n+1}. The simplest approach is to first solve (2.21) for p^{n+1}:

$$p^{n+1} = (1 + h\xi^{n+1}/2)^{-1}\left(p^{n+1/2} + \frac{h}{2}F^{n+1}\right).$$

Set $r = 1 + h\xi^{n+1}/2$; then, upon introducing the above expression into (2.22) and simplifying, we obtain a cubic equation in r,

$$r^3 + d_2r^2 + d_0 = 0,$$

with

$$d_0 = -\frac{h^2}{4\mu} w^T \mathbf{M}^{-1} w,$$

$$d_2 = -1 - \frac{h}{2}\xi^{n+1/2} + \frac{h^2}{4\mu} gk_B T,$$

$$w = p^{n+1/2} + \frac{h}{2}F^{n+1}.$$

Symplectic methods for Nosé–Poincaré

Several symplectic methods have been devised for the Nosé–Poincaré equations. The first (chronologically) was an application of the generalized leapfrog method, which for a given general Hamiltonian $G(q, p)$ is written compactly as

$$q^{n+1} = q^n + \frac{h}{2}\left(\nabla_p G(q^n, p^{n+1/2}) + \nabla_p G(q^{n+1}, p^{n+1/2})\right), \tag{2.23}$$

$$p^{n+1/2} = p^n - \frac{h}{2}\nabla_q G(q^n, p^{n+1/2}), \tag{2.24}$$

$$p^{n+1} = p^{n+1/2} - \frac{h}{2}\nabla_q G(q^{n+1}, p^{n+1/2}), \tag{2.25}$$

To apply this to the Nosé–Poincaré equations of motion, G here should be taken to be the time-rescaled Hamiltonian H_{NP} (2.12), q should be replaced by (q, s), and p by $(\tilde{p}, p_s)$. When implemented, this method requires the solution of a quadratic equation, but this presents no difficulty in practice. Bond *et al.* (1999) used this scheme to integrate the Nosé–Poincaré equations, as follows.

Generalized leapfrog algorithm (GLA)

Step 1. Solve for $\tilde{p}^{n+1/2}$, $p_s^{n+1/2}$:

$$\tilde{p}^{n+1/2} = \tilde{p}^n - \frac{h}{2}s^n \nabla U(q^n),$$

$$p_s^{n+1/2} = p_s^n + \frac{h}{2}\left(\frac{(\tilde{p}^{n+1/2})^T \mathbf{M}^{-1}\tilde{p}^{n+1/2}}{2(s^n)^2} - gk_B T(1 + \ln s^n)\right.$$

$$\left. - \frac{(p_s^{n+1/2})^2}{2\mu} - U(q^n) + H_N^0\right).$$

The first of these equations can be solved explicitly. The second requires an inexpensive quadratic solve for $p_s^{n+1/2}$.

Step 2. Solve for q^{n+1} and s^{n+1}:

$$q^{n+1} = q^n + \frac{h}{2}\left(\frac{1}{s^n} + \frac{1}{s^{n+1}}\right)\mathbf{M}^{-1}\tilde{p}^{n+1/2},$$

$$s^{n+1} = s^n + \frac{h}{2}(s^n + s^{n+1})p_s^{n+1/2}/\mu.$$

Step 3. Solve for $\tilde{p}^{n+1}$ and p_s^{n+1}:

$$\tilde{p}^{n+1} = \tilde{p}^{n+1/2} - \frac{h}{2}s^{n+1}\nabla U(q^{n+1}),$$

$$p_s^{n+1} = p_s^{n+1/2} + \frac{h}{2}\left(\frac{(\tilde{p}^{n+1/2})^T\mathbf{M}^{-1}\tilde{p}^{n+1/2}}{2(s^{n+1})^2} - gk_BT(1 + \ln s^{n+1})\right.$$
$$\left. - \frac{(p_s^{n+1/2})^2}{2\mu} - U(q^{n+1}) + H_N^0\right),$$

which is an explicit calculation.

Another approach is based on splitting the Hamiltonian into parts that can be integrated either explicitly or using an additional level of discretization. There are a wide variety of such splittings which would work for the Nosé–Poincaré system. A 3-term splitting with easily integrated terms was suggested by Nosé (2001):

$$H_{\text{NP}} = H_1 + H_2 + H_3,$$

where

$$H_1 = s\left(\frac{\tilde{p}^T\mathbf{M}^{-1}\tilde{p}}{2s^2} + gk_BT\ln s - H_N^0\right),$$

$$H_2 = sU(q), \quad \text{and} \quad H_3 = \frac{sp_s^2}{2\mu}.$$

Of course, other splittings are possible. We have experimented with moving the sH_N^0 from H_1 to H_2. In either of these splitting methods, the first Hamiltonian depends only on $\tilde{p}$ and s and is trivially integrable. The second involves only q and s and is again trivially integrable. The third gives equations

$$\frac{\mathrm{d}p_s}{\mathrm{d}t} = -\frac{p_s^2}{2\mu},$$

$$\frac{\mathrm{d}s}{\mathrm{d}t} = \frac{sp_s}{\mu},$$

with solution

$$p_s(t) = p_s(0)\left(1 + \frac{p_s(0)t}{2\mu}\right)^{-1},$$

$$s(t) = s(0)\left(1 + \frac{p_s(0)t}{2\mu}\right)^{2}.$$

An algorithm can be constructed based on the usual Trotter scheme, which can be written in the simple form

$$e^{\frac{h}{2}H_3}e^{\frac{h}{2}H_2}e^{hH_1}e^{\frac{h}{2}H_2}e^{\frac{h}{2}H_3},$$

where e^{tH} represents the flow map (solution map taking a point of phase space to its evolution through t units of time) of a Hamiltonian system with Hamiltonian H and the product of exponentials is understood to represent composition of maps. The details of this method are given below.

A 3-term Hamiltonian splitting method (HSP)

Step 1. Solve H_3 for a step of size $h/2$:

$$s^{n+1/2} = s^n\left(1 + \frac{p_s^n h}{4\mu}\right)^{2},$$

$$p_s^a = p_s^n\left(1 + \frac{p_s^n h}{4\mu}\right)^{-1}.$$

Step 2. Solve H_2 for a step of size $h/2$:

$$\tilde{p}^{n+1/2} = \tilde{p}^n + \frac{h}{2}s^{n+1/2}F^n,$$

$$p_s^b = p_s^a - \frac{h}{2}U(q^n).$$

Step 3. Solve H_1 for a step of size h:

$$q^{n+1} = q^n + h(s^{n+1/2})^{-1}\mathbf{M}^{-1}\tilde{p}^{n+1/2}$$

$$p_s^c = p_s^b - h\left(-\frac{(\tilde{p}^{n+1/2})^T\mathbf{M}^{-1}\tilde{p}^{n+1/2}}{2(s^{n+1/2})^2} + gk_BT(1 + \ln s^{n+1/2}) - H_N^0\right).$$

Step 4. Solve H_2 for a step of size $h/2$:

$$\tilde{p}^{n+1} = \tilde{p}^{n+1/2} + \frac{h}{2}s^{n+1/2}F^{n+1},$$

$$p_s^d = p_s^c - \frac{h}{2}U(q^{n+1}).$$

Step 5. Solve H_3 for a step of size $h/2$:

$$s^{n+1} = s^{n+1/2} \left(1 + \frac{p_s^d h}{4\mu} \right)^2,$$

$$p_s^{n+1} = p_s^d \left(1 + \frac{p_s^d h}{4\mu} \right)^{-1}.$$

2.7. Experiment: seven-atom chain

To illustrate the use of these schemes, the system (1.1) was simulated for 40,000 time-steps, with stepsize $h = 0.01$. To obtain an initial condition, the system was first started at a global minimum of energy, the last configuration shown in Figure 1.1. Random initial velocities were applied to each atom, corresponding approximately to a temperature (average kinetic energy per degree of freedom) of $T = 0.5$. The initial velocities were normalized so that the total linear momentum was zero. Snapshots of the (constant energy) dynamics show how the chain evolves over this time interval. At each of nine equally spaced times, the graphs show a few hundred time-steps of the positional motion of each atom. Note that the chain unfolds and refolds itself back to the vicinity (the *basin*) of a global minimum of potential.

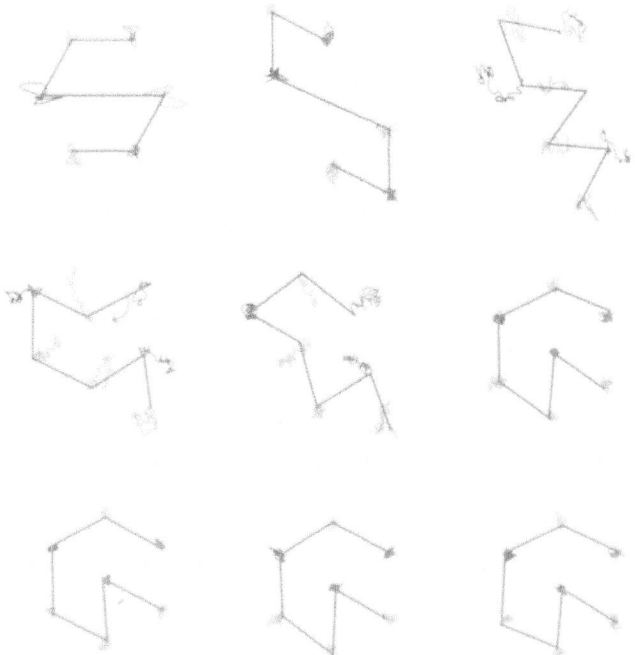

Figure 2.6. Computed configurations of the seven-atom chain as snapshots of the dynamics at equally spaced points in time.

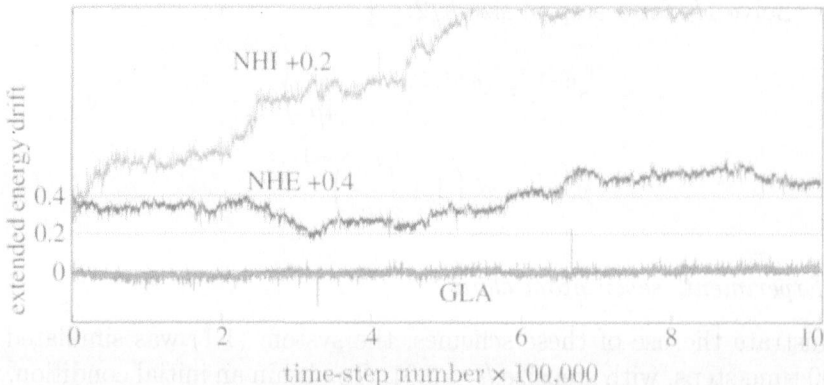

Figure 2.7. Illustration of advantage of symplectic integrator
based on Nosé–Poincaré formulation compared to Nosé–Hoover
schemes. Energy errors along trajectories of GLA: symplectic
generalized leapfrog method for Nosé–Poincaré as suggested in
Bond *et al.* (1999). NHI: implicit Nosé–Hoover method given in
Frenkel and Smit (1996). NHE: explicit Nosé–Hoover method of
Holian *et al.* (1990).

It is curious to observe that atom 4 (numbering sequentially from one
end), initially at the centre of the structure, is eventually forced away from
that point. When the system re-folds, it is atom 6 which has moved to the
centre. The lack of initial energy applied to the central atom is apparent in
the lack of motion of that atom in the first few frames. This is an 'unlikely'
state from the perspective of statistical mechanics. We say that the system
was poorly equilibrated. As the system 'equilibrates' all the atoms show
approximately the same range of motion.

We compared the behaviour of three methods, two implementations of
Nosé–Hoover and one of Nosé–Poincaré (the one used in Bond *et al.* (1999)),
based on the generalized leapfrog algorithm. Particularly for large stepsizes,
the symplectic approach is clearly superior. Sample energetic evolution for
the three methods are shown in Figure 2.7, for a million-time-step simulation
with stepsize $h = 0.03$, just below the observed Verlet stability threshold
for this problem. This result mirrors observations of Bond *et al.* (1999), as
well as subsequent studies for various generalizations (Sturgeon and Laird
2000, Hernández 2001).

3. Modified distributions and the error of numerically computed averages

The concept of a geometric (*e.g.*, symplectic) integrator has obvious appeal to the student of classical mechanics. It seems intuitively correct to
demand that a numerical method should mimic as much as possible the

available structure of the dynamical system, when this is possible to do without incurring large computational overheads in the process. So much the better that discrete dynamical systems designed in this way appear to be related to modified continuous dynamics with appropriate properties. This observation is purely aesthetic.

The real and practical importance of molecular dynamics comes in the computation of statistical mechanics using trajectories. The purpose of this section is to demonstrate that, by using the principles of geometric integration and the backward error analysis, it is possible to compute effective estimates of the error in numerically computed averages, and even to correct those averages by a straightforward reweighting.

Suppose we apply an sth-order numerical method, Ψ_h, to approximate the flow of the vector field f. Using backward error analysis (as described in Section 1.5) we can derive a modified vector field, $\bar{f}_r$, for which Ψ_h is an rth-order approximation with $r > s$. Now suppose f has an invariant distribution, ρ, which solves the corresponding Liouville equation (2.1). It is natural to ask if $\bar{f}_r$ has an invariant distribution, $\bar{\rho}_h$, corresponding to ρ. To derive an equation for $\bar{\rho}_h$, suppose

$$\bar{f} = f + h^s g, \quad \text{and} \quad \bar{\rho}_h = \rho \omega,$$

where $\omega = 1 + \mathcal{O}[h^s]$. Inserting $\bar{f}$ and $\bar{\rho}_h$ in the Liouville equation, (2.1), results in

$$\frac{D\bar{\rho}_h}{Dt} + \bar{\rho}_h \nabla_z \cdot \bar{f} = 0,$$

or

$$\omega\left(\frac{D\rho}{Dt} + \rho \nabla_z \cdot f\right) + \rho\left(\frac{D\omega}{Dt} + h^s \omega \nabla_z \cdot g\right) = 0.$$

The first term is zero, since ρ is an invariant distribution of f. Assuming $\rho > 0$, which it is for Nosé–Hoover, we can conclude

$$\frac{D\omega}{Dt} + h^s \omega \nabla_z \cdot g = 0.$$

Assuming $\omega > 0$ and integrating with respect to t results in

$$\bar{\rho}_h = \frac{1}{C} \rho \omega = \frac{1}{C} \rho \exp\left[-h^s \int \nabla \cdot g \, dt\right], \tag{3.1}$$

where C is a normalizing constant. We can use (3.1) to obtain modified distributions of any order, once we have the corresponding modified vector field. Substituting $h^s f_{[s]} + \cdots + h^r f_{[r]}$ for $h^s g$ in (3.1) results in the desired modified distribution.

In general, we expect the error in trajectory averages to depend on both truncation and sampling errors,

$$\left|\langle A\rangle_{\text{Num}} - \langle A\rangle_{\text{Exact}}\right| = \mathcal{O}[h^s] + \mathcal{O}\left[t^{-1/2}\right],$$

where t is the simulation time, and s is the order of the method (Cancès et al. 2004, 2005). For integrable systems, the convergence with respect to sampling time is faster, and can be accelerated using filtering techniques (Cancès et al. 2004, 2005). Similar filtering techniques can be applied to non-integrable systems, although the improvement is less dramatic (see Cancès et al. 2004). Here we will ignore the sampling problem, and focus on the truncation error, which (for ergodic systems) will be described by the modified distribution. In the following subsections, we derive the modified distribution for the Nosé–Hoover and Nosé–Poincaré methods.

3.1. Nosé–Hoover

The explicit Nosé–Hoover method (NHE) given in Section 2.6 can be viewed as an application of the second-order, Lobatto IIIa–IIIb, partitioned Runge–Kutta method,

$$z_a^{n+1/2} = z_a^n + \frac{h}{2} f_a\big(z_a^n, z_b^{n+1/2}\big), \tag{3.2}$$

$$z_b^{n+1/2} = z_b^n + \frac{h}{2} f_b\big(z_a^n, z_b^{n+1/2}\big), \tag{3.3}$$

$$z_a^{n+1} = z_a^{n+1/2} + \frac{h}{2} f_a\big(z_a^{n+1}, z_b^{n+1/2}\big), \tag{3.4}$$

$$z_b^{n+1} = z_b^{n+1/2} + \frac{h}{2} f_b\big(z_a^{n+1}, z_b^{n+1/2}\big), \tag{3.5}$$

to the partitioned system of differential equations,

$$\frac{dz_a}{dt} = f_a(z_a, z_b), \quad \text{and} \quad \frac{dz_b}{dt} = f_b(z_a, z_b),$$

with $z_a = (q, \xi)$ and $z_b = (p, \eta)$.

Calculating the terms of the modified vector field for a numerical method is a tedious (but straightforward) task. Since the methods are second-order accurate, the modified vector field can be written in partitioned form as

$$\bar{f}_{a,r} = f_a + h^2 f_{a,[2]} + \cdots + h^r f_{a,[r]},$$

$$\bar{f}_{b,r} = f_b + h^2 f_{b,[2]} + \cdots + h^r f_{b,[r]},$$

for which it can be shown that the second-order terms are

$$f_{a,[2],i} = \frac{1}{12}\left[\frac{\partial^2 f_{a,i}}{\partial z_{a,j}z_{a,k}} f_{a,j}f_{a,k} + \frac{\partial f_{a,i}}{\partial z_{a,j}}\frac{\partial f_{a,j}}{\partial z_{a,k}} f_{a,k} + \frac{\partial f_{a,i}}{\partial z_{a,j}}\frac{\partial f_{a,j}}{\partial z_{b,k}} f_{b,k} \right.$$

$$- \frac{\partial^2 f_{a,i}}{\partial z_{a,j}\partial z_{b,k}} f_{a,j}f_{b,k} - \frac{1}{2}\frac{\partial^2 f_{a,i}}{\partial z_{b,j}\partial z_{b,k}} f_{b,j}f_{b,k}$$

$$\left. - 2\frac{\partial f_{a,i}}{\partial z_{b,j}}\frac{\partial f_{b,j}}{\partial z_{a,k}} f_{a,k} + \frac{\partial f_{a,i}}{\partial z_{b,j}}\frac{\partial f_{b,j}}{\partial z_{b,k}} f_{b,k} \right], \tag{3.6}$$

$$f_{b,[2],i} = \frac{1}{12}\left[\frac{\partial^2 f_{b,i}}{\partial z_{a,j}\partial z_{a,k}}f_{a,j}f_{a,k} + \frac{\partial f_{b,i}}{\partial z_{a,j}}\frac{\partial f_{a,j}}{\partial z_{a,k}}f_{a,k} + \frac{\partial f_{b,i}}{\partial z_{a,j}}\frac{\partial f_{a,j}}{\partial z_{b,k}}f_{b,k}\right.$$

$$- \frac{\partial^2 f_{b,i}}{\partial z_{a,j}\partial z_{b,k}}f_{a,j}f_{b,k} - \frac{1}{2}\frac{\partial^2 f_{b,i}}{\partial z_{b,j}\partial z_{b,k}}f_{b,j}f_{b,k}$$

$$\left. - 2\frac{\partial f_{b,i}}{\partial z_{b,j}}\frac{\partial f_{b,j}}{\partial z_{a,k}}f_{a,k} + \frac{\partial f_{b,i}}{\partial z_{b,j}}\frac{\partial f_{b,j}}{\partial z_{b,k}}f_{b,k}\right]. \tag{3.7}$$

Here, summation is implied for terms with repeated indices. Applying (3.6)–(3.7) to the partitioning of the Nosé–Hoover vector field for NHE, $z_a = (q, \xi)$ and $z_b = (p, \eta)$ results in

$$f_{a,[2]} = \frac{1}{12}\left[\begin{array}{c} 2M^{-1}U''\dot{q} + 2\dot{\xi}\dot{q} - \xi M^{-1}\dot{p} \\ -\dot{p}^T M^{-1}\dot{p}/\mu + 4\dot{q}^T U''(q)\dot{q}/\mu + 4p^T\dot{q}\dot{\xi}/\mu - 2\xi\dot{q}^T\dot{p}/\mu \end{array}\right],$$

$$f_{b,[2]} = \frac{-1}{12}\left[\begin{array}{c} U'''\{\dot{q},\dot{q}\} + U''M^{-1}\dot{p} + 2\dot{q}^T\dot{p}p/\mu + 2\xi U''\dot{q} - \dot{\xi}\dot{p} + 2\xi\dot{\xi}p - \xi^2\dot{p} \\ -2\dot{q}^T\dot{p}/\mu \end{array}\right].$$

where $\dot{q}$, $\dot{p}$, $\dot{\xi}$, correspond to the vector field of the unmodified Nosé–Hoover system.

Solving for a modified invariant distribution requires computing

$$\bar{\rho}_{2,h}^{NHE} \propto \rho_{NH}\exp\left[-h^2\int \nabla \cdot f_{[2]}\, dt\right], \tag{3.8}$$

where ρ_{NH} is an invariant distribution of the unperturbed Nosé–Hoover vector field. A similar procedure can be followed to find a modified distribution for the implicit Nosé–Hoover method (NHI).

3.2. Nosé–Poincaré methods

As discussed in Section 1.5, when a symplectic integrator is applied to a Hamiltonian system, we can derive a modified Hamiltonian corresponding to the modified vector field. Hamiltonian vector fields are divergence-free, and hence the Liouville equation simply states that the invariant distribution is constant. This corresponds to the microcanonical distribution (ensemble)

$$\rho \propto \delta[H(z) - E].$$

From this we conclude that the modified distribution for the Nosé–Poincaré methods should be the microcanonical ensemble corresponding to the modified Hamiltonian,

$$\bar{\rho} \propto \delta[\bar{H}_r - \bar{E}_0],$$

where H_r is the rth-order modified Hamiltonian for a Nosé–Poincaré method and $\bar{E}_0$ is the initial value of $\bar{H}_r$. We would like to derive a marginal modified distribution, $\bar{\rho}(q,p)$, such that

$$\bar{\rho}(q,p)\,\mathrm{d}p\,\mathrm{d}q = \frac{1}{C}\iint\limits_{s\ p_s} \delta\big[\bar{H}_r(q,s,\tilde{p},p_s) - \bar{E}_0\big]\,\mathrm{d}\tilde{p}\,\mathrm{d}q\,\mathrm{d}p_s\,\mathrm{d}s, \qquad (3.9)$$

where the integration is over the extended variables s and p_s, the 'real momenta' $p = \tilde{p}/s$, and C is a normalizing constant.

Bond *et al.* (1999) demonstrated that if $\bar{H}_r$ is the unmodified Nosé–Poincaré Hamiltonian, the resulting marginal distribution is the canonical distribution in (2.2). Following this proof for the unperturbed Hamiltonian, we can compute an approximation to the modified marginal density, $\bar{\rho}$. Assuming the underlying Nosé–Poincaré numerical method is symplectic and second-order accurate, we can write the second-order modified Hamiltonian as

$$\bar{H}_{\mathrm{NP},2} = s\bigg(H(q,\tilde{p}/s) + \frac{p_s^2}{2\mu} + gk_BT\ln s - H_N^0\bigg) + h^2\,s\,G(q,s,\tilde{p}/s,p_s),$$

where

$$H(q,p) = \frac{1}{2}p^T\mathbf{M}^{-1}p + U(q), \quad sG(q,s,\tilde{p}/s,p_s) := H_{[2]}(q,s,\tilde{p},p_s),$$

and $H_{[2]}$ is the first term of the modified Nosé–Poincaré Hamiltonian. Using this expression in (3.9) results in

$$\bar{\rho}(q,p)\,\mathrm{d}p\,\mathrm{d}q = \frac{1}{C}\iint\limits_{s\ p_s} \delta\big[s\big(H_N - H_N^0 + h^2G\big)\big]\,\mathrm{d}\tilde{p}\,\mathrm{d}q\,\mathrm{d}p_s\,\mathrm{d}s.$$

Note that we have assumed the initial value of the modified Hamiltonian is zero (within the order of the expansion). Although this can be achieved by adding a constant to the modified Hamiltonian, the remaining analysis becomes significantly more complicated. An alternative, which we adopt here, is to make an order h^2 modification to the value of H_N^0. This can be viewed as the value of H_N^0 for which the modified Nosé–Poincaré Hamiltonian, $\bar{H}_{\mathrm{NP},2}$, is a Poincaré time transformation of a modified Nosé Hamiltonian, $H_N + h^2G$.

Introducing a change of variables, $p \leftarrow \tilde{p}/s$ and $\eta \leftarrow \ln s$, results in

$$\bar{\rho} = \frac{1}{C}\iint\limits_{p_s\ \eta} e^{N_f\eta}\delta\bigg[e^\eta\bigg(H(q,p) + \frac{p_s^2}{2\mu} + gk_BT\eta + h^2G(q,e^\eta,p,p_s) - H_N^0\bigg)\bigg]\,\mathrm{d}p_s\,\mathrm{d}\eta,$$

where N_f is the number of degrees of freedom (effective dimension of p). The integral over η involves a delta function of the form $\delta[r(\eta)]$. Assuming that $r(\eta)$ is differentiable and has a single, simple root η_0, we can apply the identity $\delta[r(\eta)] = \delta[\eta - \eta_0]/|r'(\eta_0)|$, and integrate over η:

$$\bar{\rho} = \frac{1}{C} \int\limits_{p_s} e^{N_f \eta_0} \left| g k_B T + h^2 \frac{\partial}{\partial \eta} G(q, e^{\eta}, p, p_s) \right|_{\eta=\eta_0}^{-1} dp_s.$$

The root of $r(\eta)$, denoted by η_0, is implicitly defined by

$$\eta_0 = \frac{-1}{g \, k_B \, T} \left(H(q, p) + \frac{p_s^2}{2\,\mu} + h^2 G(q, e^{\eta_0}, p, p_s) - H_N^0 \right),$$

which can solved to any power of h using a series expansion. Before we can integrate over p_s, we will derive the first term of the modified Hamiltonian for the Nosé–Poincaré methods.

For the generalized leapfrog algorithm (GLA) in (2.23)–(2.25), we note that the second-order Lobatto IIIa–IIIb partitioned Runge–Kutta method in (3.2)–(3.5) applied to a Hamiltonian system results in GLA. Substituting $\partial H/\partial p$ and $-\partial H/\partial q$ for f_a and f_b in (3.6)–(3.7) results in partial differential equations for $H_{[2]}$, which can be solved in the first term of the modified Hamiltonian for GLA:

$$H_{[2]}(q, p) = \frac{1}{24} \sum_j \sum_k \left(2 H_{q_j q_k} H_{p_j} H_{p_k} + 2 H_{q_j p_k} H_{p_j} H_{q_k} - H_{p_j p_k} H_{q_j} H_{q_k} \right).$$

Here, $H(q, p)$ is the original Hamiltonian system, h is the time-step size, and subscripts indicate partial derivatives. It can be shown that generalized leapfrog preserves $\bar{H}_2 = H + h^2 H_{[2]}$ to fourth-order accuracy.

To derive the modified Hamiltonian for Nosé–Poincaré GLA, we apply the above formula to the Nosé–Poincaré Hamiltonian, which yields

$$H_{[2]}(q, s, \tilde{p}, p_s) = \frac{s}{12} \left[\frac{p_s}{\mu} \frac{\tilde{p}^T \mathbf{M}^{-1}}{s} \nabla_q U(q) + \frac{\tilde{p}^T \mathbf{M}^{-1}}{s} U''(q) \frac{\mathbf{M}^{-1} \tilde{p}}{s} + \frac{2 p_s^2}{\mu^2} g k_B T \right.$$

$$\left. - \frac{1}{2} \nabla_q U(q)^T \mathbf{M}^{-1} \nabla_q U(q) - \frac{1}{2\mu} \left(\frac{\tilde{p}^T \mathbf{M}^{-1} \tilde{p}}{s^2} - g k_B T \right)^2 \right]. \qquad (3.10)$$

Here we have discarded all $H_N - H_N^0$ terms since they are zero to order h^2.

Inserting (3.10) in the definition of G, we obtain

$$G(q, s, p, \pi_s) = \frac{1}{12} \left[\frac{p_s}{\mu} p^T \mathbf{M}^{-1} \nabla_q U(q) + p^T \mathbf{M}^{-1} U''(q) \mathbf{M}^{-1} p + \frac{2 p_s^2}{\mu^2} g k_B T \right.$$

$$\left. - \frac{1}{2} \nabla_q U(q)^T \mathbf{M}^{-1} \nabla_q U(q) - \frac{1}{2\mu} (p^T \mathbf{M}^{-1} p - g k_B T)^2 \right], \qquad (3.11)$$

which is not a function of s, and hence not a function of η. Inserting G in the expression for the modified distribution and integrating over p_s results in

$$\bar{\rho} = \frac{\rho_c}{\bar{C}} \exp\left\{ -\frac{h^2}{24 k_B T} \left[\sum_j \sum_k \frac{2 p_j p_k U_{q_j q_k}}{m_j m_k} \right.\right.$$
$$\left.\left. - \sum_j \frac{U_{q_j}^2}{m_j} - \frac{1}{\mu} \left(\sum_j \frac{p_j^2}{m_j} - g k_B T \right)^2 \right] \right\},$$

where $\bar{C}$ is a constant, ρ_c is the canonical distribution, and we have assumed $g = N_f$. Hence, we have shown that the marginal modified distribution for the Nosé–Poincaré GLA can be written as

$$\bar{\rho}(q, p) = \rho_c(q, p) \omega(q, p) + \mathcal{O}[h^4],$$

where ω is a 'reweighting' factor,

$$\omega \propto \exp\left\{ -\frac{h^2}{24 k_B T} \left[\sum_j \sum_k \frac{2 p_j p_k U_{q_j q_k}}{m_j m_k} - \sum_j \frac{U_{q_j}^2}{m_j} - \frac{1}{\mu} \left(\sum_j \frac{p_j^2}{m_j} - g k_B T \right)^2 \right] \right\}.$$

To derive the first term of the modified Hamiltonian for the Hamiltonian splitting method (HSP), Nosé (2001) used the Baker–Campbell–Hausdorff formula for a 3-term splitting,

$$H_{[2]} = -\frac{1}{24} \left[\{\{H_1, H_2\}, H_2\} + 2\{\{H_1, H_2\}, H_1\} \right.$$
$$\left. + \{\{H_1 + H_2, H_3\}, H_3\} + 2\{\{H_1 + H_2, H_3\}, H_1 + H_2\} \right],$$

where $\{\cdot, \cdot\}$ is the Poisson bracket. Inserting the particular H_1, H_2, and H_3 used in the splitting of the Nosé–Poincaré Hamiltonian results in

$$H_{[2]}(q, s, \tilde{p}, p_s) = \frac{s}{12} \left[-2 \frac{p_s}{\mu} \frac{\tilde{p}^T \mathbf{M}^{-1}}{s} \nabla_q U(q) + \frac{\tilde{p}^T \mathbf{M}^{-1}}{s} U''(q) \frac{\mathbf{M}^{-1} \tilde{p}}{s} \right.$$
$$- \frac{p_s^2}{4\mu^2} \left(\frac{\tilde{p}^T \mathbf{M}^{-1} \tilde{p}}{s^2} + 3 g k_B T - \frac{p_s^2}{2\mu} \right) - \frac{1}{2} \nabla_q U(q)^T \mathbf{M}^{-1} \nabla_q U(q)$$
$$\left. + \frac{1}{\mu} \left(\frac{\tilde{p}^T \mathbf{M}^{-1} \tilde{p}}{s^2} - g k_B T + \frac{p_s^2}{2\mu} \right)^2 \right].$$

Here we have discarded all $H_N - H_N^0$ terms since they are zero to order h^2. The corresponding marginal modified distribution can be derived for this modified Hamiltonian using the process outlined earlier in this section.

Given an expression for the reweighting factor, ω, we can use it to reduce truncation error in averages. If both the exact and numerical dynamics are ergodic, the ensemble (or distribution) average will be the same as the time average. To obtain a better approximation of the exact (canonical) average, we can reweight by dividing by ω:

$$\langle A \rangle_c = \frac{\langle A/\omega \rangle_{\text{Num}}}{\langle 1/\omega \rangle_{\text{Num}}} + \mathcal{O}[h^4],$$

where $\langle \cdot \rangle_c$ and $\langle \cdot \rangle_{\text{Num}}$ correspond to the exact canonical and numerical averages respectively.

3.3. Numerical experiment using Nosé–Poincaré

To verify the backward error theory outlined in the previous sections, we performed a constant temperature molecular dynamics simulation of a system of a 256 particle Lennard–Jones gas with periodic boundary conditions. The density and temperature in reduced units (Frenkel and Smit 2002) were set to 0.95 and 1.5 respectively. The time-step size was varied over a range from 0.012 to 0.0001 resulting in as many as 2 million total time-steps for the longest simulation.

In Figure 3.1, the instantaneous temperature is shown as a function of time for a typical simulation. The instantaneous temperature is defined

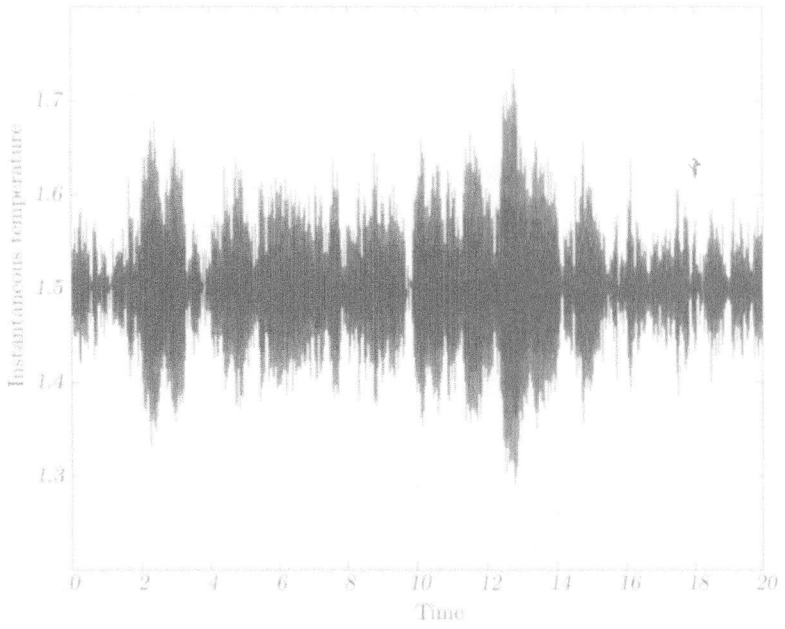

Figure 3.1. The instantaneous temperature is shown as a function of time using a standard per particle kinetic energy estimator.

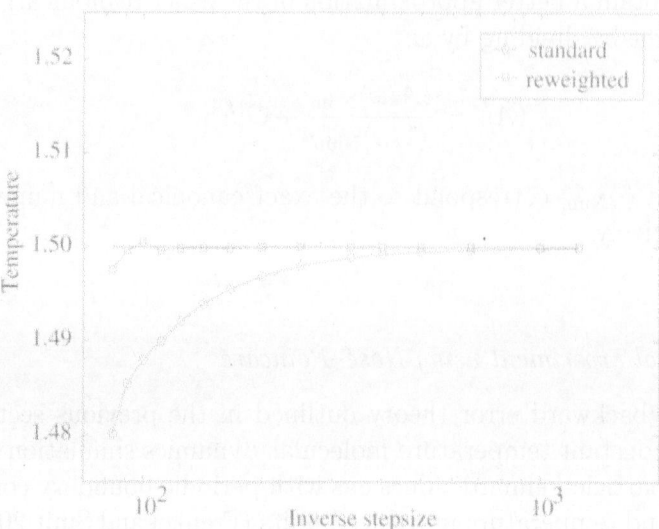

Figure 3.2. The dependence of average
instantaneous temperature on time-step size
is plotted using two different estimators.

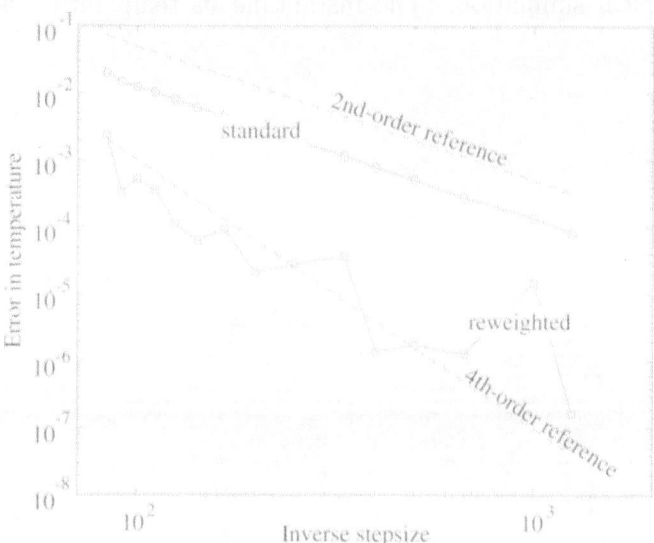

Figure 3.3. Fourth-order accuracy of the
corrected temperatures (after reweighting).

through the per particle kinetic energy,

$$k_B T_{\text{inst}} := \frac{1}{N_f} \sum_j \frac{p_j^2}{m_j},$$

where N_f is the number of degrees of freedom. One can show that for a constant temperature simulation, the instantaneous temperature is not constant, but instead is nearly normally distributed with mean equal to the temperature.

In Figure 3.2, the average of instantaneous temperature is shown as a function of inverse stepsize using a standard average and the proposed reweighting technique. As expected, the standard average converges quadratically to the correct value since the method is second-order accurate. What is remarkable is how insensitive the reweighted average is to stepsize. It converges extremely rapidly, and achieves the correct value to within statistical error for stepsizes near the stability limit of the numerical algorithm.

We examined the error in the computed temperatures *after reweighting*. We found the computation of these errors extremely challenging, with a high variance and the requirement of extremely long integrations, but ultimately we were able to verify the anticipated fourth-order accuracy in the reweighted averages (see Figure 3.3).

4. Open questions

In this section, we consider a variety of issues related to the implementation and enhancement of the Hamiltonian-based Nosé dynamics framework, and the application of these techniques in complicated systems.

4.1. What is 'thermostatted molecular dynamics'?

So far, we have avoided a precise definition of what we mean by thermostatted molecular dynamics, being content to study the properties of certain differential equation models that are often referred to by this title. This is in fact a complex question. Colloquially, a thermostat is a device that regulates the temperature of a molecular system. In order to discuss the thermostat precisely, we need to bring into the discussion the dimension of the system, and with it the notion that a given system in thermal equilibrium is effectively a part of an infinite system with which it exchanges energy.

The confusion comes from the fact that the canonical ensemble is just identified with a precise probabilistic interpretation (equilibrium statistical mechanics). However, it is evidently not enough to say that thermostatted molecular dynamics is just an arbitrary canonical sampling process, for then it would be unnecessary for the dynamics to have anything in common whatsoever with realistic dynamical motions. With that limited interpretation,

we may as well refer to Monte Carlo simulation as a 'thermostatted molecular dynamics'. In the physics literature there appears to be an assumption of
a closer connection between molecular dynamics and thermostatted molecular dynamics. The natural way to characterize this correspondence appears
to be in terms of an asymptotical approximation with the dimension of the
system. Based on this, let us attempt a definition which is not quite precise
and whose verification would be at best difficult for any standard system.

Definition 4.1. Assume a family of molecular models defined for a sequence of numbers of degrees of freedom $\{N_i\}$, $N_1 < N_2 < \cdots$ with microscopic Hamiltonian description $H(w = (q, p); N)$. We also consider a family
of bulk (intensive) properties, defined as averages of microscopic quantities
f_N (functions of the corresponding $2N$-dimensional space) whose canonical
averages converge in the limit of increasing particle number to the average
of a limiting observable f.

Suppose that for each $H(w; N)$ there is a modified dynamics, which generates trajectories $w_N(t) = (q_N(t), p_N(t))$ as follows.

(1) For almost all initial values, $w_N(t)$ is a canonical sampler, *i.e.*,

$$\lim_{S \to \infty} \frac{1}{S} \int_0^S f_N(w_N(t)) \, \mathrm{d}t = \int f_N(w) e^{-\frac{1}{kT} H(w; N)} \, \mathrm{d}w,$$

where the latter integration is performed over the phase space of the
N-degrees-of-freedom model.

(2) Temporal correlation functions computed from dynamics of the thermostatted molecular model approximate corresponding temporal correlations of the microcanonical system, as $N \to \infty$. By this we mean that
a function of the form C_N defined, for almost any trajectory w_N of the
modified dynamics, by

$$C_N(\tau) = \lim_{S \to \infty} S^{-1} \int_0^S w_N(t + \tau)^T B w_N(t) \, \mathrm{d}t,$$

for some arbitrary matrix B, converges asymptotically to its microcanonical equivalent in the large N limit.

Under these conditions we refer to the modified dynamics as a thermostatted
molecular dynamics.

As we have indicated, this is probably not a good practical definition, since
it is quite difficult to verify condition (2), except in special circumstances.
In the context of Nosé dynamics, condition (1) is immediately verified by
Nosé's original paper.

4.2. What is the role of numerical error in sampling dynamics?

On the surface this appears to be a simple question: for sufficiently small steps, we assume the method provides improving accuracy on any given time interval. This is typically made rigorous by numerical analysis which provides error bounds and estimates for computational methods, or at least for idealized model problems, or, in our case, by the backward error analysis which was the subject of the previous sections. However, the strong constraint we work under in molecular simulation is the need to push methods to their limits by increasing time-step size. This means that, in real-world molecular simulation, we generally allow some error to be introduced into the computation of a trajectory.

It is commonly supposed that any explicit numerical method with fixed time-step is destabilized by a sufficiently large velocity of any particular atom. When we generate sampling dynamics, we expect the momenta to be drawn from a normal distribution, which is not compactly supported. Although the tails of the normal distribution are small, the practical consequence of sampling dynamics is that the stability threshold must depend on temperature. We could imagine that, in any sufficiently long sampling trajectory computation, if we are not too far below the threshold, there will be occasional events in which the stepsize is sufficiently large that the tenets of backward error analysis fail, or, at least, that the constraint introduced by backward error analysis is not strong.

In the context of thermostatted molecular dynamics, this raises two possible scenarios. First, the success of molecular dynamics may hinge on limitation of the sampling. This could even be introduced in a practical and explicit way by adding some sort of dynamic restraint to prevent certain types of excursions which lead to instability. The second scenario is that the thermostat acts as a sort of reservoir for numerical error. This could

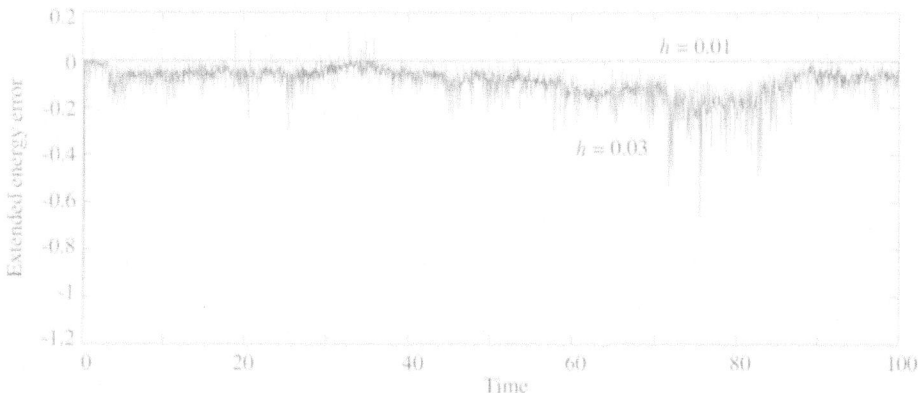

Figure 4.1. Energy errors *vs* time for two different stepsizes.

explain why the NHI scheme, which appears to show obvious instability in Figure 2.7, is a widely used and popular scheme. The method controls the temperature of the physical variables well, even as the auxiliary thermostat variables show large and growing deviations from the initial state.

To illustrate, we include here a graph (Figure 4.1) showing the extended energy error at two stepsizes using a Nosé–Poincaré method (based on generalized leapfrog) for the seven-atom chain model. In simulation, the growth of energy error is associated to the collisional dynamics of individual pairs of atoms. At isolated points along the trajectory, collisions (close approaches of the atoms) are observed which cause an increase in energy. When the stepsize is very large (or the temperature is very large), these energetic collisions may result in a catastrophic chain reaction, with the isolated strong collision causing one after another. However, there is often, as here, a substantial grey area (a large range of stepsizes) where this type of explosion in energy is never observed, but where the tenets of backward error analysis nevertheless fail to hold.

In our example, backward error analysis does not appear to constrain error growth at $h = 0.03$, as we observe an erratic drift in the extended energy. However, temperature is well controlled in both simulations, and the simulations result in good sampling of the energy landscape. If we removed the graph for $h = 0.03$, and took a closer look at the graph for $h = 0.01$, we would observe the same jumps/drift as for $h = 0.03$, just with substantially smaller magnitude and a slower growth rate. In practice, the simulation with $h = 0.01$ may well be no better than that for $h = 0.03$, for practical purposes, and, since larger stepsizes mean that longer intervals can be covered in the same amount of wall clock time, it is not unlikely that an experimenter would opt for the larger stepsize. If the key feature of thermostatted MD is control of temperature, and this can be obtained as well or better by use of Nosé–Hoover methods, or other methods, the importance of symplectic methods and of a perturbed Hamiltonian expansion is in doubt.

This example, which is hardly atypical, raises important questions regarding the basis for relying on symplecticness as a key criterion for molecular dynamics integrators. On the other hand, at present, nothing is known regarding the statistics of jumps observed in the energy (or rather the shadow energy) in molecular simulation in the case that backward error analysis fails.

Finally, we mention that there are a variety of different geometric properties that are associated to molecular models: first integrals, time-reversal (TR) symmetry and symplectic structure. Although it seems, from practical experience, that maintaining first integrals alone is not sufficient to allow long-term simulations to be performed with sufficient accuracy for sampling, it is far less obvious to which extent TR symmetry is an appropriate foundation for method building for highly chaotic molecular systems.

There is numerical evidence that TR symmetry does allow, at some sufficiently small stepsize and for some problems, long-term simulations to proceed with good energy conservation compared to non-time-reversible and non-symplectic methods, just as a similar statement can be made regarding symplectic methods. The situation is complicated by the fact that time-reversible methods are, in many cases, easier to construct than their symplectic counterparts and sometimes more efficient. Moreover, there are examples of efficient numerical methods that are not only non-symplectic, but are also not time-reversible, volume- or integral-preserving, but which appear to give good long-term averages in molecular simulations (Leimkuhler, Legoll and Noorizadeh 2007), although it is clear by now from vast numerical experience that most of the popular molecular dynamics integrators are symplectic, or at least time-reversible. It may well be that these properties make it easier for an integrator to be effective, but are not essential.

Leaving aside these weighty concerns, we ask the more practical question: Can the stability of the methods mentioned for simulating Nosé dynamics be improved upon by attention to heuristic considerations?

4.3. Can we enhance stability in Nosé dynamics simulations?

Designing effective schemes for Nosé dynamics and variants is obviously crucial to implementation as a practical tool. The importance of this is most dramatic in the case of the generalized thermostat chains mentioned above. For example, in recent work, recursive multiple thermostats were applied to simulate an alanine dipeptide model (Barth, Leimkuhler and Sweet 2005). It was found that enhanced ergodicity was possible compared to more traditional Nosé–Hoover chains (Martyna, Tuckerman, Tobias and Klein 1996, Jang and Voth 1997), but the results were disappointing in the sense that numerical stability was clearly compromised and small time-steps were needed.

Since the cost per time-step is similar for all the methods of interest, and we are mostly limited by stability rather than accuracy, the time-step size restriction is essentially the measure of efficiency of a method. In what follows we describe some preliminary ideas to increase the stability threshold.

In Nosé–Hoover, we always work with a physical momentum variable p. On the other hand, in each of the symplectic methods mentioned above, the momentum $\tilde{p}$ and the thermostat variable s are computed at staggered time points. This appears to raise the possibility of an instability when s approaches zero compared to the situation where $\tilde{p}$ and s are computed simultaneously within the method at the same time level. In that case we can view the scheme as evolving, instead of $\tilde{p}$, the physical momentum $p = \tilde{p}/s$. There appears to be no obvious way to generate a symplectic method that works with the physical momentum using generalized leap-frog, since if we rearrange the equations in applying this method or a variant thereof, it will cease to be symplectic. On the other hand, it is possible to adjust the symplectic splitting of H so that p is evolved instead of $\tilde{p}$ by using a Hamiltonian splitting. All we need to do is to make sure that the terms

$$H^{p_s} = s\frac{p_s^2}{2\mu},$$

and

$$H^q = sU(q),$$

which are the only terms of H_N that directly involve p_s and q, are evolved simultaneously. As one example, we could use

$$H_1 = H^{p_s} + H^q = s\frac{p_s^2}{2\mu} + sU(q)$$

as the basis of our splitting, and either set $H_2 = H - H_1$ or further split this term. It is also necessary to assume that only H_1 involves q and p_s. (Technically these variables could be reintroduced in other terms of the splitting, but it is unlikely to be advantageous in any case.)

Balanced methods
A potential problem exists with all the splitting methods mentioned so far. Let us reformulate the integrator in terms of the physical momentum p and $\xi = p_s/\mu$ (the same variables as used in Nosé–Hoover). The differential equations corresponding to H_1 can be written as

$$\frac{dp}{dt} = F - \xi p, \tag{4.1}$$

$$\frac{ds}{dt} = s\xi, \tag{4.2}$$

$$\mu\frac{d\xi}{dt} = -\frac{\xi^2}{2} \tag{4.3}$$

(q constant). This system is similar in appearance to the Nosé–Hoover formulation, although the force F is here fixed and the control law (4.3) is different. H_2 gives rise to

$$\frac{dq}{dt} = \mathbf{M}^{-1}p,$$

$$\mu\frac{d\xi}{dt} = -\frac{\partial H_2}{\partial s}$$

$$= \frac{p^T\mathbf{M}^{-1}p}{2} - gk_BT(1 + \ln s), \tag{4.4}$$

when written in terms of the physical momentum.

In the Nosé–Hoover equations of motion, the 'force' acting on the thermostat variable has zero mean and is close to zero when the average kinetic energy is close to the target temperature, *i.e.*, when the system is large and near thermal equilibrium. In the Nosé–Poincaré method, this equation is slightly perturbed, but by a term which typically remains small $(O(h^2))$ on exponentially long time intervals (or, anyway, throughout a typical molecular simulation, as can be verified in retrospect by monitoring the extended energy error ΔH_N). Under discretization, it is desirable that the thermostat momentum be updated from an equation that maintains this special feature, in order to limit oscillations of the thermal variable (see Figure 4.2). The methods proposed so far for Nosé–Poincaré do not retain this feature.

To correct the physical momentum method mentioned above, it suffices to use instead the following splitting terms:

$$H_1^b = s\frac{p_s^2}{2\mu} + sU(q) + gk_BTs\ln s - sH_N^0 - \frac{1}{2}sk_BT,$$

$$H_2^b = \frac{\tilde{p}^T\mathbf{M}^{-1}\tilde{p}}{2s} + \frac{1}{2}sk_BT$$

It is easily verified that this is a valid splitting of H. After writing the

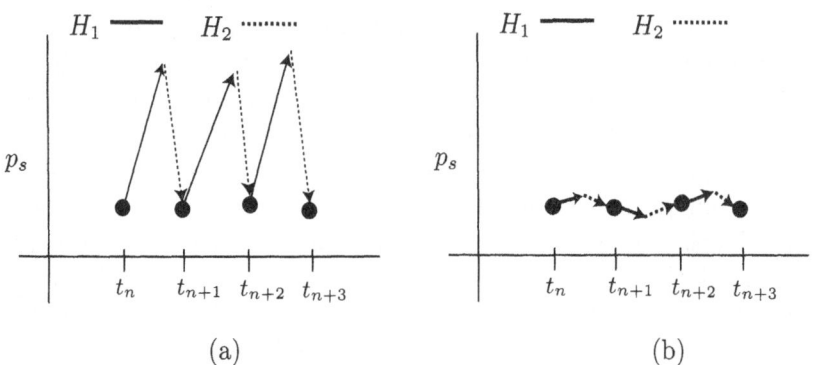

Figure 4.2. Illustration of balanced method. (a) In an unbalanced scheme, relatively large oscillations in the internally computed thermostat momentum are introduced which must cancel to make a small correction. In a balanced method (b), the oscillations are controlled in each substep.

differential equations on this Hamiltonian we obtain, for H_1^b, the system

$$\frac{d\tilde{p}}{dt} = sF,$$

$$\frac{ds}{dt} = sp_s/\mu,$$

$$\frac{dp_s}{dt} = -\frac{p_s^2}{2\mu} - U(q) - gk_BT(1 + \ln s) + H_N^0 - \frac{1}{2}k_BT.$$

However, we may rewrite the latter equation in the form

$$\mu\frac{d\xi}{dt} = \frac{p^T M^{-1} p}{2} - \frac{1}{2}k_BT + \Delta H_N,$$

where we have used the previous definitions of ξ and p. Along a symplectic numerical trajectory (*i.e.*, generated by iterating a symplectic integrator), we see that when the Hamiltonian system is near to thermal equilibrium, then the thermostat variable will be subject to small perturbations.

H_2^b gives rise to the equations of motion

$$\frac{dq}{dt} = M^{-1}p,$$

$$\frac{dp_s}{dt} = \frac{1}{2}(p^T M^{-1} p - gk_BT),$$

which is, once again, a balanced system in the sense introduced above.

This splitting is a little different from those typically suggested, in that the H_1^b term is not easily integrated (it is formally integrable, but the resolution of the motion with time is not trivial). A more common approach is to look for splittings of the equations into pieces which are either trivialized or which are easy to integrate in the sense that they are low-dimensional and can be treated inexpensively by a general-purpose symplectic method such as (2.23)–(2.25). Such a method could of course be employed to solve H_1^b. A conceptually simpler approach is not to concern oneself with obtaining exactly integrable problems, but rather to suppose the existence of an accurate 'black box' ordinary differential equation solver which is capable of producing solutions to some part of the split system to available numerical precision (*i.e.*, to rounding error). After all, Bessel functions and Jacobi elliptic functions are examples of solutions to differential equations we are perfectly comfortable to compute using such accurate numerical codes provided in the form of numerical libraries.

This approach could have stability benefits and it still produces a symplectic integrator (to within the tolerance of the numerical solver used) if accuracy is controllable and the cost of solving the subsystems accurately is contained. It should be recalled that in typical large-dimensional molecular dynamics simulations, the cost of computing a handful of thermostat variables, even very accurately, will quickly pale into insignificance compared to the cost of computing molecular dynamics forces which grow rapidly with system size. Many numerical methods are available to integrate the small-dimensional problem, including a variety of high-order variable stepsize methods, as discussed in Hairer, Nørsett and Wanner (1987).

In limited experiments, however, it appears that the balanced methods have only shown modest improvement in usable stepsize for general MD models. More extensive experimentation is required to fully clarify this picture.

4.4. Can we improve sampling efficiency?

There are two distinct types of situations in which the Nosé dynamics methods may be employed. First they may be used for sampling near the equilibrium state of an ergodic system such as a Lennard–Jones liquid. Then the task of the thermostat is typically to maintain by small corrections the equilibrium state and/or to allow adjustment of the temperature from an arbitrary initial condition under a relatively slow change while maintaining the system near equilibrium. The second use of thermostats is for studying systems which are far from equilibrium, e.g., because of inadequate strong coupling or poor choice of initial data. In this case the thermostat (or more likely, a chain of thermostats) is expected to add the necessary ergodicity so that the system can achieve equilibrium.

One case where this occurs is when the system is dominated by harmonic interactions. For example, biomolecules with stiff chemical bonds will present such difficulties. In materials science, it is not uncommon to compute the absolute free energy (the amount of thermodynamic energy which can be converted to work) using the technique of *alchemical free energy perturbation* or *thermodynamic integration* in which a given reference system with known free energy is morphed into another more interesting system by a smoothly parametrized change. For solids the starting state frequently used is an 'Einstein crystal', consisting of purely harmonic interatomic potentials, and much of the simulation evolves with a 'nearly harmonic' model (Kaczmarski, Rurali and Hernández 2004).

It should be clear from examination of the effective potential (considered in Section 2) that it is unlikely that Nosé dynamics alone would enhance ergodicity. It was understood by Nosé himself from his first work on the subject that Nosé (or Nosé–Hoover) dynamics is unable to thermalize the harmonic oscillator. The only tunable parameter in Nosé dynamics is the thermal mass μ. It is not even clear that there is a 'best' choice of this parameter, in any practical sense, although for harmonic models we might imagine that such a choice exists. Until recently the question was not properly addressed even in the case of a single harmonic oscillator. Leimkuhler and Sweet (2005) made a careful study of the impact of the mass parameter on the resulting combined dynamics, using a technique to estimate the envelope of the accessible phase space. By predicting the onset of chaos in the thermostatted harmonic oscillator, the authors were able to bracket the optimal thermal mass to a certain interval. Unfortunately, even with the best choice, identified from numerical experiments, the thermalized trajectories obtained from Nosé dynamics do not fill in the accessible phase space. The problem is associated to the presence of KAM tori, which can be seen as arriving from the effective perturbation of the integrable system (*i.e.*, the harmonic oscillator, associated to infinite thermal mass). Recently, the KAM approach was used to analyse Nosé dynamics by Legoll, Luskin and Moeckel (2007).

The idea of dynamical thermostatting requires a more complex underlying model to achieve ergodicity in such an application. An alternative is to introduce a more complicated thermal bath, say with several, rather than just one additional degree of freedom.

In the most general form, these thermostats can be described by a Hamiltonian of the form

$$H_{GN} = H(q, \tilde{p}/S) + H_G(s_1, s_2, \ldots, s_m, p_{s_1}, p_{s_2}, \ldots, p_{s_m}),$$

where $S = \Pi_{\alpha \in A} s_\alpha$ is taken over a subset A of the indices $1, 2, \ldots, m$, and the 'bath' H_G is chosen so that the canonical density can be obtained through integration over the entire set of thermostatting variables and their momenta:

$$\int \cdots \int \left(\delta \left[H_{GN} - H_{GN}^0 \right] \mathrm{d}\tilde{p}_1 \, \mathrm{d}\tilde{p}_2 \cdots \mathrm{d}\tilde{p}_{3N} \right) \mathrm{d}s_1 \, \mathrm{d}s_2 \cdots \mathrm{d}s_m \, \mathrm{d}p_{s_1} \, \mathrm{d}p_{s_2} \cdots \mathrm{d}p_{s_m}$$

$$= \exp\left(-\frac{1}{k_B T} H(q, p) \right) \mathrm{d}p_1 \, \mathrm{d}p_2 \cdots \mathrm{d}p_{3N}.$$

Examples of this type of thermostatting bath are developed and applied to physical and chemical systems in Laird and Leimkuhler (2003), Leimkuhler and Sweet (2004, 2005), Barth *et al.* (2005), Jia and Leimkuhler (2006) and Gill, Jia, Leimkuhler and Cocks (2006). One of the key challenges that has become evident through this work is that all of the dynamical thermostat models contain a large number of parameters which must be chosen carefully to ensure efficient and complete sampling. Barth *et al.* (2005) described a heuristic for automatically determining the optimal thermostatting parameters, but it seems evident that these parameters will need to be chosen using an adaptive method. The choice of parameters is complicated by the fact that their selection influences not only ergodicity, but also numerical stability.

4.5. Are there effective generalizations of Nosé dynamics for other molecular ensembles?

In many applications, it is necessary to modify the Nosé dynamics scheme to make it useful in a physical setting. For example, it is often necessary to control not only the temperature of a simulation, but also the pressure. To model such a situation, we allow the simulation cell size to vary, following a suggestion of Andersen (1980). In some cases, for example for homogeneous liquids, we can use a cubic simulation cell. The cell volume is assumed to be V and the box side $L = V^{1/3}$. We rescale coordinates,

$$q_i = L\hat{q}_i,$$

and simulation is performed within the unit cube. Transforming coordinates in this way means that momenta should be rescaled by

$$p_i = L^{-1}\hat{p}_i.$$

The Nosé–Andersen equations of motion (Nosé 1984b) are then derived from an extended Hamiltonian that includes both thermostatting and barostatting terms:

$$H_{\text{NTP}} = \frac{\hat{p}^T M^{-1}\hat{p}}{2s^2 V^{2/3}} + \frac{p_s^2}{2\mu V^{2/3}} + \frac{p_V^2}{2\nu s^2}$$
$$+ U(V^{1/3}\hat{q}) + PV + gk_BT\ln s,$$

where P is the external pressure. A temporal rescaling must be introduced (now by $sV^{1/3}$). The resulting equations of motion can be treated symplectically using a generalized leapfrog discretization, as demonstrated by Sturgeon and Laird (2000), with similar benefit in stability as observed for Nosé–Poincaré.

The ideas of Nosé–Poincaré sampling can also be applied to simulate a system with an irregular shaped box, or one that allows variation of the relative dimensions (Hernández 2001). Artificial ensembles which have a density which can be viewed as a smooth function of the Hamiltonian can be treated using dynamic thermostats based on ideas similar to Nosé–Poincaré (Barth, Laird and Leimkuhler 2003). Jia and Leimkuhler (2006) introduced methods for isothermal/isobaric simulation in only one part of a multiple scale model.

Even more exotic ensembles are of interest, however, including the grand canonical ensemble in which the particle number N is allowed to fluctuate during simulation. A numerical method for this type of simulation was proposed by Lynch and Pettitt (1997). There is no current understanding of how to treat this type of system using a geometric integration method.

4.6. What is the role of the thermostat in non-equilibrium modelling?

This article has been mostly concerned with equilibrium simulation in the canonical ensemble for standard autonomous Hamiltonian systems. In recent years there has been a rise in interest in simulating pre-equilibrium or transient dynamics at the molecular level of detail, and there is a need for methods for equilibrium simulations which are driven (slowly or rapidly) in time.

Partial thermostats
As an illustration, we mention the use of thermostats that are applied to only a part of the system. This can be achieved relatively easily within the context of Nosé–Hoover dynamics (Nosé 1991), but it is also possible using a partial thermostatting technique based on Nosé–Poincaré, as described in a recent article of Jia and Leimkuhler (2006). Let us label the two groups of variables of the system q, p and Q, P and assume an underlying Hamiltonian of the form

$$H = \frac{P^T \mathbf{M}^{-1} P}{2} + \frac{p^T m^{-1} p}{2} + U(q, Q).$$

The idea arrived at in Jia and Leimkuhler (2006) is to work with a dynamical model which couples Newtonian dynamics in Q, P with a Nosé–Poincaré thermostatted subsystem:

$$\dot{Q} = \mathbf{M}^{-1} P,$$

$$\dot{P} = -\nabla_Q U(q, Q),$$

$$\dot{q} = \frac{m^{-1} p}{s},$$

$$\dot{p} = -s \nabla_q U(q, Q),$$

$$\dot{s} = s \frac{\pi}{\mu},$$

$$\dot{p}_s = \frac{p^T m^{-1} p}{s^2} - g_f k_B T - \Delta \mathcal{H},$$

where

$$\Delta \mathcal{H} = \frac{P^T \mathbf{M}^{-1} P}{2} + \mathcal{H}_{\text{Nosé}}^{[f]} - \mathcal{H}_0,$$

$$\mathcal{H}_{\text{Nosé}}^{[f]} = \frac{p^T m^{-1} p}{2} + \frac{p_s^2}{2\mu} + U(q, Q) + g_f k_B T \ln s,$$

g_f is equal to the number of degrees of freedom of light particles, and $\mathcal{H}_0$ is given by

$$\frac{P^T \mathbf{M}^{-1} P}{2} + \mathcal{H}_{\text{Nosé}}^{[f]}$$

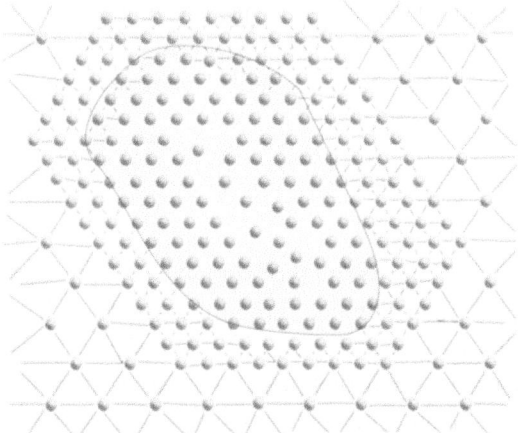

Figure 4.3. Coarse-grained molecular dynamics:
the simulation involves both fully atomistic
dynamics (*e.g.*, in the vicinity of a developing
defect) and a natural finite-element discretization
based on a coarsened atomic lattice.

at initial values. In Dupuy, Tadmor, Miller and Phillips (2005) and Gill
et al. (2006), thermostatted simulation of a partially coarse-grained molecular dynamics model was considered. Figure 4.3 presents an illustration of
the model whereby a defect region is to be treated with a coarse-grained
boundary domain. To illustrate, suppose that we have a homogeneous system, consisting of N similar atoms interacting in a uniform potential energy:

$$H(q,p) = \sum_{i=1}^{N} \frac{\|p_i\|^2}{2m} + \sum_{i<j} \phi(\|q_i - q_j\|).$$

The assumption is that in some region of space, we may define a natural
discretization based on $2\times$, $4\times$, ..., spacing of the approximate atomic lattice. For this purpose, we simply remove intermediate atoms and view the
configuration as being determined by a collection of *representative* atoms.
Between the atoms the interaction becomes the free energy of the representative atoms, averaging over the motion of the constituent atoms of each
element. Assuming that these intermediate atoms are near atomic lattice
sites, we can directly compute the free energy for the harmonic approximation. For example, in one dimension, we are simply removing successive
atoms between two neighbours. Assuming nearest neighbour interactions
only, the approximate energy of element i is found to be

$$\frac{1}{2} k_B T (n_i - 1) \ln \left(\left. \frac{\partial^2 \phi}{\partial r^2} \right|_{r=r_i} \right),$$

where $n_i - 1$ is the number of intermediate atoms removed from the element and r_i is the mean spacing within the element. This results in an effective coarse-grained potential for the representative atoms and a Hamiltonian of the form

$$H(q, p, Q, P) = \sum_{i=1}^{N_a} \frac{\|p_i\|^2}{2m} + \sum_{i=1}^{N_b} \frac{\|P_i\|^2}{2M_i} + V_{\text{eff}}(q, Q), \qquad (4.5)$$

where Q represents positions of the N_b coarse-grained element representative atoms and q the N_a atoms which are untouched. The goal is accurate dynamics of the fully atomistic region using a sampled force field associated to the coarse-grained part. Various possibilities exist for the masses assigned to the representative atoms in the coarse-grained part: because they are only used to provide sampling in the coarse-grained region, their choice should not affect dynamics in the atomistic domain. Most authors have used lumped masses, assigning to the representative atoms also the mass of the eliminated atoms.

The challenge then is to develop effective thermostatting methods for (4.5) which preserve the dynamical evolution in the atomistic region. One approach is to use a generalized bath as described in Section 4.4, applied only to the coarse-grained part of the system (Gill *et al.* 2006).

In Jia and Leimkuhler (2006), a non-equilibrium partial thermostatting method somewhat similar to this was proposed and tested for simulating systems with an artificial thermal gradient (two temperatures in one simulation).

Time-dependent models
As an illustration of the treatment of a time-dependent model, consider the physical process of *annealing*, which is used commonly to strengthen metallic alloys. By slowly cooling a material, it often enables a more perfect crystal structure to form which has stronger material properties. In *simulated annealing* (Kirkpatrick, Gelatt and Vecchi 1983), a schedule of temperature decay is introduced into Monte Carlo simulation. The result is a method that samples an ever-lower value of energy. It can be shown that if the temperature schedule is sufficiently slow, *e.g.*, proportional to $1/\log(t)$, then the system is forced to find a global minimum (Hajek 1988). For many applications it is necessary to forgo the theoretical foundation and use a much faster temperature reduction, so that a minimum is reached in a shorter period of time.

Annealing can also be implemented in an MD setting by introducing a time-dependent temperature parameter. The idea is then to implement a temperature control mechanism that allows temperature to vary with time. The theory that applies to simulated annealing can be used essentially unmodified to justify this dynamical approach, which is effectively a

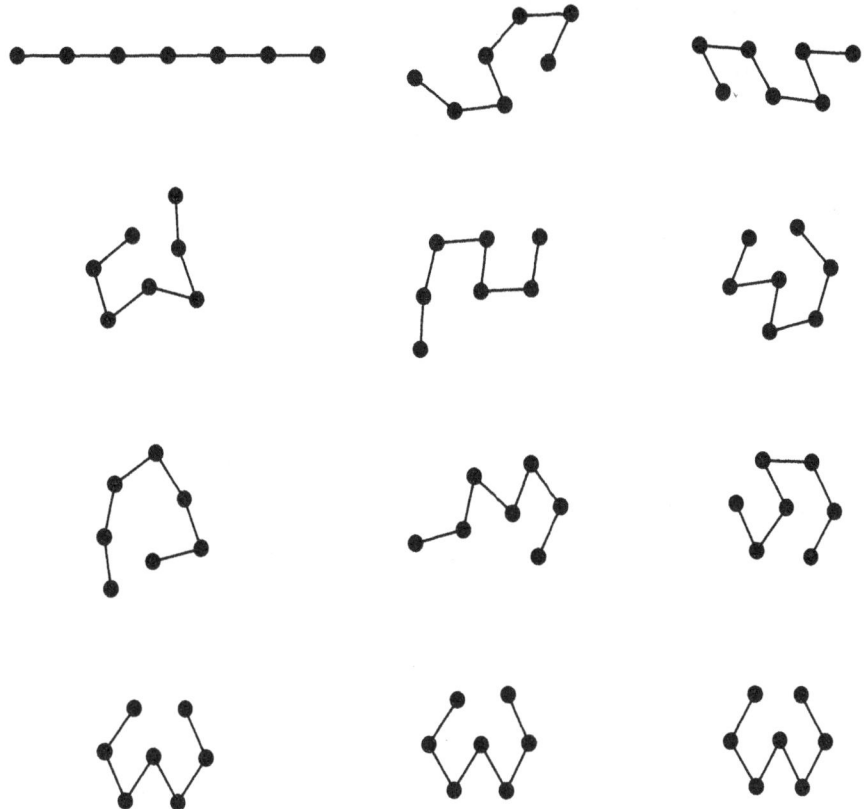

Figure 4.4. Annealing simulation of the seven-atom chain: snapshots at equally spaced points in time show progression to the vicinity of a low-energy state.

microscopic formulation of annealing. The natural question is: How can we formulate (high-quality) numerical algorithms in such a setting? One idea is simply to modify an existing algorithm for constant temperature simulation so that the temperature is allowed to vary.

For example, we have implemented the GLA scheme allowing $T(t)$ to vary from step to step. We applied this to simulate the folding of the seven-atom chain from an extended configuration, with the results shown in Figure 4.4. The temperature schedule used was just $T_0/(1 + \log(t))$, where $T_0 = 1$.

Despite the fact that it can minimize at least a simple chain model, this algorithm has a serious flaw. In Nosé–Poincaré, the rescaling $H \to s(H - H_0)$ is only correct if $H - H_0 = 0$. However, if $T = T(t)$, we have $H = H(t)$ and this will no longer hold. This means that the time variable is incorrect in this approach.

To correct the algorithm we assume that the temperature variation can be regarded as adiabatic compared to the evolution of the physical variables and thermostat. Then we write

$$\tilde{H} = s\left[H(q, \tilde{p}/s) + \frac{p_s^2}{2\mu} + gk_B T(t) \ln s - H_N\right] = s\Delta.$$

This yields equations of motion

$$\frac{dq}{d\tau} = \mathbf{M}^{-1}\tilde{p}/s, \qquad (4.6)$$

$$\frac{d\tilde{p}}{d\tau} = -s\nabla_q H(q, \tilde{p}/s), \qquad (4.7)$$

$$\frac{ds}{d\tau} = sp_s/\mu, \qquad (4.8)$$

$$\frac{dp_s}{d\tau} = s\left[s^{-2}\tilde{p} \cdot \nabla_p H(q, \tilde{p}/s) - T(t)\right] - \Delta. \qquad (4.9)$$

This is in exactly the same form as the Nosé–Poincaré method. The difference is the time-dependent temperature and also the fact that H_N (in Δ), and t are viewed as dependent variables. As H_N should be regarded as conjugate to time, we have

$$\frac{dH_N}{d\tau} = gkT'(t)s \ln s, \qquad \frac{dt}{d\tau} = s,$$

which is an additional pair of equations that must be solved in tandem with the physical variables. A discretization for this system can be derived by application of the generalized leapfrog method.

In several recent articles, Michel Cuendet has employed a similar technique for time-dependent Hamiltonians together with Jarzynski's relation (Jarzynski 1997a, 1997b) to relate non-equilibrium work averages and thermodynamic free energy differences, where paths are computed using Nosé–Hoover or Nosé–Poincaré thermostatted dynamical trajectories (Cuendet 2006a, 2006b). The precise details of numerical treatment and the benefit of symplectic methods in such time-dependent systems are still unclear.

Acknowledgements

The authors wish to thank Ruslan Davidchack and Nana Arizumi for reading a draft of this article and providing many useful comments.

REFERENCES

E. Akhmatskaya and S. Reich (2005). The targetted shadowing hybrid Monte Carlo (TSHMC method), in *New Algorithms for Macromolecular Simulation*, Springer, Berlin, pp. 141–153.

B. Alder and T. E. Wainwright (1957), 'Phase transition for a hard sphere system', *J. Chem. Phys.* **27**, 1208–1209.

M. P. Allen and D. J. Tildesley (1987), *Computer Simulation of Liquids*, Oxford Science, Oxford.

H. C. Andersen (1980), 'Molecular dynamics simulations at constant pressure and/or temperature', *J. Chem. Phys.* **72**, 2384–2393.

J. Barnes and P. Hut (1986), 'A hierarchical $O(N \log N)$ force-calculation algorithm', *Nature* **324**, 446–449.

E. Barth, B. Laird and B. Leimkuhler (2003), 'Generating generalized distributions from dynamical simulation', *J. Chem. Phys.* **118**, 5759–5768.

E. Barth, B. Leimkuhler and C. R. Sweet (2005), Approach to thermal equilibrium in biomolecular simulation, in *New Algorithms for Macromolecular Simulation* (B. Leimkuhler, C. Chipot, R. Elber, A. Laaksonen, A. Mark, T. Schlick, C. Schütte and R. Skeel, eds), Vol. 49 of *Lecture Notes in Computational Science and Engineering*, Springer, pp. 125–140.

G. Benettin and A. Giorgilli (1994), 'On the Hamiltonian interpolation of near to the identity symplectic mappings with application to symplectic integration algorithms', *J. Statist. Phys.* **74**, 1117–1143.

S. Bond, B. Laird and B. Leimkuhler (1999), 'The Nosé–Poincaré method for constant temperature molecular dynamics', *J. Comput. Phys.* **151**, 114–134.

A. T. Brunger, C. L. Brooks and M. Karplus (1984), 'Stochastic boundary conditions for molecular dynamics simulations of ST2 water', *Chem. Phys. Lett.* **105**, 495–500.

E. Cancès, F. Castella, P. Chartier, E. Faou, C. Le Bris, F. Legoll and G. Turinici (2004), 'High-order averaging schemes with error bounds for thermodynamical properties calculations by molecular dynamics simulations', *J. Chem. Phys.* **121**, 10346–10355.

E. Cancès, F. Castella, P. Chartier, E. Faou, C. Le Bris, F. Legoll and G. Turinici (2005), 'Long-time averaging for integrable Hamiltonian dynamics', *Numer. Math.* **100**, 211–232.

R. Car and M. Parinello (1985), 'Unified approach for molecular dynamics and density functional theory', *Phys. Rev. Lett.* **22**, 2471–2474.

D. Chandler (1987), *Introduction to Modern Statistical Mechanics*, Oxford University Press, New York.

M. Cuendet (2006a), 'The Jarzynski identity derived from general Hamiltonian or non-Hamiltonian dynamics reproducing NVT or NPT ensembles', *J. Chem. Phys.* **125**, # 144109.

M. Cuendet (2006b), 'Statistical mechanical derivation of the Jarzynski identity for non-Hamiltonian thermostated dynamics', *Phys. Rev. Lett.* **96**, # 120602.

T. Darden, D. York and L. Pedersen (1993), 'Particle mesh Ewald: An $N \cdot \log(N)$ method for Ewald sums in large systems', *J. Chem. Phys.* **98**, 10089–10092.

R. De Vogelaere (1956), Methods of integration which preserve the contact transformation property of the Hamiltonian equations, Technical Report 4, Department of Mathematics, University of Notre Dame, IN.

C. P. Dettmann (1999), 'Hamiltonian reformulation for a restricted isoenergetic thermostat', *Phys. Rev. E* **60**, 7376–7377.

C. P. Dettmann and G. P. Morriss (1997), 'Hamiltonian reformulation and pairing of Lyapunov exponents for Nosé–Hoover dynamics', *Phys. Rev. E* **55**, 3693–3696.

S. Duane, A. D. Kennedy, B. J. Pendleton and D. Roweth (1987) 'Hybrid Monte Carlo', *Phys. Lett. B* **195**, 216–222.

A. Dullweber, B. Leimkuhler and R. McLachlan (1997), 'Symplectic splitting methods for rigid body molecular dynamics', *J. Chem. Phys.* **107**, 5840–5851.

L. M. Dupuy, E. B. Tadmor, R. E. Miller and R. Phillips (2005), 'Finite temperature quasicontinuum: Molecular dynamics simulations without all the atoms', *Phys. Rev. Lett.* **95**, # 060202.

R. D. Engle, R. D. Skeel and M. Drees (2005), 'Monitoring energy drift with shadow Hamiltonians', *J. Comput. Phys.* **206**, 432–452.

D. Frenkel and B. Smit (1996), *Understanding Molecular Simulation*, Academic Press, New York.

D. Frenkel and B. Smit (2002), *Understanding Molecular Simulation*, 2nd edn, Academic Press, New York.

B. García-Archilla, J. M. Sanz-Serna and R. D. Skeel (1998), The mollified impulse method for oscillatory differential equations, in *Numerical Analysis 1997* (D. F. Griffiths and G. A. Watson, eds), pp. 111–123.

S. Gill, Z. Jia, B. Leimkuhler and A. Cocks (2006), 'Rapid thermal equilibration in coarse-grained molecular dynamics', *Phys. Rev. B* **73**, # 184304.

L. Greengard (1994), 'Fast algorithms for classical physics', *Science* **265**, 909–914.

E. Hairer (1994), 'Backward analysis of numerical integrators and symplectic methods', *Ann. Numer. Math.* **1**, 107–132.

E. Hairer and C. Lubich (1997), 'The lifespan of backward error analysis for numerical integrators', *Numer. Math.* **76**, 441–462.

E. Hairer and C. Lubich (2000), 'Long-time energy conservation of numerical methods for oscillatory differential equations', *SIAM J. Numer. Anal.* **38**, 414–441.

E. Hairer, C. Lubich and G. Wanner (2002), *Geometric Numerical Integration*, Springer.

E. Hairer, C. Lubich and G. Wanner (2003), 'Geometric numerical integration illustrated by the Störmer–Verlet method', in *Acta Numerica*, Vol. 12, Cambridge University Press, pp. 399–450.

E. Hairer, S. Nørsett and G. Wanner (1987), *Solving Ordinary Differential Equations I: Nonstiff Problems*, Springer, Berlin.

B. Hajek (1988), 'Cooling schedules for optimal annealing', *Math. Oper. Res.* **13**, 311–329.

J. P. Hansen and I. R. McDonald (1986), *Theory of Simple Liquids*, 2nd edn, Academic Press, New York.

E. Hernández (2001), 'Metric-tensor flexible-cell algorithm for isothermal-isobaric molecular dynamics simulations', *J. Chem. Phys.* **115**, 10282–10290.

M. Hochbruck and C. Lubich (1999), 'A Gautschi-type method for oscillatory second-order differential equations', *Numer. Math.* **83**, 403–426.

B. L. Holian, A. J. D. Groot, W. G. Hoover and C. G. Hoover (1990), 'Time-reversible equilibrium and nonequilibrium isothermal-isobaric simulations with centered-difference Stoermer algorithms', *Phys. Rev. A* **41**, 4552–4553.

J. A. Izaguirre and S. S. Hampton (2004) 'Shadow hybrid Monte Carlo: An efficient propagator in phase space of macromolecules', *J. Chem. Phys.* **200**, 581–604.

J. Izaguirre, S. Reich and R. Skeel (1999), 'Longer time steps for molecular dynamics', *J. Chem. Phys.* **110**, 9853–9864.

S. Jang and G. A. Voth (1997), 'Simple reversible molecular dynamics algorithms for Nosé–Hoover chain dynamics', *J. Chem. Phys.* **107**, 9514–9526.

C. Jarzynski (1997*a*), 'Equilibrium free-energy differences from nonequilibrium measurements: A master equation approach', *Phys. Rev. E* **56**, 5018–5035.

C. Jarzynski (1997*b*), 'Nonequilibrium equality for free energy differences', *Phys. Rev. Lett.* **78**, 2690–2693.

Z. Jia and B. Leimkuhler (2006), 'A projective thermostatting dynamics technique'.

M. Kaczmarski, R. Rurali and E. Hernández (2004), 'Reversible scaling simulations of the melting transition in silicon', *Phys. Rev. B* **69**, # 214105.

S. Kirkpatrick, C. D. Gelatt and M. P. Vecchi (1983), 'Optimization by simulated annealing', *Science* **220**, 621–680.

A. Kol, B. B. Laird and B. J. Leimkuhler (1997), 'A symplectic method for rigid-body molecular simulation', *J. Chem. Phys.* **107**, 2580–2588.

R. Krasny and Z.-H. Duan (2002), Treecode algorithms for computing nonbonded particle interactions, in *Computational Methods for Macromolecules: Challenges and Applications* (T. Schlick and H. H. Gan, eds), Vol. 24 of *Lecture Notes in Computational Science and Engineering*, Springer, pp. 359–380.

B. Laird and B. Leimkuhler (2003), 'Generalized dynamical thermostatting technique', *Phys. Rev. E* **68**, # 016704.

F. Legoll, M. Luskin and R. Moeckel (2007), 'Non-ergodicity of the Nosé–Hoover thermostatted harmonic oscillator', *Arch. Ration. Mech. Anal.*, to appear.

B. Leimkuhler (2002), 'A separated form of Nosé dynamics for constant temperature and pressure simulation', *Comput. Phys. Comm.* **148**, 206–213.

B. Leimkuhler and S. Reich (2001), 'A reversible averaging integrator for multiple time-scale dynamics', *J. Comput. Phys.* **171**, 95–114.

B. Leimkuhler and S. Reich (2005), *Simulating Hamiltonian Dynamics*, Cambridge University Press, Cambridge.

B. Leimkuhler and C. Sweet (2004), 'The canonical ensemble via symplectic integrators using Nosé and Nosé–Poincaré chains', *J. Chem. Phys.* **121**, 108–116.

B. Leimkuhler and C. Sweet (2005), 'A Hamiltonian formulation for recursive multiple thermostats in a common timescale', *SIAM J. Appl. Dyn. Syst.* **4**, 187–216.

B. J. Leimkuhler and R. D. Skeel (1994), 'Symplectic numerical integrators in constrained Hamiltonian systems', *J. Comput. Phys.* **112**, 117–125.

B. Leimkuhler, F. Legoll and E. Noorizadeh (2007), 'Temperature regulated microcanonical dynamics', in preparation.

B. J. Leimkuhler, S. Reich and R. D. Skeel (1996), Integration methods for molecular dynamics, in *IMA Volumes in Mathematics and its Applications*, Vol. 82, Springer, New York, pp. 161–186.

G. Lynch and B. Pettitt (1997), 'Grand canonical ensemble molecular dynamics simulations: Reformulation of extended system dynamics approaches', *J. Chem. Phys.* **107**, 8594–8610.

J. A. McCammon and S. C. Harvey (1987), *Dynamics of Proteins and Nucleic Acids*, Cambridge University Press, Cambridge.

J. A. McCammon, B. R. Gelin and M. Karplus (1977), 'Dynamics of folded proteins', *Nature* **267**, 585–590.

R. McLachlan (1993), 'Explicit Lie–Poisson integration and the Euler equations', *Phys. Rev. Lett.* **71**, 2043–3046.

R. McLachlan and A. Zanna(2005), 'The discrete Moser–Veselov algorithm for the free rigid body, revisited', *Found. Comput. Math.* **5**, 87–123.

D. A. McQuarrie (1976), *Statistical Mechanics*, Harper and Row, New York.

G. J. Martyna, M. E. Tuckerman, D. J. Tobias and M. L. Klein (1996), 'Explicit reversible integrators for extended systems dynamics', *Mol. Phys.* **87**, 1117–1157.

A. I. Neishtadt (1984), 'The separation of motions in systems with rapidly rotating phase', *J. Math. Mech.* **48**, 133–139.

S. Nosé (1984a), 'A molecular-dynamics method for simulations in the canonical ensemble', *Mol. Phys.* **52**, 255–268.

S. Nosé (1984b), 'A unified formulation of the constant temperature molecular-dynamics methods', *J. Chem. Phys.* **81**, 511–519.

S. Nosé (1991), 'Constant temperature molecular dynamics methods', *Prog. Theor. Phys. Supp.* **103**, 1–46.

S. Nosé (2001), 'An improved symplectic integrator for Nosé–Poincaré thermostat', *J. Phys. Soc. Japan* **70**, 75–77.

G. R. W. Quispel and C. Dyt (1997), Solving ODE's numerically while preserving symmetries, Hamiltonian structure, phase space volume, or first integrals, in *Proc. 15th IMAC World Congress* (A. Sydow, ed.), Vol. 2, Wissenschaft & Technik, Berlin, pp. 601–607.

A. Rahman (1964), 'Correlations in the motion of atoms in liquid argon', *Phys. Rev.* **136**, A405–A411.

A. Rahman and F. Stillinger (1971), 'Molecular dynamics study of liquid water', *J. Chem. Phys.* **55**, 3336–3359.

A. Rappe and W. A. Goddard (1991), 'Charge equilibration for molecular dynamics simulations', *J. Phys. Chem.* **95**, 3358–3363.

S. Reich (1994), 'Momentum conserving symplectic integrators', *Physica D* **76**, 375–383.

S. Reich (1999), 'Backward error analysis for numerical integrators', *SIAM J. Numer. Anal.* **36**, 1549–1570.

S. W. Rick, S. J. Stuart and B. J. Berne (1994), 'Dynamical fluctuating charge force fields: Application to liquid water', *J. Chem. Phys.* **101**, 6141–6156.

R. D. Ruth (1983), 'A canonical integration technique', *IEEE Trans. Nuclear Science* **30**, 2669–2671.

J. P. Ryckaert, G. Ciccotti and H. J. C. Berendsen (1977), 'Numerical integration of the Cartesian equations of motion of a system with constraints: Molecular dynamics of n-alkanes', *J. Comput. Phys.* **23**, 327–341.

J. M. Sanz-Serna (1997), Geometric integration, in *The State of the Art in Numerical Analysis* (I. S. Duff and G. A. Watson, eds), Clarendon Press, Oxford, pp. 121–143.

J. M. Sanz-Serna and M. P. Calvo (1994), *Numerical Hamiltonian Problems*, Chapman and Hall, New York.

T. Schlick (2002), *Molecular Modeling and Simulation*, Springer.

R. Skeel and D. Hardy (2001), 'Practical construction of modified Hamiltonians', *SIAM J. Sci. Comput.* **24**, 1172–1188.

M. Sprik and M. L. Klein (1988), 'A polarizable model for water using distributed charge sites', *J. Chem. Phys.* **89**, 7556–7560.

J. Sturgeon and B. Laird (2000), 'Symplectic algorithm for constant-pressure molecular dynamics using a Nosé–Poincaré thermostat', *J. Chem. Phys.* **112**, 3474–3482.

T. Takai, C. Lee, T. Halicioglu and W. A. Tiller (1990), 'A model potential function for carbon systems: Clusters', *J. Phys. Chem.* **94**, 4480–4482.

M. Toda, R. Kubo and N. Saitô (1991), *Statistical Physics I*, 2nd edn, Springer, New York.

P. Tupper (2007), 'A conjecture about molecular dynamics', *Proc. Abel Symposium*, to appear.

W. F. van Gunsteren and H. J. C. Berendsen (1977), 'Algorithms for macromolecular dynamics and constraint dynamics', *Mol. Phys.* **34**, 1311–1327.

W. F. van Gunsteren and H. J. C. Berendsen (1982), 'Algorithms for Brownian dynamics', *Mol. Phys.* **45**, 637–647.

R. van Zon and J. Schofield (2007), 'Numerical implementation of the exact dynamics of free rigid bodies', *J. Comput. Phys.*, to appear.

L. Verlet (1967), 'Computer experiments on classical fluids I: Thermodynamical properties of Lennard–Jones molecules', *Phys. Rev.* **159**, 98–103.

C. Verlinde and W. Hol (1994), 'Structure-based drug design: Progress, results, and challenges', *Structure* **2**, 577–587.

A. F. Voter, F. Montalenti and T. C. Germann (2002), 'Extending the time scale in atomistic simulation of materials', *Annu. Rev. Mater. Res.* **32**, 321–346.

P. J. Whittle and T. L. Blundell (1994), 'Protein structure-based drug design', *Annu. Rev. Biophys. Biomol. Struct.* **23**, 349–375.

Acta Numerica (2007), pp. 67–154
doi: 10.1017/S0962492906290019

Modelling atmospheric flows

Mike Cullen

Met Office,

Fitzroy Road,

Exeter EX1 3PB, UK

E-mail: mike.cullen@metoffice.gov.uk

This article demonstrates how numerical methods for atmospheric models can be validated by showing that they give the theoretically predicted rate of convergence to relevant asymptotic limit solutions. This procedure is necessary because the exact solution of the Navier–Stokes equations cannot be resolved by production models. The limit solutions chosen are those most important for weather and climate prediction. While the best numerical algorithms for this purpose largely reflect current practice, some important limit solutions cannot be captured by existing methods. The use of Lagrangian rather than Eulerian averaging may be required in these cases.

CONTENTS

1. Introduction

This article is a review of the mathematical basis of the numerical methods used in production atmosphere models. Many of the results can also be applied to ocean models. A recent review of ocean modelling issues is given by Higdon (2006). Atmospheric models are routinely used in weather prediction for time-scales of a few hours up to a few seasons, and for climate predictions for hundreds of years ahead. These predictions require modelling not just of the atmosphere, but also of the ocean and the rest of the 'Earth system' such as the vegetation. They also require modelling of the interactions between the atmosphere, ocean and land surface. This is difficult because of the very different time-scales of the separate systems.

In atmospheric predictions, the state of the system at an initial time has to be determined from observations. The techniques used are usually called 'data assimilation', where information from new observations is blended with a first guess computed from observations made at previous times. The data assimilation problem is not discussed in this article. A good introduction is given in Kalnay (2003). However, knowledge of the dynamics of the system is very important in data assimilation, and some discussion of how the knowledge is used is given. In addition, modern data assimilation methods such as 'four-dimensional variational data assimilation' (see Courtier, Thépaut and Hollingsworth (1994)) require a model trajectory to be fitted to the observations. Practical ways of doing this require a linearization of the numerical algorithm used in the model, together with an adjoint. The ability to create an accurate linearization of the numerical method used in the nonlinear model is thus important, and is discussed in the article. However, there are many other numerical issues associated with the data assimilation problem which are not discussed.

In production atmosphere models, the affordable resolutions are many orders of magnitude coarser than those required to solve the equations accurately. As discussed in Section 2.2, this would require a grid-length of the order of 1 mm. In practice, the equations are averaged in space and time, and thus depend on implicit or explicit sub-grid modelling assumptions. The problem is therefore to show that the numerical solution stays close to the averaged solution. It is, of course, not possible to write down a set of equations which describe the averaged behaviour exactly; and it is also not possible to estimate the difference between the solution of the averaged equations and the average of the true solution.

The article concentrates on numerical methods for the averaged equations governing the atmosphere, rather than the choice of sub-grid models. The choice of averaging methods and the design of sub-grid models are very large subjects, and exploit detailed observational and modelling studies of the small-scale behaviour. Only those issues which cannot be separated from the design of numerical methods are discussed. Some more information about this very diverse subject is given in Garratt (1992) and Smith (1997). The atmospheric circulation is forced directly by radiation and indirectly through boundary fluxes. The effects of phase changes involving water vapour is very important. In this article, we assume very simple representations of these processes. In production models, a large proportion of the computer code and execution time is spent on modelling them.

The ideal in a mathematical analysis of the numerical methods used in prediction models is to prove that the numerical solution stays close to the exact solution for large times. While this can be attempted if the sub-grid model is assumed to be exact, such estimates are of limited use because of the unknown error of the sub-grid model itself. It is therefore more useful to analyse the accuracy of the approximation to asymptotic limit solutions of the governing equations, whose accuracy is known and which can be well represented using the space- and time-averaging scales that can be afforded. This is the method used in this article. Production numerical models can describe a large variety of asymptotic regimes of the governing equations; and ideally it is necessary to demonstrate the accuracy of the numerical method in all of them. In practice only the most important regimes are considered.

While this article describes how to optimize numerical methods for the most important regimes being modelled; this is only one aspect of the subject of numerical methods for the atmosphere. More comprehensive descriptions of other aspects of the subject can be found in Durran (1998).

In Section 2 we introduce some of the most important asymptotic regimes and discuss their properties relevant to numerical approximation. In Section 3, we discuss the properties of numerical methods in representing these regimes with reference to methods used in current operational models. In Section 4, we illustrate these procedures using the Met Office Unified Model. We demonstrate that the numerical solutions converge to solutions of asymptotic limit equations appropriate to large scales in three cases which are simple enough for accurate numerical solutions to be possible, but still physically relevant. We also demonstrate that the solutions converge to asymptotic limit solutions governing smaller-scale behaviour.

2. Asymptotic limits of the equations of motion

2.1. Basic equations

The starting point is that the atmosphere can be regarded as a fluid continuum which obeys the basic physical laws of dynamics and thermodynamics. The study of fluid dynamics recognizes that fluids exhibit a wide range of different behaviour under different circumstances. These are characterized as asymptotic regimes by identifying dimensionless parameters that control the flow, and choosing appropriate ranges of values of these parameters. A comprehensive survey is given in Batchelor (1967).

A number of simplifying assumptions are universally made when modelling the atmosphere. Current research is exploring whether some of these should be relaxed as the availability of more powerful computers enables more accurate solutions. A more detailed account of these issues is given in White (2002).

The Earth is assumed to rotate with angular velocity $\boldsymbol{\Omega}$ on an axis through the coordinate poles. The acceleration due to gravity and the centrifugal acceleration due to the Earth's rotation are combined, and act normally to geopotential surfaces. The geopotential surfaces are then approximated by spherical surfaces. The equations are defined in spherical polar coordinates (λ, ϕ, r), with origin at the centre of the Earth. The Earth's surface is then assumed to be a spherical surface with radius a with perturbations due to orography. It is defined by the equation $r = r_0(\lambda, \phi)$. The combined gravitational and centrifugal acceleration are assumed to be towards the origin, with a constant magnitude g.

The atmosphere is assumed to consist of a compressible ideal gas with pressure, density and temperature p, ρ, T which are functions of position and time. It contains a mixing ratio q of water vapour. It moves with a vector velocity $\mathbf{u} = (u, v, w)$. The evolution is described by the compressible Navier–Stokes equations, the first law of thermodynamics and the equation of state for an ideal gas, all written in a frame of reference rotating with the Earth's angular velocity. These are

$$\frac{D\mathbf{u}}{Dt} + 2\boldsymbol{\Omega} \times \mathbf{u} + \frac{1}{\rho}\nabla p + g\hat{\mathbf{r}} = \nu\nabla^2\mathbf{u} + \frac{1}{3}\nu\nabla(\nabla \cdot \mathbf{u}),$$

$$\frac{\partial\rho}{\partial t} + \nabla \cdot (\rho\mathbf{u}) = 0,$$

$$C_v\frac{DT}{Dt} - \frac{RT}{\rho}\frac{D\rho}{Dt} = \kappa_h\nabla^2 T + S_h + LP, \qquad (2.1)$$

$$\frac{Dq}{Dt} = \kappa_q\nabla^2 q + S_q - P,$$

$$p = \rho RT.$$

Here the Lagrangian derivative D/Dt is a shorthand for $\partial/\partial t + \mathbf{u} \cdot \nabla$ and $\hat{\mathbf{r}}$ is a unit vector in the radial direction. R is the gas constant and C_v the specific heat of air at constant volume. ν is the kinematic viscosity. All of these are assumed constant. S_h and S_q are the total heat and moisture sources. P is the rate of conversion of water vapour to liquid water or ice, with L the associated latent heat. $\rho\kappa_h$ and κ_q are the thermal conductivity and moisture diffusivity, also assumed constant. In production atmospheric modelling the thermodynamic parameters are allowed to be functions of atmospheric composition. The true viscosity and thermal conductivity are invariably superseded by sub-grid models. The representation of phase changes and forcing terms is very complex, but only the leading order effects will be discussed in this article.

These equations form a system of seven equations for the unknowns $(\mathbf{u}, p, \rho, T, q)$. The obvious physical boundary conditions are that $\mathbf{u} = 0$ at $r = r_0$ and that $p, \rho \to 0$ as $r \to \infty$. While the no-slip condition at the Earth's surface is standard, the issue of the correct mathematical upper boundary condition for an unbounded atmosphere is open. Fluxes of heat and moisture are specified at the lower boundary. Fluxes of momentum are discussed in the next section.

Suppose there are no dissipation and source terms, so that the right-hand side terms of the first four equations of (2.1) vanish. Solve equations (2.1) in a closed time-independent region Γ with boundary conditions $\mathbf{u} \cdot \mathbf{n} = 0$, where $\mathbf{n}$ is a vector pointing outward from the boundary. Then the energy integral

$$E = \int_{\Gamma} \rho \left(\frac{1}{2}(u^2 + v^2 + w^2) + C_v T + gr \right) r^2 \cos\phi \, d\lambda \, d\phi \, dr \qquad (2.2)$$

is conserved. The requirement that the upper boundary be rigid can be removed most conveniently, while retaining energy and angular momentum conservation, by reformulating the equations in 'mass' coordinates (Wood and Staniforth 2003).

It is convenient to rewrite the first law of thermodynamics in terms of the potential temperature $\theta = T(p/p_{\text{ref}})^{-R/C_p} \equiv T/\Pi$, where C_p is the specific heat of air at constant pressure, p_{ref} is a constant reference pressure equal to a typical pressure at the Earth's surface, and Π is the Exner pressure. This gives, noting $R = C_p - C_v$,

$$\frac{D\theta}{Dt} = \frac{1}{C_p \Pi}(\kappa_h \nabla^2 T + S_h + LP). \qquad (2.3)$$

This form of the equation is particularly useful in situations where the right-hand side terms can be neglected. The equation of state can now be rewritten as

$$p_{\text{ref}} \Pi^{\frac{1}{\gamma-1}} = \rho R \theta \qquad (2.4)$$

where $\gamma = C_p/C_v$.

We can also rewrite the momentum equations by using the definition of θ and the equation of state (the last equation of (2.1)). After some algebraic manipulations we obtain

$$\frac{D\mathbf{u}}{Dt} + 2\mathbf{\Omega} \times \mathbf{u} + C_p \theta \nabla \Pi + g\hat{\mathbf{r}} = \nu\nabla^2\mathbf{u} + \frac{1}{3}\nu\nabla(\nabla \cdot \mathbf{u}). \qquad (2.5)$$

In the absence of dissipation and source terms, equations (2.1) imply a conservation law for the *Ertel potential vorticity*

$$Q = \frac{1}{\rho}(\nabla \times \mathbf{u} + 2\mathbf{\Omega}) \cdot \nabla\theta \qquad (2.6)$$

in the form

$$\frac{DQ}{Dt} = 0. \qquad (2.7)$$

2.2. Methods of averaging

As noted in the Introduction, the equations have to be averaged before they are solved numerically. This article only discusses a few of the key issues in the choice of averaging method. A much fuller discussion is given by Ferziger (1998). The averaging is much coarser than that which is usual in large eddy modelling. The kinematic viscosity of air is about $10^{-5}\,\mathrm{m^2\,s^{-1}}$ near the Earth's surface. The resolution required for a direct numerical simulation has to be chosen to make the Reynolds number based on the grid-length $O(1)$. This is difficult to estimate. An estimate based on dissipation rates given by Gill (1982, p. 79) gives a resolution of about 1 mm. The resolution used in even the finest-scale production models is about 1 km. Thus the averaging has to reduce the resolution required by a factor of 10^9. In practice, therefore, the general ideas of turbulence modelling based on the maintenance of an inertial range have to be supplemented by specialized sub-grid models, which use additional knowledge about the behaviour of the atmosphere on scales which cannot be resolved.

The averaging has to be carried out in time as well as space. The ratio of the space- and time-averaging scales is important. The optimal choice will depend on the type of motion being modelled, and is quite different for sound waves and for solutions which move with the flow speed; see Browning and Kreiss (1994). It is therefore difficult to separate the issue from the identification of important asymptotic limits, which is discussed in the following sections.

A basic issue is the choice between Eulerian and Lagrangian averaging. In principle, the state of the atmosphere can either be described in terms of space-time averages at particular locations, or in terms of the behaviour of fluid parcels of a finite size. The latter is very appealing in terms of the observed physics. Descriptive accounts of meteorology often talk in terms of 'air-masses' with particular characteristics. The boundaries between different air-masses can be quite sharp, and spatial averaging would not then give satisfactory results. The mathematical theory underlying Lagrangian averaging is set out in Andrews and McIntyre (1978). That paper illustrates that it is technically much harder to work with than Eulerian averaging.

The coarseness of the averaging required means that the optimum technique may well depend on the asymptotic regime being modelled. For instance, we would expect Eulerian averaging to be appropriate to describe flow over hills, since the hills are fixed in space. However, Lagrangian averaging would be more appropriate for treating moving air-mass boundaries. In the rest of the article, we therefore discuss the optimum averaging for particular regimes alongside the discussion of the numerical method. In particular, it is possible to build the averaging into the numerical method, thus obtaining an implicit turbulence model as discussed in Fureby and Grinstein (2002). However, many production models use an analytic turbulence model, which is added to the equations.

In designing averaging methods, it is essential to note that the prognostic variables in equations (2.1) are observed to be bounded quantities, and therefore their space-time averages will be smooth on the averaging scale. If analytic Eulerian averaging is used to produce modified versions of equations (2.1), it is essential that their solutions can be proved to be smooth on the averaging scale. The modified equations will therefore need to include terms which prevent the growth of unresolvable gradients in the prognostic variables. However, some terms in the governing equations will become insignificant in the presence of the averaging, and can be removed. Thus there is never any point in including the real kinematic viscous or thermal conduction terms. Some other approximations which are often made for coarser averaging scales are noted in the next section. Alternatively, if implicit Eulerian averaging is included in the numerical methods, it is essential that the computed solutions vary smoothly on the averaging scale.

A particular issue with the averaging is the treatment of the no-slip boundary conditions. The real viscous sub-layers are much too thin to resolve explicitly, even if the vertical grid is stretched to give increased resolution near the boundaries. Specialized sub-grid models are therefore used to represent the momentum fluxes near the boundaries. A review of this topic can be found in Garratt (1992).

Another issue is the treatment of instabilities which occur in the real system on scales much too small to be resolved in production models. If these instabilities are not removed by the sub-grid modelling, unstable circulations are likely to develop on resolved scales which may be orders of magnitude larger than the real ones. This can lead to results which are too inaccurate to be useful. It is normal to develop specialized sub-grid models to deal with this. In the atmosphere, such instabilities are often triggered by moisture phase changes, and thus correspond to significant weather events.

2.3. Important asymptotic limits

In this section we discuss some of the important asymptotic regimes for production atmosphere and ocean models. A more comprehensive treatment is given in textbooks such as Pedlosky (1987) and Gill (1982).

Figure 2.1 shows a diagram due to Smagorinsky (1974) which plots atmospheric phenomena as a function of horizontal scale and time-scale. The phenomena traditionally associated with extra-tropical weather forecasting are grouped along the diagonal, indicating that their horizontal scale L is typically proportional to $T^{\frac{3}{2}}$, where T is their time-scale, so they can all be characterized by a horizontal velocity $U = LT^{-1}$ between about

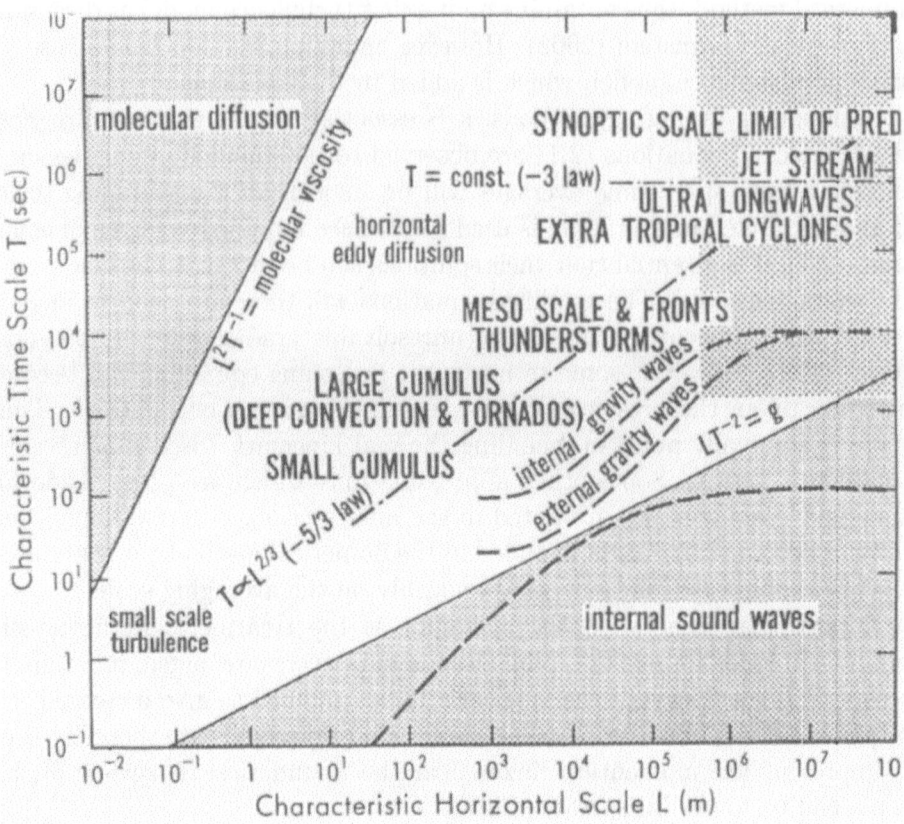

Figure 2.1. Typical space- and time-scales of atmospheric phenomena (following Smagorinsky (1974)).

1 and $10 \, \text{m s}^{-1}$. There are also other classes of motion, such as gravity waves, which have time-scales shorter by an order of magnitude for a given length-scale, and sound waves with a time-scale several orders of magnitude shorter. Molecular diffusion acts on much smaller horizontal scales, as discussed earlier.

Though not shown in Figure 2.1, the typical vertical scale is also important. The effect of surface friction dominates in a boundary layer, which has a thickness of order $1 \, \text{km}$. The troposphere, which contains nearly all the moisture, and where most weather systems are confined, is about $10 \, \text{km}$ deep. Above the troposphere is the stratosphere which is much more stably stratified. It is therefore necessary to consider a range of aspect ratios H/L, where H is the vertical scale. The requirement that the vertical velocity W has a similar time-scale to the horizontal velocity gives that $W/U \simeq H/L$.

Large horizontal scales

We now consider some of the important limits in more detail. We first consider large horizontal scales, where 'large' will be specified below but is intended to correspond to the scale of extra-tropical weather systems. This means that the viscous and thermal conductivity terms can be neglected.

The first step is to recognize that the acceleration due to gravity, g, is much larger than the acceleration of the air in a weather system. However, this acceleration is largely compensated by a vertical pressure gradient. We therefore define a time-independent reference state at rest, which satisfies equations (2.1), with uniform potential temperature θ_0 and with pressure p_0 and density ρ_0 depending only on the radial coordinate. This is given by an Exner pressure $\Pi_0(r)$ satisfying

$$C_p\theta_0\frac{\mathrm{d}\Pi_0}{\mathrm{d}r} + g = 0,$$

$$\theta_0 = \text{constant}, \qquad (2.8)$$

$$\Pi_0 = 1 \text{ at } r = a.$$

Subtract this state from (2.5). The equation becomes

$$\frac{\mathrm{D}\mathbf{u}}{\mathrm{D}t} + 2\boldsymbol{\Omega} \times \mathbf{u} + C_p\theta\nabla\Pi' - g\frac{\theta'}{\theta_0}\hat{\mathbf{r}} = 0 \qquad (2.9)$$

where $\theta' = \theta - \theta_0$ and $\Pi' = \Pi - \Pi_0$.

We next separate the cases where the horizontal scale is comparable to the radius of the Earth (so that the spherical geometry has to be considered), and smaller horizontal scales which can be studied in plane geometry. In the large-scale case, the next step is to recognize that the atmosphere is 'shallow'. The restriction of the vertical scale to the depth of the troposphere means that the aspect ratio for such flows is less than 0.01 and therefore the ratio of the vertical to the horizontal velocities is also less than 0.01. Under these conditions it can be shown that only the horizontal components of the Coriolis force $2\boldsymbol{\Omega} \times \mathbf{u}$ need be considered, and that the radial coordinate r can be replaced by a wherever it appears undifferentiated. This approximation is discussed in detail by White, Hoskins, Roulstone and Staniforth (2005). The result is that the components of the Coriolis force can be written as $(-fv, fu, 0)$ where the Coriolis parameter $f = 2\Omega\sin\phi$. The shallow atmosphere approximation is very accurate, and is used in many operational weather forecasting and climate models.

The next step in analysing large-scale flow is to show that it is hydrostatic. In the vertical component of equation (2.9), the term $g\theta'/\theta_0$ is typically about $1\,\mathrm{m\,s^{-2}}$, since the horizontal variations of θ are typically about 10% of the mean value. Since horizontal and vertical velocities have similar time-scales, $\mathrm{D}w/\mathrm{D}t \simeq (H/L)\mathrm{D}u/\mathrm{D}t$. Given the aspect ratio of 0.01 discussed above, if $\mathrm{D}w/\mathrm{D}t \simeq 1\,\mathrm{m\,s^{-2}}$, then $\mathrm{D}u/\mathrm{D}t \simeq 100\,\mathrm{m\,s^{-2}}$. This is far larger than observed. The implication is that the vertical component of equation (2.9) can be replaced by a statement of hydrostatic balance.

$$C_p\theta\frac{\partial\Pi'}{\partial r} - g\frac{\theta'}{\theta_0} = 0. \tag{2.10}$$

This approximation is very accurate on large scales, and is also used in many weather forecasting and climate models. Its relation to other possible large-scale approximations is discussed by White *et al.* (2005).

We summarize the resulting reduced version of (2.1), (2.3), (2.4) and (2.5), writing ∇_r for the horizontal components of the gradient operator and $\mathbf{u}_r = (u, v)$:

$$\frac{D\mathbf{u}_r}{Dt} + (-fv, fu) + C_p\theta\nabla_r\Pi' = 0,$$

$$C_p\theta\frac{\partial\Pi'}{\partial r} - g\frac{\theta'}{\theta_0} = 0,$$

$$\frac{\partial\rho}{\partial t} + \nabla\cdot(\rho\mathbf{u}) = 0,$$

$$\frac{D\theta}{Dt} = \frac{1}{C_p\Pi}(S_h + LP), \tag{2.11}$$

$$\frac{Dq}{Dt} = S_q - P,$$

$$p_{\text{ref}}\Pi^{\frac{1}{\gamma-1}} = \rho R\theta.$$

The hydrostatic approximation means that there is no explicit evolution equation for w. The next step is to deduce w from the other equations. The hydrostatic relation (2.10) and the equation of state together give two constraints between the thermodynamic variables Π, ρ and θ. Consistency of the separate evolution equations for ρ and θ yields 'Richardson's equation' (see White (2002)):

$$\gamma\frac{\partial}{\partial r}\left\{p\left(\frac{\partial w}{\partial r} + \nabla_r\cdot\mathbf{u} - \frac{1}{C_p\Pi}(S_h + LP)\right)\right\} = \frac{\partial p}{\partial r}\nabla_r\cdot\mathbf{u} - \frac{\partial\mathbf{u}_r}{\partial r}\cdot\nabla_r p. \tag{2.12}$$

It is also shown in White (2002) that equations (2.11) with $S_h = P = 0$ yield a potential vorticity conservation law of the form (2.7) with potential vorticity

$$Q = \frac{1}{\rho}(\hat{\mathbf{r}}\times\nabla\mathbf{u}_r + f\hat{\mathbf{r}})\cdot\nabla\theta. \tag{2.13}$$

Classification of large-scale flows

We now classify the various types of large-scale flow further. The approximations made in the previous subsubsection filter sound waves which have a component propagating in the vertical. We now carry out an analysis of equations (2.10) to (2.12) linearized about a state of rest with a basic state θ which varies with r. This enables us to identify the different types of motion and the important flow parameters. These are then used to define asymptotic regimes.

As in Thuburn, Wood and Staniforth (2002a), a simple analytic treatment requires a choice of basic state where the squared speed of sound $c^2 = C_p\theta\Pi(\gamma-1)$ and the Brunt–Väisälä frequency $N = \sqrt{\frac{g}{\theta}\frac{\partial\theta}{\partial r}}$ are constants. Set $(p,\rho,\theta) = (p_1(r),\rho_1(r),\theta_1(r))$ with $C_p\theta_1\partial\Pi_1/\partial r + g = 0$. Setting $(p,\rho,\theta) = (p_1,\rho_1,\theta_1) + (p',\rho',\theta')$ in (2.11) and (2.12) yields the system

$$\frac{\partial u}{\partial t} - fv + \frac{C_p\theta_1}{a\cos\phi}\frac{\partial\Pi'}{\partial\lambda} = 0,$$

$$\frac{\partial v}{\partial t} + fu + \frac{C_p\theta_1}{a}\frac{\partial\Pi'}{\partial\phi} = 0,$$

$$\frac{\partial\theta}{\partial t} + w\frac{\partial\theta_1}{\partial r} = 0, \qquad (2.14)$$

$$C_p\theta_1\frac{\partial\Pi'}{\partial r} - g\frac{\theta'}{\theta_1} = 0,$$

$$(\gamma - 1)\frac{\partial p_1}{\partial r}\nabla_r\cdot\mathbf{u} + \gamma\frac{\partial p_1}{\partial r}\frac{\partial w}{\partial r} + p_1\gamma\frac{\partial}{\partial r}\nabla_r\cdot\mathbf{u} = 0.$$

This can be analysed in plane geometry by writing $x = a\lambda\cos\phi$, $y = a\phi$, $z = r$ and considering solutions proportional to $\exp^{i(kx+ly-\omega t)}$. Write the vertical structure function as $Z(z)$. Following Thuburn $et\ al.$ (2002a), this can be shown to yield non-trivial solutions if $\omega = 0$ or

$$-Zc^2\frac{(k^2 + l^2)}{(\omega^2 - f^2)} - Z + \frac{c^2}{N^2}\left(\frac{\mathrm{d}}{\mathrm{d}z} + \frac{N^2}{g}\right)\left(\frac{\mathrm{d}}{\mathrm{d}z} + \frac{g}{c^2}\right)Z = 0. \qquad (2.15)$$

Equation (2.15) represents the dispersion equation for inertia-gravity waves. The boundary conditions on w imply that

$$\left(\frac{\mathrm{d}}{\mathrm{d}z} + \frac{g}{c^2}\right)Z = 0 \qquad (2.16)$$

at $z = 0$, z_{top}. This is an eigenvalue problem for $1 + c^2\frac{(k^2+l^2)}{(\omega^2-f^2)}$. ω can then be calculated for given horizontal wave-numbers. This equation is identical in form to equation (4.9) of Thuburn $et\ al.$ (2002a) with a different constant multiplying Z. The solutions therefore take the form given in Thuburn $et\ al.$'s equations (4.13) and (4.15). There are external modes, with

$$Z \propto \exp\left(-\frac{gz}{c^2}\right), \qquad (2.17)$$

and internal modes

$$Z \propto (\Gamma \sin(mz) - m \cos(mz)) \exp\left(-\frac{\gamma g z}{2c^2}\right), \qquad (2.18)$$

where $\Gamma = (1/2)(g/c^2 - N^2/g)$.

Consider the external modes first. Then

$$\left(\frac{d}{dz} + \frac{g}{c^2}\right) Z = 0$$

and ω is given by

$$\omega^2 = f^2 + c^2(k^2 + l^2). \qquad (2.19)$$

This corresponds to a horizontally propagating sound wave, modified by rotation. The sound wave speed is $c = \sqrt{\gamma p_1/\rho_1}$ which is of order $300 \, \mathrm{m \, s^{-1}}$. The associated external Froude number Fr is the ratio of the typical horizontal velocity U to c, and is usually less than 0.1 in the atmosphere. The frequency associated with the horizontal velocity is U/L. This is much less than ω if either $U \ll c$, so that Fr is small, or $U \ll fL$, so that the Rossby number Ro $= U/fL$ is small. The two conditions coincide for a length-scale equal to the external *Rossby radius of deformation* L_R, in this case equal to c/f. This is about $3000 \, \mathrm{km}$ in the atmosphere, corresponding to a wave-length of $20\,000 \, \mathrm{km}$, or a planetary wave-number 2.

For the internal waves, the eigenvalue equation takes the form

$$c^2 \frac{(k^2 + l^2)}{(\omega^2 - f^2)} + 1 + \Lambda(m) = 0,$$

where $\Lambda(m)$ is a quadratic polynomial in m with leading term $c^2 m^2/N^2$. For large m, this gives

$$\omega^2 \simeq f^2 + N^2 \frac{(k^2 + l^2)}{m^2}. \qquad (2.20)$$

Given a characteristic horizontal velocity U, the associated frequency is U/L. Equation (2.20) shows that this frequency is significantly less than the inertia-gravity wave frequency ω if either $U/L \ll NH/L$, so the internal Froude number $U/(NH) \ll 1$, or $U/L \ll f$, so the Rossby number $U/(fL) \ll 1$. The two conditions coincide if $H/L = f/N$. This defines a characteristic aspect ratio. Equivalently, we can state that the horizontal scale is equal to the internal Rossby radius NH/f. If $f = 10^{-4} \, \mathrm{s^{-1}}$, which is characteristic of the extra-tropics, $N = 10^{-2} \, \mathrm{s^{-1}}$, which is characteristic of the troposphere, and $H = 10 \, \mathrm{km}$, which is the tropospheric depth, then the internal Rossby radius is $1000 \, \mathrm{km}$, corresponding to a wave-length of $6000 \, \mathrm{km}$. The aspect ratio f/N is 0.01. In the stratosphere $N = 3 \times 10^{-1} \, \mathrm{s^{-1}}$, giving an internal Rossby radius of $3000 \, \mathrm{km}$ and $f/N = 0.001$. If $U = 10 \, \mathrm{m \, s^{-1}}$, then the frequency U/L is much less than the inertia-gravity wave frequency in the troposphere if either $H \gg 1 \, \mathrm{km}$ or $L \gg 100 \, \mathrm{km}$.

The aspect ratio f/N is important when understanding the behaviour of atmospheric flows with frequencies less than that given by (2.20). Large extra-tropical disturbances tend to develop with aspect ratios close to f/N because this configuration allows efficient conversion of potential to kinetic energy, as first established by Charney (1948). If the aspect ratio is less than f/N, the potential vorticity defined by (2.13) is approximately $\frac{1}{\rho}f\hat{\mathbf{r}}\cdot\nabla\theta$ and is thus primarily a measure of static stability, together with the variation of the planetary vorticity $f\hat{\mathbf{r}}$. The energy of disturbances is primarily potential rather than kinetic energy. If the aspect ratio is greater than f/N, but smaller than 1, which is required for the validity of the shallow atmosphere hydrostatic equations, then the potential vorticity is approximately $\frac{1}{\rho}(\hat{\mathbf{r}}\times\nabla\mathbf{u}_r + f\hat{\mathbf{r}})\cdot\nabla\theta_1$, where $\theta_1(r)$ is a reference state such as used in (2.14). Thus potential vorticity variations are determined by vorticity variations. The energy of disturbances is primarily kinetic energy. These behaviours are analysed by Gill (1982), and also hold for disturbances independent of height when the threshold becomes the external Rossby radius L_R. The case of aspect ratios of order 1 is discussed in the next subsection. We will show in Section 3 that the behaviour of the solutions of (2.11) is qualitatively different as the aspect ratio changes through f/N. This is important in considering optimal numerical methods or data assimilation techniques.

If the frequency of forcing terms is comparable to the inertia-gravity wave frequency, then the inertia-gravity wave response will be important. This happens on large scales in the tropics where f tends to zero. Since f changes rapidly away from zero away from the equator, tropical waves tend to have a very asymmetric structure with a larger scale along the equator (Gill 1982), which determines the response to localized forcing on time-scales less than a few days. Tidal motions occur because the frequency (2.17) determined by the horizontal sound speed is comparable to the diurnal frequency of radiative forcing, so a resonant response is possible.

We now show how this analysis can be exploited in the nonlinear equations (2.11) and (2.12) by deriving a second-order wave equation for the evolution of the horizontal divergence. This will be needed when applying the analysis to production models. The horizontal momentum equations and the thermodynamic equation from (2.11) can be rewritten in component form as

$$\frac{\partial u}{\partial t} - fv + \frac{C_p\theta}{a\cos\phi}\frac{\partial\Pi'}{\partial\lambda} = A_1,$$

$$\frac{\partial v}{\partial t} + fu + \frac{C_p\theta}{a}\frac{\partial\Pi'}{\partial\phi} = A_2, \qquad (2.21)$$

$$\frac{\partial\theta}{\partial t} + w\frac{\partial\theta}{\partial r} = A_3.$$

where A_1, A_2 and A_3 represent all the remaining terms. The choice of terms retained on the left-hand side of equations (2.21) is suggested by the linear analysis above.

The first two of equations (2.21) can be combined into an equation for the evolution of the horizontal divergence $\Delta = \nabla_r \cdot \mathbf{u}_r$. The terms resulting from differentiating f and θ are transferred to the right-hand side:

$$\frac{\partial \Delta}{\partial t} - \frac{f}{a \cos \phi} \left(\frac{\partial v}{\partial \lambda} - \frac{\partial}{\partial \phi}(u \cos \phi) \right) + C_p \theta \nabla_r^2 \Pi' = A_4, \qquad (2.22)$$

where $\nabla_r^2 \equiv \nabla_r \cdot (\nabla_r)$. Now calculate the second time derivative, substituting for $\partial u/\partial t$ and $\partial v/\partial t$ using the first two equations of (2.21), and again transferring some terms to the right-hand side:

$$\frac{\partial^2 \Delta}{\partial t^2} + f^2 \Delta + C_p \theta \nabla_r^2 \frac{\partial \Pi'}{\partial t} = A_5. \qquad (2.23)$$

The final step is to differentiate with respect to r, to use the hydrostatic relation in the form $C_p \theta \partial \Pi/\partial r + g = 0$ and the time-independence of the basic state to substitute for $\partial^2 \Pi'/\partial r \partial t$ in terms of $\partial \theta/\partial t$, and to use the third equation of (2.21) for $\partial \theta/\partial t$. This gives

$$\frac{\partial^2}{\partial t^2} \left(\frac{\partial \Delta}{\partial r} \right) + f^2 \frac{\partial \Delta}{\partial r} + N^2 \nabla_r^2 w = A_6, \qquad (2.24)$$

where the Brunt–Väisälä frequency $N = \sqrt{\frac{g}{\theta} \frac{\partial \theta}{\partial r}}$. Using (2.12) it is possible to eliminate w in favour of Δ by integrating in r, thus obtaining the desired wave equation for the horizontal divergence. It takes the generic form

$$\frac{\partial^2 \Delta}{\partial t^2} + \mathbf{L}\Delta = A \qquad (2.25)$$

where $\mathbf{L}$ is a positive definite linear operator. This operator will reduce to that derived from (2.14) if the atmospheric state is given by the reference state used in (2.14) and plane geometry is assumed.

If $\mathbf{L}\Delta$ is larger than the right-hand side terms A, then equation (2.25) describes forced linear inertia-gravity waves. If $\mathbf{L}\Delta$ and A are of similar magnitude, and the natural frequencies of the waves are large compared with those contained in A, then the response to the 'forcing' terms A can be expressed as the 'slow' equation

$$\mathbf{L}\Delta = A. \qquad (2.26)$$

This equation will apply in the cases identified above where ω as given by (2.19) or (2.20) is much larger than U/L. When this happens $\mathbf{L}$ will have large eigenvalues compared with the frequencies implied by A.

Smaller-scale behaviour

Now consider smaller-scale behaviour. We characterize this by assuming an aspect ratio of $O(1)$, so that the horizontal scale is of order $10\,\mathrm{km}$. The hydrostatic approximation was justified by the assumption of a small aspect ratio, so is no longer applicable. For typical wind speeds, this horizontal scale implies a time-scale less than f^{-1}, so that the Rossby number $\mathrm{Ro} > 1$. The internal Froude number Fr may be greater or less than 1 according to the vertical scale and the strength of the stratification. This scale is, however, still sufficiently large for the viscous and thermal conductivity terms in (2.1) to be neglected. On these scales we can use Cartesian coordinates (x, y, z). Therefore consider equations (2.1) with (2.3), (2.4) and (2.5), omit the viscous and conductivity terms, and subtract the reference state (2.8) to give (2.9). We also omit the moisture equation and other forcing terms. This yields the equations

$$\frac{D\mathbf{u}}{Dt} + 2\boldsymbol{\Omega} \times \mathbf{u} + C_p \theta \nabla \Pi' - g \frac{\theta'}{\theta_0} \hat{\mathbf{r}} = 0,$$

$$\frac{\partial \rho}{\partial t} + \nabla \cdot (\rho \mathbf{u}) = 0,$$

$$\frac{D\theta}{Dt} = 0,$$

$$p_{\mathrm{ref}} \Pi^{\frac{1}{\gamma-1}} = \rho R \theta. \tag{2.27}$$

As in the large-scale case, we first study the behaviour of (2.27) linearized about the hydrostatic reference state defined before (2.14). We assume that rotation is unimportant on this scale, so the equations are

$$\frac{\partial \mathbf{u}}{\partial t} + C_p \theta_1 \nabla \Pi' - g \frac{\theta'}{\theta_1} \hat{\mathbf{r}} = 0,$$

$$\frac{\partial \rho}{\partial t} + \nabla \cdot (\rho_1 \mathbf{u}) = 0,$$

$$\frac{\partial \theta}{\partial t} + w \frac{\partial \theta_1}{\partial z} = 0,$$

$$\frac{1}{\gamma-1} \frac{\Pi'}{\Pi_1} = \frac{\rho'}{\rho_1} + \frac{\theta'}{\theta_1}. \tag{2.28}$$

We assume rigid upper and lower boundary conditions. These equations are a subset of those analysed by Thuburn, Wood and Staniforth (2002b). Consider solutions proportional to $\exp^{i(kx+ly+mz)}$. The dispersion relation for the external mode can then be deduced from Thuburn *et al.*'s equation (4.4) as

$$\omega^2 = c^2(k^2 + l^2). \tag{2.29}$$

This corresponds to a horizontally propagating sound wave. The dispersion relation for the internal modes is deduced from Thuburn *et al.*'s equation (4.5) as

$$(\omega^2 - c^2(k^2 + l^2))(\omega^2 - N^2) - c^2\left(m^2 + \frac{1}{4}\left(\frac{g}{c^2} - \frac{N^2}{g}\right)^2\right)\omega^2 = 0. \quad (2.30)$$

This equation describes pairs of oppositely propagating gravity and sound waves. The sound waves propagate with speed c. For small vertical scales, the gravity wave dispersion relation is approximately $\omega^2 m^2 = N^2(k^2 + l^2)$.

Observed atmospheric winds are always smaller than the speed of sound, usually by an order of magnitude. It is thus safe only to consider the case where $U \ll c$. The internal Froude number is defined as the ratio of the wind speed to the internal gravity wave speed, this is given approximately by $\mathrm{Fr} = U/NH$. This can be either greater or less than 1 according to the circumstances. Observations of flow over hills show that lee waves occur if $\mathrm{Fr} > 1$ and flow blocking, possibly with downstream hydraulic jumps, occurs if $\mathrm{Fr} < 1$ (Gill 1982). Since the qualitative nature of the solutions to (2.27) is different in these cases, the choice of optimal numerical methods and data assimilation techniques may also be different, as discussed in Section 3.

We now seek an analogue of the derivation of (2.26) from (2.25) by exploiting the knowledge that $U \ll c$. By analogy with the derivation of (2.21), rewrite equations (2.27) in the form

$$\frac{\partial \mathbf{u}}{\partial t} + C_p \theta \nabla \Pi' - g\frac{\theta'}{\theta_0}\hat{\mathbf{k}} = B_1,$$

$$\frac{\partial \rho}{\partial t} + \nabla \cdot (\rho \mathbf{u}) = 0,$$

$$\frac{\partial \theta}{\partial t} + w\frac{\partial \theta_1}{\partial z} = B_3, \quad (2.31)$$

$$p_{\mathrm{ref}}\Pi^{\frac{1}{\gamma - 1}} = \rho R\theta,$$

where $\hat{\mathbf{k}}$ is a unit vector in the $z-$ direction. Take the divergence of the momentum equations and transfer the terms in $\partial \theta/\partial r$ and $\partial \theta'/\partial r$ to the right-hand side to give

$$\frac{\partial (\nabla \cdot \mathbf{u})}{\partial t} + C_p \theta \nabla^2 \Pi' = B_4. \quad (2.32)$$

Differentiating the last equation of (2.31) with respect to time, the result can be written

$$\frac{\rho}{\gamma - 1}\frac{\partial \Pi}{\partial t} - \Pi\frac{\partial \rho}{\partial t} = B_5. \quad (2.33)$$

Combining (2.33), (2.32) and the continuity equation from (2.31) gives

$$\frac{\partial^2 (\nabla \cdot \mathbf{u})}{\partial t^2} - C_p \theta \Pi(\gamma - 1)\nabla^2(\nabla \cdot \mathbf{u}) = B_6. \quad (2.34)$$

This is a forced second-order wave equation for $\nabla \cdot \mathbf{u}$. Now $C_p \theta \Pi (\gamma - 1) = \gamma R \theta \Pi = \gamma p / \rho = c^2$, so we see that if $U \ll c$, (2.34) can be approximated by the slow equation

$$-c^2 \nabla^2 (\nabla \cdot \mathbf{u}) = B_6. \tag{2.35}$$

Note that this is an equation for the three-dimensional divergence, rather than the two-dimensional divergence which is determined by (2.26). This is an analogue of (2.26) which can be used on small scales.

2.4. Use of asymptotic limit solutions

We assume that the governing equations can be averaged, as discussed in Section 2.2. The viscous and thermal conductivity terms can be removed, and additional terms inserted on the right-hand sides of the equations representing the sub-grid model. After subtracting the reference state (2.8) to give (2.9), equations (2.1) with (2.3), (2.4) and (2.5) can be written as

$$\frac{D\mathbf{u}}{Dt} + 2\mathbf{\Omega} \times \mathbf{u} + C_p \theta \nabla \Pi' - g \frac{\theta'}{\theta_0} \hat{\mathbf{r}} = F_{\mathbf{u}},$$

$$\frac{\partial \rho}{\partial t} + \nabla \cdot (\rho \mathbf{u}) = 0,$$

$$\frac{D\theta}{Dt} = \frac{1}{C_p \Pi} (S_h + LP) + F_\theta, \tag{2.36}$$

$$\frac{Dq}{Dt} = S_q - P + F_q,$$

$$p_{\mathrm{ref}} \Pi^{\frac{1}{\gamma - 1}} = \rho R \theta.$$

Suppose that the averaging is on a horizontal scale L, a vertical scale H and a time-scale T. The solutions of the equations will then vary smoothly on these space- and time-scales, either in an Eulerian or a Lagrangian sense. The sub-grid terms may be introduced implicitly by the method of discretization. We can then apply analyses such as those set out in Section 2.3 on the assumption that the solutions of the equations only contain scales larger than L, H and T.

For example, if $L \gg 10\,\mathrm{km}$, the aspect ratio of the solutions must be small, and the shallow atmosphere hydrostatic equations, (2.11) will be satisfied accurately on all scales that are permitted by the averaging. They can then be used in the predictive equations. The right-hand side of (2.11) should now include the sub-grid model terms, giving

$$\frac{D\mathbf{u}_r}{Dt} + (-fv, fu) + C_p \theta \nabla_r \Pi' = F_{\mathbf{u}_r},$$

$$C_p \theta \frac{\partial \Pi'}{\partial r} - g \frac{\theta'}{\theta_0} = F_w,$$

$$\frac{\partial \rho}{\partial t} + \nabla \cdot (\rho \mathbf{u}) = 0,$$

$$\frac{D\theta}{Dt} = \frac{1}{C_p \Pi}(S_h + LP) + F_\theta, \qquad (2.37)$$

$$\frac{Dq}{Dt} = S_q - P + F_q,$$

$$p_{ref}\Pi^{\frac{1}{\gamma-1}} = \rho R\theta.$$

Note that the sub-grid term must even appear in the hydrostatic relation, since it is this term that ensures that Dw/Dt is always small. Equations (2.37) are widely used in operational prediction models, though the F_w term is not normally included. For instance, the model currently used by the European Centre for Medium Range Weather Forecasts (ECMWF) uses these equations with an averaging scale of about 50 km.

Now consider the approximation of (2.25) by (2.26), which filters inertia-gravity waves. This is not valid on all scales permitted by current production models. This is because the typical T is less than 1 hour, while the inertial period $2\pi f^{-1}$ is at least 12 hours. However, the period of some inertia-gravity waves is much less than 1 hour. Thus (2.26) will be satisfied on some, but not all scales that are represented. This is discussed in detail in Section 3.2. Selective application of (2.26) is useful in data assimilation. It is also useful for ensuring that time-varying forcing is included in a way which ensures the response to it is on an appropriate time-scale. A further application is in validating model solutions.

The inclusion of the sub-grid terms is important when considering what conservation properties are important. Equation (2.11) conserves the potential vorticity (2.13) if the source and sink terms are zero. However, the inclusion of the sub-grid terms to give (2.37) will destroy potential vorticity conservation. Thus there is no longer a justification for enforcing it in discrete models.

In order to demonstrate that the sub-grid terms are correct, it is necessary to ensure that (2.36) or (2.37) have solutions which are smooth on the scales L, H and T. For example, Cao and Titi (2005) analyse a slightly simplified form of equations (2.37), and show that they can be solved if viscosity is included. Their proof depends critically on the viscosity, as it shows that the problem can be reduced to the three-dimensional Burgers equation $D\mathbf{u}/Dt = 0$. This describes colliding particle trajectories and can only be solved if there is sufficient viscosity to prevent such collisions. Equations (2.37) can thus be solved if the sub-grid term $F_{\mathbf{u}_r}$ incorporates sufficient viscosity. A similar result was obtained by Lions, Temam and Wang (1992a), where the importance of including the term F_w in equations similar to (2.37) is demonstrated. With horizontal averaging scales of order 100 km, this level of viscosity may lead to inaccurate solutions, since it is unlikely that the averaged behaviour of the atmosphere is like that of a fluid with a low enough Reynolds number to give smooth behaviour on this scale.

The limitation of the averaging approach is that, because of the nonlinearity of the equations, it is not possible to write down sub-grid models which will ensure that the solution of the modified governing equations will accurately represent the average of the solution of the original equations. An alternative approach is to prove that the behaviour of the unaveraged governing equations stays close to that of suitable asymptotic limit equations for 'long' time periods, typically of order the length of time that is being modelled. This requires proving the existence and uniqueness of solutions to the asymptotic limit equations for sufficiently large times to describe the phenomena of interest. The results can then be extended to numerical approximations to the governing equations. Typically, the proof of existence depends on particular conservation properties of the limit equations, and some analogue of these conservation properties should then be enforced on the numerical approximation to the governing equations.

In choosing useful asymptotic limit equations, it is desirable that existence of solutions can be proved without introducing extra regularization which is not appropriate for the regime being considered. If such terms have to be included, they may degrade the accuracy and usefulness of the resulting estimate. An example is provided by the viscosity that has to be included to solve (2.37). We show in Section 3.6 that this is probably much larger than the real viscosity, and will lead to solutions that suggest that large-scale circulations are more dissipative than they really are.

3. Numerical methods for particular asymptotic regimes

3.1. Introduction

In this section we consider a selection of the most important asymptotic regimes of the atmosphere introduced in the previous section. It does not attempt to be comprehensive, since almost all types of fluid flow can occur somewhere in the atmosphere. We discuss the structure of the appropriate limit equations, and identify the conservation properties which would be required to integrate the limit equations for long periods. We then discuss how these could be applied in integrations of the basic equations (2.1). We expect that it will be necessary to do this in order to ensure that the numerical solution of (2.1) will stay as close to the limit solution as it should.

3.2. Regimes with 'fast' inertia-gravity waves

Consider the case where the advection frequency U/L is much less than the inertia-gravity wave frequency ω calculated in (2.15). This is often called the 'balanced' regime, but this term is ambiguous and so better avoided.

We derive limit equations called the nonlinear balance equations from the shallow atmosphere hydrostatic equations (2.11) by using $\varepsilon = U/(\omega L)$ as the small parameter. For internal waves with small vertical scale, where ω is given by (2.20), ε can be related to the Rossby and internal Froude numbers by $\varepsilon^2 = (\text{Ro}^{-2} + \text{Fr}^{-2})^{-2}$. A similar equation applies for external waves with ω given by (2.19). There are many versions of this type of analysis in the literature; for instance, see McWilliams, Yavneh, Cullen, and Gent (1999), Warn, Bokhove, Shepherd and Vallis (1995), Lynch (1989), Holm (1996), Ford, McIntyre and Norton (2000) and Mohebalhojeh and Dritschel (2001).

Equations (2.11) can be rewritten in terms of the vertical component of the vorticity, ζ, and the horizontal divergence, Δ, defined by

$$\zeta = \frac{1}{a\cos\phi}\left(\frac{\partial v}{\partial\lambda} - \frac{\partial u\cos\phi}{\partial\phi}\right),$$

$$\Delta = \frac{1}{a\cos\phi}\left(\frac{\partial u}{\partial\lambda} + \frac{\partial v\cos\phi}{\partial\phi}\right). \tag{3.1}$$

The horizontal momentum equations become

$$\frac{\partial\zeta}{\partial t} + \frac{u}{a\cos\phi}\frac{\partial(\zeta+f)}{\partial\lambda} + \frac{v}{a}\frac{\partial(\zeta+f)}{\partial\phi} + w\frac{\partial\zeta}{\partial r} + (\zeta+f)\Delta + \tag{3.2}$$

$$\frac{1}{a\cos\phi}\frac{\partial w}{\partial\lambda}\frac{\partial v}{\partial r} - \frac{1}{a}\frac{\partial w}{\partial\phi}\frac{\partial u}{\partial r} = 0,$$

$$\frac{\partial\Delta}{\partial t} + \frac{u}{a\cos\phi}\frac{\partial\Delta}{\partial\lambda} + \frac{v}{a}\frac{\partial\Delta}{\partial\phi} + w\frac{\partial\Delta}{\partial r} + \Delta^2 - 2J(u,v) + \tag{3.3}$$

$$\frac{1}{a\cos\phi}\frac{\partial w}{\partial\lambda}\frac{\partial u}{\partial r} + \frac{1}{a}\frac{\partial w}{\partial\phi}\frac{\partial v}{\partial r} + \nabla_r\cdot C_p\theta\nabla_r\Pi' - \nabla_r\cdot(fv,-fu) = 0,$$

where ∇_r again denotes a horizontal derivative and $(u,v) \equiv \mathbf{u}_r$. We now follow the steps used to derive (2.24), but making explicit some of the most important terms in A_6. This gives

$$-\frac{\partial^2}{\partial t^2}\frac{\partial^2 w}{\partial r^2} - f\frac{\partial}{\partial r}\left((\zeta+f)\frac{\partial w}{\partial r}\right) - \frac{g}{\theta}\nabla_r^2\left(w\frac{\partial\theta}{\partial r}\right) = \tag{3.4}$$

$$\frac{g}{\theta}\nabla_r^2\left(\mathbf{u}_r\cdot\nabla_r\theta - \frac{1}{C_p\Pi}(S_h+LP)\right)$$

$$+ \frac{\partial}{\partial r}\left(-f\mathbf{u}_r\cdot\nabla_r(\zeta+2f) + 2\frac{\partial}{\partial t}J(u,v)\right) + \text{remainder}.$$

The linearization of this equation describes inertia-gravity waves with the frequency ω calculated in (2.15). Consider a regime with velocity-scales and length-scales U and L, such that $\varepsilon = U/(L\omega) \ll 1$, and depth-scale H such that $H \ll L$, so that (3.2) and (3.4) are valid. Then the argument that was used to derive (2.26) from (2.25) shows that equation (3.4) can be approximated by

$$f\frac{\partial}{\partial r}\left((\zeta + f)\frac{\partial w}{\partial r}\right) + \nabla_r^2(N^2 w) = \tag{3.5}$$

$$-\frac{g}{\theta}\nabla_r^2\left(\mathbf{u}_r \cdot \nabla_r\theta - \frac{1}{C_p\Pi}(S_h + LP)\right)$$

$$-\frac{\partial}{\partial r}\left(-f\mathbf{u}_r \cdot \nabla(\zeta + 2f) + 2\frac{\partial}{\partial t}J(u, v)\right) + \text{remainder}.$$

It can be shown that it is also consistent to neglect many but not all of the terms in 'remainder'. Equation (2.26) shows that $\Delta = \mathbf{L}^{-1}A$ and is thus small under our assumptions. The second equation of (3.2) can then be approximated by

$$-2J(u, v) + \frac{1}{a\cos\phi}\frac{\partial w}{\partial\lambda}\frac{\partial u}{\partial r} + \frac{1}{a}\frac{\partial w}{\partial\phi}\frac{\partial v}{\partial r} + \nabla_r \cdot C_p\theta\nabla_r\Pi' - \nabla_r \cdot (fv, -fu) = 0.$$
$$\tag{3.6}$$

Equation (3.5) is exactly of the form (2.26), and is thus used to define $\mathbf{L}$. The system of equations comprising (3.1), (3.2), the diagnostic relations (2.10) and (2.12), the definition of Π', and equations (3.5) and (3.6) form eight equations for $u, v, w, \zeta, \Delta, \theta, p, \Pi'$. These form the nonlinear balance equations. After some manipulations, further consistent approximations, and the use of potential temperature as a coordinate, these reduce to the equations used in McWilliams et al. (1999), subject to changes in notation. It is also shown there that the resulting equations retain the potential vorticity conservation law (2.13).

Solvability of these equations is shown by McWilliams et al. (1999) to depend on the conditions

(i) $\partial\theta/\partial r$ does not change sign,

(ii) $\zeta + f$ does not change sign,

(iii)

$$(\zeta + f)^2 - \left(\frac{\partial^2\psi}{\partial x^2} - \frac{\partial^2\psi}{\partial y^2}\right)^2 - 4\left(\frac{\partial^2\psi}{\partial x\partial y}\right)^2 > 0, \tag{3.7}$$

where the stream-function ψ is defined by $\nabla_r^2\psi = \zeta$, with ζ defined by (3.1). The product of conditions (i) and (ii) requires that the potential vorticity does not change sign. Since (2.13) still holds, spontaneous violations of condition (i) are only possible if both (i) and (ii) are violated simultaneously. Condition (iii) is liable to spontaneous violations as it is not a constant of the motion. Computations (McWilliams and Yavneh 1998) show that these do indeed occur. If the initial data has small Rossby number, the velocity gradients have magnitude $U/L \ll f$ and condition (3.7) will be satisfied. If the initial data has small internal Froude number $U/(NH)$ but large Rossby number, (3.7) may not be satisfied even though $H \ll L$. The solvability conditions result from writing the equations as a fully coupled system. If the Froude number is small, but the Rossby number is large, the eigenvalue of the system associated with $\partial\theta/\partial r$ is much larger than the others. A good approximation to the system can then be solved by decoupling it, as described in Section 3.5.

The nonlinear balance equations approximate equations (2.11) to an accuracy $O(\varepsilon^2)$. This was achieved by neglecting the first and second time derivatives of Δ in (3.3) and (3.4). In Mohebalhojeh and Dritschel (2001) it is shown that approximations of any polynomial order to (2.11) could be generated by neglecting successively higher time derivatives in deriving approximations to equations (3.3) and (3.4). If the resulting equations could be solved for long time periods, it would be possible to prove that solutions of (2.11) with suitable initial data remained close to the solutions of the nonlinear balance equations for long times, and the inertia-gravity waves would thus be restricted to very small amplitudes. The apparent lack of solvability means that there is no such thing as a solution of the nonlinear balance equations for long times which can be compared with the solution of the original equations, and so no such conclusion can be drawn. In McWilliams and Yavneh (1998) it is suggested that spontaneous violations of condition (3.7) correspond to local bursts of inertia-gravity wave activity.

The discussion in Section 2.4 suggests that it would be of interest to regularize the equations by including sub-grid terms in a way that ensured solvability. The approximations made in deriving the nonlinear balance equations are derived in an Eulerian frame. Start from the averaged shallow atmosphere hydrostatic equations (2.37), which requires assuming a horizontal averaging scale $L \gg 10\,\mathrm{km}$. Assume additionally that L is large enough for ε to be small. Following the derivation used to obtain equation (3.4) gives

$$-\frac{\partial^2}{\partial t^2}\frac{\partial^2 w}{\partial r^2} - f\frac{\partial}{\partial r}\left((\zeta+f)\frac{\partial w}{\partial r}\right) - \frac{g}{\theta}\nabla_r^2\left(w\frac{\partial\theta}{\partial r}\right) = \tag{3.8}$$

$$\frac{g}{\theta}\nabla_r^2\left(\mathbf{u}_r\cdot\nabla_r\theta - \frac{1}{C_p\Pi}(S_h + LP)\right)$$

$$+\frac{\partial}{\partial r}\left(-f\mathbf{u}_r\cdot\nabla_r(\zeta+2f) + 2\frac{\partial}{\partial t}J(u,v)\right) + \frac{\partial^2}{\partial t\partial r}(\nabla\cdot F_{\mathbf{u}_r}) + \text{remainder}.$$

A sub-grid term $\frac{\partial^2}{\partial t\partial r}(\nabla\cdot F_{\mathbf{u}_r})$ has been introduced.

The assumption $\varepsilon \ll 1$ means that (3.8) can be approximated by

$$f\frac{\partial}{\partial r}\left((\zeta + f)\frac{\partial w}{\partial r}\right) + \nabla_r^2(N^2 w) = \qquad (3.9)$$

$$-\frac{g}{\theta}\nabla_r^2\left(\mathbf{u}_r \cdot \nabla_r\theta - \frac{1}{C_p\Pi}(S_h + LP)\right)$$

$$-\frac{\partial}{\partial r}\left(-f\mathbf{u}_r \cdot \nabla_r(\zeta + 2f) + 2\frac{\partial}{\partial t}J(u, v)\right) - \frac{\partial^2}{\partial t\partial r}(\nabla \cdot F_{\mathbf{u}_r}),$$

which is derived by including the sub-grid term in (3.5). Similarly, a sub-grid term has to be added to the right-hand side of (3.6), giving

$$-2J(u, v) + \frac{1}{a\cos\phi}\frac{\partial w}{\partial\lambda}\frac{\partial u}{\partial r} + \frac{1}{a}\frac{\partial w}{\partial\phi}\frac{\partial v}{\partial r} \qquad (3.10)$$

$$+\nabla_r \cdot C_p\theta\nabla_r\Pi' - \nabla_r \cdot (fv, -fu) = \nabla \cdot (F_{\mathbf{u}_r}).$$

Since (3.7) is satisfied under these conditions, the modified form (3.9) and (3.10) of the nonlinear balance equations can be solved, and the solutions will be close to those of equations (2.37).

Sub-grid terms will now appear in the potential vorticity equation. Potential vorticity conservation will then no longer apply. An example is given by Ziemianski and Thorpe (2003), though with a different interpretation. In their computations using a numerical solution of (2.1), large sources of potential vorticity appear. These are in a frontal zone, where there is organized unresolved small-scale activity in the solution, and so potential vorticity, which is nonlinear, would not be conserved under the Eulerian averaging implied by the model. This is represented by the sub-grid term in the potential vorticity equation and would be likely to explain the results shown in that paper. The effect of Lagrangian averaging on potential vorticity conservation for a frontal zone is discussed in Section 4.3.

Now, given $L \gg 10\,\mathrm{km}$, we achieve the condition that ε is small by choosing a sufficiently large vertical averaging scale H. This choice of L is not sufficient to ensure that (3.7) is satisfied. In this case (3.9) and (3.10) will hold, but the time evolution cannot be calculated using the nonlinear balance equations. An alternative procedure which does work is described in Section 3.5.

The discussion above shows that the diagnostic conditions (3.9) and (3.10) are useful, but can only be applied selectively. This is exploited in the semi-implicit method of time integration introduced by Robert (1981). This is widely used in operational models, such as the ECMWF model (Ritchie, Temperton, Simmons, Hortal, Davies, Dent and Hamrud 1995), and the Met Office Unified Model (UM) (Davies, Cullen, Malcolm, Mawson, Staniforth, White and Wood 2005). Robert's motivation was to increase the efficiency of operational models by not resolving the time evolution of fast inertia-gravity waves. Thus the time-step is chosen such that $U\delta t \leq 1$, where U is the maximum velocity for which advection has to be treated accurately. Inertia-gravity waves will not then be resolved in time in cases where $\varepsilon \ll 1$. In such cases, the solution will satisfy the nonlinear balance conditions (3.9), (3.10) accurately. We therefore seek a time discretization of equations (2.37) which ensures that the solution satisfies the nonlinear balance conditions whenever $\varepsilon \ll 1$, while noting that this condition will not be satisfied everywhere. We illustrate the method by writing a time discretization of equation (2.21) as

$$\delta_t u - f\overline{v}^t + \frac{C_p \overline{\theta}^t}{a \cos \phi} \frac{\partial \overline{\Pi}'^t}{\partial \lambda} = \overline{A}_1^t,$$

$$\delta_t v + f\overline{u}^t + \frac{C_p \overline{\theta}^t}{a} \frac{\partial \overline{\Pi}'^t}{\partial \phi} = \overline{A}_2^t, \tag{3.11}$$

$$\delta_t \theta + \overline{w}^t \frac{\partial \overline{\theta}^t}{\partial r} = \overline{A}_3^t,$$

where $\delta_t u = (u(t+\delta t) - u(t))/\delta t$ and $\overline{v}^t = (v(t+\delta t) + v(t))/2$.

We can now go through the same steps as were used to derive (2.25) to give a second-order equation for $\Delta(t+\delta t)$. The result is

$$\delta_{tt}\Delta + \overline{\mathbf{L}\Delta}^{tt} = \overline{A}^{tt}. \tag{3.12}$$

Assuming A is proportional to $e^{i\nu t}$, and replacing $\mathbf{L}$ by its eigenvalue ω^2, there is a solution of (3.12) proportional to $e^{i\nu t}$ which is

$$-\sin^2(\tfrac{1}{2}\nu\delta t)\Delta + \delta t^2 \omega^2 \cos^2(\tfrac{1}{2}\nu\delta t)\Delta = A\delta t^2 \cos^2(\tfrac{1}{2}\nu\delta t). \tag{3.13}$$

For non-zero A, since $\omega\delta t \gg 1$, this reduces to $\omega^2 \Delta = A$ as desired. This is because of the consistent time averaging on both sides of the equation. If an equation of the form (3.12) were derived from a three-time-level approximation to (3.11), then the time averaging of A would not be required for numerical stability, and (3.13) would reduce to $\omega^2 \cos^2(\tfrac{1}{2}\nu\delta t)\Delta = A$, thus degrading the accuracy of the balanced solution. The frequency of free inertia-gravity waves (obtained by setting $A = 0$), is reduced to $\omega \cos(\omega\delta t)/\sin(\omega\delta t) \ll \omega$. The inertia-gravity waves in the solution are thus no longer accurately treated.

This illustrates that, if $\omega^2 \delta t^2 \gg 1$, (3.12) can be approximated by

$$\overline{\mathbf{L}\Delta}^{tt} = \overline{A}^{tt} \tag{3.14}$$

which is a discretization of (2.26). This approximation is exactly that used to go from (3.4) to (3.5), which defines the nonlinear balance equations. Moreover, the presence of the first term on the left-hand side of (3.12) will ensure solvability if δt is small enough, even though (3.14) may not be solvable. The use of (3.12) therefore achieves the aim of enforcing nonlinear balance selectively where $\omega^2 \delta t^2 \geq 1$.

It is important that the spatial approximations maintain the condition $|\mathbf{L}|\delta t^2 \geq 1$. Some numerical discretizations will allow the eigenvalues of $\mathbf{L}$ to become zero on the smallest spatial scales. Examples of this are given in Section 3.3, and detailed analysis is given in Durran (1998). It is important to note that the operator $\mathbf{L}$ which appears in (2.26) is a variable-coefficient operator. This is necessary to ensure that the nonlinear balance equations are valid on large scales, where the coefficients may have O(1) variations. Most implementations of the semi-implicit method have used a constant-coefficient operator, which will thus not ensure consistency with the balanced solution. The UM uses a variable-coefficient operator for this reason. If equation (3.14) is solved iteratively, with a constant-coefficient approximation to $\mathbf{L}$ as a preconditioner, then the accuracy of the balanced solution will be degraded if the convergence of the iteration is insufficient.

The enforcement of balance can be strengthened by using a decentred time averaging of the terms on the left-hand side of (3.11). If $\mathbf{L}$ is time-independent, this yields

$$\delta_{tt}\Delta + \mathbf{L}(\alpha^2 \Delta(t + \delta t) + 2\alpha(1 - \alpha)\Delta(t) + (1 - \alpha)^2 \Delta(t - \delta t)) = \overline{A}^{tt}, \quad (3.15)$$

as an approximate replacement for (3.14). The condition $\alpha \geq \frac{1}{2}$ is required for stability. The decentering will not change the potential vorticity conservation law, except for the change to the advection of the potential vorticity by the divergent wind. The inertia-gravity waves which are not well-resolved in time will be damped, but this is desirable as they are no longer travelling at the correct speed.

When considering the effect of decentred time integration in production atmospheric models, note that invariably $f\delta t \ll 1$, so that the rotational contribution to ω as defined in (2.20) will not be significant on scales where $\omega \delta t \gg 1$. In the global forecast version of the UM, the horizontal grid-length is about 40 km, and the time-step 15 minutes. The Courant number $U\delta t/\delta x$ will thus be 1 for a speed U of about 45 m s^{-1}. The condition $\omega \delta t \geq 1$ is satisfied for the external gravity wave which has speed approximately 300 m s^{-1} if the spatial wave-length is less than 300 km, and thus decentering will suppress the waves with wave-lengths up to about 1000 km. Internal gravity waves with speeds greater than 45 m s^{-1} will be suppressed on progressively smaller scales. An example of the effect of decentering with explicit time integration is described and analysed by Fox-Rabinowitz (1996), while the effect of decentred implicit time-differencing was pointed out by A. Staniforth (personal communication); see Section K.5.6 of Staniforth, White, Wood, Thuburn, Zerroukat and Cordero (2002).

While decentering is useful in preventing rapidly oscillating solutions, because $f\delta t \ll 1$ it is far short of selecting out the meteorologically significant motions which are rotation-dominated. The model will thus resolve many other motions, including external gravity waves with a wave-length greater than about 1000 km. The use of decentred time integration may be particularly useful when forcing terms with high time-frequency are added to the equations, since their effect is then time-averaged to fit the time-resolution of the model. It may also be useful in four-dimensional variational data assimilation, where observation increments are added in a way which may not be compatible with the diagnostic relations (3.5) and (3.6). This was the motivation for the paper by Fox-Rabinowitz (1996).

Equation (3.11) represents a fully implicit time-discretization of (2.21). The time-average on the right-hand side generates a nonlinear implicit problem for the values at time $t + \delta t$. As noted by Staniforth, the procedure introduced by Fox-Rabinowitz (1996) formed the first two iterations of an implicit procedure for solving this system. It should also be noted that the Heun advection scheme used in the original UM, Cullen and Davies (1990) takes the same form. We illustrate by displaying the iterations for solving the first equation of (3.11):

$$
\frac{(u^* - u(t))}{\delta t} - f\bar{v}^* + \frac{C_p\theta(t)}{a\cos\phi}\frac{\partial\overline{\Pi}'^*}{\partial\lambda} = A_1(t),
$$

$$
\delta_t u - f\bar{v}^t + \frac{C_p\bar{\theta}^*}{a\cos\phi}\frac{\partial\overline{\Pi}'^t}{\partial\lambda} = \overline{A}_1^*,
$$

$$(3.16)$$

where the notation $\bar{v}^*$ indicates $\frac{1}{2}(v(t)+v^*)$. Each iteration requires solution of an implicit problem of the form (3.12). Yeh, Côté, Gravel, Méthot, Patoine, Roch and Staniforth (2002) used this scheme in the Canadian GEM model, and Cullen (2001) demonstrated its use in the ECMWF model. In four-dimensional variation data assimilation, it is necessary to integrate a linearized version of the model forward in time. An obvious method is to use the first iteration of the procedure (3.16) as the basis for the linear model. The right-hand side terms will now represent perturbations to A_1, and therefore be linear in the evolution variables, Lawless, Nichols and Ballard (2003).

An example using the UM is shown in Figures 3.1 and 3.2. The model solves equations (2.1) by a semi-implicit scheme as described in Davies *et al.* (2005). The balance constraint is applied by using backward time-weighting in the semi-implicit scheme as in (3.12). The operational UM has $\alpha = 0.7$ and the modified version illustrated has $\alpha = 1$. Figure 3.1 shows that the decentering has a very small effect on even the small-scale structure of the potential vorticity Q, while Figure 3.2 shows a significant reduction in the divergence tendency $\partial\Delta/\partial t$, which should be small if $\varepsilon \ll 1$. The larger values in the operational forecast probably mainly reflect the effect of insufficiently constrained initial data and insufficiently smooth representations of

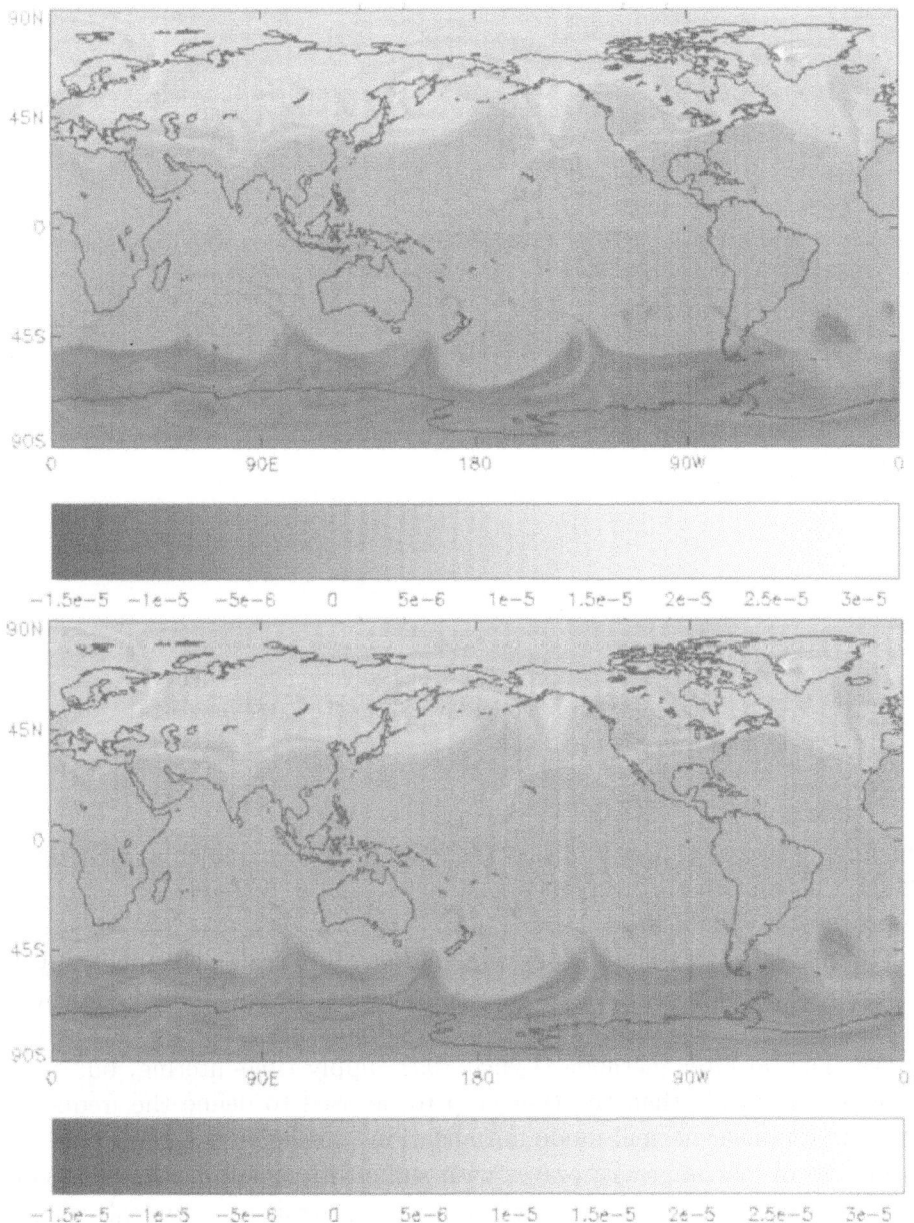

Figure 3.1. Ertel potential vorticity at $\theta = 330\,\mathrm{K}$ predicted by the UM for 0900UTC 24 January 2006 from data at 0900UTC on 22 January 2006. *Top panel*: $\alpha = 0.7$. *Bottom panel*: $\alpha = 1.0$.

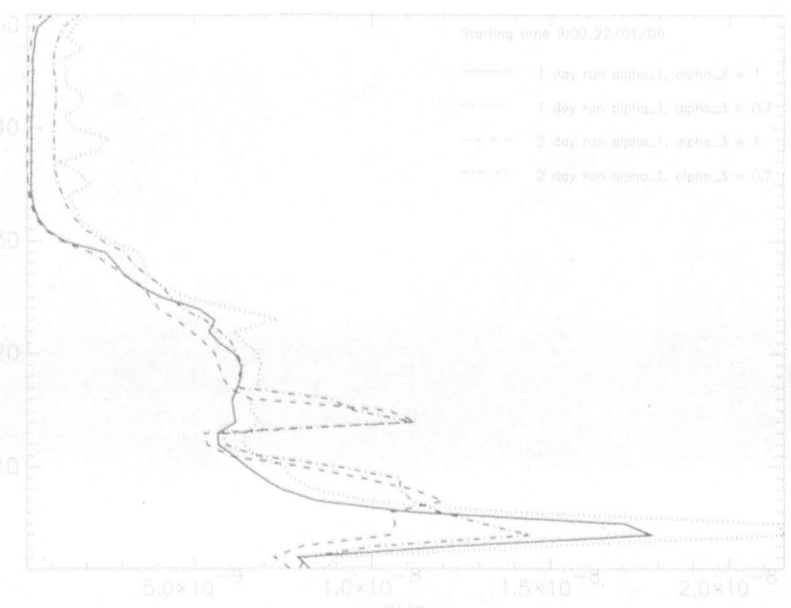

Figure 3.2. Vertical profile of r.m.s. divergence tendency
(s^{-2}) plotted against model level from the same forecasts
as shown in Figure 3.1.

the forcing terms. However, it is also shown by Mohebalhojeh and Dritschel
(2004) that unrealistic divergence tendencies can be created by a poor choice
of prognostic variables.

The constraints (3.4) and (3.5) are enforced in a scale-dependent way in
other methods of initialization. The bounded derivative method of Browning and Kreiss (1994) is a systematic way of doing this, based on the analysis
above. Digital filter methods (Lynch 1997) apply time-filtering, but allow
a longer time-scale than the time-step to be used to define the frequency
cut-off. Nonlinear normal-mode initialization (Machenhauer 1977) enforces
(3.14) for all inertia-gravity waves with sufficiently small m, thus applying
a speed constraint rather than a frequency constraint. A typical value is
about $50\,\mathrm{m\,s^{-1}}$. These methods are usually modified to ensure that the
diurnal tidal signal is not filtered out.

Since the nonlinear balance equations, including the sub-grid terms, can
be written as prediction of potential vorticity, Q, together with diagnostic relations to recover the other variables, it is natural to try and exploit
this structure in numerical methods. It was shown by Mohebalhojeh and
Dritschel (2004) that in order to achieve the best approximations to the nonlinear balance solutions, the model should be formulated in variables which

separate the potential vorticity from variables describing inertia-gravity waves. Recent work by D. Devlin and D. Dritschel (personal communication) suggests that decentering is more effective at preserving nonlinear balance when these variables are used. Exploiting this requires inversion of the operator $\mathbf{L}$ in order to make the change of variables. This is difficult because of the scale-dependence of the inertia-gravity wave frequency ω, equation (2.15). This means that $\mathbf{L}$ will have a large condition number which may be made worse by a poor choice of discretization, as discussed in Section 3.3. It is therefore expensive to invert it, which is why these transformations are not yet in widespread use. The potential benefit of using these variables for data assimilation is illustrated by Cullen (2003).

A number of the lessons for numerical methods have been pointed out in this section, in particular the need to preserve large values of $|\mathbf{L}|$ in the spatial discretization, the need to maintain the condition $\mathbf{L}\Delta = A$ in the large time-step limit, and the need to preserve accuracy in the advection of potential vorticity. These issues will be analysed further in Section 3.3 in a constant-coefficient context. It is not, however, possible to identify what properties are needed to preserve long-time accuracy for cases where $\epsilon \ll 1$ from this analysis because of the non-existence of long-time solutions to the nonlinear balance equations. In order to make further progress, we split the regime $\varepsilon \ll 1$ into parts where $H/L \simeq f/N$, $H/L < f/N$, and $H/L > f/N$, as discussed in Section 2.3. These cases are discussed respectively in Sections 3.3, 3.4 and 3.5. It is then possible to find solvable systems of equations appropriate to each case and thus possible in principle to prove long-time estimates for the errors in numerical methods.

3.3. The quasi-geostrophic regime

This regime is defined by the requirement that the Rossby and internal Froude numbers Ro and Fr are small and equal. It was introduced by Charney (1948). This assumption requires the aspect ratio to be small, typically 10^{-2} in the troposphere and 10^{-3} in the stratosphere. Under these conditions $\varepsilon = \sqrt{2}\,\mathrm{Ro} = \sqrt{2}\,\mathrm{Fr}$ and is thus small, so the nonlinear balance approximation can be made. It is thus appropriate to start from (3.1), (3.2), (2.10), (2.12), (3.5) and (3.6).

We seek a system of equations which is valid in this regime and can be solved for large times. The quasi-geostrophic equations are thus derived as a leading-order approximation in ε which is chosen so that energy is conserved, which aids the proof of solvability. A higher-order version is derived in Bourgeois and Beale (1994), but this cannot be solved for arbitrarily large times. The equations are given below. The justification for the various approximations is quite complex and details of the analysis are given by Pedlosky (1987). In particular, in some terms θ is replaced by a hydrostatic reference state value $\theta_1(r)$, chosen such that the reference value N_0^2 of N^2 is a constant as in Section 2.3. Also f is replaced by a reference value f_0 except where it is differentiated. Thus we obtain

$$\frac{\partial \zeta}{\partial t} + \frac{u}{a\cos\phi}\frac{\partial(\zeta+f)}{\partial\lambda} + \frac{v}{a}\frac{\partial(\zeta+f)}{\partial\phi} + f_0\Delta = 0,$$

$$\frac{\partial\theta}{\partial t} + \frac{u}{a\cos\phi}\frac{\partial\theta}{\partial\lambda} + \frac{v}{a}\frac{\partial\theta}{\partial\phi} + w\frac{\partial\theta_1}{\partial r} = \frac{1}{C_p\Pi}(S_h + LP),$$

$$C_p\theta_1\frac{1}{a\cos\phi}\frac{\partial}{\partial\lambda}\Pi' - f_0 v = 0,$$

$$C_p\theta_1\frac{1}{a}\frac{\partial}{\partial\phi}\Pi' + f_0 u = 0, \qquad (3.17)$$

$$C_p\theta_1\frac{\partial\Pi'}{\partial r} - g\frac{\theta'}{\theta_1} = 0,$$

$$C_p\theta_1\frac{\partial\Pi_1}{\partial r} + g = 0,$$

$$\frac{\partial}{\partial r}\left(\frac{w}{\theta_1}\right) + \frac{\Delta}{\theta_1} = 0.$$

Here ζ and Δ are still defined by (3.1) and $\theta' = \theta - \theta_1$, $\Pi' = \Pi - \Pi_1$. Consistency between equations (3.17) implies that w satisfies

$$f_0^2\frac{\partial^2}{\partial r^2}\left(\frac{w}{\theta_1}\right) + N_0^2\nabla_r^2\left(\frac{w}{\theta_1}\right) = \qquad (3.18)$$

$$-\frac{g}{\theta_1^2}\nabla_r^2\left(\mathbf{u}_r \cdot \nabla_r\theta - \frac{1}{C_p\Pi}(S_h + LP)\right) + f_0\frac{\partial}{\partial r}\left(\frac{1}{\theta_1}\mathbf{u}_r \cdot \nabla(\zeta+f)\right).$$

In the absence of source terms, these equations after some manipulations imply the potential vorticity conservation law

$$\frac{\partial Q}{\partial t} + \frac{u}{a\cos\phi}\frac{\partial Q}{\partial\lambda} + \frac{v}{a}\frac{\partial Q}{\partial\phi} = 0,$$

$$Q = \frac{1}{\theta_1}\left(N_0^2(\zeta+f) + f_0\frac{\partial}{\partial r}\left(\frac{g\theta'}{\theta_1}\right)\right). \qquad (3.19)$$

The horizontal velocities and potential temperature can be calculated from Q by first using the third, fourth and fifth equations of (3.17) to give a Poisson equation for Π' given Q:

$$f_0 Q = \frac{N_0^2 f f_0}{\theta_1} + C_p\left(N_0^2\nabla_r^2\Pi' + f_0^2\frac{\partial^2\Pi'}{\partial r^2}\right). \qquad (3.20)$$

The horizontal velocities are then calculated from the third and fourth equations of (3.17) and the potential temperature from the fifth equation. w can then be calculated from the final equation.

Equation (3.20) can be solved for Π' given suitable boundary conditions. Neumann boundary conditions on Π' imply that θ' is given on horizontal boundaries and u or v on lateral boundaries. The proofs of solvability in Bourgeois and Beale (1994), which use plane geometry, assume periodic boundary conditions in the horizontal and constant values of θ' at the upper and lower boundaries. Other proofs are given in Majda (2003) and Bennett and Kloeden (1981, 1982).

The proofs cited above show that solvability of equations (3.17) depends on the fact that equation (3.19) cannot generate any values of Q outside the range of the initial values. The constant-coefficient elliptic equation (3.20) is used to find u, v and θ' from Q, and this is shown in Bourgeois and Beale (1994) to give a bound on the gradients of (u, v) in terms of bounds on Q and its derivatives:

$$\| \nabla \mathbf{u} \|_{L^\infty} + \| \nabla \theta' \|_{L^\infty} \leq C \| Q \|_{L^\infty} \left(1 + \log^+ \frac{\| Q \|_{H^s}}{\| Q \|_{L^\infty}} \right) \qquad (3.21)$$

where $s \geq 2$. The L^∞ norm is the maximum norm, and H^s measures the L^2 norm of all derivatives up to order s. The $+$ superscript indicates that only positive values are used. C is a constant. Precise definitions are given in Bourgeois and Beale (1994). The maximum value of Q is independent of time, so it is necessary to control the mean-square value of the derivatives of Q. It is shown by differentiating (3.19) that the rate of increase of the gradient of Q is controlled by the gradients of $\mathbf{u}$, which can be estimated from (3.21). This gives an estimate for any time t that

$$\| Q(t) \|_{H^s} \leq C \| Q(0) \|_{H^s} \exp\big(C(e^{Ct\|Q(0)\|_{H^s}} - 1)\big). \qquad (3.22)$$

This can be calculated for any t given the initial data for Q, and provides the required long-time estimate.

In order to ensure that a numerical approximation to (3.17) remains bounded for large times, in other words is nonlinearly stable, it is necessary to be able to derive finite-dimensional analogues of the estimates (3.21) and (3.22).

We first have to maintain the bound on Q by its initial values. If Q is a prognostic variable, this can be achieved by using a monotonicity-preserving advection scheme. There is a very large literature on this topic, though much of it deals with enforcing monotonicity in high-speed flow problem with shocks. The method has the disadvantage that monotonicity enforcement inevitably involves diffusion and thus energy loss. This can be limited by use of sufficiently selective filters. For instance, recent work by Zerroukat, Wood and Staniforth (2005) demonstrates an appropriate selective filter for cases where non-monotonicity can only be generated by errors in upstream interpolation. The bound on Q can also be maintained by a fully Lagrangian method such as one based on contour dynamics, as in Mohebalhojeh and Dritschel (2004). Such methods are not inherently diffusive.

The prescription above could be followed in solutions of (2.1) or (2.11) if the potential vorticity is used as a prognostic variable as in Mohebalhojeh and Dritschel (2004). This will be a good approximation to Q in the quasi-geostrophic regime. We will show how a similar requirement may be achievable without introducing new prognostic variables in Section 3.4.

Another approach is to note that (3.19), together with the non-divergence of the advecting velocity (u, v) resulting from the third and fourth equations of (3.17), implies that all moments $\int Q^n$ are conserved. A numerical method can only conserve a finite number of moments, and a popular method is to conserve only the first and second moments, following Sadourny (1975). This method has the disadvantage that, even if the distribution of Q is well-resolved at the initial time, it may not stay well-resolved. Conservation of the integral of Q^2 may thus result in variance being incorrectly retained in resolved scales.

The derivation of (3.21) involves controlling the second derivatives of Π' resulting from the solution of the elliptic equation (3.20). This requires the approximation to the elliptic operator to satisfy a discrete maximum principle. This is a standard requirement in numerical solution of elliptic equations, and is discussed, for instance, by Ganzha (1996, p. 216). It leads to the standard five-point discrete approximation to the Laplacian in two dimensions. The condition will be satisfied if the three-dimensional Laplacian in (3.20) is approximated by a seven-point stencil. However, if the derivation of (3.20) from (3.17) is carried out at the discrete level, a seven-point stencil is only obtained if u is held at grid-points staggered from Π in the ϕ-direction, v at grid-points staggered from Π in the λ-direction, and θ at grid-points staggered from Π in the r-direction. This is illustrated in Figure 3.3. This horizontal staggering is referred to as the Arakawa D-grid (Arakawa and Lamb 1977), and the vertical staggering as the Charney–Phillips grid (Arakawa and Konor 1996). If any of the other grids shown in Figure 3.3 is used to solve (3.17), the solution of (3.20) is likely to oscillate in space, leading to instability. While it may be possible to control this using filtering, significant inaccuracy will result.

We now need to consider the conditions for a numerical solution of (3.17) to stay close to a solution of (2.11). In Section 2.2 we showed that this required the solution of (3.12) to stay close to that of (3.14). Equation (3.13) showed that this depends on the condition $\omega^2 \delta t^2 \gg 1$. When the spatial discretization is considered, the eigenvalue ω^2 of $\mathbf{L}$ will be approximated by an eigenvalue ω_h; so the condition becomes $\omega_h^2 \delta t^2 \gg 1$. In the quasi-geostrophic case, this can be analysed because $\mathbf{L}$ is a constant-coefficient operator. Such analyses are given in Arakawa and Lamb (1977) and Durran (1998) for various spatial discretizations. If the horizontal grids B or D

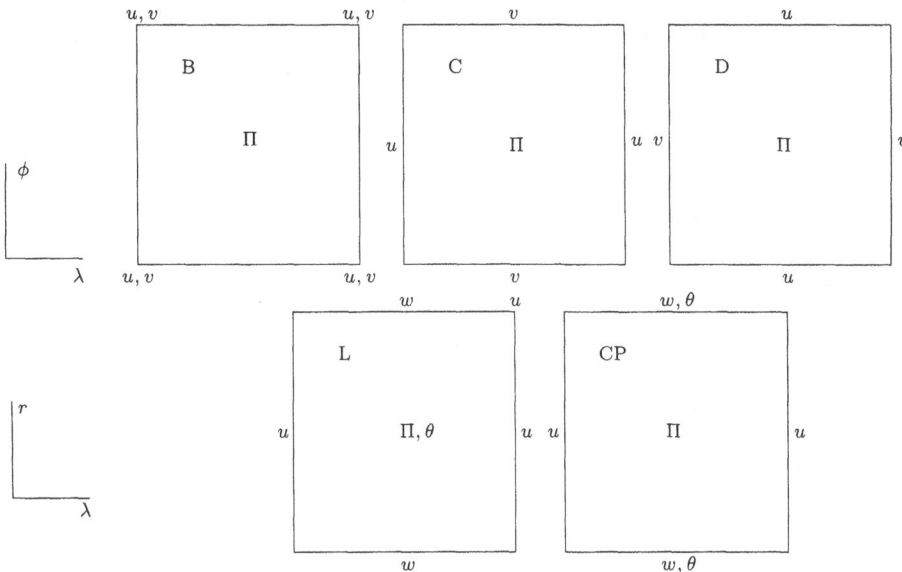

Figure 3.3. *Top line*: Various horizontal staggerings of variables for solving (3.17). These are referred to as Arakawa grids B, C and D. *Bottom line*: Two vertical staggerings, the Lorenz (L) and Charney–Phillips (CP).

shown in Figure 3.3 are used, the smallest-scale inertia-gravity waves become stationary so that $\omega_h = 0$, and the condition for (3.17) to be accurate will be violated. Only grid C gives satisfactory behaviour. In the vertical, the required condition is satisfied if the Charney–Phillips grid is used (Thuburn 2006).

There is thus a clash between the requirements on horizontal staggering of variables between the need to ensure that the discrete **L** is large when it should be, and the need to obtain stable solutions of (3.20). The Charney–Phillips grid, however, is optimal in the vertical on both counts. This clash indicates that the use of semi-implicit time integration is not sufficient to ensure accurate treatment of the limit solution, but spatial differencing issues must also be considered. Various methods have been used to address the issue. In Lin and Rood (1997) the C- and D-grids are both used for the parts of the calculation for which they are suited. However, interpolation between the grids at some point is unavoidable. A compromise horizontal arrangement called the B-grid is shown in Figure 3.3. Analysis, such as in Bryan (1989), shows that the B-grid is superior when the horizontal grid-length of the model is greater than the Rossby radius, but the C-grid is superior when the grid-length is less than the Rossby radius. In the atmosphere the internal Rossby radius for the most energetic parts of the flow is more like 1000 km, and is well-resolved. Thus the C-grid is more appropriate. The importance of preventing incorrectly small values of ω_h

is greatest in semi-implicit schemes, where the left-hand side of (3.12) has to be inverted. In explicit schemes, the problems can be managed by using filtering procedures. Thus the original Met Office Unified Model (Cullen and Davies 1990) used split-explicit finite differencing on the B-grid in the horizontal and the Lorenz grid in the vertical, while the current version (Davies *et al.* 2005) uses semi-implicit time integration on a C-grid in the horizontal and the Charney–Phillips grid in the vertical.

The difficulties in establishing an optimal horizontal representation can be resolved by using vorticity and divergence, rather than u and v, as prognostic variables. This is expensive in finite-difference methods, because of the need to solve elliptic equations to recover the velocity components. However it is a natural choice when using a spectral horizontal representation, which is very popular for global models. The elliptic problems then become trivial in spectral space. Under the assumptions made in this section, the operator $\mathbf{L}$ is separable in the horizontal and vertical. It has coefficients that depend only on the vertical. If equation (3.12) is discretized in the vertical, $\mathbf{L}$ becomes a matrix. Transforming variables to diagonalize this matrix means that the solution procedure can be decoupled into a set of constant-coefficient two-dimensional problems for each vertical mode: see Durran (1998, p. 387). A spectral model is very well-suited to solving these. The same transformation will reduce (3.20) to a set of two-dimensional problems. In practice, the scale-dependence of ω shown in (2.15) means that $\mathbf{L}$ has a large condition number. Standard iterative solvers will find it difficult to invert $\mathbf{L}$ accurately. The method of transforming the variables, followed by a solution in spectral space, will invert it to machine precision, which may be an important advantage. It should be noted that this method can be used to solve semi-implicit discretizations of (2.1): it does not depend on the approximations made in deriving (3.17). The spectral method is thus well-suited to maintaining the accuracy of (3.17) and to solving (3.20).

The advantages of spectral methods discussed above have to be balanced against the computational overhead of the transforms between grid-point and spectral space. They also maintain quadratic conservation properties, such as those discussed by Sadourny (1975). However, they are not well-suited to maintaining the bounds on Q, since there is no way of enforcing a monotonicity property. The combination of a spectral representation with semi-Lagrangian advection, which can be made to enforce monotonicity, has therefore become popular, as in the ECMWF model (Ritchie *et al.* 1995).

3.4. Large-scale flows

For this purpose we consider large-scale as a horizontal scale large relative to the Rossby radius, so that Ro < Fr, and we also assume Ro to be small. It is then appropriate to make the shallow atmosphere and hydrostatic assumptions, as in (2.11), and the geostrophic approximation, but the assumptions that the static stability and Coriolis parameter are close to reference values are not made. In particular, this means that the geostrophic wind is not non-divergent, so it is better to work directly from (2.11) rather than introduce the vorticity and divergence equations (3.1). The semi-geostrophic equations introduced by Hoskins (1975) are then appropriate, but note that f is not assumed constant. The system is

$$\frac{D\mathbf{u}_g}{Dt} + (-fv, fu) + C_p \theta \nabla_r \Pi' = 0,$$

$$\frac{D\theta}{Dt} = \frac{1}{C_p \Pi}(S_h + LP),$$

$$\frac{\partial \rho}{\partial t} + \nabla \cdot (\rho \mathbf{u}) = 0,$$

$$\frac{Dq}{Dt} = S_q - P,$$

$$p_{\text{ref}} \Pi^{\frac{1}{\gamma-1}} = \rho R\theta,$$

$$\frac{C_p \theta}{a \cos \phi} \frac{\partial}{\partial \lambda} \Pi' - f v_g = 0,$$

$$\frac{C_p \theta}{a} \frac{\partial}{\partial \phi} \Pi' + f u_g = 0,$$

$$C_p \theta \frac{\partial \Pi'}{\partial r} - g \frac{\theta'}{\theta_0} = 0.$$

(3.23)

In these equations, $\mathbf{u}_g = (u_g, v_g)$ is the geostrophic wind, defined by the sixth and seventh equations. Comparing (3.23) with (2.11) shows that the momentum has been approximated by its geostrophic value, but no other approximations have been made. In particular, the trajectory is not approximated, so that $\frac{D}{Dt} = \frac{\partial}{\partial t} + \mathbf{u} \cdot \nabla$. It can be shown that the resulting equations conserve energy in the absence of forcing terms (Cullen, Norbury, Purser and Shutts 1987). The conserved energy is

$$E = \int \rho \left(\tfrac{1}{2}(u_g^2 + v_g^2) + C_v T + gr \right) a^2 \cos \phi \, dr \, d\lambda \, dr. \qquad (3.24)$$

This differs from the energy (2.2) conserved by equations (2.1) by the replacement of the kinetic energy by its geostrophic value.

Most analyses of (3.23) have been carried out for constant f, which allows the equations to be solved by the geostrophic coordinate transformation, Hoskins (1975). However, this assumption is inappropriate for large-scale flow. An analysis of the spherical case is given by Cullen, Douglas, Roulstone and Sewell (2005). A review of the mathematical theory of these equations is given by Cullen (2006).

Equations (3.23) differ in an important respect from the nonlinear balance equations considered in Section 3.2. They contain two different velocity fields, representing the trajectory and the momentum respectively. This looks strange as a direct approximation to (2.11). However, it arises naturally if the equations are considered as Lagrangian averages of (2.11). The analysis of Andrews and McIntyre (1978) shows that the trajectory should represent the Lagrangian-averaged velocity, but the momentum-like quantity is a 'pseudo-momentum' which includes the effects of the waves excluded by the averaging. Such equations arise naturally when deriving asymptotic limit equations from Hamilton's principle. A systematic review is given in Holm, Marsden and Ratiu (2002). This idea has been exploited in sub-grid models, with the aim of modelling unresolved motions without introducing energy dissipation. Examples are the Gent–McWilliams parametrization of oceanic eddies (Gent and McWilliams 1996), and the 'alpha' model of turbulence (Foias, Holm and Titi 2001). Equations (3.23) can be interpreted as a Lagrangian average of (2.11) under the assumption that the pseudo-momentum is geostrophic. This is not the same as assuming that the momentum is geostrophic, but still assumes rotation-dominated flow.

The structure of these equations can be understood by rewriting them in the form used by Schubert (1985). This gives

$$\mathbf{Q}\begin{pmatrix} u \\ v \\ w \end{pmatrix} + C_p\theta\frac{\partial}{\partial t}\nabla\Pi' = \begin{pmatrix} f^2 u_g \\ f^2 v_g \\ \frac{g}{C_p\overline{\Pi}\theta}(S_h + LP) \end{pmatrix}, \tag{3.25}$$

$$\mathbf{Q} = \begin{pmatrix} f^2 + \frac{f\theta}{a\cos\phi}\frac{\partial}{\partial\lambda}\left(\frac{v_g}{\theta}\right) + \frac{fu_g\tan\phi}{a} & \frac{f\theta}{a}\frac{\partial}{\partial\phi}\left(\frac{v_g}{\theta}\right) & f\theta\frac{\partial}{\partial r}\left(\frac{v_g}{\theta}\right) \\ -\frac{f\theta}{a\cos\phi}\frac{\partial}{\partial\lambda}\left(\frac{u_g}{\theta}\right) + \frac{fv_g\tan\phi}{a} & f^2 - \frac{f\theta}{a}\frac{\partial}{\partial\phi}\left(\frac{u_g}{\theta}\right) & -f\theta\frac{\partial}{\partial r}\left(\frac{u_g}{\theta}\right) \\ \frac{g}{a\theta\cos\phi}\frac{\partial\theta}{\partial\lambda} & \frac{g}{a\theta}\frac{\partial\theta}{\partial\phi} & \frac{g}{\theta}\frac{\partial\theta}{\partial r} \end{pmatrix}.$$

Equation (3.25) can be rewritten as an elliptic equation for $\partial\Pi/\partial t$ by using the last three equations of (3.23) to give

$$\frac{1}{1-\gamma}\frac{\rho}{\Pi}\frac{\partial\Pi}{\partial t} + \frac{\rho C_p\theta}{g}\frac{\partial}{\partial r}\left(\frac{\partial\Pi}{\partial t}\right) + \nabla\cdot\left[C_p\rho\theta\mathbf{Q}^{-1}\nabla\left(\frac{\partial\Pi}{\partial t}\right)\right] = \tag{3.26}$$

$$\nabla\cdot\rho\mathbf{Q}^{-1}\begin{pmatrix} f^2 u_g \\ f^2 v_g \\ \frac{g}{\theta}(S_h + LP) \end{pmatrix}.$$

This is an elliptic equation for $\partial\Pi/\partial t$ if $\mathbf{Q}$ is positive definite. Since $\mathbf{Q}$ is purely a function of Π, this is a constraint on the pressure field. It is shown by Shutts and Cullen (1987) that this corresponds to *symmetric stability*, that is, stability of the resulting flow to parcel displacements, neglecting perturbation pressures arising from the displacement. They do this by showing that a geostrophic and hydrostatic state with $\mathbf{Q}$ positive definite corresponds to a minimum energy state with respect to such parcel displacements. The semi-geostrophic approximation is only relevant for such flows, since flows which are unstable in this sense will evolve on a faster time-scale than f^{-1}. The manifestation of this in the horizontal is inertial stability, and in the vertical is static stability. Shutts and Cullen also show that the neglect of perturbation pressures is justified in the large-scale regime where semi-geostrophic theory is appropriate.

If (3.26) can be solved, the trajectory $\mathbf{u}$ can be deduced from (3.25). However, it is not obvious that the positive-definiteness of $\mathbf{Q}$ can be maintained during the time evolution. In order to study this, following Cullen and Feldman (2006), we introduce the Lagrangian map $F(t, \mathbf{x})$, where $\mathbf{x} = (\lambda, \phi, r)$ is a shorthand for the three-dimensional space coordinates. This map takes initial parcel positions to positions at time t, so that in particular $F(0, \mathbf{x}) = \mathbf{x}$. The map is assumed to conserve mass. This means that the Jacobian of $F(t, \mathbf{x})$ is $\rho(t, \mathbf{x})/\rho(0, \mathbf{x})$. We use the notation $F(t, \mathbf{x}) \# \rho(0, \mathbf{x}) = \rho(t, \mathbf{x})$ as shorthand for this property. Now write equations (3.23), omitting the moisture equation and forcing terms, in the Lagrangian form

$$\partial_t Z(t, \mathbf{x}) - f \partial_t F(t, \mathbf{x}) = f J Z,$$

$$F(t, \mathbf{x}) \# \rho(0, \mathbf{x}) = \rho(t, \mathbf{x}),$$

$$p_{\mathrm{ref}} \Pi^{\frac{1}{\gamma-1}} = \rho R \theta, \tag{3.27}$$

$$Z(0, \mathbf{x}) = f^{-1} C_p \theta^0 \nabla \Pi^0(\mathbf{x}).$$

Here, $Z(t, \mathbf{x}) = f^{-1} C_p \theta \nabla \Pi'(t, F(t, \mathbf{x}))$, the first equation of (3.27) corresponds to the first two equations of (3.23) and the second equation corresponds to the third equation of (3.23).

We now use this formulation to define an energy minimizer with respect to infinitesimal parcel displacements, assumed to take place over a virtual time δ. Given a state $\tilde{\Pi}(0, \mathbf{x}), \tilde{u}(0, \mathbf{x}), \tilde{v}(0, \mathbf{x}), \tilde{\theta}(0, \mathbf{x})$, with $\tilde{\rho}(0, \mathbf{x}), \tilde{T}(0, \mathbf{x})$ calculated from $\tilde{\Pi}(0, \mathbf{x})$ using the equation of state and the thermodynamic relations, define the energy $\tilde{E}$ by

$$\tilde{E} = \int \tilde{\rho} \left(\frac{1}{2} (\tilde{u}^2 + \tilde{v}^2) + C_v \tilde{T} + gr \right) d\mathbf{x}. \tag{3.28}$$

Define the displacement by the Lagrangian map $F(\delta, \mathbf{x})$. The assumption of no perturbation pressure due to the displacement means that

$$Z(\delta, F(\delta, \mathbf{x})) - Z(0, F(0, \mathbf{x})) = f(F(\delta, \mathbf{x}) - F(0, \mathbf{x})),$$

$$F(\delta, \mathbf{x}) \# \tilde{\rho}(0, \mathbf{x}) = \tilde{\rho}(\delta, \mathbf{x}), \tag{3.29}$$

$$Z(0, \mathbf{x}) = (\tilde{v}, -\tilde{u}).$$

We calculate the change to the energy by substituting $\tilde{\rho}(\delta, \mathbf{x}), \tilde{u}(\delta, \mathbf{x}), \tilde{v}(\delta, \mathbf{x})$

and $\tilde{T}(\delta, \mathbf{x})$ into (3.28). Then Theorem 4.1 of Cullen (2006) states that $\tilde{E}$ is stationary with respect to these variations if $Z = f^{-1}C_p\tilde{\theta}\nabla\tilde{\Pi}$. Thus a solution of (3.27) corresponds to a stationary energy state. The requirement that the solution is a minimizer, rather than just stationary, comes from the physical consistency discussed above. The mathematical theory of the semi-geostrophic equations shows that it is always possible to find an energy minimizer, and thus a physically consistent solution of the equations. Other possible solutions are ignored. The effect is that it is always possible to find a time evolution such that $\mathbf{Q}$ remains positive definite, given initial data such that $\mathbf{Q}$ is positive definite. One result of this is that there can be no horizontal pressure gradients along the equator, as discussed by Cullen *et al.* (2005). Details of the theorems are reviewed in Cullen (2006).

The situation is thus different from the nonlinear balance equations, (3.1), (3.2), (2.10), (2.12), (3.5) and (3.6). These also reduce to a variable-coefficient equation but whose ellipticity cannot be guaranteed in the time evolution. There are no solvability issues with the quasi-geostrophic equations because they reduce to solving a constant-coefficient elliptic equation, (3.20). However, no system of limit equations valid on large scales can be reduced to a constant-coefficient elliptic problem.

Another feature of the solutions of (3.23) is that they can be discontinuous. This does not conflict with the idea that the equations represent a Lagrangian average. There is no reason why air parcels with different physical properties should not come close together in a (near-)discontinuity. An example is discussed in Section 4.3, where the ability to capture such a discontinuity in a conventional numerical method is explored.

The condition for positive-definiteness of $\mathbf{Q}$ is harder to maintain in the vertical if moist effects are allowed for. This is because we can write the LP term that appears in the second equation of (3.23) as

$$LP = L\frac{\mathrm{d}q_{\mathrm{sat}}}{\mathrm{d}t} \simeq L\frac{\mathrm{d}q_{\mathrm{sat}}}{\mathrm{d}T}\frac{\mathrm{d}T}{\mathrm{d}t} \simeq L\Gamma w \tag{3.30}$$

where Γ now represents the moist adiabatic lapse rate (Houghton 2002, p. 21). Since this term is proportional to w, it should be transferred to the left-hand side of (3.25). The effect is to reduce the diagonal term in $\mathbf{Q}$ and make it easier to violate the condition for $\mathbf{Q}$ to be positive definite. Though no rigorous mathematical treatment has been given in this case, computations by Holt (1990) suggest that (3.25) can still be solved for general data. These computations demonstrate that the solutions may involve discontinuous mass transport when the condition $q \geq q_{\mathrm{sat}}$ is only satisfied in parts of the domain. This is still compatible with the interpretation of the equations as a Lagrangian average of (2.1). Representing this type of process in a sub-grid model to be used with equations (2.1) has been a major research challenge; see Smith (1997).

A priori estimates of the difference between solutions of (3.23) and (2.11) or (2.1) given in Cullen (2006) show that the difference is $O(\mathrm{Ro}_L(\mathrm{Ro}/\mathrm{Fr})^2)$ for small Ro_L. Ro_L is the Lagrangian Rossby number defined as the ratio of D/Dt and f. It is small if the rate of change of wind direction following a fluid particle is small compared with f. At latitude $45°$ that requires the direction of a trajectory to change by less than $45°$ in 24 hours. The effect of this is illustrated in Figure 3.4. Trajectories are plotted from a UM forecast for a very active period of weather in January 2005. Making allowance for the Mercator projection, the condition on the trajectory direction is only violated in the trajectory starting furthest west off the Californian coast, and in the lowest level trajectory as it crosses the Rockies. This suggests that the approximation will be valid most of the time for scales represented by this model, which has an averaging scale of about 120 km.

We now consider conventional finite-difference approximations to (3.26). In the case of small disturbances, $\mathbf{Q}$ is well approximated by the diagonal matrix

$$\begin{pmatrix} f^2 & 0 & 0 \\ 0 & f^2 & 0 \\ 0 & 0 & N^2 \end{pmatrix}.$$

Equation (3.26) is naturally solved at pressure points. Then u_g and v_g are naturally held at the positions occupied by u and v on the D-grid in Figure 3.3 and θ will be held as indicated on the Charney–Phillips grid in Figure 3.3; u, v and w are naturally held on the C-grid and Charney–Phillips grid. As in the quasi-geostrophic case, this arrangement is not ideal in the horizontal, because the accurate treatment of the right-hand side requires u_g at v_g-points and *vice versa*.

In general, the structure of equation (3.26) suggests that it could be incorporated in a semi-implicit discretization of (2.1), analogous to the derivation of equation (3.12). This is difficult to achieve, because of the presence of the two different velocity fields $\mathbf{u}$ and $\mathbf{u}_g$. In addition, it is found that in circumstances where (3.23) has discontinuous solutions, conventional finite-difference methods attempt to maintain smoothness by allowing $\mathbf{Q}$ not to be positive definite. This then results in computational instability (see Cullen (2006, Section 5.3.3)). In such cases, it is not even clear if the Eulerian form of advection used to write (3.25) makes sense.

To make further progress, we therefore need to base the discretization on the Lagrangian formulation of the equations (3.27), which means finding the Lagrangian map $F(t, \mathbf{x})$ by using all the constraints in equations (3.27). This requires methods where advection is treated in a Lagrangian manner. When solving (2.11) in circumstances when (3.23) is accurate, we then seek a Lagrangian map F close to that which satisfies (3.27).

We first have to establish a condition on Π such that the matrix $\mathbf{Q}$ derived from it is positive definite. Equations (3.23) and the definition of the reference state, (2.8), show that

$$C_p \nabla \Pi = \left(\frac{f v_g}{\theta}, -\frac{f u_g}{\theta}, -\frac{g}{\theta} \right). \tag{3.31}$$

Then it can be seen from (3.25) that, in the case of Cartesian geometry where f is constant, $\mathbf{Q}$ is the matrix of second derivatives of Π with all terms multiplied by θ and the additional terms f^2 on the diagonal. If variations

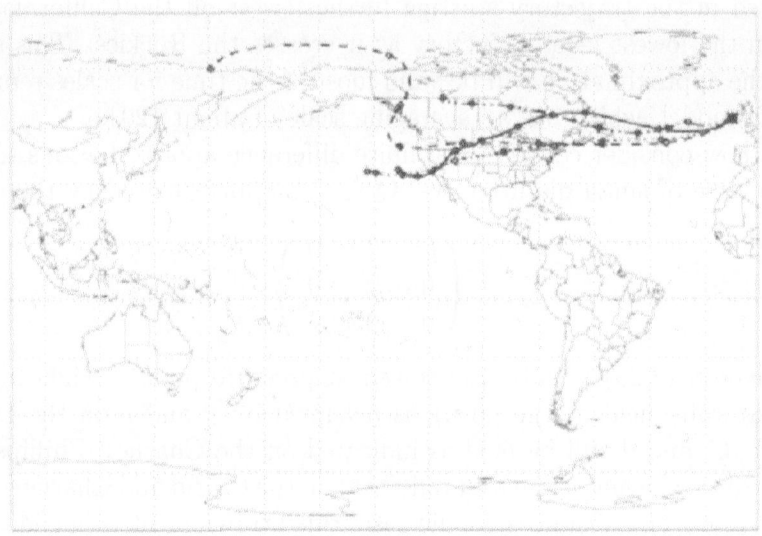

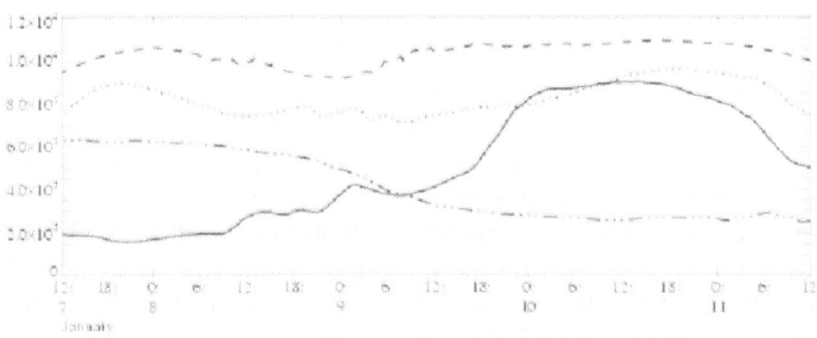

Figure 3.4. *Top*: Back trajectories from Mace Head for 12UTC on 11 January 2005. The total time covered is 96 hours, with points marked at 12 hour intervals. The trajectories were computed from 3-hourly data. *Bottom*: The vertical position of the trajectories (m). Source: Atmospheric Dispersion Group, Met Office.

of the multiplying factor θ are ignored, the condition for $\mathbf{Q}$ to be positive definite is that $C_p\Pi + \frac{1}{2}(x^2 + y^2)$ is convex, where x and y are Cartesian coordinates. When variations of θ are included, the equivalent condition is derived by Shutts and Cullen (1987). In spherical geometry, the condition is derived by Cullen *et al.* (2005), and is referred to as *involutivity*. In both these cases, it can be shown that the condition prevents oscillatory behaviour of Π, in the same way that a convex function cannot oscillate.

Another important feature of the solutions of (3.27) is that only integrals of fluid properties are well-defined. This allows the solutions to be singular. Physically it means that the behaviour of infinitesimally small volumes of the fluid is irrelevant. Thus we choose a function φ which is very smooth, and define a 'weak' solution of (3.27) on a time interval $(0, \tau)$ by requiring that, for any $\varphi \in C^\infty([0, \tau] \times \Omega : \mathcal{R}^3)$,

$$\int_{\Omega \times (0,\tau)} \left[Z(t, \mathbf{x}) \cdot \partial_t \varphi(t, \mathbf{x}) - fF(t, \mathbf{x})\partial_t\varphi(t, \mathbf{x}) + fJZ(t, \mathbf{x}) \cdot \varphi(t, \mathbf{x}) \right] \mathrm{d}t\,\mathrm{d}\mathbf{x}$$

$$+ \int_\Omega f^{-1}\theta^0\nabla\Pi^0(\mathbf{x}) \cdot \varphi(0, \mathbf{x})\,\mathrm{d}\mathbf{x} = 0. \qquad (3.32)$$

Suppose that we are given initial data $\Pi^0(\mathbf{x})$, assumed to be an involutive bounded non-negative function defined on an open set $\Omega \subset \mathcal{R}^3$. Let ρ^0 and θ^0 be calculated from Π^0 by the hydrostatic relation. In the case where f is constant, Cullen and Feldman (2006) proved that we can find for any $t \in (0, \tau)$ a Lagrangian map $F_t = F(t, \mathbf{x}) : \Omega \to \Omega$ satisfying $F_t\#\rho^0 = \rho(t, \mathbf{x})$ and such that (3.32) is satisfied. This map also has an inverse F_t^* such that $F_t^* \circ F_t(\mathbf{x}) = \mathbf{x}$ for almost all $\mathbf{x}$ (*i.e.*, it may not exist for infinitesimal masses of fluid). This property means that the Lagrangian map F satisfies the 'flow' property that $F_{t_1+t_2} = F_{t_1} \circ F_{t_2}$. This is required for the solution to make physical sense, and not to depend on an arbitrary discretization of the time interval. This proof has not yet been extended rigorously to the case of variable f. Cullen *et al.* (2005) describe the formal arguments which show that the result is expected to hold in that case.

Now consider numerical methods for approximating the Lagrangian map. The most widely used Lagrangian-based methods are the semi-Lagrangian methods introduced by Robert (1982). A review of these is given by Staniforth and Côté (1991). They are used in both the UM and the ECMWF model, as well as many others. They have the advantage that the data are mapped onto an Eulerian grid at each time-step, thus allowing computations of terms other than advection to be carried out easily. They also have the advantage of being unconditionally stable for time integration unless the trajectories cross; see Durran (1998, Chapter 6). This was the original reason for their introduction. Enforcement of the condition $F(t, \mathbf{x})\#\rho(0, \mathbf{x}) = \rho(t, \mathbf{x})$ requires the discrete Lagrangian map to be mass-preserving. This is the motivation for conservative semi-Lagrangian methods such as Zerroukat, Wood and Staniforth (2004). However, such methods are hard to use and have not yet been implemented in production models. They are discussed in more detail in Section 3.6. The proof that (3.32) can be solved depends on the fact that a Lagrangian map cannot create new values of the quantity to which it is applied, as in the advection of Q in the quasi-geostrophic case. In the present case the condition has to be applied separately to the momentum components and the potential temperature. This again favours the use of monotone interpolation schemes, as discussed in Section 3.3. However, the remapping to grid-points every time-step means that strictly monotone schemes will be too diffusive, and more selective schemes such as that of Zerroukat et al. (2005) are needed.

The trajectory is derived implicitly by the solution procedure. This suggests that an implicit calculation of the trajectory departure points is desirable. An iteration like (3.16) can be used for this purpose, as in Yeh et al. (2002) and Cullen (2001). In Cullen (2001), the increased accuracy of doing this is demonstrated. In White (2003) the dynamical consistency resulting from this choice is discussed. The application of this scheme to the UM is described in Diamantakis, Davies and Wood (2007). An example is shown in Figure 3.5. The top two panels use the operational version of the UM with the departure point calculation extrapolated in time to obtain second-order accuracy. They show that the vertical motion is greatly exaggerated in a run with a 30-second time-step as compared to a run with a 10-second time-step. If the departure-point calculation is iterated once in time, as in Cullen (2001), then the results shown in the bottom panel are obtained using a 40-second time-step. The results are close to that obtained with the operational scheme and a 10-second time-step, showing that the time iteration is beneficial and cost-effective.

The proof of existence of solutions to (3.23) depends critically on the involutivity of Π, which prevents oscillatory behaviour and thus gives considerable stability to the time evolution, as demonstrated in Cullen (2002a). This is why large-scale disturbances persist. However, on smaller scales, where semi-geostrophic theory is not applicable, there is no reason why solutions of (2.1) or (2.11) should obey this condition and oscillatory and unstable behaviour is readily observed. Section 4.4 illustrates such a case, where the instability is triggered by orography. The direct solution of (2.1) is quite successful, though it has clearly not converged. This is, however, a rather simple case. In real cases, satisfactory solutions to (2.1) in regions where $\mathbf{Q}$ is not positive definite can usually only be obtained by using a horizontal grid-length of no more than 1 km. Operational averaging scales are normally much greater than this, except for local models. In order to get satisfactory performance with coarser averaging scales, the sub-grid model has to enforce the involutivity of Π in order to prevent unrealistic unstable

Figure 3.5. Cross-sections of vertical velocity $(\mathrm{m\,s^{-1}})$ from UM integration over the Alpine region plotted against height (m) and latitude relative to the centre of the domain (degrees). *Top*: Explicit departure-point calculation, time-step 30 s, decentering parameter $\alpha = 0.7$. *Middle*: As above with time-step 10 s and $\alpha = 0.6$. *Bottom*: Iterated departure-point calculation, time-step 40 s, $\alpha = 0.6$.

circulations developing. Since typical time-averaging scales in the atmosphere are much less than f^{-1}, but greater than N^{-1}, it is usual to enforce only the condition that $\partial\theta/\partial r \geq 0$, modified for the effects of moisture as shown in (3.30). Recent collections of papers in this area are given in ECMWF (2002) and ECMWF (2005).

The long-term behaviour of solutions to (3.23) was studied in the shallow water case by Cullen (2002a). The cascades of vorticity to small scales that are a feature of classical two-dimensional turbulence, Leith and Kraichnan (1972), are inhibited, and disturbances persist for long times. The long-lived anomalous circulations observed in the extra-tropical atmosphere thus represent the natural dynamics of rotation-dominated flows. This is therefore the most appropriate regime in which to study the requirements for maintaining long-time accuracy in numerical solutions.

This behaviour means that, if flows on a scale larger than the Rossby radius are initially well-resolved, then they will stay well-resolved. Numerical methods such as those of Sadourny (1975) may then be beneficial in preserving long-term accuracy. In particular, it is tempting to suggest that schemes which conserve $\int Q^2$, where Q is the Ertel potential vorticity, would be appropriate. However, fronts can form in three-dimensional flow. As demonstrated in Section 4.3, these are associated with apparent potential vorticity sources in an Eulerian sense, even though potential vorticity is still conserved in a Lagrangian sense. These solutions could not develop if schemes like those of Sadourny (1975) were used. It is therefore a major research question how to preserve long-term accuracy in this regime.

3.5. Small-scale flows with $\varepsilon \ll 1$

In this case we assume that the aspect ratio may be O(1), but that the horizontal extent is sufficiently small for plane geometry to be used. The assumption that $\varepsilon \ll 1$ for such aspect ratios means that U/NH has to be small. The solvability condition for the nonlinear balance equations is not necessarily satisfied, as noted in Section 3.2. The difficulty is caused by the flow-dependent coefficients in the elliptic problem that has to be solved. If only small-scale flows are considered, the flow-dependent coefficients are not required. Consider the equation for the evolution of θ:

$$\frac{\partial\theta}{\partial t} + \mathbf{u}_r.\nabla_r\theta + w\frac{\partial\theta}{\partial r} = \frac{1}{C_p\Pi}(S_h + LP). \tag{3.33}$$

If the right-hand side terms of (3.33) are ignored, then it can be shown that the assumption $U/NH \ll 1$ means that $\nabla_r\theta \ll \frac{H}{L}\frac{\partial\theta}{\partial r}$. We can therefore assume a hydrostatic reference state $\theta_1(r)$ with a reference value N_0^2 of N^2 as in Section 3.3. Comparison of the terms in equation (3.33) then shows that $w/\mathbf{u}_r \ll H/L$, so that the vertical advection can be neglected in the momentum equations of (2.11). Majda (2003, Chapter 6) shows that an energetically consistent system can then be obtained by approximating the pressure gradient term and replacing the continuity equation by $\nabla_r \cdot \mathbf{u} = 0$. The resulting system is

$$\frac{D_r \mathbf{u}_r}{Dt} + (-fv, fu) + C_p \theta_1 \nabla_r \Pi' = 0,$$

$$C_p \theta \frac{\partial \Pi'}{\partial r} - g \frac{\theta'}{\theta_0} = 0,$$

$$\nabla_r \cdot (\mathbf{u}) = 0, \qquad (3.34)$$

$$\frac{D_r \theta}{Dt} + w \frac{\partial \theta_1}{\partial r} = \frac{1}{C_p \Pi}(S_h + LP),$$

$$\Pi = (p/p_{\text{ref}})^{-R/C_p},$$

$$p = \rho R \theta \Pi.$$

The first and third equations in (3.34) reduce to the equations for two-dimensional incompressible flow. The nonlinear balance relation, (3.6), reduces to

$$-2J(u, v) + \nabla_r \cdot C_p \theta_1 \nabla_r \Pi' - \nabla_r \cdot (fv, -fu) = 0. \qquad (3.35)$$

These equations can be solved by integrating the first equation of (3.34) forward in time, with the pressure gradient terms determined implicitly by using (3.35) to enforce the incompressibility constraint, the third equation of (3.34). Equation (3.35) together with the hydrostatic relation (the second equation of (3.34)) can be used to determine Π and θ. It is then possible to derive w from the evolution equation for θ.

Solvability of these equations follows from the solvability of the equations for two-dimensional incompressible flow; *e.g.*, Chemin (2000). This depends on the fact that the vorticity ζ defined by (3.1) is bounded by its initial values, and that the elliptic equation (3.35) for Π' satisfies second derivative estimates analogous to (3.22). This ensures that the horizontal velocity fields stay smooth for smooth initial data. However, the vorticity typically cascades to small scale, so an initially well-resolved solution will not stay well-resolved. The remaining steps are explicit calculations.

A discrete version of the first requirement can be achieved in numerical models by using ζ as a prognostic variable, and advecting it with a monotonicity-preserving scheme. The alternative method, based on Sadourny (1975), of enforcing conservation of ζ^2 has the disadvantage of forcing the variance of ζ to be conserved on resolved scales, which is inappropriate in the presence of cascades of variance to small scales. The second requirement can be met if the elliptic equation (3.35) for Π' satisfies a discrete maximum principle. As in (3.19), this favours the use of the D-grid for the velocity components. The explicit calculation of θ from Π needs the

Charney–Phillips grid, as the discrete version of this calculation on the Lorenz grid is impossible because of the computational mode (Arakawa and Konor 1996).

The behaviour of this system is analysed in Majda (2003). The difficulty is that the first two equations of (3.34) evolve independently at each value of r. There is thus no control over the vertical scale of the solutions. In particular, values of $\partial\Pi/\partial r$ may become very large, leading to unrealistic values of θ and hence w. The problem is that the solutions violate the assumption $U/NH \ll 1$ as H is reduced. This is called 'zigzag' instability. In Majda (2003) it is shown that the problem can be resolved either by vertical viscosity, which cannot be physically justified in this strongly stratified case, or by rotation. As the vertical scale reduces, it may be that the condition $NH = fL$ becomes valid while the Froude number is still small. In that case the quasi-geostrophic equations derived in Section 3.3 apply, and the vertical coupling is restored.

As will be illustrated in Section 4.2, the assumption $\nabla_r \cdot \mathbf{u}_r = 0$ is very restrictive. A scheme with higher-order accuracy, but avoiding the solvability condition (3.7), involves iterating the solution procedure above. Thus, the calculated w is used via Richardson's equation (2.12) to generate a value of $\nabla_r \cdot \mathbf{u}_r$. This is then used along with ζ to calculate $\mathbf{u}$. In Ford et $al.$ (2000) such an iterative procedure is developed in a rather more general context. Numerical solutions described by McIntyre and Norton (2000) show its effectiveness in shallow water integrations. These did not use sufficiently high resolution for (3.7) to be violated. The convergence of such an iteration in cases where (3.7) is violated is uncertain.

The solution procedure above is very attractive for numerical modelling, because it only involves solving two-dimensional rather than three-dimensional elliptic problems. However, the instability of this regime means that such methods will be of limited applicability.

3.6. Small-scale flows with $\varepsilon \gg 1$

In this case we assume that the aspect ratio may be O(1), but that the horizontal extent is sufficiently small for plane geometry to be used. We thus use Cartesian coordinates (x, y, z) and assume that f is a constant. Equations (2.27), with the moisture equation and forcing terms restored, become

$$\frac{D\mathbf{u}}{Dt} + f(-v, u, 0) + C_p\theta\nabla\Pi' - g\frac{\theta'}{\theta_0}\hat{\mathbf{r}} = 0,$$

$$\frac{\partial\rho}{\partial t} + \nabla \cdot (\rho\mathbf{u}) = 0,$$

$$\frac{D\theta}{Dt} = \frac{1}{C_p\Pi}(S_h + LP), \qquad (3.36)$$

$$\frac{Dq}{Dt} = S_q - P,$$

$$p_{\text{ref}}\Pi^{\frac{1}{\gamma-1}} = \rho R\theta.$$

We showed in Section 2.3 that the characteristic velocity $U \ll c$ for all cases of interest. This means that equation (2.34) can be approximated by (2.35). Energetically consistent approximations to (3.36) that achieve this include the anelastic approximations of Ogura and Phillips (1962) and Lipps and Hemler (1982). In these approximations, the continuity equation is replaced by the constraint $\nabla \cdot (\rho_0 \mathbf{u}) = 0$, where ρ_0 is given by either the reference state (2.8) or the reference state used in (3.34). This reference state has also to be used in the momentum equations in order to retain energetic consistency. Issues with the effect of this approximation on convergence to the anelastic limit are discussed below. It is assumed that the scale is still large enough for the real viscous and conductive terms to be neglected.

With either choice of reference state, these approximations reduce equations (3.36) to

$$\frac{D\mathbf{u}}{Dt} + f(-v, u, 0) + C_p \nabla(\theta_0 \Pi') - g\frac{\theta'}{\theta_0}\hat{\mathbf{r}} = 0,$$

$$\nabla \cdot (\rho_0 \mathbf{u}) = 0,$$

$$\frac{D\theta}{Dt} = \frac{1}{C_p\Pi}(S_h + LP), \qquad (3.37)$$

$$\frac{Dq}{Dt} = S_q - P,$$

$$p_{\text{ref}}\Pi^{\frac{1}{\gamma-1}} = \rho R\theta.$$

The conserved energy density is

$$\rho_0\left(\frac{1}{2}\mathbf{u}^2 + gr + C_p\Pi_0\theta\right). \qquad (3.38)$$

Since, in effect, this approximation makes the speed of sound infinite, it is not valid on large scales where the finite sound speed matters; see Davies, Staniforth, Wood and Thuburn (2003).

In the case where $f = g = 0$, (3.37) shows that $\nabla \cdot \theta_0 \nabla \Pi$ is of order $|\mathbf{u}|^2$. Thus the variations in Π will be of the same order as the variations in θ, with an additional variation of $O(|\mathbf{u}|^2)$. The equation of state then shows that ρ will have the same order of magnitude of variations as Π. In the special case of uniform θ, this means that the error in using the reference state value of ρ is of order (Mach number)^2. This is the case where convergence of compressible flow to incompressible flow has been proved (Majda 1984). There will also be an error related to the departures in θ from its assumed reference value. In general, this error will increase with the size of the variations in θ in the initial data. It will not be related to the Mach number. The error will be reduced in strongly stratified flows where θ remains close to a reference state, even though that reference state has large variations in θ. This was the case treated in Section 3.5 where the dynamics is hydrostatic and can be described by (3.34).

We therefore consider the case where $U/(NH)$ is large, so that the stratification is not dominant. This will be particularly true of the circulations in convective clouds, for instance. We avoid the difficulties caused by the nonlinearity by using a Lagrangian form of the equations, as in Section 3.4. Thus write $\mathbf{x} = (x, y, z)$ and define a Lagrangian map $F(t, \mathbf{x})$ with $F(0, \mathbf{x}) = \mathbf{x}$. The conserved mass density of a fluid element is now $\rho_0(z)$. The continuity equation is expressed by the condition $F(t, \mathbf{x}) \# \rho_0 = \rho_0$. Define $Z(t, \mathbf{x}) = (-fv(t, F(t, \mathbf{x})), fu(t, F(t, \mathbf{x})), -g\theta'(t, F(t, \mathbf{x}))/\theta_0)$ and $Y(t, \mathbf{x}) = C_p \nabla(\theta_0 \Pi)(t, F(t, \mathbf{x}))$. The Lagrangian form of (3.37) and the initial conditions is then

$$\partial_t F(t, \mathbf{x}) = \mathbf{u}(t, F(t, \mathbf{x})),$$

$$\partial_t \mathbf{u}(t, F(t, \mathbf{x})) + Z(t, \mathbf{x}) + Y(t, \mathbf{x}) = 0,$$

$$F(t, \mathbf{x}) \# \rho_0 = \rho_0,$$

$$\partial_t \theta(t, F(t, \mathbf{x})) = \frac{1}{C_p \Pi(t, F(t, \mathbf{x}))}(S_h + LP),$$

$$\partial_t q(t, F(t, \mathbf{x})) = S_q - P, \tag{3.39}$$

$$\mathbf{u}(0, \mathbf{x}) = \mathbf{u}^0(\mathbf{x}),$$

$$\theta(0, \mathbf{x}) = \theta^0(\mathbf{x}),$$

$$q(0, \mathbf{x}) = q^0(\mathbf{x}),$$

$$F(0, \mathbf{x}) = \mathbf{x}.$$

As in Section 3.4, the omission of viscosity and thermal conductivity means that there is no reason why the solutions should be smooth. Air parcels with different properties can 'tangle' as much as desired. We therefore again seek weak solutions, where the behaviour of infinitesimally small fluid volumes is ignored, and work only with integrals. We thus multiply (3.39) by smooth test functions φ and ϖ and seek solutions in the sense that, for any $\varphi \in C^\infty([0, \tau] \times \Omega : \mathcal{R}^3)$, $\varpi \in C^\infty([0, \tau] \times \Omega : \mathcal{R})$:

$$\int_{\Omega \times (0, \tau)} \left[F(t, \mathbf{x}) \cdot \partial_t \varphi(t, \mathbf{x}) + \mathbf{u}(t, F(t, \mathbf{x})) \cdot \varphi(t, \mathbf{x}) \right] dt \, d\mathbf{x} \tag{3.40}$$

$$+ \int_\Omega \mathbf{x} \cdot \varphi(0, \mathbf{x}) \, d\mathbf{x} = 0,$$

$$\int_{\Omega \times (0,\tau)} \left[\mathbf{u}(t, F(t, \mathbf{x})) \cdot \partial_t \varphi(t, \mathbf{x}) + Z(t, \mathbf{x}) \cdot \varphi(t, \mathbf{x}) \right.$$

$$\left. + Y(t, \mathbf{x}) \cdot \varphi(t, \mathbf{x}) \right] dt \, d\mathbf{x} + \int_{\Omega} \mathbf{u}^0(\mathbf{x}) \cdot \varphi(0, \mathbf{x}) \, d\mathbf{x} = 0,$$

$$\int_{\Omega \times (0,\tau)} \left[\theta(t, F(t, \mathbf{x})) \partial_t \varpi(t, \mathbf{x}) \frac{1}{C_p \Pi} (S_h + LP) \varpi(t, \mathbf{x}) \right] dt \, d\mathbf{x}$$

$$+ \int_{\Omega} \theta^0(\mathbf{x}) \varpi(0, \mathbf{x}) \, d\mathbf{x} = 0,$$

$$\int_{\Omega \times (0,\tau)} \left[q(t, F(t, \mathbf{x})) \partial_t \varpi(t, \mathbf{x}) - (S_q - P) \varpi(t, \mathbf{x}) \right] dt \, d\mathbf{x}$$

$$+ \int_{\Omega} q^0(\mathbf{x}) \varpi(0, \mathbf{x}) \, d\mathbf{x} = 0.$$

The existence of such solutions is an open question. However, a time-discretized version of (3.37) can be solved by a method based on Brenier (1991). It is described in Cullen (2002b, Section 3.3). Given data at $t = 0$, construct a 'solution' at time δt by the following procedure. The forcing terms are omitted. Define a first guess Lagrangian map F^+ by

$$x_c(\mathbf{x}) = x + \frac{v^0(\mathbf{x})}{f},$$

$$y_c(\mathbf{x}) = y - \frac{u^0(\mathbf{x})}{f},$$

$$F^+(\delta t, \mathbf{x}) = \left(x_c + (x - x_c)\cos(f\delta t) + (y - y_c)\sin(f\delta t), \right.$$

$$y_c + (y - y_c)\cos(f\delta t) - (x - x_c)\sin(f\delta t),$$

$$\left. z + w^0(\mathbf{x})\delta t + \frac{1}{2}g\frac{\theta'}{\theta_0}\delta t^2 \right), \tag{3.41}$$

$$u(\delta t, \mathbf{x}) = u^0(\mathbf{x})\cos(f\delta t) + v^0(\mathbf{x})\sin(f\delta t),$$

$$v(\delta t, \mathbf{x}) = v^0(\mathbf{x})\cos(f\delta t) - u^0(\mathbf{x})\sin(f\delta t),$$

$$w(\delta t, \mathbf{x}) = w^0(\mathbf{x}) + g\frac{\theta'(\mathbf{x})}{\theta_0}\delta t,$$

$$\theta'(\delta t, \mathbf{x}) = \theta^0(\mathbf{x}),$$

$$q(\delta t, \mathbf{x}) = q^0(\mathbf{x}).$$

The forcing terms can easily be included if they can be expressed in terms of $\mathbf{u}, \theta$ and q. Otherwise they have to be iterated, and a rigorous treatment may not be possible.

We now correct this first guess to enforce the condition that $F^+ \# \rho_0 = \rho_0$. This is done by using the *polar factorization* theorem of Brenier (1991). This states that under suitable non-degeneracy conditions, we can uniquely write

$$F^+(\delta t, \mathbf{x}) = F(\delta t, \mathbf{x}) \circ \nabla \Upsilon(\mathbf{x}), \tag{3.42}$$

where $F(\delta t, \cdot) \# \rho_0 = \rho_0$ and Υ is convex. If the term $C_p \nabla(\theta_0 \Pi)$ had been included in (3.41), and it took constant values on trajectories, F^+ would have been incremented by $\frac{1}{2} \delta t^2 C_p \nabla(\theta_0 \Pi)$. We can thus identify the projection $\nabla \Upsilon$ defined in (3.42) with the mapping $I + \frac{1}{2} \delta t^2 C_p \nabla(\theta_0 \Pi)$. Note that the presence of the identity map, which can be written as $\nabla(\frac{1}{2} \mathbf{x}^2)$, will ensure convexity of Υ as $\delta t \to 0$, whatever (bounded) value of Π is required for the projection.

In their proof that the viscous hydrostatic equations could be solved, Cao and Titi (2005) showed that equations (2.11) could be reduced to the 'pressure-less' equations which are solved to define the first guess trajectory (3.41). In these equations trajectories will usually intersect. They then used the theory of the three-dimensional Burgers equation to show that the equations could be solved by including viscosity. In the present case, we have shown that the pressure gradient term can be used to construct a trajectory that satisfies the incompressibility condition over a finite time interval. No viscosity is required. This is consistent with the analyses in Lions, Temam and Wang (1992b) which show that the non-hydrostatic equations (3.37) have more regular solutions than equations (2.11) when viscosity is included in both.

In the present case, it is possible that the equations can be solved without viscosity. That would require proving the existence of a sufficiently regular limit as $\delta t \to 0$. This has not yet been achieved, and existence of solutions to (3.37) remains open but possible, given that computations have so far failed to provide convincing evidence that solutions break down in finite time (Kerr 1993). This issue is discussed at length in Majda and Bertozzi (2002). If the limit of this solution procedure does not exist, it implies that trajectories lose their identity and energy dissipation is inevitable.

This solution procedure depends critically on the anelastic approximation. The projection method is analogous to that used to prove existence of solutions to the incompressible semi-geostrophic equations by Benamou and Brenier (1998). The extension of their result to the compressible case by Cullen and Maroofi (2003) suggests that the solution procedure for the anelastic equations can be generalized to at least slightly compressible flows. This would be sufficient for the atmospheric case.

In a grid-based model, the natural discrete representation of this solution procedure is the semi-Lagrangian method discussed in Section 3.4. The remapping to grid-points every time-step in this method will inevitable involve loss of identity of the trajectories, so it reasonable to infer that such a method will converge as $\delta t \to 0$, though it is not clear whether convergence will be to an inviscid energy-conserving solution or to a viscous, dissipative solution. Domaradzki, Xiao and, Smolarkiewicz (2003) illustrate how a limiting value of the dissipation can be estimated from numerical solutions. Nonlinear stability will require enforcing the constraint that transport cannot create new values of any of the primary variables (u, v, w, θ', q). This suggests the use of monotonicity-preserving schemes for all the primary variables. If this is done, the method will remain stable at a given resolution irrespective of any energy cascade to smaller scales. Thus the numerical method incorporates an implicit turbulence model, as discussed by Fureby and Grinstein (2002), but based on Lagrangian averaging. No other dissipation need be included. In the ECMWF model, monotonicity-preserving schemes are used for all variables. In the UM they are used for scalars. In the UM no other dissipation needs to be included in the spatial representation, except at the poles where the grid is highly distorted. In the ECMWF model only small amounts of extra dissipation are used. Specialized subgrid models are used in regions where there is organized small-scale flow such as the atmospheric boundary-layer and regions of convective cloud.

The other issue is the discrete version of the projection algorithm (3.42). This is needed to enforce the continuity equation. In order to maintain consistency with (3.39), the continuity equation should be written in fully Lagrangian form as

$$\frac{\rho_0(z_d)\partial(x_d, y_d, z_d)}{\rho_0(z)\partial(x, y, z)} = 1, \tag{3.43}$$

where $\mathbf{x}_d$ denotes the departure point associated with arrival point $\mathbf{x}$. Equation (3.43) is most naturally calculated with all components of $\mathbf{u}$ held together at the vertices of grid volumes, giving a three-dimensional version of the B-grid shown in Figure 3.4. This leads to a 27-point stencil for the Poisson equation for Π which will be difficult to invert. This problem does not arise if the conservative scheme is written as a product of one-dimensional schemes, as in Lin and Rood (1997) or Zerroukat et al. (2004). If an Eulerian form of the constraint is used, it takes the form of a discretization of the equation $\nabla \cdot (\rho_0 \mathbf{u}) = 0$ from (3.37). This is naturally expressed on the C-grid shown in Figure 3.4, with Π calculated by solving a discrete Poisson equation on a 7-point stencil. This satisfies a discrete maximum principle, as noted in Section 3.3, and is the basis of the scheme used in the UM (Davies et al. 2005).

3.7. Summary

In this section, we have demonstrated how numerical methods can be chosen optimally for different asymptotic regimes. The examples used were primarily nonlinear regimes, so emphasis is placed on maintaining Lagrangian

conservation properties by using spatial discretizations that preserve monotonicity. Since in most cases the asymptotic limit solutions are derived by
solving equations with time derivatives removed, the use of selective time
filtering is an important method of staying close to the asymptotic solutions.
Where the limit equations are derived by removing Lagrangian time derivatives, as in Section 3.4, the time filtering needs to act in a Lagrangian sense.

The regimes surveyed here are not comprehensive. In some cases the
behaviour is well described by linear dynamics. These cases have been extensively analysed in the literature (Arakawa and Lamb 1977, Durran 1998).
The use of monotonicity-preserving advection schemes is not optimal when
the dynamics is nearly linear and the data smooth. Long-term accuracy is
then better preserved by methods such as spectral methods, which have no
computational damping. This is another example of the dependence of the
choice of optimal numerical methods on the regime being modelled.

4. Examples using the Met Office model

4.1. The numerical scheme in the Unified Model

The issues described in Section 3 are illustrated using various simplified
configurations of the Met Office Unified Model (UM) described by Davies
et al. (2005). We summarize the equations and discretization in this section.

The equations used are the compressible Euler equations (2.1). The
momentum equation is solved in the form (2.8) and the thermodynamic
equation in the form (2.5). Sub-grid modelling terms, represented as $\mathbf{S}_u$,
which include turbulent diffusion are added to the horizontal momentum
equations. The equations are formulated in the spherical polar coordinates
(λ, ϕ, r) introduced in Section 2.1, with a lower boundary condition $\mathbf{u} = 0$
and $\partial \mathbf{u}/\partial r$ at $r = r_0(\lambda, \phi)$ and an upper boundary condition $w = 0$ at
$r = r_T$. This assumes that the terms $\mathbf{S}_u$ contain higher-order terms near
the surface consistent with a no-slip lower boundary condition, but that $\mathbf{S}_u$
does not contain second-order vertical derivatives near $r = r_T$. For convenience in applying the lower boundary condition, the radial coordinate r
is replaced by a terrain-following vertical coordinate η, such that $\eta = 0$ at
$r = r_0$ and $\eta = 1$ at $r = r_T$. The transformation from r to η is described by
Davies et al. (2005). All variables in the equations, including r and w, are
then considered as functions of (λ, ϕ, η). The terms which account for the effects of moisture are not considered in these tests and are thus omitted. For
convenience, the resulting equations are summarized in component form:

$$\frac{\mathrm{D}}{\mathrm{D}t} \equiv \frac{\partial}{\partial t} + \frac{u}{r \cos \phi} \frac{\partial}{\partial \lambda} + \frac{v}{r} \frac{\partial}{\partial \phi} + \dot{\eta} \frac{\partial}{\partial \eta},$$

$$\dot{\eta} \frac{\partial r}{\partial \eta} = w - \frac{u}{r \cos \phi} \frac{\partial r}{\partial \lambda} - \frac{v}{r} \frac{\partial r}{\partial \phi},$$

$$\frac{Du}{Dt} - \frac{uv\tan\phi}{r} + \frac{uw}{r} - 2\Omega\sin\phi v + 2\Omega\cos\phi w$$

$$+ \frac{C_p\theta}{r\cos\phi}\left(\frac{\partial\Pi}{\partial\lambda} - \frac{\partial\Pi}{\partial\eta}\left(\frac{\partial r}{\partial\eta}\right)^{-1}\frac{\partial r}{\partial\lambda}\right) = \mathbf{S}_u,$$

$$\frac{Dv}{Dt} + \frac{u^2\tan\phi}{r} + \frac{vw}{r} + 2\Omega\sin\phi u + 2\Omega\sin\phi w$$

$$+ \frac{C_p\theta}{r}\left(\frac{\partial\Pi}{\partial\phi} - \frac{\partial\Pi}{\partial\eta}\left(\frac{\partial r}{\partial\eta}\right)^{-1}\frac{\partial r}{\partial\phi}\right) = \mathbf{S}_v,$$

$$(4.1)$$

$$\frac{Dw}{Dt} - \frac{(u^2+v^2)}{r} + 2\Omega\cos\phi u + C_p\theta\frac{\partial\Pi}{\partial\eta}\left(\frac{\partial r}{\partial\eta}\right)^{-1} + g = \mathbf{S}_w,$$

$$\frac{\partial}{\partial t}\left(r^2\rho\frac{\partial r}{\partial\eta}\right) + \nabla\cdot\left(r^2\rho\frac{\partial r}{\partial\eta}(u,v,\dot\eta)\right) = 0,$$

$$\frac{D\theta}{Dt} = S_h,$$

$$\Pi = (p/p_{\text{ref}})^{-R/C_p},$$

$$p - \rho R\theta\Pi.$$

The discretization of the equations is described by Davies *et al.* (2005). The variables are held on the C-grid defined in Figure 3.4 in the horizontal and the Charney–Phillips grid in the vertical, with θ held at w-points and ρ at pressure points. It is shown in Thuburn (2006) that, given the form of the pressure gradient terms used in (4.1), this results in an optimal discretization of (4.1) in the vertical. The equations are solved by a semi-implicit scheme in which the factor $\nabla\Pi$ in the pressure gradient terms and the components of the Coriolis term proportional to $\sin\phi$ are integrated implicitly. The advection terms, including the metric terms that arise in spherical polar coordinates, are integrated explicitly by a vector semi-Lagrangian method based on that of Bates, Moorthi and Higgins (1993).

This scheme encompasses many of the features noted in Section 3 as desirable for accurate representation of important asymptotic limits. Research is being carried out into implicit calculation of the semi-Lagrangian departure points, as illustrated in Section 3.4, and into conservative semi-Lagrangian schemes, as discussed in Sections 3.4 and 3.6.

4.2. Validation for shallow water flow

The first set of tests is to demonstrate that the correct behaviour is obtained for shallow water flow. If the differences between the solutions of the exact equations and the asymptotic limit equations agree with the theoretical predictions, it is a validation of the experimental procedure, the numerical methods, and the correctness of the coding.

The tests are carried out on a sphere of radius a, to avoid boundary issues, but with the rotation terms replaced by a constant value 2Ω. This is to ensure that the Rossby radius is uniform over the domain. This formulation was used by Cullen (2002a). The water depth is written as h and the velocity as $\mathbf{u} = (u, v)$. The shallow water version of equations (4.1) is then

$$\frac{D}{Dt} \equiv \frac{\partial}{\partial t} + \frac{u}{a \cos \phi} \frac{\partial}{\partial \lambda} + \frac{v}{a} \frac{\partial}{\partial \phi},$$

$$\frac{Du}{Dt} - \frac{uv \tan \phi}{a} - 2\Omega v + \frac{g}{a \cos \phi} \frac{\partial h}{\partial \lambda} = 0,$$

$$\frac{Dv}{Dt} + \frac{u^2 \tan \phi}{a} + 2\Omega u + \frac{g}{a} \frac{\partial h}{\partial \phi} = 0,$$
$$(4.2)$$

$$\frac{\partial h}{\partial t} + \nabla \cdot (h\mathbf{u}) = 0.$$

The tests are carried out with a shallow water version of the UM code, described in Mawson (1996), and the shallow water semi-geostrophic model of Mawson (1996). The semi-geostrophic approximation to (4.2) replaces the momentum (u, v) by (u_g, v_g), where

$$2\Omega u_g = -\frac{g}{a} \frac{\partial h}{\partial \phi}, \qquad 2\Omega v_g = \frac{g}{a \cos \phi} \frac{\partial h}{\partial \lambda}. \qquad (4.3)$$

The equations can be rewritten in an analogous way to (3.25), with a matrix $\mathbf{Q}$ given by

$$\mathbf{Q} = \begin{pmatrix} 4\Omega^2 + \frac{2\Omega}{a \cos \phi} \frac{\partial v_g}{\partial \lambda} + \frac{2\Omega u_g \tan \phi}{a} & \frac{2\Omega}{a} \frac{\partial v_g}{\partial \phi} \\ -\frac{2\Omega}{a \cos \phi} \frac{\partial u_g}{\partial \lambda} + \frac{2\Omega v_g \tan \phi}{a} & 4\Omega^2 - \frac{2\Omega}{a} \frac{\partial u_g}{\partial \phi} \end{pmatrix}. \qquad (4.4)$$

We also use an incompressible version of the UM code, in which the final equation of (4.2) is replaced by

$$\nabla \cdot \mathbf{u} = 0. \qquad (4.5)$$

The experiments shown here were designed to test the effect of varying Ro/Fr for fixed Ro. Since the Froude number for shallow water flow is $U/\sqrt{gH}$ and so Ro/Fr is proportional to $\sqrt{gH}$ for a fixed length-scale, this was achieved by using the same perturbation depth field for all runs, but varying the mean value h_0 from 12800 m down to 400 m. The perturbation depth field is the same as that used in Cullen (2002a). It includes components on zonal wave-numbers ranging from 3 to 20, with amplitudes such that the total depth h is always positive and the matrix $\mathbf{Q}$, (4.4), calculated from h using the geostrophic relations is positive definite. The resulting depth field is shown in Figure 4.1. The amplitude of the variations is about ± 340 m. The horizontal velocity had a maximum value of about $16 \, \mathrm{m \, s^{-1}}$. The gravity wave speed varied from $360 \, \mathrm{m \, s^{-1}}$ to $65 \, \mathrm{m \, s^{-1}}$. The results

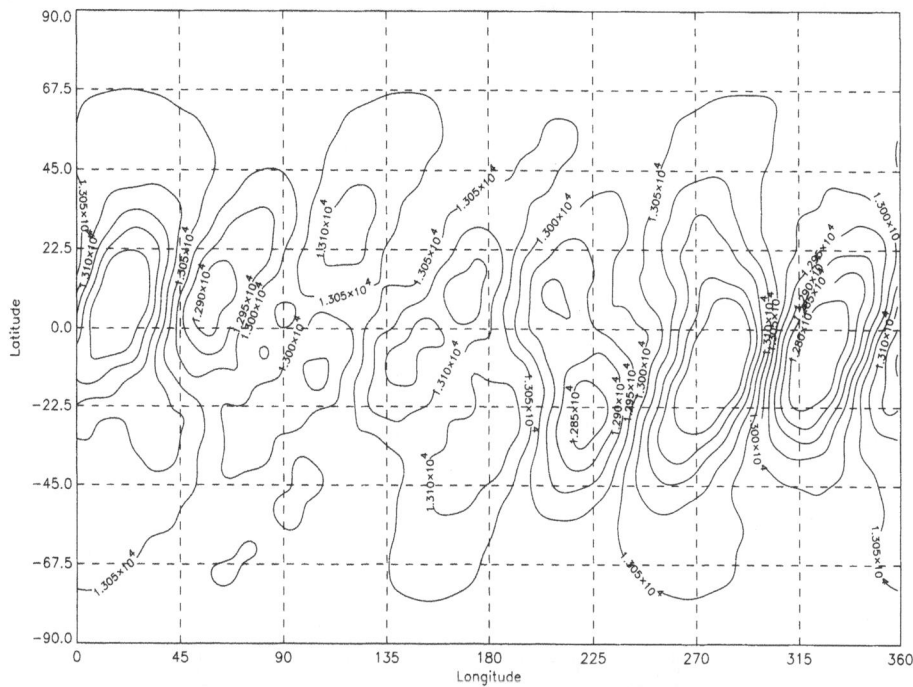

Figure 4.1. Initial depth data gh, units $\mathrm{m^2\,s^{-2}}$ for
shallow water experiments. Contour interval $50\,\mathrm{m^2\,s^{-2}}$.

quoted by Reiser (2000, p. 65) suggest a gravity wave speed of $140\,\mathrm{m\,s^{-1}}$
as giving the best match of shallow water solutions to the evolution of the
barotropic part of the atmospheric flow (*i.e.*, that part whose direction is
independent of height). The average Rossby number was thus about 0.1,
and the average Froude number ranged from 0.05 to 0.3. The Rossby radius
L_R ranged from 3600 km (about wave-number 2) to 650 km (about wave-
number 11). Thus for the largest value of L_R nearly all the variance was on
scales less than L_R, and for the smallest value much of the variance was on
scales larger than L_R. As discussed in Section 3.4, it is therefore expected
that the semi-geostrophic model should be accurate for the small values of
mean depth, but deteriorate significantly for larger values. The reverse is
expected to happen for the incompressible model.

The data are initialized for the semi-geostrophic model using the procedure described by Mawson (1996). This calculates an initial u and v consistent with the given h and the semi-geostrophic approximation to (4.2).
These data are passed to the shallow water model. The vorticity calculated
from u and v using (3.1) is used to initialize the incompressible equations.
While the incompressible solution itself is insensitive to the mean depth,
the initialization procedure means that the initial vorticity is different in

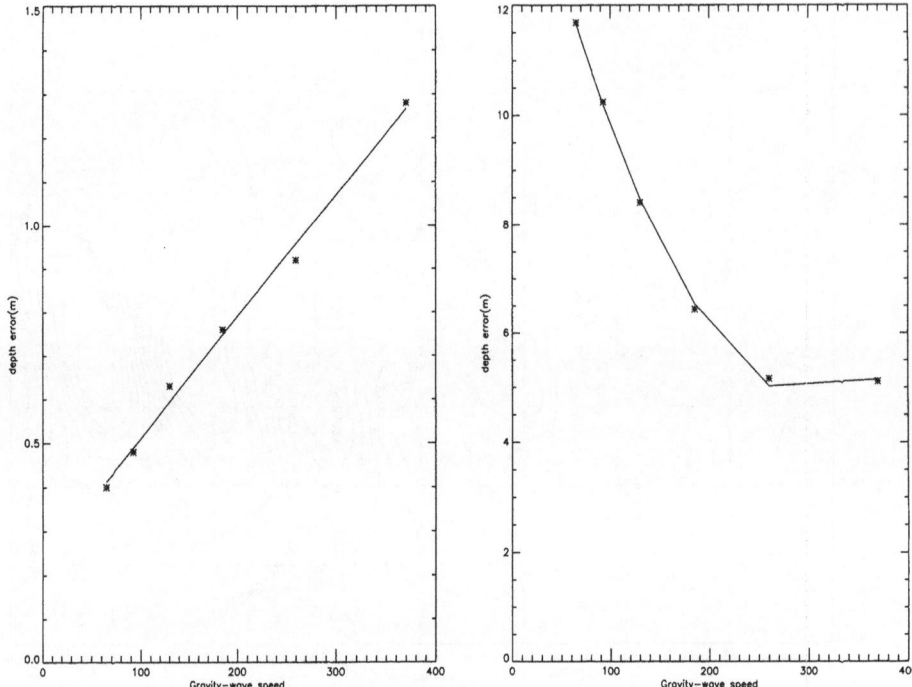

Figure 4.2. *Left*: Root-mean-square depth differences (m) between semi-geostrophic model and shallow water model with best linear fit shown. *Right*: Root-mean-square depth differences between incompressible model and shallow water model with best fit by a quadratic function. Both calculated after 48 hours and plotted against gravity-wave speed ($\mathrm{m\,s^{-1}}$).

experiments with different mean depths. The results were found to be very sensitive to the initialization procedure.

The resolution for the experiments was a latitude–longitude grid with 288 points around latitude circles and 193 points between the poles. The models are run for two days. As shown in Cullen (2002a), this is sufficiently short for the vorticity distribution to remain well resolved in the case of large mean depth. The results are compared in terms of r.m.s. errors of the primary prognostic fields, thus using the trajectory u, v from the semi-geostrophic model rather than the geostrophic winds.

The results for the depth errors plotted against gravity-wave speed are shown in Figure 4.2. The gravity wave speed is proportional to Fr^{-1}. The wind errors are shown in Figure 4.3. According to Cullen (2000), the expected error for the semi-geostrophic depth field is the greater of $O(\mathrm{Ro}(\mathrm{Ro/Fr})^2)$ and $O(\mathrm{Ro}^2)$ for small Ro and Ro $<$ Fr. The error is $O(\mathrm{Ro})$ for small Ro with Ro $\geq$ Fr and $O(\mathrm{Fr})$ for small Fr. The expected error for

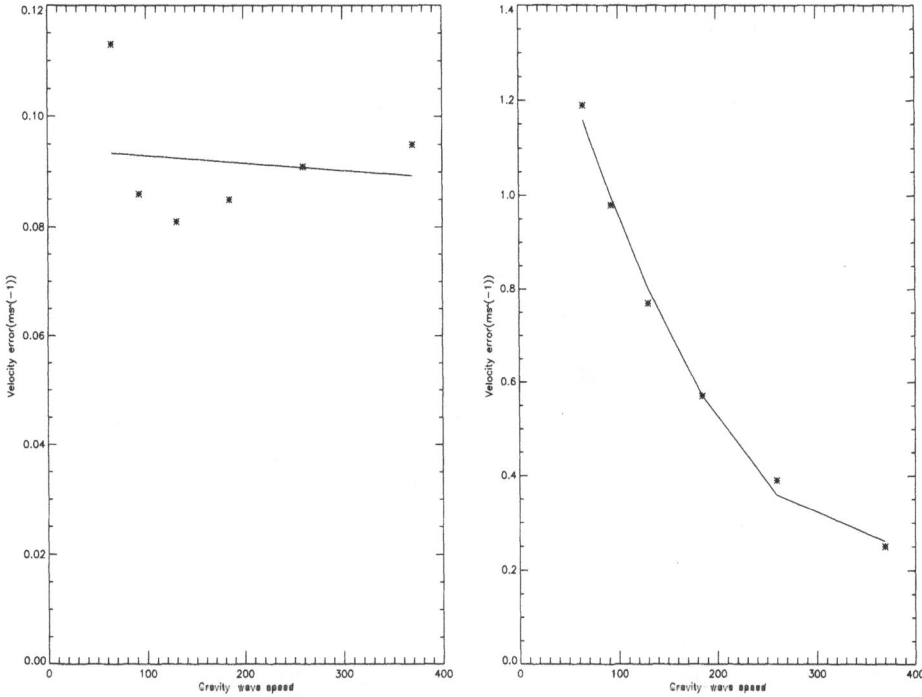

Figure 4.3. As Figure 4.2 for root-mean-square wind errors $(\mathrm{m\,s^{-1}})$.

the incompressible model is $O(\mathrm{Fr}^2)$. In the present experiment, Ro/Fr is $O(1)$, but the smallest Fr is 0.05.

Figure 4.2 shows a linear decrease of the depth error in the semi-geostrophic model as Fr increases. This is in the expected direction, but there is no quadratic convergence, unlike that shown by Cullen (2000). This reflects the fact that Ro/Fr $=O(1)$ in these experiments; the smaller values used in Cullen (2000) could not be reached using these data. The error in the incompressible model exhibits the expected quadratic behaviour, but it does not asymptote to zero. The errors in the incompressible model are much larger than those in the semi-geostrophic model. Iterated calculations allowing non-zero divergence, as described in Section 3.5, would have much smaller errors.

Figure 4.3 shows that the trajectory errors in the semi-geostrophic model are essentially independent of Fr. They are very small, about $0.1\,\mathrm{m\,s^{-1}}$. The errors in the incompressible model are quadratic in Fr, and asymptote to about $0.2\,\mathrm{m\,s^{-1}}$. These residual errors may result from initialization differences or numerical errors. To put this in perspective, if the shallow water and incompressible models were started with zero divergence, and vorticity calculated from geostrophic velocities derived from the semi-geostrophic model, the errors after 48 hours were of the order of $3\,\mathrm{m\,s^{-1}}$.

These results show that the numerical models are behaving in the expected way. We now supplement this by illustrating how schemes that are not constructed according to the principles set out in Section 3 may not give the correct asymptotic behaviour. We thus analyse discrete approximations to a linearized version of (4.2) in plane geometry, which allows the effect of different discretizations to be demonstrated. It is necessary to consider forced equations so that we can demonstrate convergence to the asymptotic limit for prescribed forcing. The equations are linearized about a state of rest with $h = h_0$. The equations solved are thus

$$\frac{\partial u}{\partial t} - 2\Omega v + g\frac{\partial h}{\partial x} = A_u,$$

$$\frac{\partial v}{\partial t} + 2\Omega u + g\frac{\partial h}{\partial y} = A_v, \tag{4.6}$$

$$\frac{\partial h}{\partial t} + h_0 \nabla \cdot \mathbf{u} = A_h.$$

We assume a domain $(-\pi L, \pi L) \times (-\pi L, \pi L) \subset \mathbb{R}^2$ and periodic boundary conditions. Assume forcing and thus solutions proportional to $e^{i(kx+ly-\nu t)}$. The inertia-gravity wave frequency ω is then $\sqrt{gh_0(k^2 + l^2) + 4\Omega^2}$. We assume that the forcing is non-resonant, so that $\nu \neq \omega$ for any choice of k and l that is an integer multiple of L^{-1}. The Froude number Fr is now the ratio of the forcing frequency ν to the gravity wave frequency $\sqrt{gh_0(k^2 + l^2)}$. The limit equations as Fr $\rightarrow 0$ are given by replacing the third equation of (4.6) by (4.5), noting that taking this limit for fixed forcing requires $A_h/h_0 \rightarrow 0$. We then eliminate h to obtain

$$\frac{\partial \zeta}{\partial t} = \frac{\partial A_v}{\partial x} - \frac{\partial A_u}{\partial y},$$

$$\zeta = \frac{\partial v}{\partial x} - \frac{\partial u}{\partial y}, \tag{4.7}$$

$$\nabla \cdot \mathbf{u} = 0.$$

Write the solution of (4.6) as $(\hat{u}, \hat{v}, \hat{h}) \exp^{i(kx+ly-\nu t)}$ and the forcing as $(\hat{A}_u, \hat{A}_v, \hat{A}_h) \exp^{i(kx+ly-\nu t)}$. Consider, for example, the case where $\hat{A}_v = \hat{A}_h = 0$. Then we can show that

$$\begin{pmatrix} \hat{u} \\ \hat{v} \\ \hat{h} \end{pmatrix} = \frac{i}{\nu\left(\frac{\nu^2}{gh_0} - (k^2 + l^2) - \frac{4\Omega^2}{gh_0}\right)} \begin{pmatrix} \frac{\nu^2}{gh_0} - l^2 \\ -\frac{2i\Omega\nu}{gh_0} - kl \\ \frac{1}{g}(-2i\Omega l + k\nu) \end{pmatrix} \hat{A}_u. \tag{4.8}$$

The limit solution for small Fr, using (4.7) and inferring $\hat{h}$ by back substitution, is given by

$$
\begin{pmatrix} \hat{u} \\ \hat{v} \\ \hat{h} \end{pmatrix} = \frac{i}{-\nu(k^2 + l^2)} \begin{pmatrix} -l^2 \\ -kl \\ \frac{1}{g}(-2i\Omega l + k\nu) \end{pmatrix}. \tag{4.9}
$$

Since $\mathrm{Fr}^2 \propto 1/(gh_0)$, we can see from (4.8) and (4.9) that the solution of (4.6) converges to the limit solution at a rate $O(\mathrm{Fr}^2)$. The accuracy of the limit solution increases as k and l increase, consistent with the dependence on $\mathrm{Ro/Fr}$ demonstrated above.

Now consider a discretization of (4.6) using the C-grid and implicit time integration as used in the UM. This can be written using the notation introduced in Section 3.2 as

$$
\delta_t u - 2\Omega \overline{v}^t + g\delta_x \overline{h}^t = \overline{A}_u^t,
$$

$$
\delta_t v + 2\Omega \overline{u}^t + g\delta_y \overline{h}^t = \overline{A}_v^t, \tag{4.10}
$$

$$
\delta_t h + h_0(\delta_x \overline{u}^t + \delta_y \overline{v}^t) = \overline{A}_h^t.
$$

Assuming grid-lengths $\delta x, \delta y$, write

$$
\tilde{k} = 2\sin(\tfrac{1}{2}k\delta x)/\delta x, \qquad \tilde{l} = 2\sin(\tfrac{1}{2}l\delta y)/\delta y.
$$

Write $\tilde{\nu} = 2\tan(\tfrac{1}{2}\nu\delta t)$ and write $\tilde{A}_u$ for the spatially discretized version of the forcing term A_u, with similar notation for the other forcing terms. Then the discrete version of (4.8) becomes

$$
\begin{pmatrix} \hat{u} \\ \hat{v} \\ \hat{h} \end{pmatrix} = \frac{i}{\tilde{\nu}\left(\frac{\tilde{\nu}^2}{gh_0} - (\tilde{k}^2 + \tilde{l}^2) - \frac{4\Omega^2}{gh_0}\right)} \begin{pmatrix} \frac{\tilde{\nu}^2}{gh_0} - \tilde{l}^2 \\ -\frac{2i\Omega\tilde{\nu}}{gh_0} - \tilde{k}\tilde{l} \\ \frac{1}{g}(-2i\Omega\tilde{l} + \tilde{k}\tilde{\nu}) \end{pmatrix} \tilde{A}_u \tag{4.11}
$$

and (4.9) becomes

$$
\begin{pmatrix} \hat{u} \\ \hat{v} \\ \hat{h} \end{pmatrix} = \frac{i}{-\tilde{\nu}(\tilde{k}^2 + \tilde{l}^2)} \begin{pmatrix} -\tilde{l}^2 \\ -\tilde{k}\tilde{l} \\ \frac{1}{g}(-2\Omega\tilde{l} + \tilde{k}\tilde{\nu}) \end{pmatrix} \tilde{A}_u. \tag{4.12}
$$

Since $k\delta x, l\delta y, \nu\delta t$ are all less than or equal to π for any resolved function, $\tilde{k}, \tilde{l}$ and $\tilde{\nu}$ are all non-zero. Therefore convergence of (4.11) to (4.12) will occur at the predicted rate $O(\mathrm{Fr}^2)$.

If the spatial discretization is instead performed on the B-grid defined in Figure 3.3, then

$$
\tilde{k} = 2\sin(\tfrac{1}{2}k\delta x)\cos(\tfrac{1}{2}l\delta y)/\delta x, \qquad \tilde{l} = 2\sin(\tfrac{1}{2}l\delta y)\cos(\tfrac{1}{2}k\delta x)/\delta y.
$$

These are both smaller than their values on the C-grid, so the accuracy of the approximation to the limit solution will be degraded. $\tilde{k}$ and $\tilde{l}$ are both zero if $k\delta x = l\delta y = \pi$. In this case, (4.11) reduces to

$$\begin{pmatrix} \hat{u} \\ \hat{v} \\ \hat{h} \end{pmatrix} = \frac{i}{\tilde{\nu}\left(\frac{\tilde{\nu}^2}{gh_0} - \frac{4\Omega^2}{gh_0}\right)} \begin{pmatrix} \frac{\tilde{\nu}^2}{gh_0} \\ -\frac{2i\Omega\tilde{\nu}}{gh_0} \\ 0 \end{pmatrix} \tilde{A}_u = \frac{i}{(\tilde{\nu}^2 - 4\Omega^2)} \begin{pmatrix} \tilde{\nu} \\ -2i\Omega \\ 0 \end{pmatrix} \tilde{A}_u. \quad (4.13)$$

This represents a forced inertial oscillation. This cannot be a solution of (4.7) because the latter equation has no dependence on Ω. The B-grid analogue of (4.7) becomes completely degenerate in this case and is satisfied by any $(\hat{u}, \hat{v})$, leading to accumulation of noise and possible computational instability. We thus see that an inappropriate choice of discretization can lead to the limiting behaviour of the solution being incorrect. Note that in the analytic case, the manipulations leading to (4.7) do not make sense if $k = l = 0$ and so in this case (4.7) is not the correct limit of (4.6) as Fr $\to 0$. The B-grid discretization creates this situation unphysically because of numerical errors.

4.3. Validation for baroclinic waves and fronts

The tests described in this subsection use the Eady model of frontogenesis (Gill 1982, p. 556) and the Boussinesq incompressible forms of (4.14) and (4.15). This allows the fundamental mechanism by which extra-tropical weather systems evolve to be studied in two-dimensional vertical slice geometry. Thus all variables are assumed to be independent of one horizontal direction except for basic state variations in pressure and potential temperature which are in hydrostatic balance. We use Cartesian coordinates (x, y) in the horizontal. Since we will use the same model for tests of flow over ridges, we retain the terrain-following coordinate from (4.1). Write

$$\Pi = \Pi_1(y, \eta) + \Pi(x, \eta), \quad \theta = \theta_1(y) + \theta(x, \eta),$$

with $C_p\theta_1(\partial\Pi_1/\partial\eta)(\partial z/\partial\eta)^{-1} + g = 0$. Equations (4.1) then reduce to

$$\frac{D}{Dt} \equiv \frac{\partial}{\partial t} + u\frac{\partial}{\partial x} + \dot{\eta}\frac{\partial}{\partial \eta},$$

$$\dot{\eta}\frac{\partial z}{\partial \eta} = w - u\frac{\partial z}{\partial x},$$

$$\frac{Du}{Dt} - fv + C_p\theta\left(\frac{\partial\Pi}{\partial x} - \frac{\partial\Pi}{\partial \eta}\left(\frac{\partial z}{\partial \eta}\right)^{-1}\frac{\partial z}{\partial x}\right) = 0,$$

$$\frac{Dv}{Dt} + fu + C_p\theta\frac{\partial\Pi_1}{\partial y} = 0,$$

$$\frac{Dw}{Dt} + C_p\theta\frac{\partial\Pi}{\partial \eta}\left(\frac{\partial z}{\partial \eta}\right)^{-1} + g = 0, \quad (4.14)$$

$$\frac{\partial}{\partial t}\left(\rho\frac{\partial z}{\partial \eta}\right) + \nabla \cdot \left(\rho\frac{\partial z}{\partial \eta}(u, \dot{\eta})\right) = 0,$$

$$\frac{D\theta}{Dt} + v\frac{\partial \theta_1}{\partial y} = 0,$$

$$\Pi = (p/p_{\text{ref}})^{-R/C_p},$$

$$p = \rho R\theta\Pi.$$

The use of vertical slice geometry means that the semi-geostrophic model is the natural approximation when rotation is important. This is because all quantities are assumed independent of y, and the frontal scaling introduced by Hoskins (1975) is appropriate. The semi-geostrophic approximation to equations (4.14) is

$$-fv + C_p\theta\left(\frac{\partial \Pi}{\partial x} - \frac{\partial \Pi}{\partial \eta}\left(\frac{\partial z}{\partial \eta}\right)^{-1}\frac{\partial z}{\partial x}\right) = 0,$$

$$\frac{Dv}{Dt} + fu + C_p\theta\frac{\partial \Pi_1}{\partial y} = 0,$$

$$C_p\theta\frac{\partial \Pi}{\partial \eta}\left(\frac{\partial z}{\partial \eta}\right)^{-1} + g = 0, \tag{4.15}$$

$$\frac{\partial}{\partial t}\left(\rho\frac{\partial z}{\partial \eta}\right) + \nabla \cdot \left(\rho\frac{\partial z}{\partial \eta}(u, \dot{\eta})\right) = 0,$$

$$\frac{D\theta}{Dt} + v\frac{\partial \theta_1}{\partial y} = 0,$$

$$\Pi = (p/p_{\text{ref}})^{-R/C_p},$$

$$p = \rho R\theta\Pi.$$

The first of these equations states that v is geostrophic, showing that these equations are consistent with (3.23) under the assumption of no y-dependence. As with the derivation of (3.26), these equations can be solved by reducing them to a single elliptic equation for $\partial \Pi/\partial t$. The method is described in detail in Cullen (2007). However, it is not clear if this form of the equations has well-posed solutions.

Cullen and Maroofi (2003) showed that these equations are well-posed if written in geostrophic and isentropic coordinates and then solved as a transport equation for the mass density. The coordinates are defined by

$$X = x + f^{-1}v, \ Z = \theta. \tag{4.16}$$

The second and fifth of equations (4.15) then become

$$\frac{DX}{Dt} + C_p\theta\frac{\partial \Pi_1}{\partial y} = 0,$$

$$\frac{DZ}{Dt} + f(X - x)\frac{\partial \theta_1}{\partial y} = 0. \tag{4.17}$$

The UM equations (4.14) are solved by the methods described in Section 4.1. Thus u is staggered in the horizontal from Π, but, because of the slice geometry, v is held at the same points as Π. The algorithm for solving (4.15) uses the same vertical arrangement of variables. However, in order to represent geostrophic balance, v is staggered in the horizontal from Π. The matrix $\mathbf{Q}$, defined as in (3.25), which appears in the elliptic equation (3.26) for $\partial\Pi/\partial t$, is evaluated at pressure points. This means that the finite differences on the diagonal are calculated over a single grid-length, but the off-diagonal terms involve additional averaging. This has the desirable property of increasing the diagonal dominance of the system. The reduction of the problem to a single equation for Π is needed in order to ensure that v and θ represent an exactly geostrophic and hydrostatic state. As discussed in Section 3.4, physically relevant solutions of SG are characterized by the matrix $\mathbf{Q}$ being positive definite. This condition is also required for (3.26) to be solvable. This condition is not naturally enforced by the numerical method. It is therefore necessary to correct the data at the end of each time-step so that $\mathbf{Q}$ is positive definite. This is done by a variational method described in Cullen (2007).

Previous work on this problem by Cullen and Roulstone (1993) used a geometric algorithm to solve the Boussinesq incompressible form of (4.15) and showed that the solutions represent a sequence of energy-conserving life-cycles in which disturbances develop and decay. The solutions are discontinuous when the disturbances are fully developed and are shown to be predictable for many life-cycles. These results were confirmed by a more efficient geometric algorithm due to R. J. Purser; see Cullen (2006, Section 5.3.2).

We illustrate the main points of the solutions obtained using the geometric algorithm. The equations are solved on a domain $\Gamma : [-L, L] \times [0, H]$ in the (x, z) plane with periodic boundary conditions in x and rigid wall conditions $w = 0$ on $z = 0, H$. The initial data are represented on a set of fluid elements, on each of which X and Z as defined in (4.16) are constant. Two resolutions are used, of 21×13 and 40×26 elements. Solutions of the Boussinesq incompressible form of (4.15) for initial data of the form

$$\theta' = N_0^2 \theta_0 z/g + B \sin\big(\pi(x/L + z/H)\big), \qquad (4.18)$$

where N_0^2 and B are positive constants, are illustrated. The data used are taken from Nakamura (1994), so that

$$L = 1000\,\mathrm{km}, \quad H = 10\,\mathrm{km}, \quad N_0^2 = 2.5 \times 10^{-5}\,\mathrm{s}^{-2}, \quad f = 10^{-4}\,\mathrm{s}^{-1},$$
$$g = 10\,\mathrm{m\,s}^{-1}, \quad \theta_0 = 300\,\mathrm{K}, \quad \partial\theta_1/\partial y = 3 \times 10^{-6}\,\mathrm{m}^{-1}\,\mathrm{K}.$$

These data correspond to an unstable mode of the linearized equations derived from (4.15). As discussed in Gill (1982, p. 556), if the isentropes have a negative slope $\mathrm{d}x/\mathrm{d}z$, then v_g will increase with z, and the evolution

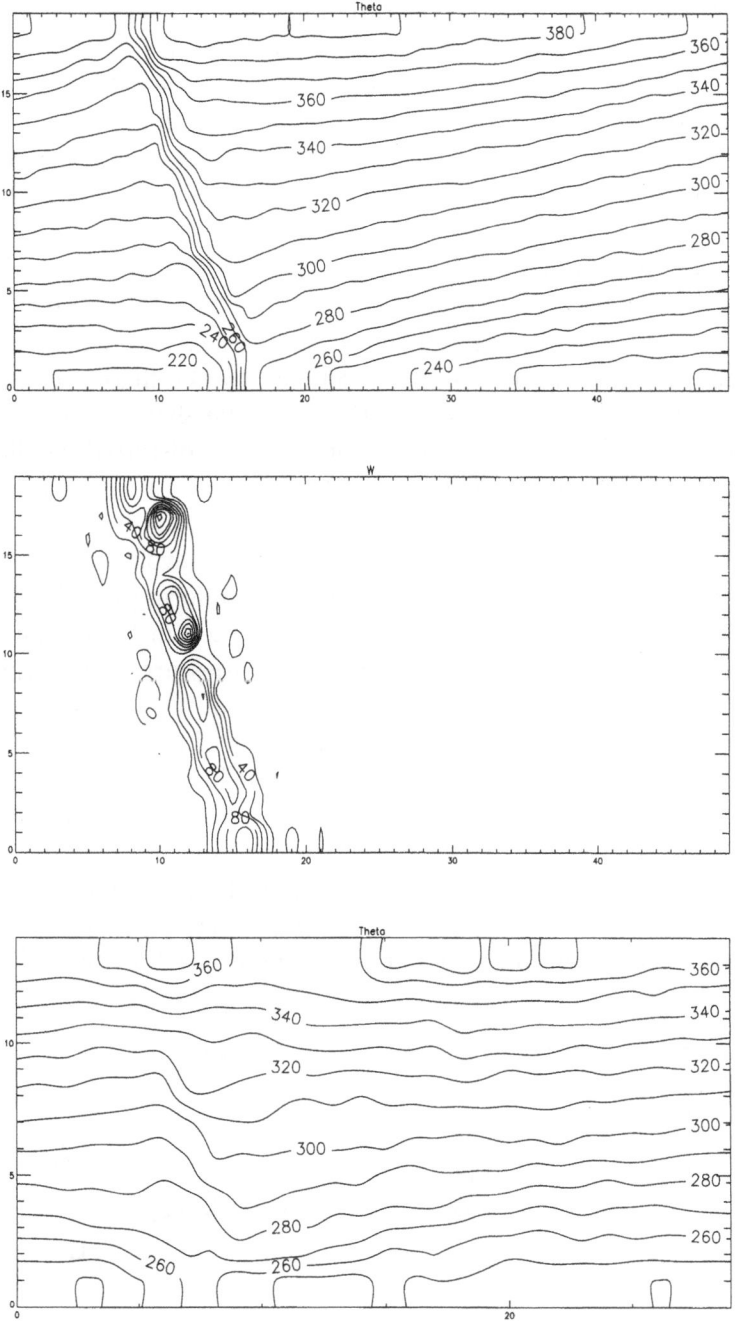

Figure 4.4. *Top*: Plot of potential temperature after 8 days (degrees K, contour interval 10 K). *Middle*: Plot of potential vorticity scaled by $f^2 N_0^2$ after 8 days (contour interval 20). *Bottom*: Plot of potential temperature after 15 days (degrees K, contour interval 10 K). From Cullen (2006).

equation for θ' will increase the vertical gradient of θ, giving a positive feed-back. It represents conversion of potential energy from the infinite reservoir implied by the imposed basic state $\partial\theta_1/\partial y$ into kinetic energy.

We show the solutions after 8 days in Figure 4.4, at which point there is strong frontogenesis, as illustrated by the potential temperature plot. The potential vorticity at the same time is also shown. There is a large increase in the value in the frontal zone. The semi-geostrophic potential vorticity is defined as the determinant of the matrix $\mathbf{Q}$ defined in (3.25). As discussed in Cullen (2006, Chapter 3), it represents the Jacobian of the transformation between physical coordinates and the coordinates (X, Z) defined in (4.16). During the time evolution, the area occupied by the fluid is conserved in both (x, z) and (X, Z) coordinates. However, the shape of the region oc-cupied by the fluid in (X, Z) space becomes highly distorted, as illustrated schematically in Figure 4.5. The total potential vorticity calculated in (x, z) space would be the ratio of the area of the convex hull of Σ to the area of Γ. This ratio will not be conserved during the time evolution.

The potential vorticity is initially equal to the uniform value $f^2 N_0^2$. At the front it becomes a Dirac mass, which is represented in Figure 4.4 as large values by the plotting software. The irregularities are due to the use of piecewise constant data and thus the irregularity of the boundaries between elements. The small negative values are artifacts of the plotting.

The formation of the fronts at the upper and lower boundaries destroys the normal mode property of the initial data, and the vertical shear in the basic state reverses the slope of the isentropes by day 10. The front is then destroyed. The mean potential vorticity remains larger than its initial value, represented by layers of enhanced static stability near the upper and lower boundaries as shown in Figure 4.4. Further life-cycles then take place, with

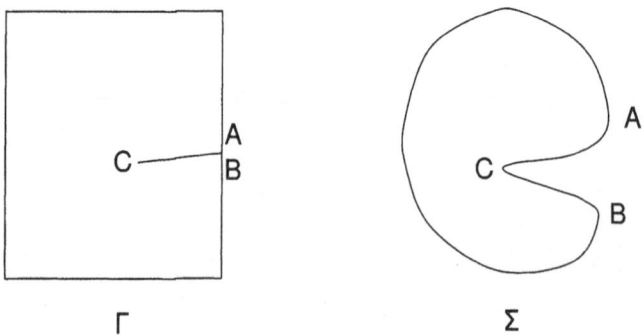

Figure 4.5. Mapping a region in (X, Z) space (Σ) to a rectangular region in (x, z) (Γ). The points A, B and C indicate corresponding points in Γ and Σ. From Cullen (2006).

the disturbances confined to the region of uniform static stability in the centre of the domain.

A graph of the kinetic energy against time is shown in Figure 4.6. This shows that after 8 days the maximum kinetic energy is reached. It then reduces, and smaller amplitude life-cycles follow. The periodic oscillations continue to day 30. The graphs for the two resolutions are almost identical, showing that the solution is highly predictable, despite the formation of fronts. The Boussinesq incompressible form of (4.15) has a natural period equal to $2Lf\theta_1(gH\partial\theta_1/\partial y)$, the length of time between when features at the upper and lower boundaries come back into phase under the action of the basic state wind. With the data chosen, the difference in the basic state wind between the boundaries is $10\,\mathrm{m\,s^{-1}}$, giving a period of about 2.3 days. This is much shorter than the period observed, reflecting the fact that the vertical shear is impeded during the growth phase. It also shows that the prediction of the same period by two different discretizations is a non-trivial achievement.

Solutions using conventional numerical methods have been obtained by Nakamura and Held (1989) and Nakamura (1994). They lose predictability after a single life-cycle. Nakamura and Held (1989), using solutions of the Boussinesq incompressible form of (4.14), show that the loss in predictability is associated with the need to include a form of artificial viscosity in order to capture the frontogenesis. The solution after the first life-cycle

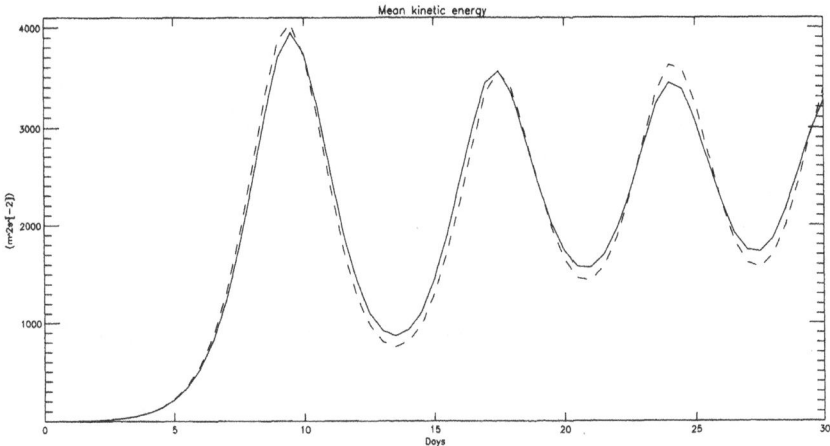

Figure 4.6. Graph of domain-averaged meridional kinetic energy $\mathrm{m^2\,s^{-2}}$ against time for the solution of the Boussinesq incompressible form of (4.15) by the geometric method. Solid line = 40×26 elements, dashed line = 21×13 elements. From Cullen (2006).

is strongly dependent on the form of artificial viscosity used. Nakamura (1994) describes solutions of the Boussinesq incompressible form of (4.15) using Eulerian numerical methods based on potential vorticity, and using artificial viscosity to capture the front. There is a large increase in potential vorticity generated by the artificial viscosity, which again results in layers of enhanced static stability near the boundaries.

In this section we illustrate solutions obtained using the UM discretization scheme and a solution of the semi-geostrophic equations in real variables, based on (4.15). These are compared qualitatively with the solutions of the Boussinesq incompressible equations discussed above, but the additional approximations mean that quantitative comparisons are not appropriate. The integration domain is the same as that used above, with the same flow parameters. In order to ensure that the initial potential vorticity, as defined by (2.6), is uniform, N_0^2 is set to $2.5 \times 10^{-5} \rho_0(0)/\rho_1(z)$, where $\rho_1(z)$ is an initial reference profile in hydrostatic balance calculated for an isentropic state using (2.8). The basic state $\partial \Pi_1/\partial y$ is chosen to vary linearly in z, so that the initial u in geostrophic balance with it varies from -5 to $5\,\mathrm{m\,s}^{-1}$ over the depth of the domain, as in the Boussinesq incompressible case. The basic state $\partial \theta_1/\partial y$ is calculated from $\partial \Pi_1/\partial y$ using the hydrostatic relation.

The mean meridional kinetic energy from the two integrations is illustrated in Figure 4.7. The mean potential vorticity, scaled by its initial value, is shown in Figure 4.8. The semi-geostrophic integration produces an earlier growth of the initial disturbance. The UM integration gives an initial growth on the same time-scale as the results of Nakamura (1994). After the initial growth, the semi-geostrophic integration reduces the meridional kinetic energy to very small values while the UM retains a value close to the maximum reached. Both integrations show a large increase in potential vorticity as the disturbance grows, and retain high values for the rest of the integration. Further life-cycles occur in the UM integration, but with much smaller amplitude than those given by the geometric model and even those shown by Nakamura (1994).

Further study of the results (diagrams not shown) shows that, after the initial growth, both solutions are dominated by a wave in the v-field which is almost independent of height and a geostrophically balanced pressure. The potential temperature signal is rather weak. In the UM, the wave then propagates slowly through the domain. It is completely different from the solutions shown in Nakamura (1994). However, the compressible formulation of the UM means that it has solutions not available to the models of Nakamura (1994). After the initial life-cycle, the semi-geostrophic solution becomes almost independent of height, and the amplitude of the wave in the v-field then slowly decays. The increased potential vorticity is represented by layers of enhanced static stability at both boundaries. The latter feature can also be seen in the geometric model solutions shown in Figure 4.4.

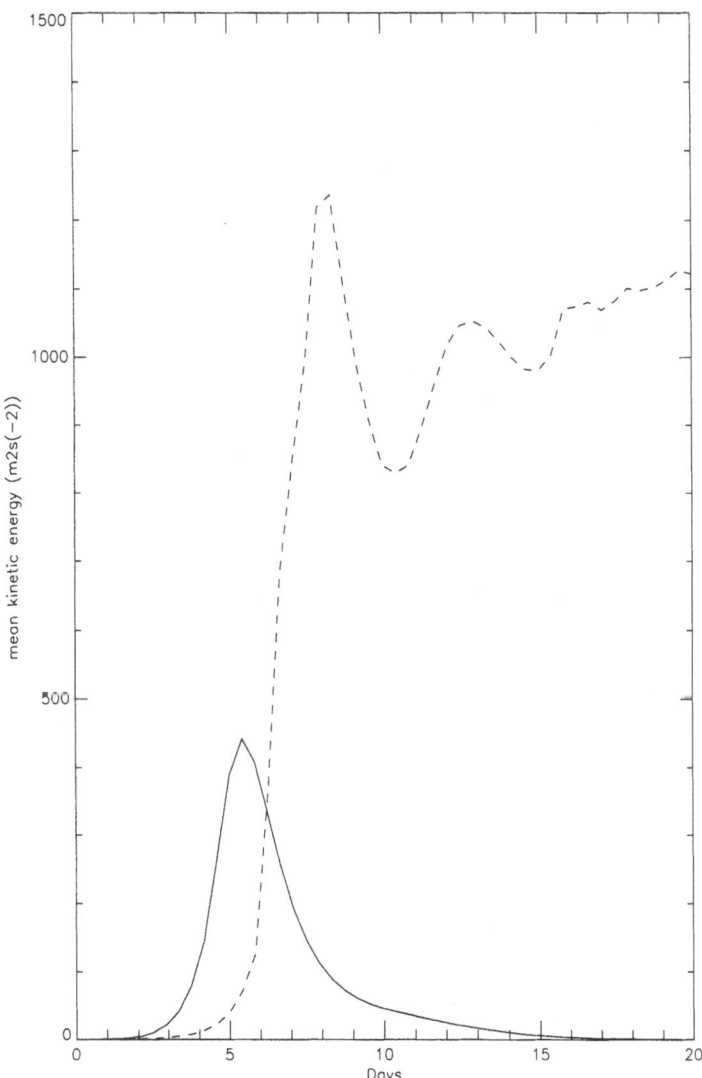

Figure 4.7. Graph of domain-averaged meridional kinetic energy $m^2 s^{-2}$ against time. Solid line = solution of (4.15), dashed line = solution of (4.14).

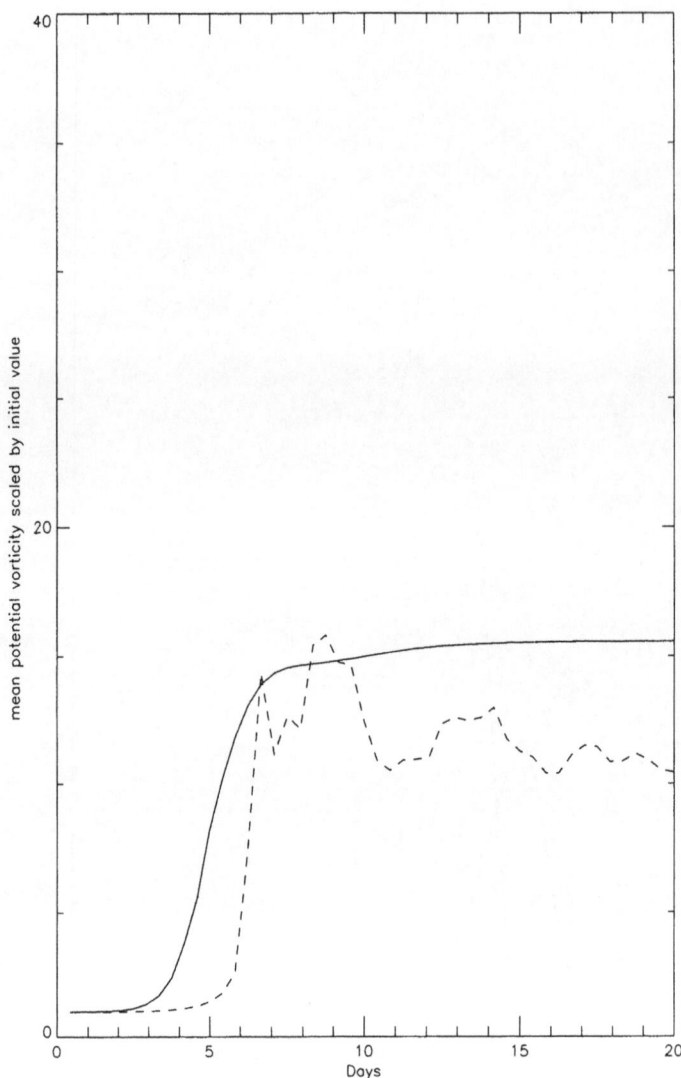

Figure 4.8. Graph of domain-averaged potential vorticity, scaled by its initial value, against time. Solid line = solution of (4.15), dashed line = solution of (4.14).

The symmetry of the solution in the vertical shown from the geometric model is lost because of the large basic-state density variation.

These results suggest that neither model is able to capture the qualitative behaviour shown in the geometric model. The reason is that the state left after the initial life-cycle is strongly dependent on the numerical algorithm. The geometric method is fully Lagrangian, and thus much more naturally suited to the problem. Fully Lagrangian methods have not yet been considered as practical for production models, though the contour advection method of Mohebalhojeh and Dritschel (2004) has been successful in idealized problems. The normal formulation of this method would be unsuitable in the present case because of the lack of potential vorticity conservation integrated over the physical domain. It would be necessary to apply the method in geostrophic and isentropic coordinates, which is only practicable even in principle for a semi-geostrophic model.

4.4. Validation for large-scale flow over a ridge

In this section we demonstrate in vertical-slice geometry that solutions of the UM equations (4.14) converge to the solutions of the semi-geostrophic (SG) equations (4.15) at the expected rate in the presence of orography. These experiments are described in more detail in Cullen (2007), where the tests are carried out in a wider range of cases.

The UM equations (4.14) and SG equations (4.15) are solved as described in Section 4.3. The forcing pressure gradient $\partial \Pi_0 / \partial y$ is chosen to be uniform in z. The initial horizontal pressure gradient is set to zero. Because of the orography, the values of Π have to be calculated so that

$$\frac{\partial \Pi}{\partial x} - \frac{\partial \Pi}{\partial \eta} \left(\frac{\partial z}{\partial \eta} \right)^{-1} \frac{\partial z}{\partial x} = 0 \qquad (4.19)$$

in a discrete sense. This can only be achieved if $\partial^2 \Pi / \partial z^2$ is uniform, because otherwise the finite difference expressions for the terms in (4.19) do not cancel up to rounding error. We therefore use initial and boundary pressure data satisfying this condition, and modify the finite differencing in Davies et al. (2005) to extrapolate θ to the lower boundary assuming hydrostatic balance and constant $\partial^2 \Pi / \partial z^2$. In the SG model, the matrix $\mathbf{Q}$ defined in (3.25) is calculated in (x, z) coordinates rather than (x, η) coordinates, and the positive-definiteness condition is enforced in these coordinates. The matrix is then transformed to (x, η) coordinates as described in Cullen (2007).

Assume a length-scale l, which will be the half-width of the ridge, and a geostrophic velocity u_g in the x-direction. The Froude number Fr is defined as $u_g / (Nh)$, where N is the Brunt–Väisälä frequency and h is the ridge height. The Rossby number Ro is defined as u_g/fl. The Rossby radius $L_R = NH/f$. The error estimates for SG theory were given in Section 4.2. We test the limit Ro $\to$ 0 for fixed L_R by letting $u_g \to 0$ and leaving all other parameters fixed. In Cullen (2007) the limit Ro $\to$ 0 with Ro $= O(Fr^2)$ was also tested.

We choose a domain of length 2000 km and height 10 km. The ridge is centred at $x =$ 750 km, and has width 150 km and height 2400 m. The ridge profile is an isolated cosine-squared hill. The Brunt–Väisälä frequency is 0.01 and the Coriolis parameter 10^{-4}. The ridge is thus slightly narrower than the radius of deformation. Since u_g is determined from $\partial\Pi_0/\partial y$, which is height-independent, u_g increases slowly with z because of the increase of θ. Values of u_g at $z = 0$ are chosen to range from $0.625\,\mathrm{m\,s^{-1}}$ to $10\,\mathrm{m\,s^{-1}}$, giving Ro ranging from 0.083 to 1.33.

The integrations start from very simple initial data with uniform u and zero w. A rapid initial adjustment will take place to create an approximately non-divergent flow. A further adjustment will take place on the inertial time-scale to establish approximate geostrophic balance of the wind parallel to the ridge. The experiments are therefore run for several inertial periods to allow this to happen. The solutions that are verified will be close to steady-state. Since it is unlikely that either model actually gives steady-state solutions to this problem, there will be a sampling issue in the error calculations. Results are therefore shown for two times, corresponding to 1.875 and 2.5 days for $u_g = 10\,\mathrm{m\,s^{-1}}$. Two spatial resolutions are also shown for the UM integrations, of 121×61 and 201×121 points respectively. The SG solutions are only shown for 121×61 resolution, since they have essentially converged by this point.

We note that the SG solution is independent of u_g in the sense that the pressure, and fields derived directly from it, only depend on $u_g T$ where T is the period of integration. The velocities u, $\dot{\eta}$ and w will be proportional to u_g. Thus, in order to compare results with different values of u_g, we choose a total integration time inversely proportional to u_g. This means that the same SG states can be used for comparison of UM runs with all values of u_g.

The limit solution for $u_g = 0$ has the same p, v_g and θ as the solution for finite u_g, and has $u = \dot{\eta} = 0$. Since this represents an exactly geostrophic and hydrostatic state, it is also a solution of the UM equations. Convergence should therefore be possible. The limit solution depends on the history of the problem. If the limit $u_g \to 0$ is taken with negative values of u_g, the SG solution will be reflected in $x = 0$, and thus be different.

The main characteristics of the solution are discussed in Cullen (2006, Chapter 6) and Shutts (1998). Some of these can be seen from the cross-section of v from the UM shown in Figure 4.9. A particular feature is the blocking of cold air near the surface on the upstream side of the ridge and the associated barrier jet. The effect is to extend the influence of the ridge

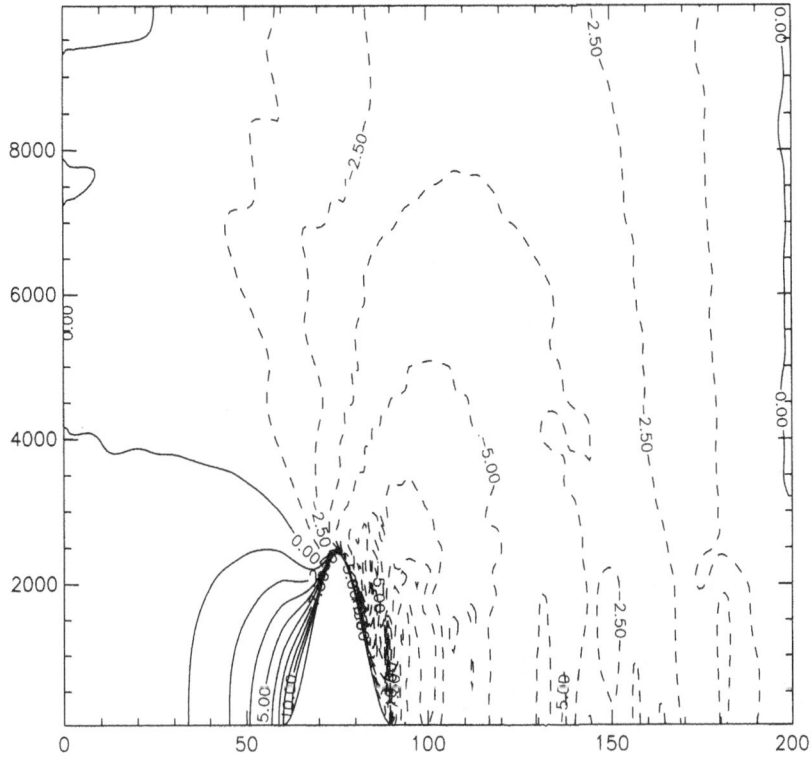

Figure 4.9. UM solution for v at 40 days (solid), with
Ro $= 0.083$ and 201×121 grid, plotted against
height (m) and horizontal grid-point number. Units
$\mathrm{m\,s^{-1}}$, contour interval $1.25\,\mathrm{m\,s^{-1}}$. From Cullen (2007).

upstream for a horizontal distance L_R. The solution has a pressure force
acting on the ridge, which represents the orographic drag. Air trapped on
the upstream side of the ridge will have v increasing with time, as u is
constrained to be zero. Therefore the slopes of the isentropes, which are
related to $\partial v/\partial z$, will increase with time and trapped air will reach the top
of the ridge with a large v. At this point, there will be a large negative value
of $f + \partial v/\partial x$, so that the air parcel will be unstable and 'jump' downstream
to a stable position. In reality, there would be a rapid down-slope wind not
described by semi-geostrophic theory. In the UM solutions, we therefore
expect to see a region of inertial instability downstream of the mountain,
together with a downslope wind. The overturning associated with inertial
instability is clearly seen in Figure 4.9. This would be expected to relax

to a stable state as given by the SG model, but the relaxation could take many inertial periods. In the real three-dimensional system the relaxation is likely to be faster as three-dimensional turbulence will be generated in the inertially unstable region.

Figure 4.10 shows the r.m.s. differences in v. These exhibit the expected linear convergence, with little difference between the two UM resolutions. However, the best fit lines do not actually go to zero. This is likely to reflect the failure of the UM to relax to a symmetrically stable state at very small Ro as discussed above. It is also shown in Cullen (2007) that the need to enforce positive-definiteness of $\mathbf{Q}$ in the SG model means that the maximum of the barrier jet occurs below the top of the ridge. This is because the variational adjustment which is used to enforce this condition does not know about the flow direction, and therefore makes adjustments both upstream and downstream of the ridge. In the UM, the barrier jet does reach the top of the ridge, as shown in Figure 4.9. This contributes to the error for small Ro.

Figure 4.11 shows the convergence of the drag. Again the expected linear convergence is achieved, with little sensitivity to UM resolution. There is less of an issue at small Ro than with the v-field. This is because the drag measures the total amount of flow blocking, which is determined by large-scale aspects of the problem. It is insensitive to the details of how the energy is dissipated downstream of the ridge.

Figure 4.12 shows the convergence in θ. This is less satisfactory, though the error does reduce with Ro. The convergence is better using the higher resolution UM data, suggesting that numerical errors may be significant. Difference fields (not shown) indicate that the errors are mainly upstream of the ridge, where the SG model has greater flow blocking, and in the inertially unstable region downstream. These are related to the same issues as those discussed in the context of the errors in v.

The test of the limit Ro $\to$ 0 with Ro $=O(\mathrm{Fr}^2)$ in Cullen (2007) shows a similar story. The expected higher-order rate of convergence is obtained, but the errors do not asymptote to zero. The error in the θ field is the most sensitive to numerical errors. In this limit, the SG solution becomes smooth, so issues with the removal of inertial instability are not relevant. The residuals thus probably represent accumulated computational errors.

Overall, this problem suggests reasonable agreement with theory. The residual errors are related to the technical difficulty of maintaining inertial stability in the SG model, and the issue of how fast and on what scales the UM should relax to an inertially stable solution. There are also issues in converting an accurate solution of the evolution problem into an accurate solution of the steady-state problem. Tests of accuracy in the initial evolution would be of little practical relevance because of the large transient motions set up by the initial conditions chosen.

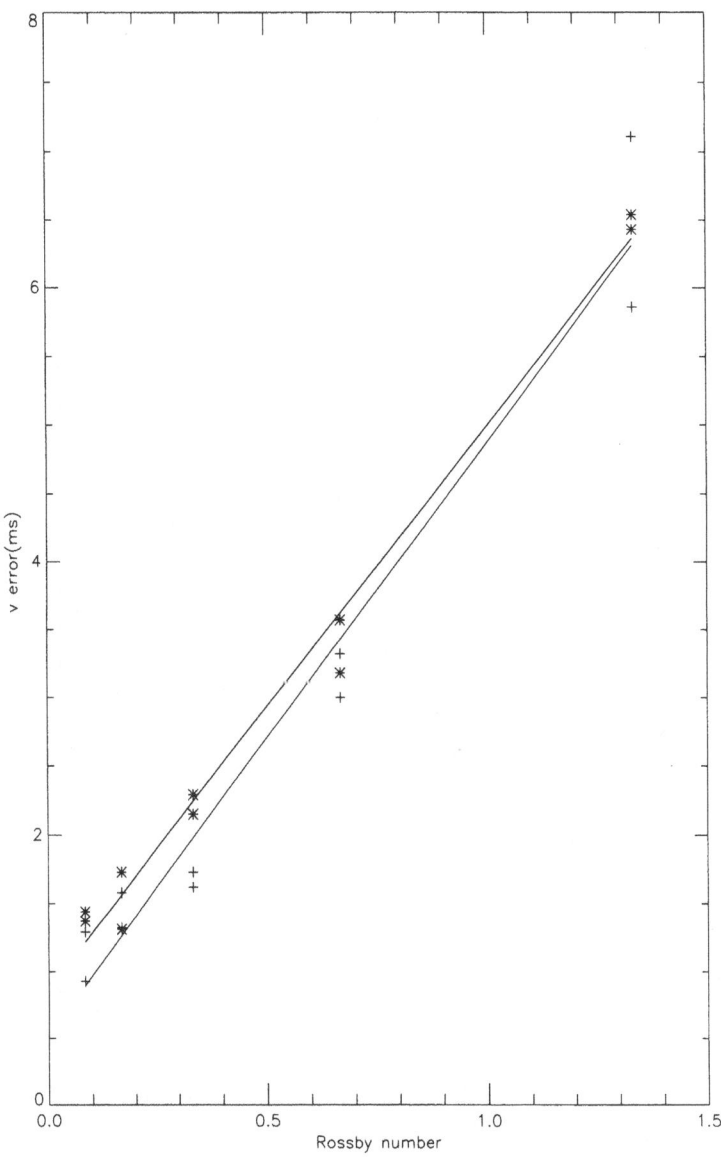

Figure 4.10. Convergence of r.m.s. differences of v (m s^{-1}) between UM and SG solutions for fixed L_R, plotted against Rossby number. SG solutions on 121×61 grid. Thin line = best linear fit in Ro to differences between UM solutions on 121×61 grid and SG solutions. Individual differences plotted as + symbols, using two verification times for each Ro. Thick line = best linear fit in Ro to differences between UM solutions on 201×121 grid and SG solutions. Individual differences plotted as asterisks, using two verification times for each Ro. From Cullen (2007).

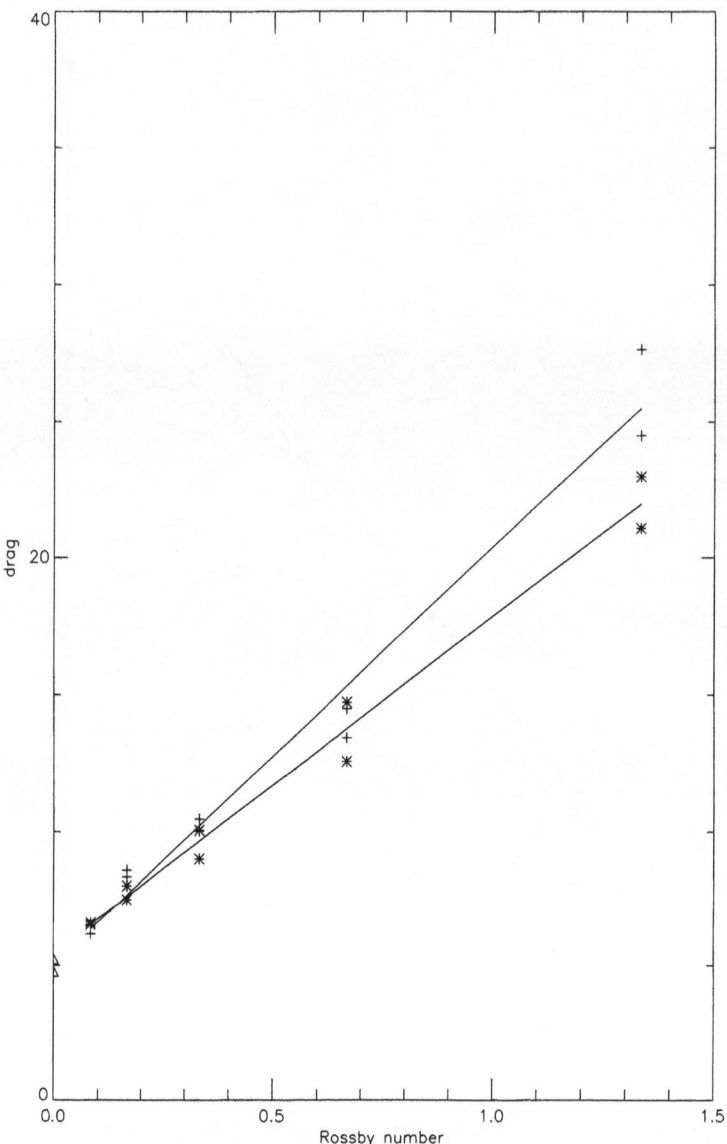

Figure 4.11. Convergence of drag on the ridge between
UM and SG solutions for fixed L_R, plotted against Rossby
number. The drag is measured per unit length in the
y-direction, units are $10^5 \, \mathrm{Pa} \, \mathrm{m}^{-1}$. SG values are plotted as
triangles at Ro = 0. Thin line = best linear fit in Ro to
UM solutions on 121 × 61 grid. Thick line = best linear fit
in Ro to UM solutions on 201 × 121 grid. Individual
differences plotted as + signs (low resolution) and asterisks
(high resolution), using two verification times for each Ro.
From Cullen (2007).

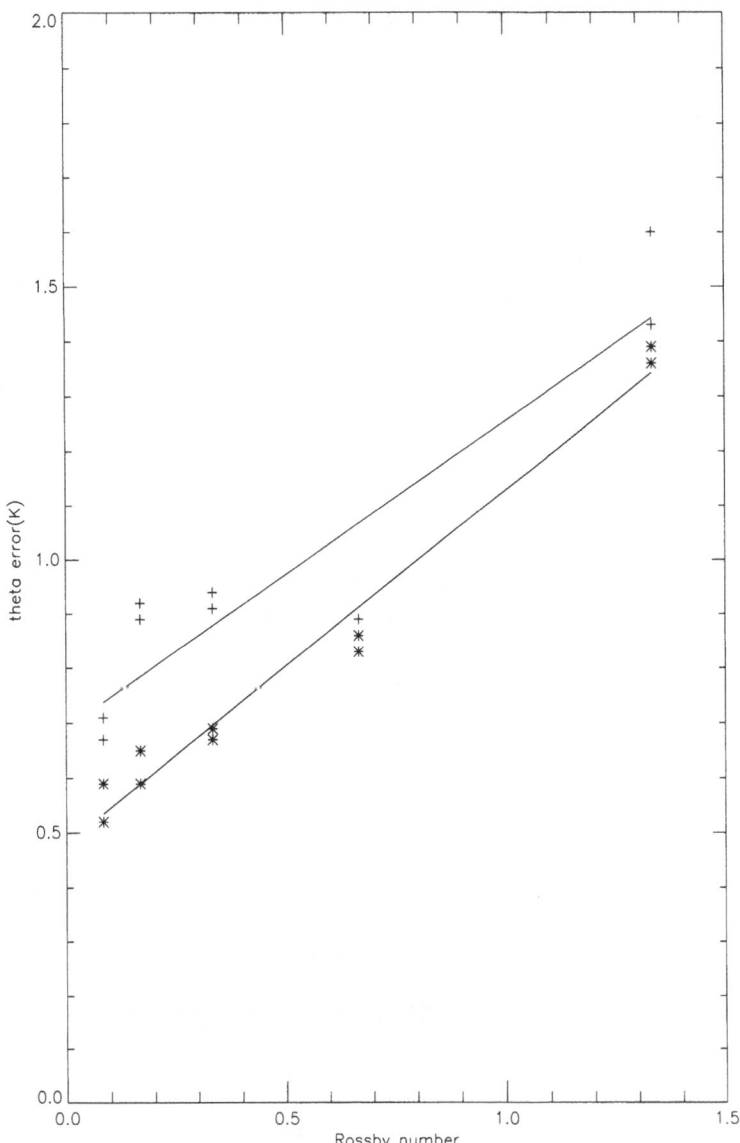

Figure 4.12. Convergence of r.m.s. differences of θ (deg K) between UM and SG solutions for fixed L_R, plotted against Rossby number. Resolutions and plotting conventions as in Figure 4.10. From Cullen (2007).

4.5. Validation for small-scale flow over a ridge

In this subsection we compare UM solutions for the flow over a ridge problem treated in Section 4.4 with solutions of an anelastic version of the UM. The aim is to demonstrate convergence of the UM to an anelastic solution at the predicted rate. The UM equations for this problem are again (4.14). The anelastic version is obtained by first defining initial data satisfying the hydrostatic equation and equation of state:

$$C_p\theta(0,\cdot)\frac{\partial\Pi(0,\cdot)}{\partial\eta}\left(\frac{\partial z}{\partial\eta}\right)^{-1} + g = 0,$$

$$\Pi(0,\cdot) = (p(0,\cdot)/p_{\text{ref}})^{-R/C_p}, \tag{4.20}$$

$$p(0,\cdot) = \rho R\theta(0,\cdot)\Pi(0,\cdot).$$

These data are chosen to have zero horizontal pressure gradient, as in Section 4.4. However, there is a non-zero gradient along the terrain-following coordinate surfaces. The resulting $\rho(0,\cdot)$ is used as a reference state, as in the derivation of (3.37). The equations are then as follows:

$$\frac{Du}{Dt} - fv + C_p\theta\left(\frac{\partial\Pi}{\partial x} - \frac{\partial\Pi}{\partial\eta}\left(\frac{\partial z}{\partial\eta}\right)^{-1}\frac{\partial z}{\partial x}\right) = 0,$$

$$\frac{Dv}{Dt} + fu + C_p\theta\frac{\partial\Pi_1}{\partial y} = 0,$$

$$\frac{Dw}{Dt} + C_p\theta\frac{\partial\Pi}{\partial\eta}\left(\frac{\partial z}{\partial\eta}\right)^{-1} + g = 0, \tag{4.21}$$

$$\nabla\cdot\left(\rho(0,\cdot)\frac{\partial z}{\partial\eta}(u,\dot\eta)\right) = 0,$$

$$\frac{D\theta}{Dt} + v\frac{\partial\theta_1}{\partial y} = 0.$$

Given data at time t, the Exner pressure Π is calculated from a Poisson equation derived by enforcing the continuity equation at time $t + \delta t$. This is a slight modification of the Helmholz equation derived by Davies et al. (2005) as part of the semi-implicit integration scheme in the UM. The resulting pressure is purely diagnostic and may not represent a realistic thermodynamic pressure in places where the anelastic approximation is inaccurate. It would be possible to derive a density from it using the equation of state, and use this density in the continuity equation (the fourth equation of (4.21)). However, this could cause unrealistic behaviour when the pressure is inaccurate, which is why a reference density has to be used in (3.37) to ensure energetic consistency.

The tests are carried out using the same geometry as in Section 4.4. The length-scale is reduced by a factor of 100 to give a small-scale regime. The ridge then has an $O(1)$ aspect ratio. The length of the domain is now 20 km, and the width of the ridge is 1.5 km. Values of u_g at $z = 0$ were chosen from $10\,\mathrm{m\,s}^{-1}$ up to $80\,\mathrm{m\,s}^{-1}$, giving Mach numbers up to 0.25. As in Section 4.4, the experiments are run for a time inversely proportional to u_g. It is not useful to measure the errors in the initial evolution because of rapid transients, so the errors have to be measured after a period of evolution. The aim is to allow time for the adjustment to a quasi non-divergent solution, but to use a time short enough for geostrophic adjustment to have a negligible effect. Results are thus shown for times corresponding to 27 and 36 minutes for $u_g = 10\,\mathrm{m\,s}^{-1}$. Two spatial resolutions were used, of 121×61 and 201×121 points.

In the first set of experiments, N^2 is set to 0.01, as in Section 4.4. Figure 4.13 shows the errors in the u field. These are consistent with quadratic convergence except for very small Mach number where the differences asymptote to a non-zero value. This does not appear to be a function of spatial resolution, though the differences at larger Mach number are sensitive to the resolution.

Figure 4.14 shows the differences in the θ-field. These are more sensitive to resolution, suggesting that numerical errors are also contributing. The results are again consistent with a second-order increase in error with Mach number, starting from a non-zero basic value.

In order to establish whether the residual error at small Mach number is due to the effect of the non-constant θ in the initial data, we repeat the experiments with $N^2 = 0.0001$. Retention of a non-zero value of N^2 allows the effect on the convergence of θ to be estimated. Figure 4.15 shows the errors in the u field, using the same scale as Figure 4.13. These are now consistent with quadratic convergence to zero even for the smallest Mach numbers plotted, and show very little sensitivity to resolution.

Figure 4.16 shows the differences in the θ-field. The scale is magnified by a factor of 100, to allow for the reduced variations in θ in the initial data. There is still a residual non-zero value of the error for small Mach number, but it is 3 times smaller than in the case with $N^2 = 0.01$. The results remain quite sensitive to the resolution, as in the more strongly stratified case.

Overall the results are consistent with the error estimate in Section 3.6. The non-convergence for non-uniform θ means that the use of anelastic equations for quantitative predictions should be treated with caution, as should numerical methods which employ a reference state, unless the effects of the reference state are removed by iteration as discussed in Section 3.2.

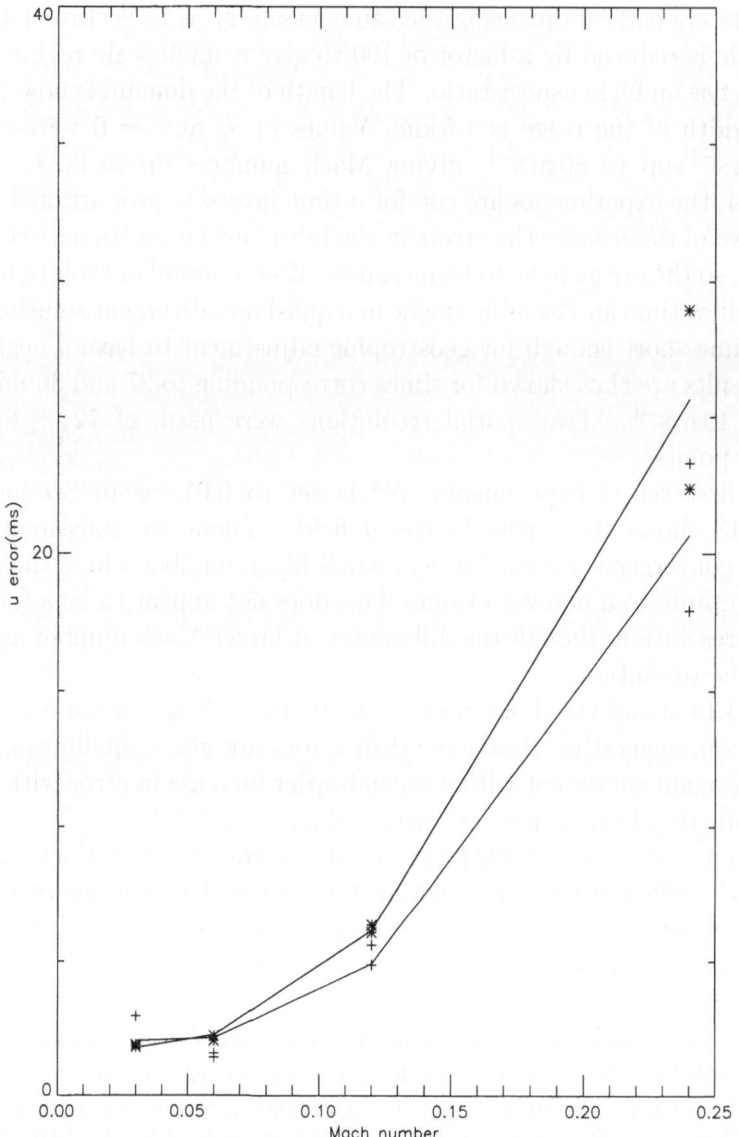

Figure 4.13. Convergence of r.m.s. differences of u $(\mathrm{m\,s}^{-1})$
between UM and anelastic solutions, plotted against Mach
number Ma. Thin line = best fit by quadratic polynomial
in Ma to differences between UM and anelastic solutions on
121×61 grid. Individual differences plotted as + symbols,
using two verification times for each Ma. Thick line = best
quadratic fit in Ma to differences between UM solutions on
201×121 grid and anelastic solutions. Individual differences
plotted as asterisks, using two verification times for each Ma.

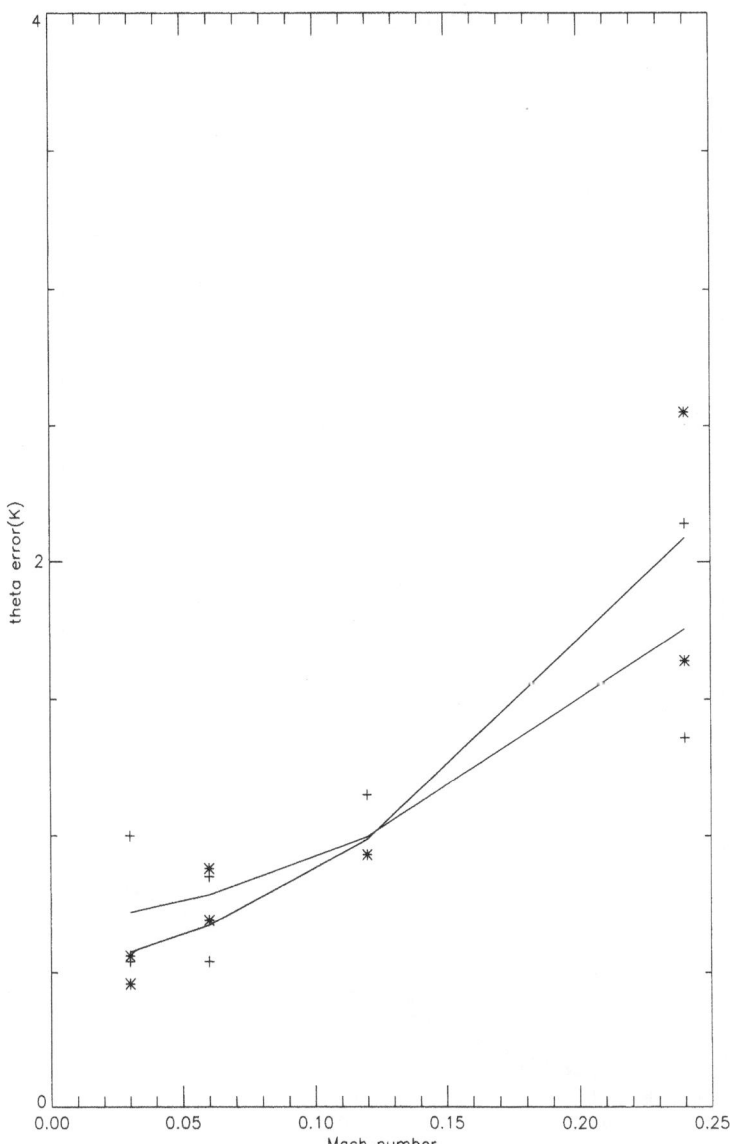

Figure 4.14. Convergence of r.m.s. differences of θ (deg K) between UM and anelastic solutions, plotted against Mach number. Resolutions and plotting conventions as in Figure 4.13.

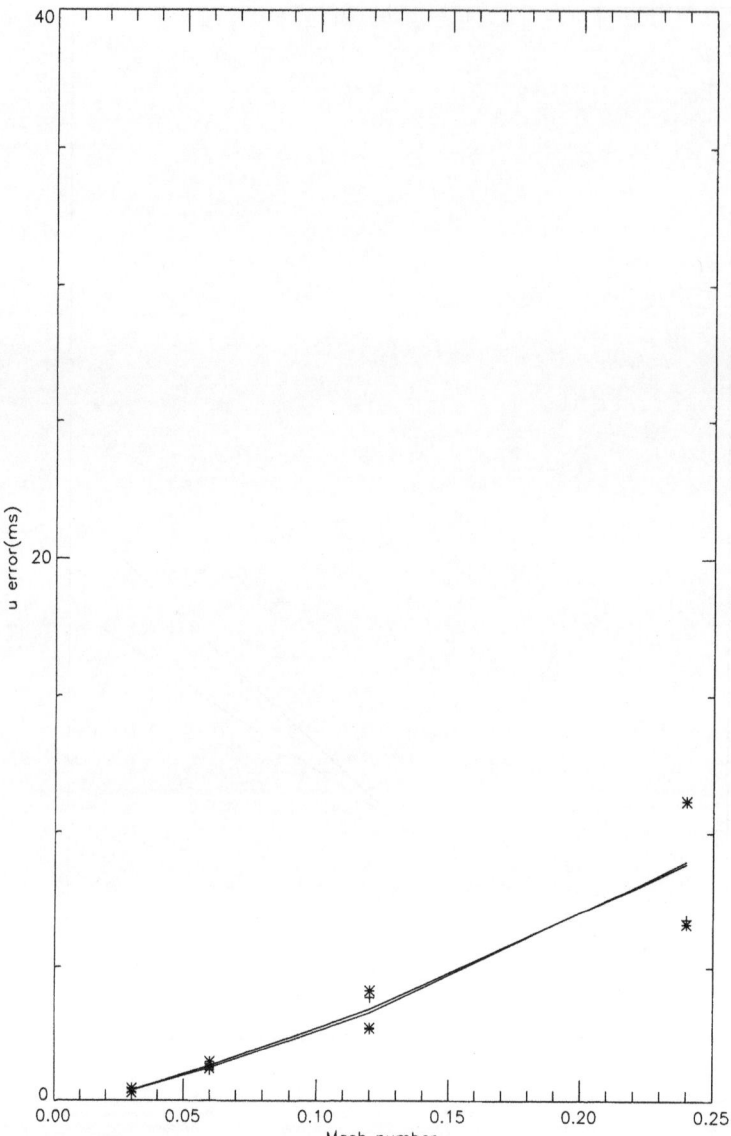

Figure 4.15. Convergence of r.m.s. differences of u $(\mathrm{m\,s^{-1}})$ between UM and anelastic solutions, with $N^2 = 10^{-4}$, plotted against Mach number Ma. Resolutions and plotting conventions as in Figure 4.13.

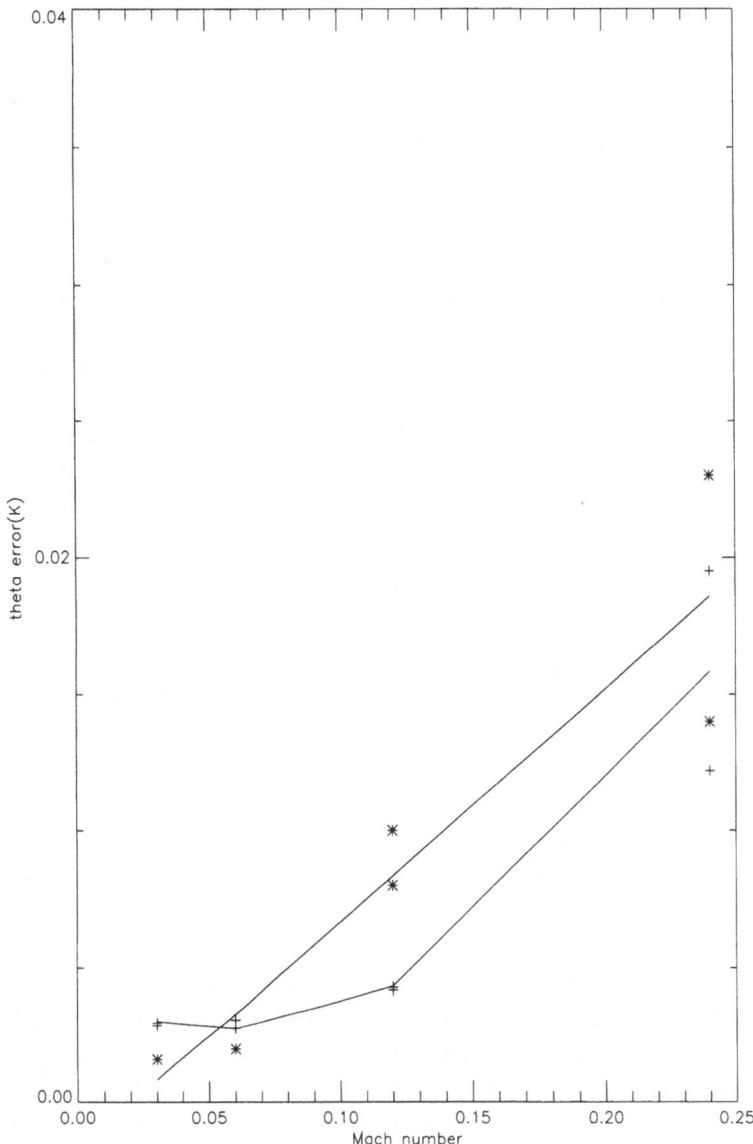

Figure 4.16. Convergence of r.m.s. differences of θ (deg K) between UM and anelastic solutions, with $N^2 = 10^{-4}$, plotted against Mach number. Resolutions and plotting conventions as in Figure 4.13.

5. Discussion

The theme of this paper is that numerical methods for atmosphere models need to be designed and evaluated using asymptotic limit solutions, rather than exact solutions of the governing equations. This is because production models fall short of adequately resolving the exact solution by several orders of magnitude. A related theme is that substantial research is needed on methods of averaging the equations. This has not been reviewed comprehensively in this paper but some key pieces of work have been discussed. An important issue is the choice between Eulerian and Lagrangian averaging.

A number of the most important asymptotic limit solutions have been described and their validity illustrated in computations. The emphasis has been on nonlinear regimes, since analysis of linear regimes is well represented in the existing literature. The related optimal numerical methods have been discussed. Not surprisingly, these mostly reflect current practice. The use of asymptotic limit solutions allows systematic justification for these choices. It should also allow long-term error estimates to be made for suitable numerical methods. There has been little work on that topic so far.

The numerical tests, with the exception of the Eady problem, show the expected convergence rates but to non-zero residuals. Some of these residuals represent dependencies on the initial data or additional parameters defining the flow. The others appear to represent a combination of numerical errors, initialization issues, and uncertainty in predicting turbulent flows. If these residuals are only significant for values of the dimensionless parameters (Rossby and Mach numbers in the examples presented) which are smaller than typical values in real data, then they will not be very significant. An issue with the larger-scale tests, which requires further study using observations, is to determine how prevalent symmetric instability actually is in the atmosphere and ocean on various scales, and whether this is correctly reflected in the solutions given by production models.

The failure of either model to predict the Eady solution after the initial life-cycle is disappointing, since this is the example most relevant to long-term prediction of weather systems. The results suggest that the Lagrangian conservation laws enforced by the geometric model are important in retaining long-term accuracy. However, it is difficult to see how they can be enforced in a production model. This remains an outstanding research challenge. The example does, however, suggest the importance of Lagrangian averaging in describing the physics of fronts.

Acknowledgements

The author would like to thank Nigel Wood for critically reading the manuscript and providing a lot of useful input. He would also like to thank John Thuburn for useful comments, and Marek Wlasak for providing Figures 3.1, 3.2 and 4.1.

REFERENCES

D. G. Andrews and M. E. McIntyre (1978) 'An exact theory of nonlinear waves on a Lagrangian-mean flow', *J. Fluid Mech.* **89**, 609–646.

A. Arakawa and C. S. Konor (1996) 'Vertical differencing of the primitive equations based on the Charney–Phillips grid in hybrid $\sigma - p$ vertical coordinates', *Mon. Weather Rev.* **124**, 511–528.

A. Arakawa and V. R. Lamb (1977) 'Computational design of the basic dynamical processes of the UCLA general circulation model', *Methods in Comput. Phys.* **17**, 173–265.

G. K. Batchelor (1967) *An Introduction to Fluid Dynamics*, Cambridge University Press.

J. R. Bates, S. Moorthi and R. W. Higgins (1993) 'A global multilevel atmospheric model using a vector semi-Lagrangian finite-difference scheme, Part I: Adiabatic formulation', *Mon. Weather Rev.* **121**, 244–263.

J.-D. Benamou and Y. Brenier (1998) 'Weak existence for the semi-geostrophic equations formulated as a coupled Monge–Ampère/transport problem', *SIAM J. Appl. Math.* **58**, 1450–1461.

A. F. Bennett and P. E. Kloeden (1981) 'The quasi-geostrophic equations: Approximation, predictability and equilibrium spectra', *Quart. J. Roy. Meteorol. Soc.* **107**, 121–136.

A. F. Bennett and P. E. Kloeden (1982) 'The periodic quasi-geostrophic equations: Existence and uniqueness of strong solutions', *Proc. Roy. Soc. Edinburgh* **91A**, 185–203.

A. J. Bourgeois and J. T. Beale (1994) 'Validity of the quasi-geostrophic model for large-scale flow in the atmosphere and ocean', *SIAM J. Math. Anal.* **25**, 1023–1068.

Y. Brenier (1991) 'Polar factorization and monotone rearrangement of vector-valued functions', *Comm. Pure Appl. Math.* **44**, 375–417.

G. Browning and H.-O. Kreiss (1994) 'Splitting methods for problems with different time-scales', *Mon. Weather Rev.* **122**, 2614–2622.

K. Bryan (1989) The design of numerical models of the ocean circulation. In *Oceanic Circulation Models: Combining Data and Dynamics* (*Proc. NATO ASI*, Les Houches, France, February 1988), pp. 465–500.

C. S. Cao and E. S. Titi (2005) Global well-posedness of the three-dimensional viscous primitive equations of large-scale ocean and atmosphere dynamics. arXiv.math AP/0503028.

J. G. Charney (1948) 'On the scale of atmospheric motions', *Geofys. Publ.* **17**, 1–17.

J.-Y. Chemin (2000) *Perfect Incompressible Fluids*, Oxford University Press.

P. Courtier, J.-N. Thépaut and A. Hollingsworth (1994) 'A strategy for operational implementation of 4D-Var, using an incremental approach', *Quart. J. Roy. Meteorol. Soc.* **120**, 1367–1387.

M. J. P. Cullen (2000) 'On the accuracy of the semi-geostrophic approximation', *Quart. J. Roy. Meteorol. Soc.* **126**, 1099–1115.

M. J. P. Cullen (2001) 'Alternative implementations of the semi-Lagrangian semi-implicit scheme in the ECMWF model', *Quart. J. Roy. Meteorol. Soc.* **127**, 2787–2802.

M. J. P. Cullen (2002*a*) 'Large scale non-turbulent dynamics in the atmosphere', *Quart. J. Roy. Meteorol. Soc.* **128**, 2623–2640.

M. J. P. Cullen (2002*b*) New mathematical developments in atmosphere and ocean dynamics, and their application to computer simulations. In *Large-Scale Atmosphere-Ocean Dynamics* (J. Norbury and I. Roulstone eds), Vol. I, Cambridge University Press, pp. 202–287.

M. J. P. Cullen (2003) 'Four dimensional variational data assimilation: A new formulation of the background error covariance matrix, based on a potential vorticity representation', *Quart. J. Roy. Meteorol. Soc.* **129**, 2777–2796.

M. J. P. Cullen (2006) *A Mathematical Theory of Large-Scale Atmospheric Flow*, Imperial College Press.

M. J. P. Cullen (2007) 'Semi-geostrophic solutions for flow over a ridge', *Quart. J. Roy. Meteorol. Soc.*, to appear.

M. J. P. Cullen and T. Davies (1990) 'A conservative split-explicit scheme with fourth-order horizontal advection', *Quart. J. Roy. Meteorol. Soc.* **117**, 993–1002.

M. J. P. Cullen and M. Feldman (2006) 'Lagrangian solutions of semi-geostrophic equations in physical space', *SIAM J. Math. Anal.* **37**, 1371–1395.

M. J. P. Cullen and H. Maroofi (2003) 'The fully compressible semi-geostrophic system from meteorology', *Arch. Rat. Mech. Anal.* **167**, 309–336.

M. J. P. Cullen and I. Roulstone (1993) 'A geometric model of the nonlinear equilibration of two-dimensional Eady waves', *J. Atmos. Sci.* **50**, 328–332.

M. J. P. Cullen, R. J. Douglas, I. Roulstone and M. J. Sewell (2005) 'Generalized semi-geostrophic theory on a sphere', *J. Fluid Mech.* **531**, 123–157.

M. J. P. Cullen, J. Norbury, R. J. Purser and G. J. Shutts (1987) 'Modelling the quasi-equilibrium dynamics of the atmosphere', *Quart. J. Roy. Meteorol. Soc.* **126**, 735–757.

T. Davies, M. J. P. Cullen, A. J. Malcolm, M. H. Mawson, A. Staniforth, A. A. White and N. Wood (2005) 'A new dynamical core for the Met Office's global and regional modelling of the atmosphere', *Quart. J. Roy. Meteorol. Soc.* **131**, 1759–1782.

T. Davies, A. Staniforth, N. Wood and J. Thuburn (2003) 'Validity of anelastic and other equation sets as inferred from normal-mode analysis', *Quart. J. Roy. Meteorol. Soc.* **129**, 2761–2775.

M. Diamantakis, T. Davies and N. Wood (2007) An iterative time-stepping scheme for the Met Office's semi-implicit semi-Lagrangian non-hydrostatic model', *Quart. J. Roy. Meteorol. Soc.*, to appear.

J. A. Domaradzki, Z. Xiao and P. K. Smolarkiewicz (2003) 'Effective eddy viscosities in implicit large eddy simulations of turbulent flows', *Phys. Fluids* **15**, 3890–3893.

D. Durran (1998) *Numerical Methods for Wave Equations in Geophysical Fluid Dynamics*, Springer.

ECMWF (2002) *Key Issues in the Parametrisation of Sub-Grid Physical Processes* (*Proc. ECMWF Seminar*, September 2001).

ECMWF (2005) *Representation of Sub-Grid Processes Using Stochastic–Dynamic Models* (*Proc. ECMWF Workshop*, June 2005).

J. H. Ferziger (1998) Direct and large eddy simulations of turbulence. In *Numerical Methods in Fluid Mechanics* (A. Vincent, ed.), Vol. 16 of *CRM Proceedings and Lecture Notes*, Université de Montreal, AMS.

C. Foias, D. D. Holm and E. S. Titi (2001) 'The Navier–Stokes-alpha model of fluid turbulence', *Physica D* **152**, 505–519.

R. Ford, M. E. McIntyre and W. A. Norton (2000) 'Balance and the slow quasi-manifold: Some explicit results', *J. Atmos. Sci.* **57**, 1236–1254.

M. S. Fox-Rabinowitz (1996) 'Diabatic dynamical initialization with an iterative time integration scheme as a filter', *Mon. Weather Rev.* **124**, 1544–1557.

C. Fureby and F. F. Grinstein (2002) 'Large eddy simulation of high-Reynolds number free and wall-bounded flows', *J. Comput. Phys.* **181**, 68–97.

V. G. Ganzha (1996) *Numerical Solutions for Partial Differential Equations*, CRC Press.

J. R. Garratt (1992) *The Atmospheric Boundary Layer*, Cambridge University Press.

P. R. Gent and J. C. McWilliams (1996) 'Eliassen–Palm fluxes and the momentum equation in non-eddy-resolving ocean circulation models', *J. Phys. Oceanog.* **26**, 2539–2546.

A. E. Gill (1982) *Atmosphere-Ocean Dynamics*, Academic Press.

R. L. Higdon (2006) Numerical modelling of ocean circulation. In *Acta Numerica*, Vol. 15, Cambridge University Press, pp. 385–470.

D. D. Holm (1996) 'Hamiltonian balance equations', *Physica D* **98**, 379–414.

D. D. Holm, J. E. Marsden and T. Ratiu (2002) The Euler–Poincaré equations in geophysical fluid dynamics. In *Large-Scale Atmosphere-Ocean Dynamics* (J. Norbury and I. Roulstone, eds), Vol. II, Cambridge University Press, pp. 251–300.

M. W. Holt (1990) 'Semi-geostrophic moist frontogenesis in a Lagrangian model', *Dyn. Atmos. Ocean.* **14**, 463–481.

B. J. Hoskins (1975) 'The geostrophic momentum approximation and the semi-geostrophic equations', *J. Atmos. Sci.* **32**, 233–242.

J. Houghton (2002) *The Physics of Atmospheres*, 3rd edn, Cambridge University Press.

E. Kalnay (2003) *Atmospheric Modelling, Data Assimilation and Predictability*, Cambridge University Press.

R. M. Kerr (1993) 'Evidence for a singularity of the three-dimensional incompressible Euler equations', *Phys. Fluids A* **5**, 1725–1746.

A. S. Lawless, N. K. Nichols and S. P. Ballard (2003) 'A comparison of two methods for developing the linearization of a shallow-water model', *Quart. J. Roy. Meteorol. Soc.* **129**, 1237–1254.

C. E. Leith and R. E. Kraichnan (1972) 'Predictability of turbulent flows', *J. Atmos. Sci.* **29**, 1041–1052

M. J. Lighthill (1978) *Waves in Fluids*, Cambridge University Press.

S.-J. Lin and R. B. Rood (1997) 'An explicit flux-form semi-Lagrangian shallow water model on the sphere', *Quart. J. Roy. Meteorol. Soc.* **123**, 2477–2498.

J.-L. Lions, R. Temam and S. Wang (1992*a*) 'New formulations of the primitive equations of atmosphere and applications', *Nonlinearity* **5**, 237–288.

J.-L. Lions, R. Temam and S. Wang (1992*b*) 'On the equations of the large-scale ocean', *Nonlinearity* **5**, 1007–1053.

F. Lipps and R. Hemler (1982) 'A scale analysis of deep moist convection and some related numerical calculations', *J. Atmos. Sci.* **29**, 2192–2210.

P. Lynch (1989) 'The slow equations', *Quart. J. Roy. Meteorol. Soc.* **115**, 201–219.

P. Lynch (1997) 'The Dolph–Chebyshev window: A simple optimal filter', *Mon. Weather Rev.* **125**, 1976–1982.

B. Machenhauer (1977) 'On the dynamics of gravity oscillations in a shallow water model with applications to normal-mode initialization', *Contrib. Atmos. Phys.* **50**, 253–271.

R. J. McCann (2001) 'Polar factorization of maps on Riemannian manifolds', *Geom. Funct. Anal.* **11**, 589–608.

M. E. McIntyre and W. A. Norton (2000) 'Potential vorticity inversion on a hemisphere', *J. Atmos. Sci.* **57**, 1214–1235.

J. C. McWilliams and I. Yavneh (1998) 'Fluctuation growth and instability associated with a singularity of the balance equations', *Phys. Fluids* **10**, 2587–2596.

J. C. McWilliams, I. Yavneh, M. J. P. Cullen and P. R. Gent (1999) 'The breakdown of large-scale flows in rotating, stratified fluids', *Phys. Fluids* **10**, 3178–3184.

A. J. Majda (1984) *Compressible Fluid Flow, and Systems of Conservation Laws in Several Space Variables*, Vol. 53 of *Applied Mathematical Sciences*, Springer, New York.

A. J. Majda (2003) *Introduction to PDEs and Waves for the Atmosphere and Ocean*, Vol. 9 of *Courant Lecture Notes*, AMS.

A. J. Majda and A. L. Bertozzi (2002) *Vorticity and Incompressible Flow*, Cambridge University Press.

M. H. Mawson (1996) 'A shallow water semi-geostrophic model on a sphere', *Quart. J. Roy. Meteorol. Soc.* **122**, 267–290.

A. R. Mohebalhojeh and D. G. Dritschel (2001) 'Hierarchies of balance conditions for the f-plane shallow water equations', *J. Atmos. Sci.* **58**, 2411–2426.

A. R. Mohebalhojeh and D. G. Dritschel (2004) 'Contour advective semi-Lagrangian algorithms for many-layer primitive-equation models', *Quart. J. Roy. Meteorol. Soc.* **130**, 347–364.

N. Nakamura (1994) 'Nonlinear equilibration of two-dimensional Eady waves: Simulations with viscous geostrophic momentum equations', *J. Atmos. Sci.* **51**, 1023–1035.

N. Nakamura and I. M. Held (1989) 'Nonlinear equilibration of two-dimensional Eady waves', *J. Atmos. Sci.* **46**, 3055–3064.

Y. Ogura and N. A. Phillips (1962) 'Scale analysis for deep and shallow convection in the atmosphere', *J. Atmos. Sci.* **19**, 173–179.

J. Pedlosky (1987) *Geophysical Fluid Dynamics*, Springer.

H. Reiser (2000) The development of numerical weather prediction in Deutsche Wetterdienst. In *Proc. 50th Anniversary of Numerical Weather Prediction*, Deutsche Meteorologische Gesellschaft, pp. 51–80.

H. Ritchie, C. Temperton, A. J. Simmons, M. Hortal, T. Davies, D. W. Dent and M. Hamrud (1995) 'Implementation of the semi-Lagrangian method in a high-resolution version of the ECMWF forecast model', *Mon. Weather Rev.* **123**, 489–514.

A. Robert (1981) 'A stable numerical integration scheme for the primitive meteorological equations', *Atmos. Ocean* **19**, 35–46.

A. Robert (1982) 'A semi-Lagrangian and semi-implicit numerical integration scheme for the primitive meteorological equations', *J. Meteorol. Soc. Japan* **60**, 319–324.

R. Sadourny (1975) 'The dynamics of finite-difference models of the shallow-water equations', *J. Atmos. Sci.* **32**, 680–689.

W. H. Schubert (1985) 'Semi-geostrophic theory', *J. Atmos. Sci.* **42**, 1770–1772.

G. J. Shutts (1998) 'Idealized models of the pressure drag force on mesoscale mountain ridges', *Contrib. Atmos. Phys.* **71**, 303–313.

G. J. Shutts and M. J. P. Cullen (1987) 'Parcel stability and its relation to semi-geostrophic theory', *J. Atmos. Sci.* **44**, 1318–1330.

J. Smagorinsky (1974) Global atmospheric modelling and the numerical simulation of climate. In *Weather and Climate Modification*, Wiley, New York, pp. 633–686.

R. K. Smith, ed., (1997) *The Physics and Parameterization of Moist Atmospheric Convection*, Kluwer, Dordrecht.

A. Staniforth and J. Côté (1991) 'Semi-Lagrangian integration scheme for atmospheric models: A review', *Mon. Weather Rev.* **119**, 2206–2223.

A. Staniforth, A. A. White, N. Wood, J. Thuburn, M. Zerroukat and E. Cordero (2002) Unified Model Documentation Paper No. 15: The Joy of UM 6.1: Model Formulation. Unpublished documentation, available from: www.metoffice.gov.uk/research/nwp/publications/papers/unified_model/index.html.

J. Thuburn (2006) 'Vertical discretizations giving optimal representation of normal modes: Sensitivity to the form of the pressure gradient term', *Quart. J. Roy. Meteorol. Soc.* **132**, 2809–2826.

J. Thuburn, N. Wood and A. Staniforth (2002a) 'Normal modes of deep atmospheres I: Spherical geometry', *Quart. J. Roy. Meteorol. Soc.* **128**, 1771–1792.

J. Thuburn, N. Wood and A. Staniforth (2002b) 'Normal modes of deep atmospheres II: $f - F$-plane geometry', *Quart. J. Roy. Meteorol. Soc.* **128**, 1793–1806.

T. Warn, O. Bokhove, T. G. Shepherd and G. K. Vallis (1995) 'Rossby-number expansions, slaving principles, and balance dynamics', *Quart. J. Roy. Meteorol. Soc.* **121**, 723–739.

T.-K. Wee and Y.-H. Kuo (2004) 'Impact of a digital filter as a weak constraint in MM5 4D-Var: An observing system simulation experiment', *Mon. Weather Rev.* **132**, 543–559.

A. A. White (2002) The equations of meteorological dynamics and various approximations. In *Large-Scale Atmosphere-Ocean Dynamics* (J. Norbury and I. Roulstone, eds), Vol. I, Cambridge University Press, pp. 1–100.

A. A. White (2003) 'Dynamical equivalence and the departure-point equation in semi-Lagrangian numerical models', *Quart. J. Roy. Meteorol. Soc.* **129**, 1317–1324.

A. A. White, B. J. Hoskins, I. Roulstone and A. Staniforth (2005) 'Consistent approximate models of the global atmosphere: Shallow, deep, hydrostatic, quasi-hydrostatic and non-hydrostatic', *Quart. J. Roy. Meteorol. Soc.* **131**, 2081–2108.

N. Wood and A. Staniforth (2003) 'The deep atmosphere Euler equations with a mass-based vertical coordinate', *Quart. J. Roy. Meteorol. Soc.* **129**, 1289–1300.

K.-S. Yeh, J. Côté, S. Gravel, A. Méthot, A. Patoine, M. Roch and A. Staniforth (2002) 'The CMC–MRB Global Environmental Multiscale (GEM) model, Part III: Non-hydrostatic formulation', *Mon. Weather Rev.* **130**, 339–356.

M. Zerroukat, N. Wood and A. Staniforth (2004) 'SLICE-S: A Semi-Lagrangian Inherently Conserving and Efficient scheme for transport problems on the sphere', *Quart. J. Roy. Meteorol. Soc.* **130**, 2649–2664.

M. Zerroukat, N. Wood and A. Staniforth (2005) 'A monotonic and positive-definite filter for a Semi-Lagrangian Inherently Conserving and Efficient (SLICE) scheme', *Quart. J. Roy. Meteorol. Soc.* **131**, 2923–2936.

M. J. Ziemianski and A. J. Thorpe (2003) 'Nonlinear balanced models for stratified fluids conserving Ertel–Rossby PV', *Quart. J. Roy. Meteorol. Soc.* **129**, 139–156.

Acta Numerica (2007), pp. 155–238
doi: 10.1017/S0962492906300013

Finite volume methods for hyperbolic conservation laws

K. W. Morton
Oxford University Computing Laboratory,
Wolfson Building, Parks Road, Oxford OX3 0DW, UK
E-mail: morton@comlab.ox.ac.uk

T. Sonar
Computational Mathematics,
TU Braunschweig, Pockelsstraße 14, D-38106 Braunschweig, Germany
E-mail: t.sonar@tu-bs.de

Finite volume methods apply directly to the conservation law form of a differential equation system; and they commonly yield cell average approximations to the unknowns rather than point values. The discrete equations that they generate on a regular mesh look rather like finite difference equations; but they are really much closer to finite element methods, sharing with them a natural formulation on unstructured meshes. The typical projection onto a piecewise constant trial space leads naturally into the theory of optimal recovery to achieve higher than first-order accuracy. They have dominated aerodynamics computation for over forty years, but they have never before been the subject of an *Acta Numerica* article. We shall therefore survey their early formulations before describing powerful developments in both their theory and practice that have taken place in the last few years.

CONTENTS

1. Introduction

The comprehensive book by Quarteroni and Valli (1994) on the numerical approximation of partial differential equations, which covers finite difference, finite element and spectral methods, devotes only its last eight pages to finite volume methods. However, they do point out that these methods are 'very popular in computational fluid dynamics' and use a terminology for their various formulations which is consistent with that which we will use, so that brief account provides a useful introduction to this survey article.

The term *finite volume method* seems to have appeared in the literature only in the early 1970s (see, *e.g.*, McDonald (1971) and Rizzi and Inouye (1973)) when it was applied to methods used to approximate the hyperbolic conservation law system corresponding to the Euler equations of gas dynamics; but the main ideas are much older. In Varga (1962) an integration method is used to derive finite difference approximations of self-adjoint elliptic equations on a non-uniform rectangular mesh, which reflected standard practice in the nuclear industry at that time and could now be regarded as a standard finite volume method. At about the same time, Preissmann (1961) was advocating a *box scheme* for approximating the St. Venant equations of hydraulic flow which we now regard as one of the basic finite volume schemes.

Consider the scalar conservation law for $u(x,t)$,

$$u_t + f(u)_x = s(x,u), \qquad (1.1)$$

which has the form of the momentum equation of the St. Venant system. In deriving a one-dimensional model of a river it is important to divide it up into sections of varying length, each with fairly uniform properties. It is also important to use an implicit time-stepping procedure because the important flood waves typically travel much more slowly than the characteristic waves that would define the CFL stability condition. So we integrate the conservation law over a rectangular box in the (x,t)-plane, use Gauss's theorem to convert the volume integral on the left to an integral along the boundary of the box shown in Figure 1.1(a), and use the trapezoidal rule to approximate the resulting integrals. Using U_j^n to denote our approximation to $u(x_j, t^n)$, we obtain the following scheme, which we consider to be the simplest form of a *cell-vertex scheme*:

$$\tfrac{1}{2}(x_{j+1} - x_j)\left[U_{j+1}^{n+1} + U_j^{n+1} - U_{j+1}^n - U_j^n\right] \qquad (1.2)$$

$$+ \tfrac{1}{2}(t^{n+1} - t^n)\left[F_{j+1}^{n+1} + F_{j+1}^n - F_j^{n+1} - F_j^n\right]$$

$$= \tfrac{1}{4}(x_{j+1} - x_j)(t^{n+1} - t^n)\left[S_{j+1}^{n+1} + S_j^{n+1} + S_{j+1}^n + S_j^n\right],$$

where we have written F_j^n for $f(U_j^n)$ with a similar notation for s.

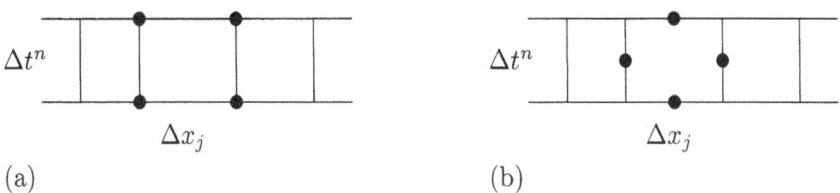

(a)   (b)

Figure 1.1. (a) The cell-vertex or Preissmann box scheme.
(b) The Godunov or cell-centre scheme.

An even earlier scheme, and one which has been the inspiration for many finite volume methods, is that due to Godunov (1959) (see also Richtmyer and Morton (1967, Section 12.15)) who developed it for application to the Euler equations of gas dynamics. If we apply it to (1.1) with the mesh as shown in Figure 1.1(b), the quantity U_j^n now represents the *cell average* of $u(x,t)$ at time level t^n in cell j, and $F_{j+1/2}^{n+1/2}$ an average between times t^n and t^{n+1} of the flux through the cell boundary at $x_{j+1/2}$. This is normally implemented as an explicit method so that, with cell length $\Delta x_j = x_{j+1/2} - x_{j-1/2}$ and time step $\Delta t^n = t^{n+1} - t^n$, we obtain

$$U_j^{n+1} = U_j^n - \Delta t^n \left[\left(F_{j+1/2}^{n+1/2} - F_{j-1/2}^{n+1/2} \right) / \Delta x_j - S_j^n \right]. \qquad (1.3)$$

To obtain the fluxes, the approximation at time level t^n can be interpreted as piecewise constant so that a Riemann problem is set up by the discontinuity at each cell boundary: this is solved exactly or approximately to give the flux. Such a scheme, when developed for two space dimensions, will be called a *cell-centre scheme*.

A scheme of the form (1.3) may seem so obvious, simple and natural that one may wonder why there have been so many alternatives in the literature – even in the class of first-order, explicit schemes. A brief explanation is in order here because it highlights the advantages of the finite volume formulation. In the absence of a source term, we can sum (1.3) over any set of contiguous cells, say $l \le j \le r$, to obtain the overall flux balance

$$\sum_{j=l}^{r} \Delta x_j (U_j^{n+1} - U_j^n) + \Delta t^n \left[F_{r+1/2}^{n+1/2} - F_{l-1/2}^{n+1/2} \right] = 0. \qquad (1.4)$$

Such a property is crucial to the correct modelling of shocks, whose structure is determined by the conservation law rather than by any differential equation derived from it. And it comes about as a result of two key choices, for the individual fluxes and the mesh length. The calculation of the flux by solving a Riemann problem at a cell boundary can be complicated, and for systems of equations a closed form solution may not exist. So it is natural to try to make use of fluxes defined as $F_j^n = f(U_j^n)$, as in a finite difference formulation in which U_j^n would be interpreted as a pointwise approximation to u and the mesh length as a difference between the corresponding mesh points. Then a simple explicit first-order scheme would make use of flux differences and take one of the following forms,

$$U_j^{n+1} = U_j^n - \Delta t^n \left[\frac{F_{j+1}^n - F_j^n}{x_{j+1} - x_j} \quad \text{or} \quad \frac{F_j^n - F_{j-1}^n}{x_j - x_{j-1}} \right]. \tag{1.5}$$

Stability, by the CFL condition, would require the choice to be determined by the sign of the characteristic speed $a(u) = f'(u)$ to give a so-called upwind scheme. But any switch between the two at a change in sign of a would preclude the cancellation and collapse of the flux sum that occurs in (1.4). Moreover, the summation on the left would not sensibly represent an integral of U except in the case of a uniform mesh.

So we are driven to the finite volume formulation as above, with some choice of the interface fluxes. The simplest such choice was given by Murman and Cole (1971) and is the scalar form of a Roe-scheme (Roe 1981): it is

$$F_{j+1/2}^{n+1/2} = \begin{cases} f(U_j^n) & \text{if } a_{j+1/2}^n \geq 0, \\ f(U_{j+1}^n) & \text{if } a_{j+1/2}^n < 0, \end{cases} \tag{1.6}$$

where $a_{j+1/2}^n = [f(U_{j+1}^n) - f(U_j^n)]/[U_{j+1}^n - U_j^n]$. This scheme deals with shocks very well; but unfortunately it treats smooth transitions, *i.e.*, expansion waves where $a(\cdot)$ is increasing from left to right, in the same way and hence gives a non-physical kink in the solution. This can be rectified, but the theoretically preferred first-order scheme is due to Engquist and Osher (1981) and takes the following form, where we write $A_j^n = a(U_j^n)$ and u_s is the *sonic point* at which $a(u_s) \equiv f'(u_s) = 0$:

$$F_{j+1/2}^{n+1/2} = \tfrac{1}{2}\big[(1 + \mathrm{sgn}A_j^n)F_j^n + (\mathrm{sgn}A_{j+1}^n - \mathrm{sgn}A_j^n)f(u_s)$$
$$+ (1 - \mathrm{sgn}A_{j+1}^n)F_{j+1}^n\big]. \tag{1.7}$$

These two schemes, (1.6) and (1.7), have very important theoretical properties which we will refer to in later sections, and which make them very important starting points for the development of higher-order schemes for systems of equations in higher dimensions.

The partial differential equation problems that we shall consider in this article will be of the general form

$$\mathbf{u}_t + \mathrm{div}\,\mathcal{F}(\mathbf{u}, \nabla\mathbf{u}) = \mathbf{s}(\mathbf{x}, t, \mathbf{u}), \quad \mathbf{u}: (\mathbf{x}, t) \in \mathbb{R}^d \times \mathbb{R}^+ \to \mathbf{u}(\mathbf{x}, t) \in \mathbb{R}^m,$$
$$\mathbf{u}(\mathbf{x}, 0) = \mathbf{u}^0(\mathbf{x}). \tag{1.8}$$

We shall concentrate on two space dimensions, and many engineering problems are steady, in which case the time t will not be involved. In the purely hyperbolic cases, such as for the Euler equations of gas dynamics, the fluxes $\mathcal{F}$ will be independent of the gradients $\nabla \mathbf{u}$ so that we have a first-order system of equations. Compressible gas dynamics is a key application area for the methods, however, so that it is important that they are readily applicable to the compressible Navier–Stokes equations through the inclusion of viscous flux terms. Another key feature of this field is that there is normally no source term $\mathbf{s}(\mathbf{x}, t, \mathbf{u})$, with the shape of the boundary being the key determinant of the flow.

The methods that we will describe will be applicable to quite general conservation laws of the form (1.8), but our discussion of them will frequently use terms and ideas that derive from fluid dynamics, as has already been the case. This is partly because this is the field with which we have most experience, but also because of the enormous influence this field has had on the development both of the mathematical models and their numerical approximation – see the beautiful historical essay on this topic by Birkhoff (1983).

Each of the finite volume schemes outlined above will meet new difficulties when applied to problems in two space dimensions; and, as we shall describe below, the extra dimension will lead to alternative variants. The system of equations generated by the Preissmann box scheme applied to the St. Venant equations describing one-dimensional river flow are generally solved by Newton iteration, exploiting the block tridiagonal form of the Jacobian system. But this does not extend to two dimensions. Hence, although the cell-vertex schemes have advantages in accuracy, the resulting algebraic systems are more difficult to solve than those generated by alternative schemes. On the other hand, the cell-centre schemes clearly provide, through their cell averages, only first-order approximations to the flow variables. Thus a very important aspect of these methods is the way in which higher-order approximations are generated from such data as cell averages. This comes within the general compass of *optimal recovery* (see Micchelli and Rivlin (1977)) and a large part of this account will be devoted to this topic. The general framework is as follows. Suppose that an unknown function is assumed to lie in a given function space and one is given the values of a set of linear functionals evaluated for the function. What then is the best estimate that one can make for the value of another linear functional? For example, how does one recover the point values of a function from its cell averages? The choice of mesh will also be crucial, especially in the neighbourhood of complicated flow features such as shocks. We shall therefore devote considerable attention to the topic of mesh adaptivity.

There is one final point that we wish to make in this Introduction. In the vast literature on finite volume methods they have sometimes been generated as finite difference schemes, and sometimes as some sort of finite element method. Given the flexibility and power of the latter methods in generating approximations on unstructured meshes, and the powerful theoretical framework in which they are formulated, we will below always regard the schemes we describe as finite element methods: in particular, we shall regard them as *Petrov–Galerkin methods*, in which the trial space may take any form but the test space is composed of piecewise constant functions. We will concentrate on algorithmic and theoretical aspects of the methods but will give sufficient numerical examples to demonstrate their power and generality.

2. Systems of conservation laws

Although the finite volume methods that we will develop should be applicable to the general conservation law form (1.8), most of their development and study has been in the context of hyperbolic equations, that is, where the fluxes $\mathcal{F}$ depend only on $\mathbf{u}$. We will therefore concentrate on these throughout this article, and begin by outlining the theoretical background; for a more detailed exposition see Godlewski and Raviart (1991) or Smoller (1983).

2.1. Hyperbolic systems

Consider the system of first-order conservation laws for $\mathbf{u}(\mathbf{x}, t) \in \mathbb{R}^m$, $(\mathbf{x}, t) \in \mathbb{R}^d \times \mathbb{R}^+$, with initial data $\mathbf{u}(\mathbf{x}, 0) = \mathbf{u}^0(\mathbf{x})$, in which $\mathcal{F} = (\mathbf{f}_1, \mathbf{f}_2, \ldots, \mathbf{f}_m)$,

$$\partial_t \mathbf{u} + \sum_{\ell=1}^{d} \partial_{x_\ell} \mathbf{f}_\ell(\mathbf{u}) = \mathbf{0}, \qquad (2.1)$$

where each flux vector $\mathbf{f}_\ell$ is a C^1-function. Then we can introduce the corresponding Jacobians of the fluxes, which we will denote by A_ℓ, so that when the solution is smooth it satisfies the quasilinear system of equations

$$\partial_t \mathbf{u} + \sum_{\ell=1}^{d} A_\ell(\mathbf{u}) \partial_{x_\ell} \mathbf{u} = \mathbf{0}. \qquad (2.2)$$

This system is *hyperbolic* in a region $G \subset \mathbb{R}^m$ of the state space if every linear combination of the Jacobians,

$$A(\boldsymbol{\nu}) := \sum_{\ell=1}^{d} \nu_\ell A_\ell(\mathbf{u}), \qquad (2.3)$$

corresponding to a unit vector $\boldsymbol{\nu} = (\nu_1, \nu_2, \ldots, \nu_d)^T \in \mathbb{R}^d$, has m real eigenvalues and associated linearly independent eigenvectors for every $\mathbf{u} \in G$. In a later section we will describe methods which make use of this property to approximate the time evolution of the solution: but for the moment we note only that it indicates how smooth data can evolve into a non-smooth solution after a finite time. Consider for example the *inviscid Burgers equation*, namely $u_t + u u_x = 0$. Its characteristics are given by $dx/dt = u$ along which u is constant, so they are straight lines; so where the initial data is a decreasing function of x it will form a front which will steepen until it breaks to form a shock.

It is therefore necessary to broaden the concept of what constitutes a solution of the PDE problem. The *weak form* of the equation is derived by multiplying (2.1) with a vector of test functions $\boldsymbol{\varphi} \in C_0^1(\mathbb{R}^d \times [0, \infty))^m$ and integrating over a $(d+1)$-dimensional sphere large enough to contain the support of the test functions. Integration by parts then results in

$$\int_{\mathbb{R}^d \times \mathbb{R}^+} \left[\mathbf{u} \cdot \partial_t \boldsymbol{\varphi} + \sum_{\ell=1}^{d} \mathbf{f}_\ell(\mathbf{u}) \cdot \partial_{x_\ell} \boldsymbol{\varphi} \right] d\mathbf{x} \, dt + \int_{\mathbb{R}^d} \mathbf{u}^0(\mathbf{x}) \cdot \boldsymbol{\varphi}(\mathbf{x}, 0) \, d\mathbf{x} = 0. \quad (2.4)$$

So a *weak solution* of (2.1) is one for which (2.4) is satisfied for all such test functions; and $L_{\text{loc}}^1(\mathbb{R}^d \times \mathbb{R}^+)^m$ seems an appropriate space for such solutions.

However, this is too large a space. For example, for Burgers' equation it would include all those containing a jump from a constant u_L on the left to a constant u_R on the right; but only those with $u_L > u_R$ are true *shocks* which would evolve from smooth data or be the limits of solutions to the viscous Burgers equation $u_t + u u_x = \mu u_{xx}$ as the viscosity $\mu \to 0$. Motivated by the equations of fluid dynamics, we therefore introduce the concept of *entropy*. An entropy, for the equation (2.1), is a convex function $\eta : \mathbb{R}^m \to \mathbb{R}$ for which there exists d scalar *entropy fluxes* q_ℓ such that the following relations hold:

$$(\nabla_{\mathbf{u}} \eta)^T A_\ell = (\nabla_{\mathbf{u}} q_\ell)^T \quad 1 \le \ell \le d, \quad (2.5)$$

for each $\mathbf{u}$. Then it is clear that a smooth solution of (2.1) will also satisfy

$$\partial_t \eta(\mathbf{u}) + \sum_{\ell=1}^{d} \partial_{x_\ell} q_\ell(\mathbf{u}) = 0; \quad (2.6)$$

that is, a further conservation law is satisfied. However, when dissipative terms are added to the equations the convexity of η ensures that the left-hand side of (2.6) is non-positive. Thus, when we take the limit as the dissipation tends to zero, we obtain the following *entropy condition* for the weak solution $\mathbf{u}$: $\forall \varphi \in C_0^1(\mathbb{R}^d \times \mathbb{R}^+)$,

$$\int_{\mathbb{R}^d \times \mathbb{R}^+} \left[\eta(\mathbf{u}) \partial_t \varphi + \sum_{\ell=1}^{d} q_\ell(\mathbf{u}) \partial_{x_\ell} \varphi \right] d\mathbf{x} \, dt \ge 0, \quad (2.7)$$

the condition corresponding to the given entropy η and its associated flux functions. A weak solution **u** is called an *entropy solution* of the PDE system if such an entropy condition is satisfied for all entropies possessed by the system of equations.

We need to introduce just one further key characterization of solutions to hyperbolic PDEs before we can state an important existence and uniqueness theorem. If g is a real function defined on the open set $\Omega \subset \mathbb{R}^d$, and $g \in L^1_{\mathrm{loc}}(\Omega)$, then its *total variation* is defined as

$$TV_\Omega(g) := \sup\left\{ \int_\Omega g \operatorname{div}\boldsymbol{\phi} \,\mathrm{d}\mathbf{x}, \ \boldsymbol{\phi} \in C^1_0(\Omega)^d, \|\boldsymbol{\phi}\|_{L^\infty(\Omega)} \le 1 \right\}. \tag{2.8}$$

Thus we introduce the notation for functions of *bounded variation*,

$$BV(\Omega) := \left\{ g \in L^1_{\mathrm{loc}}(\Omega); \ TV(\Omega(g) < \infty \right\}.$$

The fact that the existence of smooth solutions to the Navier–Stokes equations in $\mathbb{R}^3$ is one of the Millennium Grand Challenge Problems of the Clay Mathematics Institute – see Jaffe (2006) – shows that we are far from having a comprehensive theory for such PDE problems. However, there is one special case in which all the above concepts show their worth.

In the case of a scalar problem, $m = 1$ in (2.1), results obtained by Oleinik (1957) and Kružkov (1970) enable us to state the following theorem.

Theorem 2.1. If $m = 1$ and the initial data $u^0 \in L^1(\mathbb{R}^d) \cap L^\infty(\mathbb{R}^d) \cap BV(\mathbb{R}^d)$ then (2.1) has a unique entropy solution $u(\cdot, t) \ \forall \ t > 0$, for which

$$\|u(\cdot, t)\|_{L^\infty(\mathbb{R}^d)} \le \|u^0\|_{L^\infty(\mathbb{R}^d)}, \tag{2.9}$$

$$TV_{\mathbb{R}^d}(u(\cdot, t)) \le TV_{\mathbb{R}^d}(u^0). \tag{2.10}$$

This result is not only a valuable guide to the selection of numerical methods but it was also proved by taking the limit of approximations obtained by their use. The concept of a *total variation (TV)-stable* numerical scheme was introduced by Harten (1984), where he showed that for a scalar problem in one dimension any scheme that is consistent with the conservation law and its entropy inequality gives a convergent approximation if it is TV-stable. Second-order schemes with these properties were presented in that paper and in Harten (1983), where the widely used concept of *TVD (total variation diminishing)* schemes was introduced. We note that the two explicit finite volume methods introduced in Section 1, Roe's scheme (1.6) and the Engquist–Osher scheme (1.7), are TVD and stable under a very natural CFL condition.

2.2. Haar's lemma

Unfortunately, since finite volume schemes are based on using piecewise constant test functions, they use an integral form of the PDE and it is not at all clear *a priori* that this will single out the same solutions as the weak form. Suppose we introduce a *control volume* $\Omega \subset \mathbb{R}^d$ with outward normal $\mathbf{n}$ and surface measure $\mathrm{d}S$. Then, to give a form which will be useful later, we integrate the equation (2.1) over the $(d+1)$-dimensional cylinder $(t, t + \Delta t) \times \Omega$ and apply Gauss's theorem to obtain

$$\int_{\Omega} [\mathbf{u}(t + \Delta t) - \mathbf{u}(t)] \, \mathrm{d}\Omega + \int_t^{t+\Delta t} \oint_{\partial \Omega} \mathcal{F} \cdot \mathbf{n} \, \mathrm{d}S \, \mathrm{d}t = \mathbf{0}. \tag{2.11}$$

For the present purposes, however, it is convenient to work with a more general $(d+1)$-dimensional control volume Ω^e obtained by extending the $\mathbf{x}$-variable to $\mathbf{x}^e = (t, x_1, x_2, \ldots, x_d)^T$ and similarly $\mathbf{n}$ to $\mathbf{n}^e$ and $\mathrm{d}S$ to $\mathrm{d}S^e$: then we can take advantage of the $(d+1)$-dimensional divergence form of (2.1) to write the more general integral form as

$$\oint_{\partial \Omega^e} \begin{pmatrix} \mathbf{u} \\ \mathcal{F}(\mathbf{u}) \end{pmatrix} \cdot \mathbf{n}^e \, \mathrm{d}S^e = \mathbf{0}. \tag{2.12}$$

Haar's lemma is the general name given to statements which link the weak form (2.4) with this integral form of the PDE: the name derives from the early result given by Haar (1919).

The work of Morrey (1960) (see Klötzler (1970)) can be used to give the following result.

Theorem 2.2. Suppose that $\mathbf{u}$ and the d fluxes $\mathbf{f}_\ell$ are summable over the bounded region $G \subset \mathbb{R}^d \times \mathbb{R}^+$. Then

$$\oint_{\partial C} \begin{pmatrix} \mathbf{u} \\ \mathcal{F}(\mathbf{u}) \end{pmatrix} \cdot \mathbf{n}^e \, \mathrm{d}S^e = \mathbf{0},$$

for almost all cuboids $C \subset G$, if and only if

$$\int_G \begin{pmatrix} \mathbf{u} \\ \mathcal{F}(\mathbf{u}) \end{pmatrix} \cdot \nabla \boldsymbol{\varphi} \, \mathrm{d}\mathbf{x}^e \equiv \int_G \left[\mathbf{u} \cdot \partial_t \boldsymbol{\varphi} + \sum_{\ell=1}^d \mathbf{f}_\ell(\mathbf{u}) \cdot \partial_{x_\ell} \boldsymbol{\varphi} \right] \mathrm{d}\mathbf{x} \, \mathrm{d}t = 0$$

for every $\boldsymbol{\varphi}$ which vanishes on or near ∂G and is uniformly Lipschitz-continuous on $\overline{G}$.

The same result has been proved for balls instead of cuboids; and, by using a generalized divergence operator due to Müller (1957), Bruhn (1985) has extended it to quite general control volumes. Thus it is this that we exploit when developing finite volume methods by integration of the conservation laws over quite general shapes.

2.3. Euler and Navier–Stokes equations

In two space dimensions, the Euler equations for inviscid compressible gas flow have the form (2.1), expressing the conservation of mass, the two components of momentum and the total energy. We write them in terms of the density ρ, the two velocity components $\mathbf{v} := (v_1, v_2)^T$, the pressure p, the total energy per unit mass E and the enthalpy which is defined by $H := E + p/\rho$. Then we have the following definitions of $\mathbf{u}$ and the flux vectors $\mathbf{f}_\ell$:

$$
\mathbf{u} := \begin{bmatrix} \rho \\ \rho v_1 \\ \rho v_2 \\ \rho E \end{bmatrix}, \quad
\mathbf{f}_1(\mathbf{u}) := \begin{bmatrix} \rho v_1 \\ \rho v_1^2 + p \\ \rho v_1 v_2 \\ \rho v_1 H \end{bmatrix}, \quad
\mathbf{f}_2(\mathbf{u}) := \begin{bmatrix} \rho v_2 \\ \rho v_1 v_2 \\ \rho v_2^2 + p \\ \rho v_2 H \end{bmatrix}.
$$

These need to be supplemented by an equation of state giving the pressure in order to close the system: for an ideal gas this is taken to be

$$
p = (\gamma - 1)\rho\big(E - \tfrac{1}{2}|\mathbf{v}|^2\big), \tag{2.13}
$$

where γ denotes the ratio of specific heats; in the case of dry air it is taken as $\gamma = 1.4$.

As these equations have played such an important role in the development of finite volume methods we will describe their key properties in some detail. The Jacobians of the flux functions have the following form, in terms of the same variables:

$$
A_1(\mathbf{u}) = \begin{bmatrix}
0 & 1 & 0 & 0 \\
\frac{\gamma-3}{2}v_1^2 + \frac{\gamma-1}{2}v_2^2 & (3-\gamma)v_1 & (1-\gamma)v_2 & \gamma-1 \\
-v_1 v_2 & v_2 & v_1 & 0 \\
(\gamma-1)v_1|\mathbf{v}|^2 - \gamma v_1 E & \gamma E - \frac{\gamma-1}{2}(v_2^2 + 3v_1^2) & (1-\gamma)v_1 v_2 & \gamma v_1
\end{bmatrix}
$$

and

$$
A_2(\mathbf{u}) = \begin{bmatrix}
0 & 0 & 1 & 0 \\
-v_1 v_2 & v_2 & v_1 & 0 \\
\frac{\gamma-3}{2}v_2^2 + \frac{\gamma-1}{2}v_1^2 & (1-\gamma)v_1 & (3-\gamma)v_2 & \gamma-1 \\
(\gamma-1)v_2|\mathbf{v}|^2 - \gamma v_2 E & (1-\gamma)v_1 v_2 & \gamma E - \frac{\gamma-1}{2}(v_1^2 + 3v_2^2) & \gamma v_2
\end{bmatrix}.
$$

These can be written in alternative forms and have several important features. We note first that if we form a linear combination as in (2.3), corresponding to the direction $\boldsymbol{\nu} = (\nu_1, \nu_2)^T$, and introduce the *sound speed* $c := \sqrt{\gamma p/\rho}$, then $A(\boldsymbol{\nu})$ is diagonalizable with eigenvalues $\mathbf{v} \cdot \boldsymbol{\nu}$ (occurring twice), $\mathbf{v} \cdot \boldsymbol{\nu} - c$ and $\mathbf{v} \cdot \boldsymbol{\nu} + c$. Thus the system is hyperbolic, so long as the density and pressure remain positive, and we will give expressions for the eigenvectors of $A(\boldsymbol{\nu})$ below.

The first remarkable property of the Euler equations that we note is their *rotational invariance*. Using the rotation matrix $T(\mathbf{n})$, written in terms of the unit vector $\mathbf{n} = (n_1, n_2)^T$ as

$$T(\mathbf{n}) := \begin{bmatrix} 1 & 0 & 0 & 0 \\ 0 & n_1 & n_2 & 0 \\ 0 & -n_2 & n_1 & 0 \\ 0 & 0 & 0 & 1 \end{bmatrix},$$

it is easy to check that the Euler equation fluxes satisfy

$$\sum_{\ell=1}^{2} \mathbf{f}_\ell(\mathbf{u})n_\ell = T^{-1}(\mathbf{n})\mathbf{f}_1(T(\mathbf{n})\mathbf{u}).$$

We can take direct advantage of this in an integral form comparable with that in (2.11): if we do not carry out the time integration we obtain

$$\frac{\mathrm{d}}{\mathrm{d}t} \int_\Omega \mathbf{u}\,\mathrm{d}\mathbf{x} + \oint_{\partial\Omega} T^{-1}(\mathbf{n})\mathbf{f}_1(T(\mathbf{n})\mathbf{u})\,\mathrm{d}s = \mathbf{0}. \qquad (2.14)$$

Another property of the equations that is exploited by some numerical schemes is that the fluxes are *homogeneous functions of degree 1* of the conservative variables. This is obvious for some of the terms but, for example, we can write the second term in the vector $\mathbf{f}_1$ as $u_2^2/u_1 + p$ and $p = (\gamma - 1)[u_4 - \frac{1}{2}(u_2^2 + u_3^2)/u_1]$. It follows that the fluxes satisfy *Euler's relation* so that

$$\mathbf{f}_\ell(\mathbf{u}) = A_\ell(\mathbf{u})\mathbf{u}, \quad \ell = 1, 2. \qquad (2.15)$$

This means that for smooth solutions it does not matter whether the Jacobian matrices are included in the spatial differentiation in (2.2) or not: $\partial_{x_\ell}(A_\ell \mathbf{u}) = A_\ell \partial_{x_\ell} \mathbf{u}$.

Where solutions are smooth it is often convenient to write the equations in terms of the so-called *primitive variables* ρ, v_1, v_2 and p. If we form these into the vector $\mathbf{w}$ the gradient matrix defining the change of variables $\nabla_{\mathbf{w}}\mathbf{u} =: M$ is given by

$$M = \begin{bmatrix} 1 & 0 & 0 & 0 \\ v_1 & \rho & 0 & 0 \\ v_2 & 0 & \rho & 0 \\ \frac{1}{2}|\mathbf{v}|^2 & \rho v_1 & \rho v_2 & (\gamma - 1)^{-1} \end{bmatrix}. \qquad (2.16)$$

As this is lower triangular, its inverse can be written down immediately and thence the new coefficient matrices, which we denote by $B_\ell := M^{-1}A_\ell M$, can be derived to give the following:

$$B_1(\mathbf{w}) := \begin{bmatrix} v_1 & \rho & 0 & 0 \\ 0 & v_1 & 0 & \rho^{-1} \\ 0 & 0 & v_1 & 0 \\ 0 & \rho c^2 & 0 & v_1 \end{bmatrix}, \qquad B_2(\mathbf{w}) := \begin{bmatrix} v_2 & 0 & \rho & 0 \\ 0 & v_2 & 0 & 0 \\ 0 & 0 & v_2 & \rho^{-1} \\ 0 & 0 & \rho c^2 & v_2 \end{bmatrix}.$$

$$(2.17)$$

Note that we have here used the fact that the sound speed is given by $c^2 = \partial p/\partial \rho$. Clearly these matrices have a very much simpler form than for the conservative form, and make the calculation of the eigenvalue structure of the system much easier to carry out.

It is an important property of these systems, which is shared by the two-dimensional wave equation system, that the two Jacobian matrices do not commute, so that although any linear combination of them can be diagonalized they cannot be simultaneously diagonalized. With the usual notation $B(\boldsymbol{\nu}) = \nu_1 B_1 + \nu_2 B_2$, and denoting the matrix of its right eigenvectors by $R(\boldsymbol{\nu})$, we have $BR = R\Lambda$, where $\Lambda = \mathrm{diag}(\mathbf{v} \cdot \boldsymbol{\nu}, \mathbf{v} \cdot \boldsymbol{\nu}, \mathbf{v} \cdot \boldsymbol{\nu} - c, \mathbf{v} \cdot \boldsymbol{\nu} + c)$ is the diagonal matrix of eigenvalues and

$$
R(\boldsymbol{\nu}) := \begin{bmatrix} \nu_1 & 0 & \rho/2c & \rho/2c \\ 0 & \nu_2 & \nu_1/2 & -\nu_1/2 \\ 0 & -\nu_1 & \nu_2/2 & -\nu_2/2 \\ 0 & 0 & \rho c/2 & \rho c/2 \end{bmatrix}. \tag{2.18}
$$

It is then straightforward to obtain the eigenvectors of $A(\boldsymbol{\nu})$ through use of the transformation matrix M: see Hirsch (1990) for details. This form of the equations has been used in the development of finite volume evolution Galerkin methods, which will be described in Section 4.3.

Another form of the equations which is very important from both a theoretical and a practical point of view will be described in Section 6. This makes use of so-called *entropy variables* to symmetrize the equations: that is, to write them as a *symmetric hyperbolic* system, in the linearized form (2.2) but with a matrix A_0 multiplying the time derivative term, in which the three coefficient matrices are symmetric.

A key parameter in the Euler equations is the *Mach number* given by $\mathrm{Ma} := |\mathbf{v}|/c$: when and where it is less than unity the flow is subsonic, and where larger than unity it is supersonic. For steady flows in more than one space dimension, the Euler equations are elliptic where the flow is subsonic and hyperbolic where it is supersonic; and this is the source of some of the characteristic challenges posed by both the analysis of the equations and their numerical modelling. Thus, for most commercial aeroplanes in steady flight, the oncoming flow relative to the aeroplane is subsonic; but it accelerates smoothly around the leading edges to form a supersonic patch which terminates in a shock. Such a flow is termed *transonic*, an example of which is shown later in Figure 6.4.

The Euler equations result from neglecting the effects of viscosity and heat conduction in models of compressible fluid flow. Their inclusion changes the structure of the equations from being purely hyperbolic, and leads to some form of the Navier–Stokes equations. We will outline here the form of these changes; for more details the reader should consult texts such as Hirsch

(1988). With the extra flux terms the equations take the following form:

$$\partial_t \mathbf{u} + \sum_{\ell=1}^{2} \partial_{x_\ell} \left[\mathbf{f}_\ell(\mathbf{u}) - \frac{1}{\mathrm{Re}} \mathbf{g}_\ell(\mathbf{u}) \right] = \mathbf{0}, \tag{2.19}$$

where

$$g_\ell := \begin{bmatrix} 0 \\ \tau_{1,\ell} \\ \tau_{2,\ell} \\ v_1 \tau_{1,\ell} + v_2 \tau_{\ell,2} + \frac{\mu\gamma}{\mathrm{Pr}} \partial_{x_\ell} \epsilon \end{bmatrix}, \quad \ell = 1, 2.$$

Here the extra terms are controlled by the two parameters, the Reynolds number Re and the Prandtl number Pr, which is a thermodynamic property of the gas equal to 0.72 for air. The terms in the viscous stress tensor are given by $\tau_{i,j} := \mu(\partial_{x_j} v_i + \partial_{x_i} v_j) + \delta_i^j \lambda(\partial_{x_1} v_1 + \partial_{x_2} v_2)$, where δ_i^j is the Kronecker delta symbol. The coefficient μ is the viscosity and, by Stokes' hypothesis, we set $\lambda = -\frac{2}{3}\mu$. The heat conduction is given above in terms of the specific internal energy, defined by $\epsilon = E - \frac{1}{2}|\mathbf{v}|^2$. The key coefficient is that for viscosity, which is typically assumed to be given in terms of the temperature T by Sutherland's law $\mu = T^{1.5}(1 + S)/(T + S)$, where $T = \gamma(\gamma-1)(|\mathbf{v}|/c)^2 \epsilon$ and $S := 110°K/\overline{T}_\infty$ and $\overline{T}_\infty$ denotes the temperature at infinity.

The details of these formulae are unimportant for our present purposes. The points to note are the structure of the extra terms and the heavy dependence on the computed variables and their gradients. The implication is that in a finite volume method it is most important to have accurate and reliable recovery procedures which, typically from cell averages, can produce both point values and gradients of the dependent variables. In addition, many schemes will combine the inviscid and viscous fluxes, as shown in (2.19), so that they can build on the finite volume techniques developed for convection-diffusion problems: see, e.g., Morton (1996).

Well-publicized test problems have played an important role in the development of numerical models for compressible flows: examples from the early days include the one-dimensional shock tube problem of Sod (1978) and the steady transonic flow past the NACA 0012 aerofoil: see Hirsch (1990). So we conclude this section by showing the results of some Euler calculations for another widely used model problem due to Woodward and Colella (1984). This concerns the supersonic flow of a gas past a forward-facing step, the details of which will be given in a later section. Two triangular meshes are used: a coarse mesh of 2016 triangles and a finer mesh of 8064 triangles, both shown in Figure 2.1. In Figure 2.2 we show contour plots of the Mach number obtained with the coarse mesh (above) and with the fine mesh (below), using two finite volume methods: on the left the plot is obtained

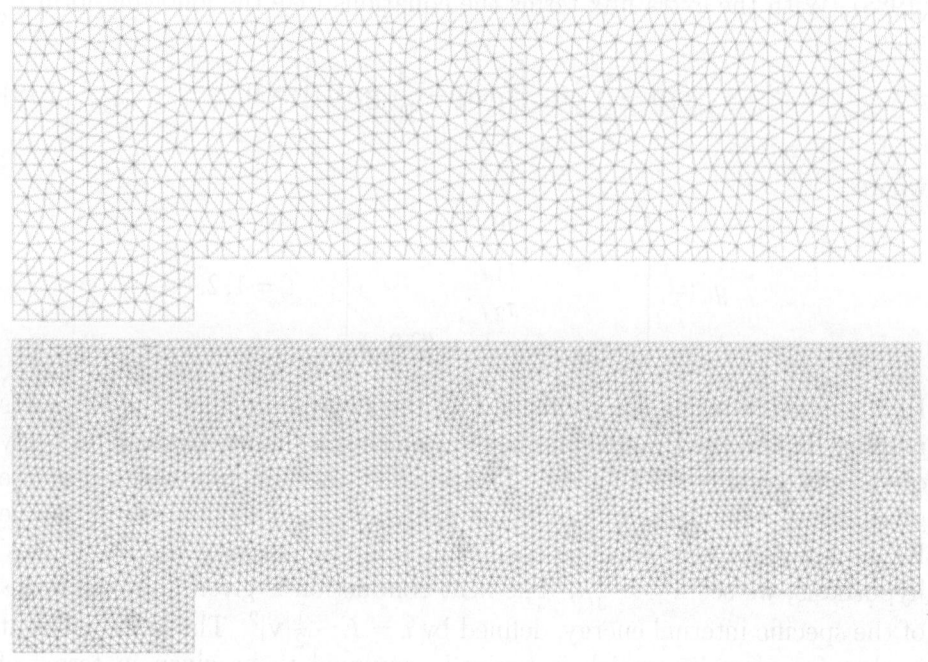

Figure 2.1. Coarse and fine grid for the Woodward and Colella test case.

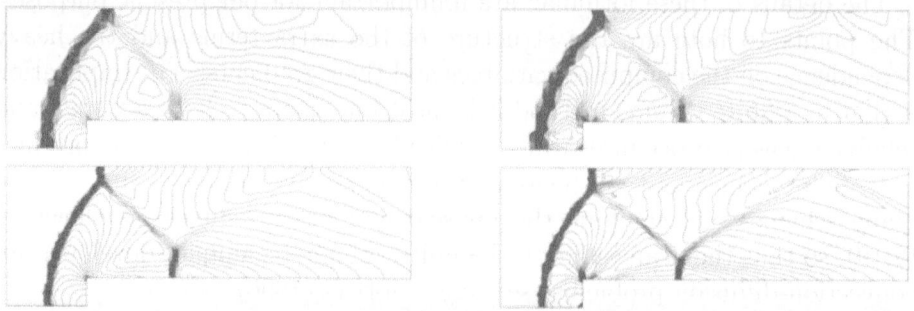

Figure 2.2. Mach number distribution on the coarse mesh (*above*) and on the fine mesh (*below*). Numerical scheme of Steger and Warming (*left*) and Osher and Solomon (*right*).

with a method due to Steger and Warming (1981), which generalizes that
shown in (1.6) by exploiting the homogeneous property of the fluxes referred
to above; and on the right the plot is obtained with a method due to Osher
and Solomon (1982) which generalizes the Engquist–Osher scheme given in
(1.7). Apart from the obvious improvement on the finer mesh, it is clear
that the second scheme captures more details of the flow, though the two
are of the same formal accuracy. We shall see later that the difference between these two schemes lies not so much in their starting points but in the
way they generalize from the scalar problem to a system of equations: the
former is an example of a *flux-vector splitting method* and the latter of a
flux-difference splitting method.

3. Finite volume formulations

There are so many finite volume schemes applied to such a variety of problems, in both the engineering and the numerical analysis literature, that we
have to be quite selective in this review. We will concentrate on formulations that are tailored to the needs of the Euler equations of aeronautical
gas dynamics because this is the field that has stimulated the most important developments; and we will focus on two space dimensions. However, we
will refer to other fields and to three space dimensions when making choices
about the methods that we describe in detail.

Many of the engineering and design problems in aeronautics concern
steady flows; and even in the unsteady problems the rates of change are
often very slow when compared with the characteristic sound speed. Thus
the approximation employed in the spatial variables is usually quite distinct
from that used for the time variable. Advantage can be taken of a finite
element formulation in the spatial variables, usually of the Petrov–Galerkin
form with the test space different from the trial space. With these points
in mind, we will first review some of the choices that have to be made.

3.1. Overall view of alternatives

Triangles vs quadrilaterals. In early two-dimensional calculations of flows
around aerofoils quadrilateral meshes were very popular: they are easy to
generate and they have the nice property that, globally, there are the same
number of vertices as cells. This has an advantage for a cell-vertex formulation, which is preferred on the grounds of accuracy on a stretched mesh
(see Morton and Paisley (1989)); and such an approach is readily extended
to approximating the Navier–Stokes equations (Crumpton, Mackenzie and
Morton 1993). However, triangular meshes are more flexible in modelling
complicated geometries and much easier to generalize to three dimensions.
So our emphasis will be on triangles, with such meshes often being referred
to as unstructured: see Figure 3.1.

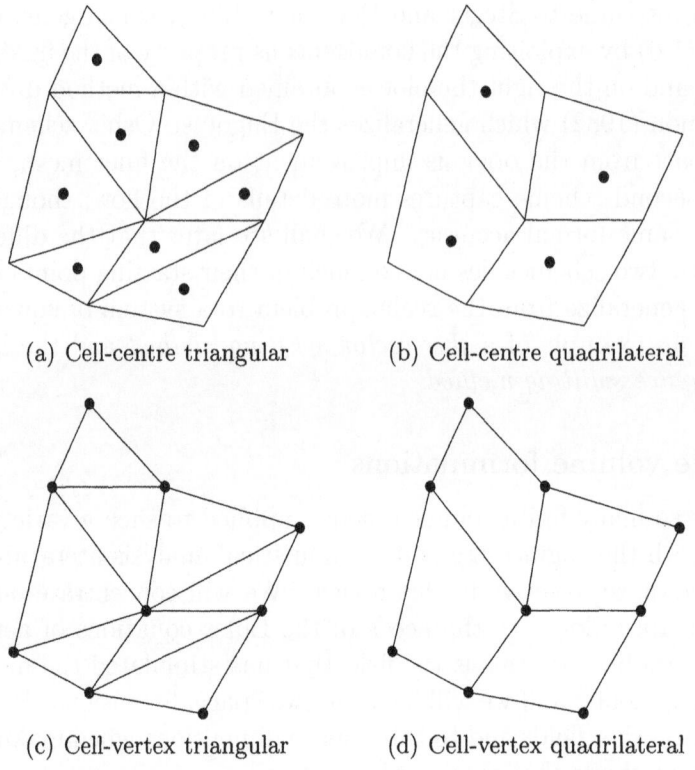

(a) Cell-centre triangular   (b) Cell-centre quadrilateral

(c) Cell-vertex triangular   (d) Cell-vertex quadrilateral

Figure 3.1. Some typical finite volume meshes.

Cell-centre vs cell-vertex. In the former the unknowns are associated with
the centres of the cells which act as the control volumes, as indicated in Figure 3.1(a) and (b); while in the latter the unknowns are associated with the
vertices of the control volumes, as indicated in Figure 3.1(c) and (d). Thus,
in a cell-vertex scheme the basic approximation would naturally be linear
on a triangle and bilinear on a quadrilateral, while from a finite element
viewpoint the test function would be piecewise constant. The local approximation can be good but difficulties can arise in setting up and solving the
overall system of equations. In a cell-centre scheme the unknowns usually
represent cell averages, so the local approximation is piecewise constant.
Thus from a finite element viewpoint the test and trial spaces are the same,
which simplifies setting up the equations: one merely has to interpret the
very low-order approximations appropriately. In this respect they can be
regarded as early forms of *discontinuous Galerkin methods*: see Cockburn,
Karniadakis and Shu (2000).

Node-centred schemes. This is a third alternative, sometimes called *box
schemes*, in which the control volumes are centred on the vertices of the
primary mesh. When the primary mesh is quadrilateral, then the new mesh
can be the same and there is then little difference from corresponding cell-

vertex schemes. But when the primary mesh is triangular, the secondary mesh control volumes are often constructed by joining the centroids to the mid-sides of the primary mesh – as shown in Figure 3.3 – although there are alternative choices for the box shape. This third choice turns out to have several advantages, which we will describe in the course of this article.

Semi-discrete vs time-integrated. In the former approach, often called the *method-of-lines* approach, approximations to only the spatial operators are sought so that an ODE solver then has to be applied to the resulting system of equations. At the other extreme, as in the box scheme of (1.2), integration over time is included in the finite volume formulation. Most commonly some intermediate approach is adopted: we will consider both extremes and relate them to some of the many alternatives.

3.2. Cell-centre schemes and node-centred schemes on triangles

We consider the approximation of the Euler (and later the Navier–Stokes) equations on a bounded open domain $\Omega \subset \mathbb{R}^2$. For the sake of simplicity we assume that the boundary $\partial\Omega := \overline{\Omega}\backslash\Omega$ is already a polygon. On $\overline{\Omega}$ we establish two types of tesselations, a primary and a secondary mesh or grid.

A *triangulation* T^h of $\overline{\Omega}$ is the set of finitely many triangular subsets $T_i \subset \overline{\Omega}$, $i = 1, \ldots, \#T$, such that the following conditions are satisfied:

- $\overline{\Omega} = \bigcup_{i\in\{1,\ldots,\#T\}} T_i$,
- every $T_i \in T^h$ is closed and non-empty,
- for two $T_i, T_j \in T^h$ with $i \neq j$ their interiors satisfy $\overset{\circ}{T}_i \cap \overset{\circ}{T}_j = \emptyset$.

A triangulation is called *conforming* if the following additional condition holds:

- every one-dimensional edge of any $T_i \in T^h$ is either a subset of $\partial\Omega$ or the edge of another T_j, $j \neq i$.

The parameter h in the notation T^h corresponds to a typical geometrical length scale of the triangulation which may be represented by the length of the longest edge.

Note that conformity ensures that there can be no *hanging nodes*, *i.e.*, vertices lying in the interior of an edge of another triangle. Although conformity is not necessary in the context of finite volume approximations, it helps to simplify nearly every algorithmic detail, especially in the case of grid adaptivity. The definition of a triangulation given here is identical to that used in finite element methods: see Ciarlet (1987). We call such a conforming triangulation T^h a *primary grid.*

A *barycentric subdivision* can be used to define a *secondary grid.* Let

$$K_{h,i} := \{T \in T^h \mid \text{node } i \text{ is vertex of } T\}$$

be the set of all triangles of a primary grid sharing node i. Denote the three edges of triangle T by $e_{T,k}$, $k = 1, 2, 3$. For each $T \in \mathcal{T}^h$ consider the following barycentric subdivision: join the barycentre or centroid of T with the mid-points of its three edges $e_{T,k}$, $k = 1, 2, 3$. This divides each triangle into three segments, as shown in Figure 3.2. The union of all those segments

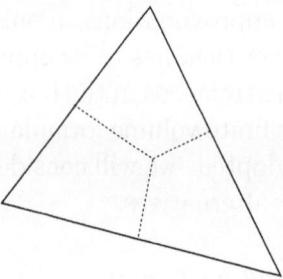

Figure 3.2. Barycentric subdivision of a triangle.

of $T \in K_{h,i}$ adjacent to node i is called the *box* B_i around node i. If node i belongs to the boundary $\partial\Omega$ the box is constructed with the two halves of the boundary edges of the boundary triangles having node i in common. The union $\mathcal{B}^h := \bigcup_{i=1,\ldots,\#B} B_i$ of all the boxes is called the *secondary grid*. An example of a primary and secondary grid is shown in Figure 3.3; this includes the situation at the boundary.

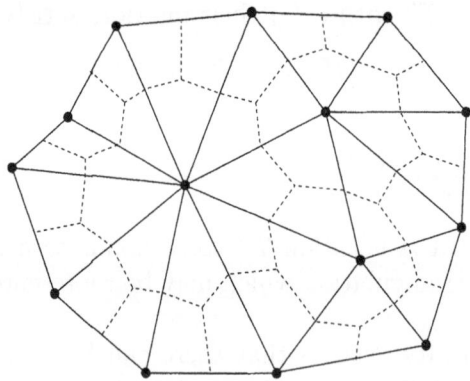

Figure 3.3. Primary and secondary grid for a node-centred scheme.

We next construct cell-centre finite volume approximations on such grids for systems of the type (2.1); we postpone consideration of the viscous fluxes to later in the section. We need the notion of the neighbourhood of a triangle T_i: so we denote the set of indices of its neighbouring triangles by

$$N(i) := \{j \in \mathbb{N} \mid T_i \cap T_j \text{ is an edge of } T_i\}.$$

Then we integrate the conservation law over the control volume formed by the triangle T_i to obtain

$$\frac{\mathrm{d}}{\mathrm{d}t}\mathcal{A}(T_i)\mathbf{u}(t) = -\frac{1}{|T_i|}\sum_{j\in N(i)}\int_{\partial T_j\cap\partial T_i}\sum_{\ell=1}^{2}\mathbf{f}_\ell(\mathbf{u})n_{ij,\ell}\,\mathrm{d}\sigma, \qquad (3.1)$$

where we denote by $\mathcal{A}(T_i)\mathbf{u}(t)$ the average of $\mathbf{u}(t)$ over the triangle. Here the outer (with respect to T_i) unit normal vector at the edge $\partial T_i\cap\partial T_j$ is denoted by $\mathbf{n}_{ij}=(n_{ij,1},n_{ij,2})^T$, and $|T_i|$ is the area of the triangle.

To allow maximum flexibility in the development of numerical schemes from this formula we replace the edge integrals by Gaussian quadrature formulae, which will assume a certain degree of smoothness. If we denote by $\mathbf{x}_{ij-}$ and $\mathbf{x}_{ij+}$ the coordinates of the vertex points of $\partial T_i\cap\partial T_j$, the edge is parametrized by $s\in[-1,1]$ such that the general point on the edge is

$$\mathbf{x}_{ij}(s) := \tfrac{1}{2}[(1-s)\mathbf{x}_{ij-}+(1+s)\mathbf{x}_{ij+}],$$

and this can be introduced in the evolution equation (3.1) on T_i. Suppose we denote the number of Gauss points on $\partial T_i\cap\partial T_j$ by n_G, the actual Gauss points by $\mathbf{x}_{ij}(s_\nu)$, $\nu=1,\ldots,n_G$, and the weights by ω_ν, then we get the system

$$\frac{\mathrm{d}}{\mathrm{d}t}\mathcal{A}(T_i)\mathbf{u}(t) = \qquad\qquad\qquad\qquad\qquad (3.2)$$

$$-\sum_{j\in N(i)}\frac{|\partial T_i\cap\partial T_j|}{2|T_i|}\left\{\sum_{\nu=1}^{n_G}\sum_{\ell=1}^{2}\omega_\nu\mathbf{f}_\ell(\mathbf{u}(\mathbf{x}_{ij}(s_\nu),t))n_{ij,\ell}+\mathcal{O}\big(h^{2n_G}\big)\right\}.$$

The lowest-order scheme corresponds to the mid-point rule and, although we will concentrate on this form of quadrature, it is clearly a simple matter to replace it by the trapezoidal rule, Simpson's rule or some other choice.

The first step in deriving a corresponding numerical approximation is to introduce $\mathbf{U}_i(t)$ as an approximation to $\mathcal{A}(T_i)\mathbf{u}(t)$. Then the crucial step is to choose a formula giving a *numerical flux function* that generalizes that given in (1.6) for the Roe scheme or (1.7) for the Engquist–Osher method. Because it takes the form of a mapping

$$(\mathbf{u}_L,\mathbf{u}_R;\mathbf{n}) \overset{\mathbf{H}}{\longmapsto} \mathbf{H}(\mathbf{u}_L,\mathbf{u}_R;\mathbf{n})\in\mathbb{R}^m$$

from two constant states to a flux, in a direction $\mathbf{n}$, it is also called an *approximate Riemann solver*. Note that we will use these terms quite generally to refer to any choices for such a mapping, even to make comparisons with, *e.g.*, Lax–Friedrichs or Lax–Wendroff difference schemes. The essential condition it has to satisfy is a consistency condition with the differential equation,

$$\forall \mathbf{u} \in \mathbb{R}^m: \quad \mathbf{H}(\mathbf{u}, \mathbf{u}; \mathbf{n}) = \sum_{\ell=1}^{2} \mathbf{f}_\ell(\mathbf{u}) n_\ell. \tag{3.3}$$

Because of this condition we can replace one of the sums over the fluxes in (3.2) by the corresponding $\mathbf{H}$ function. But more importantly, with the additional choice of a one-point quadrature rule, it leads to the following definition of the basic cell-centre finite volume scheme:

Find $\mathbf{U}_i(t), i = 1, \ldots, \#T, t \in [0, t^*], t^* > 0$, *as a solution of the system of ordinary differential equations*

$$\frac{\mathrm{d}}{\mathrm{d}t} \mathbf{U}_i(t) = -\frac{1}{|T_i|} \sum_{j \in N(i)} |\partial T_i \cap \partial T_j| \mathbf{H}(\mathbf{U}_i(t), \mathbf{U}_j(t); \mathbf{n}_{ij}), \tag{3.4}$$

$$\mathbf{U}_i(0) = \mathcal{A}(T_i)\mathbf{u}(0).$$

After a discretization in time this will form the basis of all the cell-centre finite volume schemes on triangular meshes that we shall discuss. The simplest time discretization is obtained with the explicit Euler scheme. Then we will have a direct generalization of (1.3) to systems of conservation laws in two dimensions. Such a choice will lead to a stability limit on the time step, which is in the form of a *CFL condition* (Courant, Friedrichs and Lewy 1928). In the scalar one-dimensional case, this requires that no characteristic can cross more than one cell in one time step: in the notation of Figure 1.1(b) this becomes

$$-\Delta x_{j-1} \le f'(u)\Delta t \le \Delta x_j \quad \text{for } u \text{ between } U_{j-1}^n \text{ and } U_j^n, \forall j. \tag{3.5}$$

For both the Roe flux (1.6) and the Engquist–Osher flux (1.7), it is shown in Morton (2001) that this condition is sufficient as well as necessary for stability. However, it is clear that such a condition will become more restrictive and more complicated in two dimensions on a triangular mesh. Other more sophisticated time discretizations are therefore needed, some of which will be described in Section 4.4.

One cannot expect to obtain better than first-order accuracy with a scheme that only uses a piecewise constant approximation to the unknown solution. To do better we introduce a recovery step: this produces a higher-order approximation $\widetilde{\mathbf{U}}(t)$ which preserves the cell averages,

$$\mathcal{A}(T_i)\widetilde{\mathbf{U}}(t) = \mathbf{U}_i(t), \ i = 1, \ldots, \#T;$$

and the values of this function at the quadrature points are then substituted into the calculation of the numerical flux functions in (3.4). The details of how this is done will be described in Section 5 but we will introduce some of the ideas here.

For the one-dimensional scheme (1.3), an approach that has led to the popular MUSCL algorithms introduced by van Leer (1979) makes use of discontinuous linear recovery for each variable: since the average is to be preserved in a cell, one need only choose a slope S_j for the variable in each cell. This is usually obtained by combining the divided differences D_+U_j and D_-U_j between the average in the cell and those to its right and left, but it is important to impose some restrictions on how this is done. Thus in the case of a scalar conservation law it is easily deduced from the TVD property (2.10) that monotone initial data remains monotone. Hence if a set of cell averages form a monotone sequence this property should be preserved and this makes the choice of recovery algorithm far from trivial even in this case. For example, suppose that the recovered function in cell j is given by

$$\widetilde{U}(x) = U_j + S_j(x - x_j), \tag{3.6}$$

where x_j is the centre of the cell. Then one might aim to ensure that whenever the sequence U_j is monotone increasing, then so is $\widetilde{U}(x)$; and it is easily seen that a sufficient condition for this is to have

$$0 \le S_j \le \min(D_+U_j, D_-U_j) \quad \forall j.$$

Formulae for S_j based on such considerations are referred to as *slope limiters*, with the best-known being that given by the minimum of $|D_\pm U_j|$ and called the *minmod limiter*:

$$\mathrm{minmod}(x, y) := \begin{cases} s\min(|x|, |y|) & \text{if } \mathrm{sgn}\, x = \mathrm{sgn}\, y = s, \\ 0 & \text{otherwise.} \end{cases} \tag{3.7}$$

However, such a choice is rather conservative, and could lead to clipping of local extrema, so many alternatives have appeared in the literature, a topic we will return to in Section 4.6 on ENO schemes. Meanwhile, a necessary condition for preserving a monotone increasing function that we will refer to later (and that originally suggested by van Leer) is the following:

$$U_{j-1} \le U_j - \tfrac{1}{2}S_j\Delta x_j \quad \text{and} \quad U_j + \tfrac{1}{2}S_j\Delta x_j \le U_{j+1}; \tag{3.8}$$

that is, the variation in the cell does not go beyond the averages in its neighbours.

To generalize this approach to a triangular mesh, we need to calculate a gradient for a variable in each cell from the cell averages in neighbouring cells. One way to do this is to use the general formula for obtaining an average gradient of a variable over a region Ω from values on its perimeter,

$$\mathcal{A}(\Omega)\nabla u = \frac{1}{|\Omega|} \int_{\partial\Omega} \begin{bmatrix} u\, dx_2 \\ -u\, dx_1 \end{bmatrix},$$

and applying this to the secondary grid cell around each vertex. This gives a choice of three gradients for each triangle. An alternative due to Durlofsky,

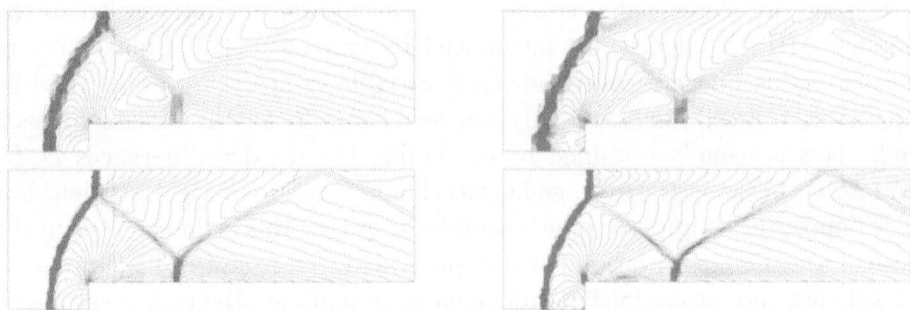

Figure 3.4. Mach number distribution on the coarse mesh (*above*) and
on the fine mesh (*below*) for the Woodward and Colella test case.
Numerical scheme of Steger and Warming (*left*) and Osher and
Solomon (*right*), both with linear recovery.

Engquist and Osher (1992) is a direct generalization of the TVD construction in one dimension. It makes use of the three neighbouring triangles, *i.e.*, those in $N(i)$, which is often called the von Neumann neighbourhood of T_i. From any pair we can construct a linear function whose averages over the pair and T_i match the corresponding values of U. Again this gives a choice of three gradients.

One needs to combine or choose between these gradients in some way, for each component of $\mathbf{U}$, and then calculate the numerical fluxes along each edge of the triangular mesh, where the variables are not only discontinuous but also non-constant; and this should be done in such a way as to utilize known properties of the differential equation system, such as monotonicity preservation or the TVD property. For example, choosing the gradient of a variable with smallest absolute value except at an extremum would be a direct generalization of the minmod limiter (3.7). However, this is rather severe and generally more sophisticated choices are made: see Section 5. These issues highlight the key principles of the recovery process: any known properties of the unknown function $\mathbf{u}(t)$ can be exploited in constructing the higher-order approximation $\widetilde{\mathbf{U}}(t)$; but its projection onto the lower-order space has to be such as to reproduce $\mathbf{U}(t)$.

We illustrate the effectiveness of such algorithms by means of the forward-facing step problem already referred to. In Figure 3.4 we show on the coarse mesh (above) and the fine mesh (below) the results obtained after linear recovery for the two schemes for which the corresponding results without recovery were shown in Figure 2.2. They clearly show the improvement due to the recovery, on both meshes.

We conclude this section by describing node-centred schemes on a triangular mesh, which we will gradually see have several advantages over the alternative cell-centre schemes, including the choice of a linear recovery

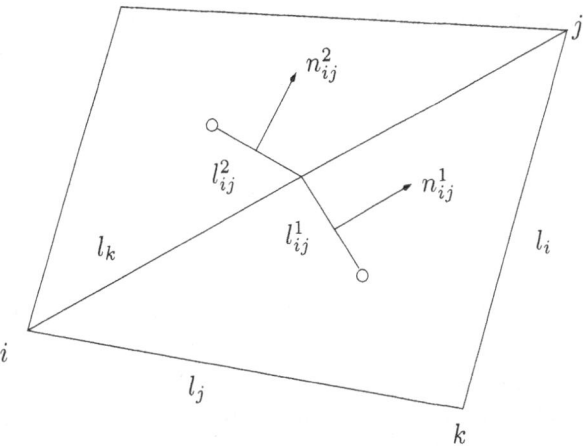

Figure 3.5. Geometry between boxes B_i and B_j.

procedure. The piecewise constant approximation in this case is given by the values $\mathbf{U}_i$ which represent averages over the boxes B_i centred on the vertices and forming the secondary grid. It is then straightforward to construct a piecewise linear function corresponding to each component variable U on each triangle by choosing the three required nodal values so that the average of the function over each box B_i corresponding to one of the nodes matches U_i. Then each triangle yields a gradient of this function; and in the box B_i we choose the gradient with the smallest magnitude from those that correspond to triangles which share the node i, i.e., are in the set $K_{h,i}$. Thus we obtain a discontinuous piecewise linear approximation to the vector of unknowns which we denote by $\widetilde{\mathbf{U}}(t)$.

The node-centred update formula corresponding to (3.4) is a little more complicated because the boundary of each box B_i has more segments: we denote the set of indices of boxes that are neighbours to B_i by

$$N^B(i) := \{j \in \mathbb{N} \mid B_i \cap B_j \text{ is edge of } B_i\};$$

and the boundary between two neighbouring boxes consists of two segments with different normals, as shown in Figure 3.5. Then, with the notation shown in the figure and with one-point Gaussian quadrature at the corresponding mid-points $\mathbf{x}_{ij}^k$ of the segments, we obtain the system of ODEs

$$\frac{d}{dt}\mathbf{U}_i(t) = -\frac{1}{|B_i|} \sum_{j \in N^B(i)} \sum_{k=1}^{2} |l_{ij}^k| \mathbf{H}\big(\widetilde{\mathbf{U}}_i(\mathbf{x}_{ij}^k, t), \widetilde{\mathbf{U}}_j(\mathbf{x}_{ij}^k, t); \mathbf{n}_{ij}^k\big), \qquad (3.9)$$

$$\mathbf{U}_i(0) = \mathcal{A}(B_i)\mathbf{u}(0). \qquad (3.10)$$

Choice of an integrator for this system completes the definition of the method.

With this scheme and its recovery procedure it is reasonably straight-forward to include the viscous fluxes and thence approximate the Navier–Stokes equations (2.19). We have linear approximations to each variable on each triangle, so that their gradients are readily computed on each section of the boundary of a box B_i; and since they are constant on each triangle they are exactly integrated by any Gaussian quadrature rule. There is just one snag. For realistic values of the Reynolds number (say, Re = $\mathcal{O}(10^6)$) it is necessary to have a highly stretched mesh near the boundary of the domain, with long thin triangles aligned with the boundary; then the normals to an edge of a box will point mainly towards and away from the boundary, rather than along it. This is clearly inappropriate for the inviscid fluxes. The remedy is to replace the barycentres of each triangle by an appropriately weighted average of its vertex positions. Such a formula is given by

$$\mathbf{x}^s = \sum_{m \in \{i,j,k\}} \alpha_m^s \mathbf{x}_m \quad \text{where } \alpha_m^s := \frac{1}{2(|l_i| + |l_j| + |l_k|)} \sum_{\substack{m' \in \{i,j,k\} \\ m' \neq m}} |l_{m'}|;$$

the effect on the mesh is illustrated in Figure 3.6.

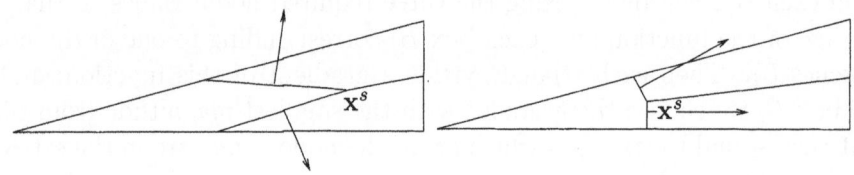

Figure 3.6. Stretched grid cell for Navier–Stokes; deformed boxes on the right.

3.3. Cell-vertex schemes on quadrilaterals

We can subdivide a bounded open domain $\Omega \subset \mathbb{R}^2$, with a polygonal boundary, into a set of quadrilaterals $Q_\alpha \subset \overline{\Omega}$, $\alpha = 1, \ldots, \#Q$ in exactly the same way as the triangulation described in the previous subsection. We will also assume it is conforming, in the same sense, and it is unnecessary to repeat here the formal detailed specification of the subdivision. However, we now use Greek subscripts $\alpha, \beta, \ldots$ to refer to the quadrilaterals and reserve Roman subscripts $i, j, \ldots$ to refer to their vertices, with which the variables $\mathbf{U}$ will be associated. In addition we include the viscous fluxes from the outset and seek a steady solution of the Navier–Stokes equations. So the formulation will be much closer to a finite element approach to a steady convection-diffusion problem: the distinction is that we limit the class of test functions to piecewise constants on the quadrilaterals.

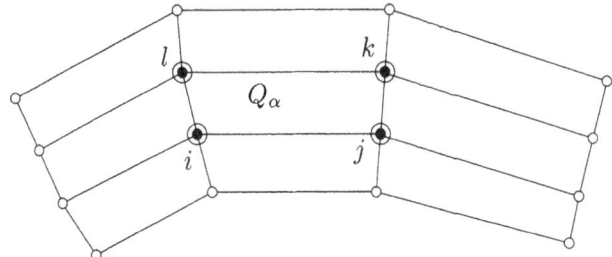

Figure 3.7. Typical quadrilateral mesh for a cell-vertex approximation to the Navier–Stokes equations, showing the vertices contributing to the cell residual: solid circles where fluxes are calculated, open circles at points needed for gradients.

A typical mesh is shown in Figure 3.7 with the cell Q_α over which the equations are to be integrated marked with solid circles. Thus, writing the integral of the hyperbolic conservation law (2.1) over this cell in a similar form to (3.1), we have

$$\frac{\mathrm{d}}{\mathrm{d}t}\mathcal{A}(Q_\alpha)\mathbf{u}(t) - -\frac{1}{|Q_\alpha|}\int_{\partial Q_\alpha}[\mathbf{f}_1\,\mathrm{d}x_2 - \mathbf{f}_2\,\mathrm{d}x_1], \qquad (3.11)$$

where we have used the usual form for the boundary integral in two dimensions; and we have a similar form for the Navier–Stokes equations (2.19).

The objective is to generate a scheme of second-order accuracy for the steady problem and this can be achieved if each quadrilateral is within $\mathcal{O}(h)$ of a parallelogram, *i.e.*, the orientations of opposite sides differ by this order, and this we will assume. Then the approximation can be regarded as a bilinear form determined by the vertex values; and a recovery procedure is needed only to calculate the gradients that appear in the viscous fluxes. This may be done in several ways, but with this assumption on the mesh it is best to do so by integrating over each cell to give the gradients at the centroids, and then interpolating between these to obtain values at the vertices.

Writing $\mathbf{U}_i$ for the approximation to $\mathbf{u}$ at the vertex $\mathbf{x}_i$, we use the more compact notation in the Navier–Stokes equations (2.19),

$$\mathbf{F}_{i,\ell} = \mathbf{f}_\ell(\mathbf{U}(\mathbf{x}_i)) - (1/\mathrm{Re})\mathbf{g}_\ell^{(r)}(\mathbf{U}(\mathbf{x}_i))$$

where the superscript in $\mathbf{g}^{(r)}$ signifies the fact that the recovered gradient has been used in the calculation of the viscous fluxes. We also approximate the boundary integrals of (3.11) by the trapezoidal rule. Then it is one of the attractive features of the cell-vertex method in two dimensions that only the components of the quadrilateral diagonals appear in the residual. So, writing $\mathbf{x}_{ij} = \mathbf{x}_i - \mathbf{x}_j$ and with the vertex lettering in Figure 3.7, to solve the steady problem we seek to satisfy the cell residual equations, which become

$$\mathbf{R}_\alpha := \tfrac{1}{2}\big[(\mathbf{F}_{i,1} - \mathbf{F}_{k,1})\mathbf{x}_{jl,2} + (\mathbf{F}_{j,1} - \mathbf{F}_{l,1})\mathbf{x}_{ki,2} \tag{3.12}$$

$$- (\mathbf{F}_{i,2} - \mathbf{F}_{k,2})\mathbf{x}_{jl,1} - (\mathbf{F}_{j,2} - \mathbf{F}_{l,2})\mathbf{x}_{ki,1}\big] = \mathbf{0}.$$

This is a difficult system to solve and, although some of the techniques that are used are properly topics for the next section, several introductory remarks are in order here.

The most direct approach to solving the system (3.12) would be to apply Newton's method, or a quasi-Newton method. And for the incompressible Navier–Stokes equations or similar systems, and with closely related finite element approximations, this is widely used very successfully: see, for example, Winters, Rae, Jackson and Cliffe (1981). In addition, if the unsteady problem were modelled by approximating in the same way the full divergence form as in (2.12), we would have a system that directly generalizes the box scheme of (1.2), which is very successfully used in conjunction with Newton's method in one-dimensional river flow modelling: see Cunge, Holly and Verwey (1980). Aerodynamic applications involving high-speed compressible flows pose much more severe problems, however, particularly in the neighbourhood of shocks. Thus, even though some progress has been made in using Newton's method, as reported in Badcock and Richards (1995), steady cell-vertex approximations have generally been obtained with iteration schemes that relate closely to time-stepping; the one important distinction is that different time steps can be used at each point in order to improve convergence rates. Nevertheless, in future developments we may expect greater use of Newton methods, and progress in this direction is discussed in Section 4.5.

In most of the flow region the inviscid flux terms dominate, so modelling the Euler equations highlights many of the difficulties. The first of these is that the discrete equations, based on the cells, do not match up with the unknowns, based on the vertices. Even having the correct number of equations depends on the careful imposition of boundary conditions. Then the resulting equations are far from diagonally dominated when linearized. Thus most iteration procedures are based on combining the residuals from the cells surrounding a vertex to form a nodal residual, which is what is actually driven to zero. So we introduce *distribution matrices* $D_{\alpha,i}$ and define nodal residuals by

$$\mathbf{N}_i(\mathbf{U}) := \frac{\sum_{\alpha=1}^{p} |Q_\alpha| D_{\alpha,i} \mathbf{R}_\alpha}{\sum_{\alpha=1}^{p} |Q_\alpha|}, \tag{3.13}$$

where p is the number of cells meeting at node i, normally 4. An important particular choice of the distribution matrices is closely related to the most widely used two-dimensional form of the Lax–Wendroff method, and corresponds to that used in the pioneering paper by Ni (1982): for node i in Figure 3.7, we have

$$D_{\alpha,i} = I + \nu_C \frac{\Delta t_\alpha}{|Q_\alpha|}[\mathbf{x}_{jl,2}A_{\alpha,1} - \mathbf{x}_{jl,1}A_{\alpha,2}], \qquad (3.14)$$

where ν_C is a cell-based global CFL number, Δt_α a local time step and $A_{\alpha,\ell}, \ell = 1, 2$, are the Jacobian matrices of the inviscid fluxes evaluated at the centre of cell Q_α. Then the basic iteration can be written as

$$\mathbf{U}_i^{n+1} = \mathbf{U}_i^n - \nu_N \Delta t_i \mathbf{N}_i(\mathbf{U}), \qquad (3.15)$$

where Δt_i is the minimum local time step from the surrounding cells and ν_N a node-based global CFL number. Such a scheme was applied successfully to solving the Navier–Stokes equations around an aerofoil in Crumpton *et al.* (1993), using a multigrid acceleration procedure based on a standard full approximation scheme. It was shown that for model problems the CFL parameters should satisfy

$$\nu_N \leq \nu_C \quad \text{and} \quad \nu_N \nu_C < 1;$$

and for the Navier–Stokes problems the role of ν_N was to control the rate of convergence of the iteration, while the value of ν_C affected the quality of the converged approximation: in particular, the extent to which the cell residuals were driven to zero rather than just the nodal residuals.

There are two further difficulties that affect these methods: the first is the presence of a spurious *chequer-board mode*; and the second is that the continuous form of the approximation is not well suited to representing shocks. The chequer-board mode arises from the averaging in the trapezoidal rule, and will be stimulated by the presence of shocks or other rapidly changing flow features to give severe oscillations in both directions of a typical mesh. These necessitate the addition of carefully chosen dissipation terms, which are critical to the success of the method. The introduction of procedures to recognize the presence of shocks, and to provide a global fit for them, can be used for simple problems such as the inviscid transonic flow around an aerofoil treated in Morton and Paisley (1989); however, this is not very feasible for general problems.

These difficulties have meant that cell-vertex methods are not as widely used at present as cell-centre and node-centred schemes. However, there has been considerable interest in the last few years in the development of cell-vertex methods on triangles. The attraction of quadrilateral meshes is that globally there are as many cells as vertices, so one can hope to drive most of the cell residuals to zero, and the nodal residuals are introduced only to achieve that end. With triangles this approach is no longer feasible. Instead, in the same way that approximate Riemann solvers use the discrepancy between two neighbouring flux values to update the two states, so the

flux residual in a triangular cell is used to update the states at its three vertices. In his influential paper Roe (1981), Roe was already expressing this
viewpoint and expanded on it in Roe (1982); subsequent collaboration with
Deconinck (Deconinck, Roe and Struijs 1993) took the ideas much further,
and a recent survey, Ricciutto, Csik and Deconinck (2005), summarizes the
present position.

4. Evolutionary algorithms

The various formulations of the finite volume approximation in the spatial
variables require differing approaches to the approximation in time. At one
extreme we have the system of ODEs in (3.4) that require a careful choice
of ODE solvers. At another we have the system of nonlinear algebraic
equations (3.12), which would be only slightly modified in an implicit time-
stepping of an unsteady problem, and this needs special solution algorithms.
And between these we have methods which use explicit time-stepping integrated into the spatial discretization. We will briefly survey each of these,
starting with the need to define numerical flux functions for the cell-centre
schemes, which leads naturally into considering first explicit, and then implicit, time-stepping algorithms.

4.1. Numerical flux functions

It is clearly not feasible to solve the generalized Riemann problem at all cell
boundaries on a triangular or quadrilateral mesh, that is, to take account
of mesh corners and the possible variation of the recovered approximation
$\widetilde{\mathbf{U}}$ along each boundary. We therefore have to consider the construction of
approximate flux functions to substitute into the typical scheme (3.4). A
useful starting point is the scalar problem and Brenier's *transport collapse
operator* (Brenier 1984). The exact solution of

$$u_t + f_1(u)_{x_1} + f_2(u)_{x_2} = 0$$

carries the initial data along the characteristics until shocks form: but in
the transport collapse operator approximation this data is carried forward
to give multivalued solutions at each point in space, and then these are combined to give the approximation. Brenier showed that repeated application
of this process over small time steps converges to the the correct solution of
the PDE as the time steps are refined; indeed, it provides one of the simplest
means of establishing Theorem 2.1. He also showed that in one dimension
it could lead directly to the Engquist–Osher scheme.

Suppose that in this one-dimensional case we have a (possibly recovered)
approximation $\widetilde{U}^n(x)$ at time t^n, and that $f'(u) = a(u)$. Let

$$y = x + a(\widetilde{U}^n(x))\Delta t$$

denote the end point of the characteristic drawn from (x, t^n) through the time step Δt. Then it was shown in Lin, Morton and Süli (1993) that the transport collapse operator can be interpreted in terms of a Riemann–Stieltjes integral along the graph $[\widetilde{U}^n, y]$, and that it is thus equivalent to a characteristic-Galerkin method. Suppose the piecewise constant basis function on the mesh of Figure 1.1(b) is denoted by

$$\chi_j(x) = H_{j+1/2}(x) - H_{j-1/2}(x)$$

in terms of two Heaviside functions. Then an update algorithm that may include a recovery step can be written in the form

$$\Delta x_j(U_j^{n+1} - U_j^n) = \int \widetilde{U}^n(x)[\chi_j(y)\,\mathrm{d}y - \chi_j(x)\,\mathrm{d}x] \tag{4.1}$$

$$= -\int \left[\int_x^{y(x)} \chi(s)\,\mathrm{d}s\right] \mathrm{d}\widetilde{U}^n.$$

To interpret this as a finite volume method we have to make use of the relationship $a\,\mathrm{d}u = \mathrm{d}f$ and carry out the integral on the right to give a difference of flux functions: this has to be done with care when crossing a sonic point, for which $a(u) = 0$. Several examples can be found in Morton (2001), including the Engquist–Osher case in which there is no recovery stage.

Even more interesting is the two-dimensional case on a rectangular mesh. It is shown in Lin, Morton and Süli (1997) that, in addition to the one-dimensional flux differences along the sides of a mesh box arising from integrals of $a_1\mathrm{d}u$ and $a_2\mathrm{d}u$, there are also corner terms arising from integrals of $a_1(u)a_2(u)\mathrm{d}u$. With no recovery, on a uniform mesh and with the CFL conditions $0 \le a_1(U^n)\Delta t \le \Delta x_1$, $0 \le a_2(U^n)\Delta t \le \Delta x_2$ satisfied, the difference scheme that results from this method has the form

$$\frac{U_{i,j}^{n+1} - U_{i,j}^n}{\Delta t} + \frac{\Delta_{-x_1} f_1(U_{i,j}^n)}{\Delta x_1} + \frac{\Delta_{-x_2} f_2(U_{i,j}^n)}{\Delta x_2} - \Delta t \frac{\Delta_{-x_1}\Delta_{-x_2} f_{12}(U_{i,j}^n)}{\Delta x_1 \Delta x_2} = 0, \tag{4.2}$$

where the corner flux is given by

$$f_{12}(u) := \int^u a_1(v)a_2(v)\,\mathrm{d}v,$$

and Δ_{-x_ℓ} is the backward difference operator in the x_ℓ direction. The scheme is stable under the given conditions, but without the extra corner term the stability limit would be given by $a_1(U^n)\Delta t/\Delta x_1 + a_2(U^n)\Delta t/\Delta x_2 \le 1$.

A key point needs to be made about these formulae, which is particularly important in the two-dimensional case. They resulted from the development of unconditionally stable methods for hyperbolic problems in which shocks were not the key phenomena: thus they were not generally put in their finite volume form and characteristics were tracked even into non-neighbouring cells before the projection was made to obtain the updated approximation. Atmospheric flows are a typical application area: see, *e.g.*, Staniforth and Côté (1991). Thus the significance of (4.2) is that the stable region in (x_1, x_2)-space is a mesh rectangle and other formulae are obtained for neighbouring rectangles so that the whole plane is covered to give an unconditionally stable scheme. This is clearly not so relevant to the general triangular and quadrilateral meshes that we are concentrating on. But what is relevant to note is that the corner terms represent an $\mathcal{O}(\Delta t)$ correction to the flux terms obtained along the edges: to omit them, as is normal with general finite volume methods, commits an error as well as limiting the stability: see LeVeque (2002) for a wider discussion on the importance of corner terms.

Formulae obtained from the transport collapse operator also provide valuable guidance on the modifications to the flux functions that arise from a recovery stage. For example, suppose the discontinuous linear recovery (3.6) is used in one dimension. Then substitution into (4.1) leads to a flux function of the same form as the Engquist–Osher flux (1.7) but with different terms: the quantities involved are obtained from the values of the recovered approximation just to the left and to the right of the interface so as to give

$$\tilde{F}_{j+1/2}^{n+1/2} = \tfrac{1}{2}\big[(1 + s_j^+)\tilde{F}_j^+ + (s_{j+1}^- - s_j^+)f(u_s) + (1 - s_{j+1}^-)\tilde{F}_{j+1}^-\big]. \qquad (4.3)$$

Here we will use the notation $\tilde{U}_j^\pm$ for the value of the recovered variable at the right and left of cell j, and $s_j^\pm$ for the signs of the corresponding characteristic speeds $a(\tilde{U}_j^\pm)$. It remains to define the two flux values. We consider only the simplest case, when the characteristic speeds are positive throughout cell j, and we need to calculate $\tilde{F}_j^+$ from all the right-moving characteristics which reach the interface from the cell in one time step. For this purpose we assume that the characteristic speed also varies linearly throughout the cell, with a slope M_j^+. Then the speed at the point which is just carried to the interface at the end of the time step is given by

$$A_j^{*+} = a(\tilde{U}_j^+)/[1 + M_j^+\Delta t]; \qquad (4.4)$$

and a short calculation gives the following flux value:

$$\tilde{F}_j^+ = \frac{A_j^{*+}f(\tilde{U}_j^+) + a(\tilde{U}_j^+)f(\tilde{U}_j^+ - S_jA_j^{*+}\Delta t)}{A_j^{*+} + a(\tilde{U}_j^+)}. \qquad (4.5)$$

This average of two flux values gives a scheme which is second-order accurate in smooth flow regions. In Morton (2001) it is shown to be TV-stable, under CFL conditions which are principally of the standard form that characteristics cross no more than one cell in one time step, and where the recovery stage satisfies only the necessary monotonicity-preserving condition (3.8).

The same framework can be used to derive a third-order accurate method by means of a continuous piecewise parabolic recovery process similar to that introduced in the *Piecewise Parabolic Method* (PPM) of Colella and Woodward (1984). In the recovery process the key step is to deduce a value $\widetilde{U}_{j+1/2}$ at each interface, the parabola in each cell then following from the requirement that the cell average is preserved. One can then show that if the recovery stage is TVD then the scheme is TV-stable under our familiar CFL conditions: see Morton (2001).

The real challenge is to carry these ideas forward to systems of conservation laws, and in particular to the Euler equations. A breakthrough was achieved by Roe (1981) by using a local linearization that ensures that, from two states $\mathbf{u}_L, \mathbf{u}_R$ and a one-dimensional flux vector $\mathbf{f}(\mathbf{u})$, a matrix $\bar{A}(\mathbf{u}_L, \mathbf{u}_R)$ is constructed so as to satisfy

$$\bar{A}(\mathbf{u}_L, \mathbf{u}_R)(\mathbf{u}_R - \mathbf{u}_L) = \mathbf{f}(\mathbf{u}_R) - \mathbf{f}(\mathbf{u}_L). \qquad (4.6)$$

The advantage of using any local linearization is that the interfacial flux can be computed straightforwardly from the waves corresponding to the eigenvalues and eigenvectors of the matrix $\bar{A}$. If these are given in the standard form $\bar{A} = R\Lambda R^{-1}$ and we define the *absolute value* of $\bar{A}$ by $|\bar{A}| = R|\Lambda|R^{-1}$, then we can write this flux very concisely as

$$\bar{F}(\mathbf{u}_L, \mathbf{u}_R) = \tfrac{1}{2}[\mathbf{f}(\mathbf{u}_R) + \mathbf{f}(\mathbf{u}_L)] - \tfrac{1}{2}|\bar{A}|[\mathbf{u}_R - \mathbf{u}_L], \qquad (4.7)$$

although this is not the way in which it is usually coded. The advantage of a linearization satisfying (4.6) is that Riemann problems whose solution is a simple discontinuity (a shock or a contact) are solved exactly, because the Rankine-Hugoniot conditions are satisfied with the shock speed given by an eigenvalue of $\bar{A}$.

The *Roe matrix* is constructed for the Euler equations by observing that both $\mathbf{u}$ and $\mathbf{f}$ can be expressed as quadratic functions of a new variable $\mathbf{z}$ given by $\rho^{1/2}(1, v, H)^T$. Then one can exploit the identity

$$2(a_1 b_1 - a_2 b_2) \equiv (a_1 + a_2)(b_1 - b_2) + (b_1 + b_2)(a_1 - a_2)$$

to introduce matrices $\bar{B}, \bar{C}$ such that

$$\mathbf{u}_R - \mathbf{u}_L = \bar{B}(\mathbf{z}_R - \mathbf{z}_L) \quad \text{and} \quad \mathbf{f}(\mathbf{u}_R) - \mathbf{f}(\mathbf{u}_L) = \bar{C}(\mathbf{z}_R - \mathbf{z}_L),$$

from which one can define $\bar{A} = \bar{C}\bar{B}^{-1}$ to satisfy (4.6); the detailed form of these matrices can be found in texts such as Hirsch (1990). For obvious reasons such methods are called *flux difference splitting methods* and are computationally quite expensive; alternatively, numerical flux functions can be computed by *flux vector splitting methods* such as that due to Steger and Warming (1981) already referred to.

The Roe matrix gives a direct generalization of the upwind scalar scheme (1.6) and has similar disadvantages in its emphasis on capturing shocks.

In particular, convergence would not necessarily be to an entropy-satisfying solution of the PDE system. Thus, when such a flux function is used it is modified by some form of *entropy fix*. Alternatively, one may seek to approximate rarefaction waves as in the method proposed in Osher and Solomon (1982) which generalizes the Engquist–Osher flux of (1.7). For a comprehensive review of schemes that generalize upwind differencing to systems of conservation laws, described in the context of the underlying theory, see Harten, Lax and van Leer (1983). The standard CFD text Hirsch (1990) also describes many widely used schemes. There are many desirable properties that numerical flux functions should have, such as ensuring that the density and pressure are always non-negative, and a discussion of such issues can be found in the recent survey by Roe (2001).

4.2. Evolution-Galerkin methods and error analysis

A useful general framework for considering the approximation of an evolutionary problem by finite difference, finite element or finite volume methods has the following form. Suppose that in a given function space V the operator $\mathcal{E}_\Delta$ represents an approximation to the true evolution operator through a time step Δt; let U^n be an approximation in some discrete subspace S^h of V to the exact solution $u(\cdot, t^n)$ at time t^n, and $\mathcal{P}$ a projection from V to that discrete space; finally, let $\mathcal{R}$ be a recovery operator giving $\tilde{U}^n = \mathcal{R}U^n$ as a recovered approximation in some larger discrete subspace of V. Then one step of an evolution-Galerkin method can be written in the alternative forms

$$U^{n+1} = \mathcal{P}\mathcal{E}_\Delta \mathcal{R}U^n \quad \text{or} \quad \tilde{U}^{n+1} = \mathcal{R}\mathcal{P}\mathcal{E}_\Delta \tilde{U}^n. \tag{4.8}$$

All of the methods we have described can be put into this form, although we have not specified the defining operators. We are interested in the error between the true solution and the recovered approximation, which we estimate by decomposing it into two parts through the introduction of a *target approximation* $u^n \in S^h$: we call $\eta^n = u(\cdot, t^n) - \mathcal{R}u^n$ the *projection error* and $\xi^n = \mathcal{R}u^n - \tilde{U}^n$ the *evolutionary error*, so that we have

$$u(\cdot, t^n) - \tilde{U}^n = [u(\cdot, t^n) - \mathcal{R}u^n] + [\mathcal{R}u^n - \tilde{U}^n] \tag{4.9}$$
$$=: \eta^n + \xi^n.$$

An appropriate choice of the target approximation can be important when comparing differing types of method on non-uniform meshes.

In order to estimate the evolutionary error, we make use of (4.8) to write

$$\xi^{n+1} \equiv \mathcal{R}u^{n+1} - \tilde{U}^{n+1} = [\mathcal{R}u^{n+1} - \mathcal{R}\mathcal{P}\mathcal{E}_\Delta \mathcal{R}u^n] + [\mathcal{R}\mathcal{P}\mathcal{E}_\Delta \mathcal{R}u^n - \mathcal{R}\mathcal{P}\mathcal{E}_\Delta \tilde{U}^n], \tag{4.10}$$

and define a *truncation error* as

$$\tilde{T}^n := (\Delta t)^{-1}(\mathcal{R}u^{n+1} - \mathcal{R}\mathcal{P}\mathcal{E}_\Delta \mathcal{R}u^n). \tag{4.11}$$

Now suppose that the method is strongly stable in the sense that

$$\|\mathcal{RPE}_\Delta \tilde{U} - \mathcal{RPE}_\Delta \tilde{V}\| \le \|\tilde{U} - \tilde{V}\|. \tag{4.12}$$

Then we have the familiar dependence of the evolutionary error on the truncation error,

$$\|\xi^{n+1}\| \le \|\xi^n\| + \|\widetilde{T}^n\|\Delta t. \tag{4.13}$$

For a well-posed problem one should expect that $\mathcal{PE}_\Delta$ will be strongly stable, so the stability assumption here is mainly a constraint on the recovery process.

The linear advection equation $u_t + au_x = 0$, with a a positive constant, is the most illuminating first test case for any proposed approximation scheme for hyperbolic problems. So it is here. Suppose that on a non-uniform mesh the upwind Roe scheme (1.3) with (1.6) is applied, using the piecewise constant approximation $U^n \equiv \{U_j^n\}$. Then, if the target approximation is based on cell averages, the truncation error will depend on the ratio of the distance between two cell centres and the length of one of the cells, so the scheme would be deemed inconsistent with the PDE, as has been observed by many researchers. On the other hand, suppose we take the target approximation $u^n \equiv \{u_j^n\}$ such that u_j^n is the value of the true solution at the upwind end of the cell, namely $u(x_{j+1/2}, t^n)$, which still gives a first-order projection error. Now the truncation error (4.11) with no recovery has the familiar form

$$T_j^n = \frac{u(x_{j+1/2}, t^{n+1}) - u(x_{j+1/2}, t^n)}{\Delta t} + \frac{a[(u(x_{j+1/2}, t^n) - u(x_{j-1/2}, t^n)]}{x_{j+1/2} - x_{j-1/2}}$$

and is clearly of first order.

In Morton (1998) this analysis is continued to show that discontinuous linear recovery gives a second-order error if the grading of the mesh and the change in solution slope are smooth; and it is also shown that continuous quadratic recovery as in the PPM scheme gives third-order accuracy under similar mesh restrictions. The target approximations in these cases are formed by truncating the Taylor expansion of the cell average of the true solution about the upwind end of the cell.

The linear advection equation also provides a convenient framework for considering both the order of accuracy best aimed for, and whether that should be attained from choice of the order of accuracy of the basic approximation U^n, or from the recovery process giving $\tilde{U}^n$. Moreover, approximating this equation on a uniform mesh means that the powerful tool of Fourier analysis is available. This shows that schemes with an even order of accuracy propagate waves with an error dominated by dispersion; while schemes with an odd order have waves dominated by dissipation. For example, the best-known second-order scheme is the Lax–Wendroff method, which notoriously suffers from a trail of oscillations when used to approximate the advection of a discontinuity; while the first-order upwind scheme suffers from severe damping for the same problem, but has surprisingly small dispersion errors; see Morton and Mayers (2005) for illustrations of these phenomena and Morton (1998) for a more general discussion.

Thus, for wave propagation problems, such as those arising in meteorology or oceanography, third-order accuracy has been advocated by many authors, such as Leonard (1991). Moreover, use of the *characteristic-Galerkin method* with the standard continuous piecewise linear finite element basis yields such a scheme (see Lesaint (1977) and Douglas and Russell (1982)); and such methods have been widely used in finite element approximations of the incompressible Navier–Stokes equations (see Pironneau (1982)). Thus, suppose we approximate the convection-dominated diffusion problem $u_t + au_x = \epsilon u_{xx}$ on a uniform mesh, with linear finite elements, in the following way. From each mesh point at the time level t^n we draw the characteristic forward to the time level t^{n+1}, and suppose that

$$\nu := a\Delta t/\Delta x = m + \hat{\nu} \quad \text{with} \quad m \in \mathbb{N}, \ 0 < \hat{\nu} \leq 1.$$

Then the projection of the approximate evolution operator, denoted by $\mathcal{PE}_\Delta$ above, is defined by carrying values along these characteristics from one time level to the next, where the diffusion is applied, and using the Galerkin projection to determine the new piecewise linear approximation. In finite difference notation, with $\mu = \epsilon\Delta t/(\Delta x)^2$, we obtain the scheme

$$[1 + (\tfrac{1}{6} - \mu)\delta^2]U_j^{n+1} = [(1 + \tfrac{1}{6}\delta^2) - \hat{\nu}\Delta_0 \qquad (4.14)$$
$$+ \tfrac{1}{2}\hat{\nu}^2\delta^2 - \tfrac{1}{6}\hat{\nu}^3\delta^3\Delta_-]U_{j-m}^n;$$

here Δ_-, Δ_0, δ^2 are the first-order backward, the first-order central and the second-order central differences, all undivided. This scheme is unconditionally stable and third-order accurate in the convection terms; and of course it is readily generalized to more space dimensions: indeed, with a triangular or quadrilateral mesh, though it would not then be expressed in difference form. It thus provides an extremely valuable yardstick against which to measure alternative schemes.

Now the piecewise constant basis function is a first-order B-spline, and on a uniform mesh higher-order B-splines are generated by a recurrence relation:

$$\chi^{(p)}(s) := \int \chi^{(p-1)}(\sigma - s)\chi(\sigma)\,d\sigma, \qquad (4.15)$$

where $\chi(\sigma) \equiv \chi^{(1)}(\sigma)$ is defined as the characteristic function of the interval $[-\tfrac{1}{2}, \tfrac{1}{2}]$; and the linear basis functions are second-order B-splines. Moreover, differentiating a spline of a given order generates splines of a lower order; and integrating the product of two splines generates higher-order splines. Thus all the terms in (4.14) could be generated from other than linear basis functions: in particular, the convection terms giving the third-order accurate scheme could be generated from using a piecewise constant approximation U^n that is recovered by quadratic splines to give $\widetilde{U}^n$.

To make this last statement more precise, we are presuming that a non-adaptive recovery process is used that defines the quadratic spline recovered approximation $\widetilde{U}^n$ by maintaining the cell averages, that is, by specifying its inner products against first-order splines; and it is these inner products exactly equalling inner products between two linear splines that leads to the equivalence of the two formulations. So in this case there is no loss of accuracy in using a piecewise constant basic approximation followed by recovery with a high-order spline, compared with using higher-order basis functions for U^n and no recovery; and this is a general conclusion. Moreover, with the former approach, which is the basis of many of our finite volume methods, we have the opportunity to make the recovery stage adaptive so as to maintain key properties of the solution, as we have already described. A fuller discussion of these points and some numerical illustrations can be found in Morton (1996).

4.3. Finite volume evolution-Galerkin methods

The linear advection equation is a reasonable model problem for devising and analysing algorithms used to solve hyperbolic problems that are dominated by a single velocity field. But it is inadequate for the equations of unsteady gas dynamics. For these the two-dimensional wave equation system is more appropriate: we write it as

$$\phi_t + c(u_x + v_y) = 0, \tag{4.16}$$
$$u_t + c\phi_x = 0, \qquad v_t + c\phi_y = 0,$$

where ϕ can be regarded as a pressure and u, v as the velocity components. The classical Kirchhoff solution of the wave equation, written as a single second-order equation, is in the form of an integral over the base of a characteristic cone with its apex at the sample point: the solution at the apex is given in terms of the data over the base. However, Butler (1960) developed an alternative form for the system (4.16), which he used to good effect in approximating the Euler equations. In his form the data on the perimeter of the base is used, but he also uses an integral over the mantle of the cone, that is, involving the solution at intermediate times. Unfortunately, his method did not make use of the data in a very consistent way and it was quickly superseded by the method of Lax and Wendroff (1960); but with the help of finite element and finite volume formulations it can be used as the basis of powerful methods.

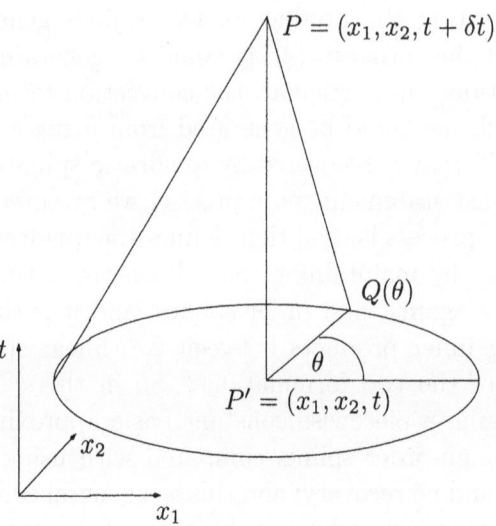

Figure 4.1. Characteristic cone for the wave equation in 2D.

Suppose one integrates a $(1, -\cos\theta, -\sin\theta)$ linear combination of the equations (4.16) along each bicharacteristic that generates the characteristic cone, centred at $P \equiv (\mathbf{x}, t + \delta t)$, and averages the result over θ; then one obtains an integral equation for ϕ that, in the notation of Figure 4.1, has the following form:

$$\phi_P = \frac{1}{2\pi} \int_0^{2\pi} [\phi_Q - u_Q \cos\theta - v_Q \sin\theta] \, \mathrm{d}\theta \qquad (4.17)$$

$$- \frac{1}{2\pi} \int_t^{t+\delta t} \int_0^{2\pi} S(t', \theta) \, \mathrm{d}\theta \, \mathrm{d}t',$$

where

$$S(t', \theta) = c[u_x \sin^2\theta - (u_y + v_x) \sin\theta \cos\theta + v_y \cos^2\theta] \qquad (4.18)$$

and (u, v) are evaluated at Q' at the intermediate time t'. Similar equations can be derived for (u_P, v_P). If the integral over t' is approximated by the rectangle rule at time level t, or the trapezoidal rule, one obtains an approximate evolution operator over a full time step by taking $\delta t = \Delta t$. This was used as the basis of various evolution Galerkin methods on a square mesh in Lukáčová-Medviďová, Morton and Warnecke (2000). The advantage of such methods was seen to lie in the possibility that, by using all characteristic directions, they would propagate wave fronts which were not distorted badly by the mesh orientation. With exact integrals over piecewise constant approximations, this hope was realized; but such methods could only be first-order accurate and did not have a good stability range.

To do better, four steps are necessary: recovery by discontinuous bilinear approximations on the square mesh, use of a more general evolution operator derived by Ostkamp (1997), adoption of a finite volume framework and better approximation of the mantle integrals. All of these steps are described in Lukáčová-Medviďová, Morton and Warnecke (2004) and were applied to the Euler equations in Lukáčová-Medviďová, Morton and Warnecke (2002). The finite volume form (2.11) was used, with the conservation form of the equations, which has the advantage that the evolution operator is needed only to evaluate the solution on the perimeter of the control volume; and in the case of the Euler equations the equations for the primitive variables were used for this purpose. To obtain a second-order accurate method one need use only the mid-point rule for the time integration, so the approximate evolution operator is needed only at time $t + \frac{1}{2}\Delta t$ and at the quadrature points used for the integral around the control volume perimeter.

To derive the general evolution operator for a hyperbolic system such as (2.2), we suppose that the linear combination $A(\boldsymbol{\nu})$ of Jacobian matrices in the direction $\boldsymbol{\nu}$, given by (2.3), has the matrix of right column eigenvectors $R(\boldsymbol{\nu})$. Then, if we apply the corresponding transformation to each individual Jacobian matrix, in general they will not all be diagonalized: writing D_ℓ for the diagonal part and B'_ℓ for the remainder, we have

$$R^{-1}A_\ell R = D_\ell + B'_\ell.$$

We also introduce the corresponding characteristic variables $\mathbf{w} \equiv \mathbf{w}(\boldsymbol{\nu})$ given by $\partial \mathbf{w} = R^{-1}\partial \mathbf{u}$. Then, operating on the differential equation with R^{-1} from the left, we get

$$\partial_t \mathbf{w} + \sum_{\ell=1}^{d} D_\ell \partial_{x_\ell} \mathbf{w} = -\sum_{\ell=1}^{d} B'_\ell \partial_{x_\ell} \mathbf{w} =: \mathbf{S}. \tag{4.19}$$

It is this equation that is integrated along the bicharacteristic corresponding to the direction $\boldsymbol{\nu}$ and the 'source' term on the right that leads to the mantle integral when the result is averaged over all directions.

For the wave equation the resulting formula for ϕ is the same as that given by Butler but that for the velocity components is different. Thus we have the following formulae, after an integration by parts in the mantle integrals over θ to remove the derivatives on the dependent variables:

$$\phi_P = \frac{1}{2\pi} \int_0^{2\pi} [\phi_Q - u_Q \cos\theta - v_Q \sin\theta] \, d\theta \tag{4.20}$$

$$-\frac{1}{2\pi} \int_0^{\delta t} \frac{1}{\tau} \int_0^{2\pi} [u_{Q'} \cos\theta + v_{Q'} \sin\theta] \, d\theta \, d\tau,$$

and

$$u_P = \frac{1}{2\pi} \int_0^{2\pi} [-\phi_Q \cos\theta + u_Q \cos^2\theta + v_Q \sin\theta \cos\theta]\, d\theta \qquad (4.21)$$

$$+ \frac{1}{2\pi} \int_0^{\delta t} \frac{1}{\tau} \int_0^{2\pi} [u_{Q'} \cos 2\theta + v_{Q'} \sin 2\theta]\, d\theta\, d\tau$$

$$+ \tfrac{1}{2} u_{P^0} - \tfrac{1}{2} c \int_0^{\delta t} \partial_{x_1} \phi_{P'}\, d\tau,$$

with a similar formula for v_P, where P^0 is the centre of the cone base. Note that in the scheme given below we take $\delta t = \frac{1}{2}\Delta t$.

The integrals at the old time level are carried out exactly for either the unrecovered piecewise constant approximation $\mathbf{U}^n$, or its discontinuous bilinear recovered counterpart $\widetilde{\mathbf{U}}^n$. It is the approximation of the mantle integrals, that involve values of the solution at intermediate times, that are crucial to both the accuracy and stability of methods based on these formulae. Fortunately, there is one case when these integrals can be evaluated exactly in terms of the known data: that is, when that data represents waves that are one-dimensional so that we can use the familiar d'Alembert formula. Lukáčová-Medviďová *et al.* (2004) used this, both for piecewise constant and continuous piecewise linear data, to relate the mantle integrals to those round the perimeter of the cone base. Substituting the result

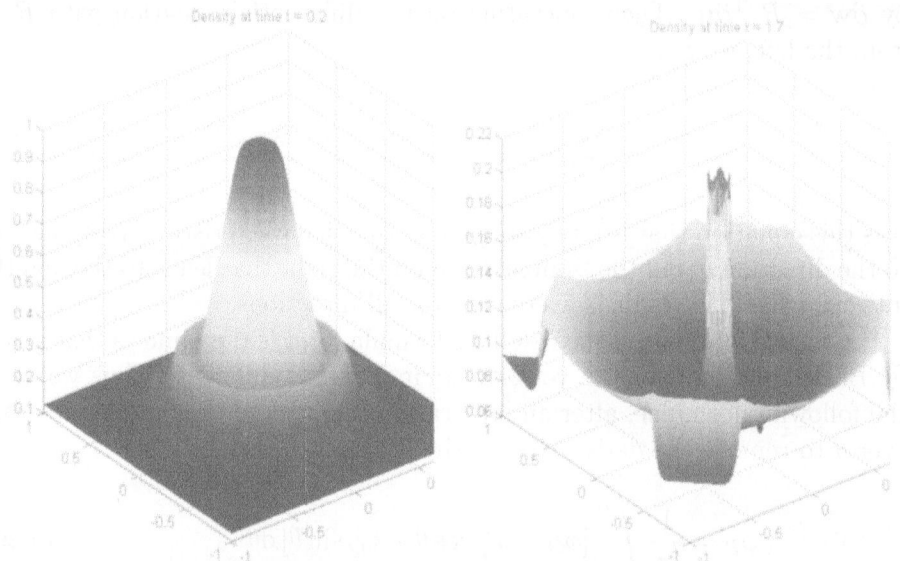

Figure 4.2. Solution to the 2D Sod test case.
Density at time $t = 0.2$ (*left*) and at $t = 1.7$ (*right*).

into the finite volume framework on a rectangular mesh, gives a scheme for updating the cell averages which corresponds to (2.11) and takes the form

$$|R_i|(\mathbf{U}_i^{n+1} - \mathbf{U}_i^n) + \Delta t \oint_{\partial R_i} \mathcal{F}(\mathcal{E}_\delta \mathcal{R} \mathbf{U}^n) \cdot \mathbf{n} \, ds = \mathbf{0}, \qquad (4.22)$$

where R_i is a mesh rectangle and $\mathcal{E}_\delta$ is the approximate evolution operator over half a time step as just described, and $\mathcal{R}$ represents the recovery operator. Using Simpson's rule for approximating the integrals around the cell perimeter, the resulting methods are second-order accurate, being some five times more accurate than the comparable Lax–Wendroff method, and have good stability properties. More importantly, when applied to the Sod-2D test problem involving cylindrically symmetric wave propagation and reflection, it preserves the symmetry very precisely: see Figure 4.2 and the results in Lukáčová-Medviďová et al. (2004).

4.4. Semi-discrete explicit time-stepping algorithms

The traditional discretization of hyperbolic equations was based on methods using finite differences in space and time, with explicit time differencing. They are simple to implement and the CFL stability limit on the time step is commonly consistent with the requirements of accuracy. However, we have already seen several situations when these arguments break down: when a steady state is sought, or the flow rate of change is much slower than important characteristic speeds; and at sharp corners in the mesh where the ideal scheme corresponding to (4.2) cannot be used and the CFL limit is drastically reduced. Moreover, when grid adaptivity is introduced in Section 6 a uniform explicit time step may be quite inappropriate. In such situations it may be preferable to consider the space discretization quite separately from that in time, not attempting in any way to have the fluxes represent averages over a time step, as is implied by the notation introduced for the Godunov method in (1.3) and which led naturally to the complications faced by the FVEG methods described in the previous subsection; instead, we seek methods to solve large systems of ODEs such as that given by (3.4).

One-step methods, such as Runge–Kutta schemes, are clearly attractive for this purpose and their use goes back to the very influential paper of Jameson, Schmidt and Turkel (1981). The behaviour of the methods, and hence the selection of the most appropriate, can be studied by considering model hyperbolic systems and applying Fourier analysis in the spatial variables. Then, in a standard stability region plot, it is clear that it is the behaviour along the imaginary axis and just to its left that is most important. So in that paper Jameson et al. used the standard explicit fourth-order Runge–Kutta scheme which is stable for a CFL number up to $2\sqrt{2}$ when upwind differencing is applied to the linear advection equation, and allows for a reasonable amount of damping to be added.

However, Shu and Osher (1988) observed that this scheme does not preserve monotonicity properties that may have been built into the spatial discretization. They therefore introduced special TVD-Runge–Kutta schemes that preserve the properties of ENO-type schemes at the cost of reduced stability ranges. If we write (3.4) as

$$\frac{\mathrm{d}}{\mathrm{d}t}\mathbf{U}_i(t) = -\mathcal{N}_i(\mathbf{U}(t)),$$

and temporarily suppress the subscripts, then a typical third-order scheme has the form

$$\mathbf{U}^{(0)} := \mathbf{U}(t), \qquad (4.23)$$
$$\mathbf{U}^{(1)} = \mathbf{U}^{(0)} - \Delta t \mathcal{N}(\mathbf{U}^{(0)}),$$
$$\mathbf{U}^{(2)} = \mathbf{U}^{(0)} - \tfrac{1}{4}\Delta t \mathcal{N}(\mathbf{U}^{(0)}) - \tfrac{1}{4}\Delta t \mathcal{N}(\mathbf{U}^{(1)}),$$
$$\mathbf{U}^{(3)} = \mathbf{U}^{(0)} - \tfrac{1}{6}\Delta t \mathcal{N}(\mathbf{U}^{(0)}) - \tfrac{1}{6}\Delta t \mathcal{N}(\mathbf{U}^{(1)}) - \tfrac{2}{3}\Delta t \mathcal{N}(\mathbf{U}^{(2)}),$$
$$\mathbf{U}(t + \Delta t) := \mathbf{U}^{(3)}.$$

But then such a scheme has a CFL limit reduced to unity for the above model problem.

4.5. Implicit time-stepping

Considerations such as those outlined in the previous subsection lead to the conclusion that implicit time-differencing is bound to play a larger role in future methods, either in the semi-discrete formulation introduced there or in a fully discrete formulation. This is despite the difficulties posed by the resulting large systems of highly nonlinear equations that such methods will lead to. There are two major hurdles to overcome: formulation of the equations to ensure convergence of the Newton or quasi-Newton iterations that are needed; and rapid solution of the linear equations at each iteration. Hyperbolicity of the equations being approximated ensures the underlying Jacobians are well behaved with finite eigenvalues and a full set of eigenvectors. So, if care is taken to reflect properly the properties of the differential equations in their discretization, attention can often be concentrated on the solution of the linear equation systems.

A natural starting point for considering these issues would seem to be provided by the Preissmann box scheme applied to the St. Venant equations for one-dimensional river flow, where solution of the global Newton system is the standard procedure for subcritical flows. However, it has long been recognized that this formulation runs into difficulties when flows develop a supercritical section. Thus we will start even more simply with the scalar problem of the inviscid Burgers equation and initial data that leads to a shock; and we consider not only the box scheme, as the simplest cell-vertex method, but also later an implicit cell-centre or node-centred scheme.

Suppose that, on the interval $0 \leq x \leq 1$, initial data $u^0(x)$ are given with $u^0(0) = u_L > 0$ and $u^0(1) = u_R < 0$. Then boundary conditions need to be imposed at both these boundaries in order to solve the problem for $t > 0$; and if these values continue to be imposed, eventually a shock will form. Now divide the interval into J cells with the points $x_0 = 0, x_1, \ldots, x_J = 1$, and suppose first that we approximate the problem with the box scheme (1.2). As there are J cells, there will be J box equations and two boundary conditions to be satisfied by the $J + 1$ unknown nodal values; clearly, something has to be sacrificed. To clarify the choice, suppose that u^0 has the constant value u_L to the left of some point $x = x_S \in (x_k, x_{k+1})$ and u_R to the right: then, in carrying out the first time step, we can set $U_0^1 = u_L$ and work from left to right using each cell residual equation to calculate successively $U_1^1, U_2^1, \ldots, U_k^1$ until the shock is reached; and we can do the same from the right for $U_J^1, U_{J-1}^1, \ldots, U_{k+1}^1$. Thus all the nodal values at the new time level could be calculated in this way without having to use the box equation for the cell containing the shock. But this would violate the basic conservation law for u and is unacceptable; and, in fact, the same problem would occur if we were to discretize the initial data.

Apart from the boundary conditions, this overall conservation property is the most important consideration. We satisfy it by means of a general algorithm which we will refer to as a *residual distribution scheme*: these ideas were originally put forward in Roe (1982) and Deconinck *et al.* (1993) in the context of explicit time-stepping algorithms, and in Crumpton *et al.* (1993) to derive iteration procedures for steady flow problems; but they are equally applicable to the tasks of discretizing the initial data or setting up implicit time-stepping equations. For a general scalar one-dimensional problem, we suppose that we have for each cell a residual $R_{j+1/2}$ and an average or representative characteristic speed $a_{j+1/2}$. Then we execute the following two steps:

- for each cell, allocate the residual $R_{j+1/2}$ to node j if $a_{j+1/2} \leq 0$ or to node $j + 1$ if $a_{j+1/2} > 0$;
- for each node, set up the appropriate nodal equation using the sum of the residuals that have been allocated to it.

For our present Burgers' equation problem, let us first consider the discretization of the initial data containing a shock by a continuous piecewise linear approximation satisfying the two boundary conditions. The integral of $u^0(x)$ over each cell gives the cell residual; and in the situation described above, we would clearly have average characteristic speeds that were positive (*i.e.*, supercritical) for the cells to the left of x_k, and negative (*i.e.*, subcritical) for the cells to the right of x_{k+1}. The only issue is with the shock cell: should its residual be combined with that from the left or the right? Let us suppose the latter. Then, by setting the residuals to zero, we would set $U_j^0 = u_L$ for $j = 0, 1, \ldots, k$ and $U_j^0 = u_R$ for $j = J, J-1, \ldots, k+2$; and from the combined residual from the cells either side of $k+1$, we obtain the following equation for U_{k+1}^0:

$$\tfrac{1}{2}(u_L + U_{k+1}^0)(x_{k+1} - x_k) + \tfrac{1}{2}(U_{k+1}^0 + u_R)(x_{k+2} - x_{k+1})$$
$$= u_L(x_S - x_k) + u_R(x_{k+2} - x_S).$$

This gives

$$U_{k+1}^0 = \frac{(2x_S - x_k - x_{k+1})u_L + (x_{k+1} + x_{k+2} - 2x_S)}{x_{k+2} - x_k}.$$

Now it is clearly important that U_{k+1}^0 should lie between u_L and u_R, a condition required to maintain the TVD property when setting up an evolution step; and this requires that

$$\tfrac{1}{2}(x_k + x_{k+1}) \leq x_S \leq \tfrac{1}{2}(x_{k+1} + x_{k+2}).$$

The first inequality implies that the shock cell residual should be combined with that from the cell to the right when the shock is closest to the boundary with that cell, at x_{k+1}; in other words, when the shock cell is dominated by u_L values and therefore considered to be supercritical. Correspondingly, the second inequality implies that these two cell residuals should be combined when the shock is in the cell (x_{k+1}, x_{k+2}) and closest to the left-hand boundary, so that it is subcritical.

These later interpretations are the key to determining how to carry out an evolution step both for this problem and more generally. For each cell we calculate the box scheme residual and allocate it to the node to its left or right according to whether it is regarded as being a subcritical or supercritical cell: and this is determined by an average characteristic speed calculated from the current solution approximation. Thus in the neighbourhood of a shock we always combine the residual of a cell which is deemed to be subcritical with one deemed supercritical.

There is also the complementary situation to consider in which the initial data, or current solution, is subcritical on the left and supercritical on the right, so that there is a *sonic point* or *critical point* at some point of the interval. Then, for example with $u_L < 0$ and $u_R > 0$, boundary conditions are not imposed on the left or the right so that now there are too few equations provided by the residuals to determine all the unknowns. The solution is to split the residual for the cell containing the sonic point at that point: for the Burgers' equation problem, either to discretize the initial data or to evolve the solution, values are obtained successively by working out from the sonic point to the boundaries. Incidentally, it is worth noting that this device corresponds to how the Engquist–Osher scheme of (1.7) breaks up the flux differences near a sonic point.

In Freitag and Morton (2007) these two techniques were applied to extend the Preissmann box scheme to solve St. Venant equation problems with a supercritical section. The mass residual was that chosen to be either split at a sonic point or combined at a shock; and the resultant system of equations was shown to be well behaved when solved by Newton's method combined with the standard Thomas algorithm applied to the resultant block tridiagonal system of linear equations.

However, these are simple problems in only one space dimension. For the compressible flow equations in two dimensions, residuals for cells crossed by shocks are in general poorly computed by the trapezoidal rule applied to the cell faces; and this can trigger violent chequer-board and washboard oscillations which are normally damped by artificial viscosity terms, of both second-order and fourth-order type. The distribution matrices, introduced in (3.13) and (3.14), to match the cell residuals with the unknowns, are also difficult to define. So, although an appeal to feedback control techniques as in Morton and Stringer (1998) offers a way forward with both these difficulties, more development is still needed for the application of cell-vertex methods to these problems.

We turn instead to cell-centre and node-centred methods, where the equations and unknowns have a more natural matching with the number of equations always equal to the number of unknowns. The node-centred method is easiest to apply to the Burgers' equation problem and the mesh described above: there are $J + 1$ unknowns corresponding to a piecewise constant representation in which the discontinuities occur at the J cell mid-points; and, for the shock problem which had ingoing characteristics at both boundaries, the end values are given by the boundary conditions, while for $j = 1, 2, \ldots, J - 1$ a typical equation will be of the Crank–Nicolson form

$$(x_{j+1/2} - x_{j-1/2})[U_j^{n+1} - U_j^n]$$
$$+ \tfrac{1}{2}(t^{n+1} - t^n)[F_{j+1/2}^{n+1} + F_{j+1/2}^n - F_{j-1/2}^{n+1} - F_{j-1/2}^n] = 0,$$

where the fluxes have to be obtained from an approximate Riemann solver, and $x_{j+1/2}$ is a cell mid-point. For a problem with an outgoing characteristic at a boundary, on the other hand, a corresponding equation is constructed over a boundary half-cell.

A Newton solver needs to be applied to this system of equations, which has some implications for the choice of Riemann solver for the fluxes: they should be smooth functions of the unknowns $\{U_j^{n+1}\}$ which yield a Jacobian matrix to which fast solvers can be applied. For our simple expository problem let us suppose we use simple upwind fluxes. Then, where the flow is supercritical, the Jacobian will have diagonal elements of the form

$$(x_{j+1/2} - x_{j-1/2}) + \tfrac{1}{2}(t^{n+1} - t^n)U_j^{n+1};$$

and where it is subcritical, of the form

$$(x_{j+1/2} - x_{j-1/2}) - \tfrac{1}{2}(t^{n+1} - t^n)U_j^{n+1}.$$

Hence these will always be increased by the flux terms; and, in general, a reasonable choice for the Riemann solver will lead to a diagonally dominant Jacobian, thus assisting with the choice of a fast solver.

Of course, the disadvantage of this approach is that the immediate outcome is only a first-order accurate approximation: a recovery procedure is needed to obtain higher-order accuracy. However, Fezoui and Stoufflet (1989) successfully used such an approach to approximating the Euler equations on a triangular mesh with simplified Jacobians, reporting almost quadratic convergence with a first-order scheme and quite acceptable results for a second-order scheme. More recently, in a series of papers (Meister and Oevermann 1996, Meister 1998, Meister and Vömel 2001), Meister and his collaborators further developed such methods and applied them successfully to the solution of both the Euler and Navier–Stokes equations. Some of the issues that need to be resolved in such an approach are as follows:

- whether to use the simple Crank–Nicolson (or more general theta-method) form of equation used above, or an implicit Runge–Kutta scheme such as a BDF method (see Hairer and Wanner (1996));
- choice of the recovery procedure, Riemann solver and approximate Jacobian;
- choice of the preconditioner and iteration scheme to solve the linearized equations.

In the papers cited above, Meister used a node-centred scheme based on a Delaunay triangulation of the flow region, and concentrated on steady flow problems. He used discontinuous linear recovery, as described below in Section 5, to obtain second-order accuracy, with numerical fluxes computed by the AUSMDV scheme due to Liou and Steffen (1993) and Wada and Liou (1994) that combines the accuracy of flux-difference splitting methods with the economy of flux-vector splitting schemes. Much of the later sections of this review will be devoted to recovery procedures; and in the next two subsections we summarize some of the other key ideas that are needed to implement such methods.

4.6. ENO and WENO schemes

In the mid-1980s it became clear that the strict imposition of Harten's TVD condition at the recovery stage of an algorithm would lead to a loss of accuracy at solution extrema; and we have already referred in Section 3.2 to the clipping of peaks when slope limiters are applied in the case of discontinuous linear recovery. New algorithms have therefore been introduced in a very influential series of papers, in which the total variation is allowed to increase to a limited extent in such a way that the order of accuracy is maintained uniformly throughout the domain: these are the *essentially non-oscillatory* or *ENO* schemes.

In the first paper, Harten and Osher (1987), the main ideas are introduced for a second-order approximation called *UNO2* to a scalar one-dimensional problem on a uniform mesh. The recovery stage uses the MUSCL discontinuous linear approximation (3.6) but the novelty lies in the choice of the slopes S_j. A non-oscillatory piecewise parabolic interpolant $Q(x; U)$ of the cell averages is constructed which, between U_j and U_{j+1}, uses either U_{j-1} or U_{j+2} (whichever gives the smaller second difference in absolute value) as the third value to determine the quadratic interpolant: it is shown that this choice results in an interpolant with no more local extrema than the set of U values. Then it is the derivatives of this function that are used to define the slopes as

$$S_j = \text{minmod}(Q(x_j - 0; U), Q(x_j + 0; U)).$$ (4.24)

The interface fluxes are calculated by approximating the constancy of the solution along its characteristics: using the same average characteristic speed $a^n_{j+1/2}$ as in the Roe scheme (1.6), when this is positive the flux given for that scheme is changed, with $\lambda = \Delta t / \Delta x$, to

$$F^{n+1/2}_{j+1/2} = f(U^n_j) + \frac{\frac{1}{2}a^n_{j+1/2}(1 - a^n_{j-1/2}\lambda)S^n_j}{1 + \lambda(a^n_{j+1/2} - a^n_{j-1/2})}.$$ (4.25)

The result is shown to be a scheme which is uniformly second-order accurate wherever the solution is smooth, including extrema and sonic points.

There are many necessary generalizations of this algorithm. Extensions to a non-uniform mesh are reasonably straightforward, since both the parabolic interpolation and the calculation of the fluxes from the discontinuous linear approximation are readily carried out on an arbitrary mesh. Even extensions to higher-order recovery procedures using Newton divided differences can be formulated in a natural way. The key step there is to define a sequence of cells running out from a given cell, such that at each stage the choice from the left or the right is made so that the resultant divided difference is the smaller in absolute value of the two available. However, Harten, Osher, Engquist and Chakravarthy (1986) showed that, at higher than second order, the process allows spurious oscillations to appear, of a magnitude limited by the order of accuracy – prompting the adoption of the general name ENO. A further point of choice is whether the cell values are regarded as point values, as in UNO2, or the cell averages are matched. The latter is more in the spirit of the optimal recovery process, and in one dimension is best done by forming the interpolation formula for the primitive function of U and then differentiating the result; but any such procedure requires more computation.

Harten, Engquist, Osher and Chakravarthy (1987) further extended these ideas to systems of equations in one dimension, in particular to the Euler equations. The recovery procedures are applied to locally defined characteristic variables; and the fluxes are calculated by using local Cauchy–Kowalewski formulae, similar to those developed by Ben-Artzi and Falcovitz (1984). In addition, it has long been recognized that *shock recovery* procedures can be applied to finite volume or characteristic-Galerkin methods in order to calculate the position and jump parameters of a shock: see, *e.g.*, Morton and Rudgyard (1988), Morton and Paisley (1989) and Childs and Morton (1990), and references therein. This is taken a step further in Harten (1989), where the ENO recovery procedures are used to detect the presence of contact discontinuities and to recover them so as to prevent the smearing that is normal for most methods. Applying all these developments leads to a very powerful set of one-dimensional schemes: and numerical experiments on standard Euler test problems, such as those from Woodward and Colella (1984), demonstrate impressive results for both second-order and fourth-order methods.

However, the major challenge is the development of such schemes on multidimensional unstructured meshes. We have already described in Section 3.2 some procedures for discontinuous linear recovery on triangular meshes; and much has been done to extend these ideas in connection with the development of ENO schemes: see Harten and Chakravarthy (1991) and Durlofsky *et al.* (1992). But it was Abgrall (1994*b*) who pointed out the advantage of using node-centred rather than cell-centre schemes in this recovery process, and we will take up this topic again in Section 5.

A problem that has generally been encountered with ENO schemes is a lack of convergence to steady flow solutions, because of oscillatory switching that can take place in the choice of recovery stencils. This has led to the development of WENO schemes by Liu, Osher and Chan (1994), in which a smoothness indicator is used to compute a weighted average of all the local recovery polynomials. Extensions to unstructured meshes have been developed by Friedrich (1998), which we will describe in Section 5.

4.7. Multigrid and Krylov subspace methods

In the thirty years in which finite volume methods have been used there have been major developments in methods for solving the large systems of algebraic equations that they generate. Two lines of development have been particularly important: one approaches the problem from the viewpoint of minimizing an appropriate function, as originated by the conjugate gradient method of Hestenes and Stiefel (1952) and generalized to unsymmetric problems in the GMRES algorithm of Saad and Schultz (1986); and the other exploits the fact that a PDE is being approximated on a mesh and the properties of the equation system depend strongly on the mesh size, as exploited in multigrid algorithms by Brandt (1977), following the earlier work of Federenko (1964) and others.

The two techniques have been applied in tandem to great effect in the solution of incompressible flow problems: see, in particular, Elman, Silvester and Wathen (2005). In such problems, the incompressibility condition brings a degree of ellipticity to the problems that enables the vast experience from Varga (1962) onwards to be built upon. There is a natural progression from discretizing the Poisson equation, and solving the resultant algebraic system, to tackling linear convection-diffusion problems, the Stokes system of equations and thence on to the incompressible Navier–Stokes equations. Algorithms such as GMRES are included in the general class of Krylov subspace algorithms: see the review by Eiermann and Ernst (2001) and the book by van der Vorst (2003). And these methods together with multigrid techniques have been progressively refined, generalized and optimized for this sequence of problems.

Compressible flow problems, with greater nonlinearity and a dominant hyperbolic character, pose greater difficulties. Thus, in early studies, point iteration methods dominated the scene. Multigrid methods were first introduced for steady transonic flow problems by Jameson (1979) and have made a massive impact on the field. The choice of appropriate smoothers is always important in multigrid applications, and the choice of alternating direction smoothers in this early study indicated how the flow direction influenced the behaviour of the methods. For a general reference on multigrid methods in the context of fluid flow problems, see Wesseling (1992); for a description of applications to compressible flow computations on unstructured meshes see Mavriplis (1995), and for a more recent review see Mavriplis (2002).

Krylov subspace methods have also been developed for these problems: see, *e.g.*, Nielsen, Anderson, Walters and Kayes (1995). In particular, their application to the finite volume methods we have described has been explored by Meister (1998). For any such algorithm, efficient preconditioning is essential and this has been developed by Meister and Vömel (2001) for the discretization of hyperbolic conservation laws.

5. Optimal recovery: theory and practice

The two preceding sections will have shown the reader the extent to which the successful development of finite volume methods depends on the recovery or reconstruction stage. Only the cell-vertex approach can give second-order accuracy without some such step, however rudimentary. We prefer the term 'recovery' because of its reference to the much more general field of *optimal recovery* which we will now outline, and because we feel that it is necessary to take advantage of this greater generality. Without such a general framework there is a natural tendency to use only local polynomials, as we have seen with the ENO methods, and these can easily lead to ill-conditioned procedures, with breakdown occurring there in the one-dimensional case when the polynomial degree exceeds six. The initial ideas for this theory are due to Golomb and Weinberger (1959) and were subsequently greatly developed in Micchelli and Rivlin (1977).

5.1. Theory of optimal recovery

Suppose we are trying to deduce some property of an unknown function u from some data given for it, and we can set up the problem in the following way. We suppose that u lies in a Hilbert space H with a bound on its norm $\|u\|_H$, and the data are given in the form of the values of a set of bounded linear functionals on that space, the *information operator*

$$\mathcal{I} := \{F_1(\cdot), F_2(\cdot), \dots, F_M(\cdot)\};$$

furthermore, what we seek is the value of another bounded linear functional $F(\cdot)$, the *feature operator*. Then Golomb and Weinberger (1959) observed that the given values of the linear functionals define a hyperplane in H and the bound on $\|u\|_H$ a hypersphere, with their intersection defining a *hypercircle* (see Synge (1957)); moreover, the centre of this hypercircle defines a function $u_c \in H$ and it gives a value of the sought-after functional $F(u_c)$ which is optimal, and this is the case independently of the information functional that is sought. We will illustrate why this is so by means of one of the most important examples of the theory.

Let $a(\cdot, \cdot)$ be a symmetric, bilinear form on $H \times H$ that defines an elliptic boundary value problem on a region Ω, with homogeneous Dirichlet boundary conditions: find $u \in H$ such that

$$a(u, v) = (f, v), \quad \forall v \in H;$$

and let the norm on H be defined by $\|v\|_H^2 := a(v, v)$. Now consider the approximation of this problem by a finite element method which uses the basis functions $\phi_1, \phi_2, \dots, \phi_M$ lying in H, and suppose we regard the data functionals to be defined by

$$F_i(u) := (f, \phi_i) = a(u, \phi_i), \quad i = 1, 2, \dots, M.$$

Indeed, whatever the given data functionals, by the Riesz representation theorem we could find their *representers* ϕ_i which would be defined by these equations; and we could continue the following construction. We define $H_M := \mathrm{span}\{\phi_i,\ i = 1, 2, \dots, M\}$; and then the centre of the hypercircle is the orthogonal projection, with respect to $a(\cdot, \cdot)$, of u onto H_M. That is, it is the familiar finite element approximation $U_M \in H_M$ to u using this set of basis functions:

$$F_i(U_M) \equiv a(U_M, \phi_i) = (f, \phi_i) \equiv F_i(u), \quad i = 1, 2, \ldots, M.$$

Then we suppose that the linear functional whose value for u we seek to estimate is

$$F(\cdot) \equiv a(\cdot, \psi),$$

where ψ is its representer.

Suppose now that ψ_M is the orthogonal projection of ψ onto H_M. It follows that

$$|F(u) - F(U_M)| = |a(u - U_M, \psi)| = |a(u - U_M, \psi - \psi_M)| \qquad (5.1)$$
$$\leq \|u - U_M\|_H \|\psi - \psi_M\|_H.$$

Moreover, let us denote by $\Delta_M u$ and $\Delta_M \psi$ the two (positive real) factors on the right and consider two alternative choices for estimating F, namely

$$u_\pm := U_M \pm \frac{\Delta_M u}{\Delta_M \psi}(\psi - \psi_M),$$

for which we can readily check that $F_i(u_\pm) = F_i(u)$, $\forall i$. Then it is easily seen that

$$F(U_M) - F(u_\pm) = \pm(\Delta_M u)(\Delta_M \psi); \qquad (5.2)$$

that is, $F(U_M)$ lies at the centre of the interval $[F(u_-), F(u_+)]$ of possible values for $F(u)$, hence giving the optimal estimate for this quantity.

It is a key property of Galerkin and Petrov–Galerkin finite element approximations that they are optimal or near-optimal approximations in an energy norm. Thus in Barrett, Moore and Morton (1988a) the framework outlined above was used to derive techniques for recovering point values of functions from their weighted L^2 best fits, using both local and global recovery procedures; while in Barrett, Moore and Morton (1988b) global recovery techniques were developed from low-order finite element approximations to ODE problems to obtain higher-order approximations, techniques which thus correspond to defect correction methods. These introduce higher-order approximations in a very similar way to those needed for the finite volume methods we will discuss below.

The theory of optimal recovery has been developed in a very general setting and the ideas applied to a wide range of problems: see Micchelli and Rivlin (1977). For finite volume methods the linear functionals that provide the data are the cell averages; and the information that is required consists of the function and derivative values from which higher-order approximations can be constructed – and, in particular, from which fluxes on the element boundaries can be computed. In the next two subsections we will describe how the ENO recovery techniques may be generalized in a very natural way to two-dimensional triangular meshes. Then we will describe some less conventional approaches to this problem, based on spline functions and radial basis functions.

5.2. Recovery on primary triangular grids

In generalizing the ENO recovery techniques to obtain higher-order approximations we will need to consider how to select the mesh for successively higher orders, what form of expansion to use for the approximations, how to solve the algebraic equations for the expansion coefficients and how each of these interact. For instance, it is well known that a linear approximation in two dimensions cannot be derived from three point values given on a straight line. So for the cell-centre scheme we would not want to use three triangles whose centroids almost lie on a line in order to generate our linear recovery approximation. In what follows we shall generally refer to the recovery of the scalar quantity U: in application to such as the Euler equations, this will denote one component of the vector of unknowns or, more commonly, chosen to be one of the primitive variables.

A very simple but surprisingly successful technique for linear recovery was described by Durlofsky *et al.* (1992). The possible node sets for recovery on the triangle T_i in Figure 5.1 consist of the barycentre of T_i and those of two of its neighbours:

$$K_1(T_i) := \{T_j, j \in \{i, i_2, i_3\}\},$$
$$K_2(T_i) := \{T_j, j \in \{i, i_3, i_1\}\},$$
$$K_3(T_i) := \{T_j, j \in \{i, i_1, i_2\}\}.$$

On each node set $K_k(T_i)$ a linear polynomial

$$\pi_i^{(k)} := a_{00}^{(k)} + a_{10}^{(k)}(x_1 - c_{i,1}) + a_{01}^{(k)}(x_2 - c_{i,2}), \quad k = 1, 2, 3,$$

where $\mathbf{c}_i$ is the barycentre of T_i, is computed by solving the linear systems

$$\mathcal{A}(T_j)\pi_i^{(k)} = U_j, \quad \forall(k, j); \tag{5.3}$$

here the range of values for (k, j) are given in the above definition of the node sets. If there is an extremum at triangle T_i with respect to its three neighbours, then none of the polynomials is chosen and the value on T_i is not recovered. Otherwise we consider the steepest of the three linear polynomials and check whether its use would result in a new extremum. If that is not the case this polynomial is taken to be the recovery on T_i. If a new extremum is created the polynomial with the next steepest gradient is considered, and so on. If all three polynomials would result in a new extremum then no recovery is used on T_i. Note that this procedure corresponds more to the classical TVD approach than to an ENO approach.

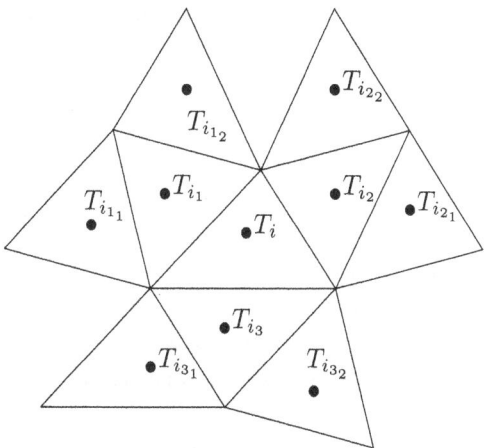

Figure 5.1. Neighbourhood of T_i.

An even simpler algorithm consists of choosing the node set that yields the linear polynomial with the gradient of smallest absolute value. This is a true generalization of the ENO idea in that small oscillations are then allowed to occur. However, this can lead to quantities such as density or pressure taking on negative values; so a remedial step would be needed. To reduce this possibility one might consider a larger number of node sets: for example, using the neighbours of the neighbours of T_i, that is, all the triangles shown in Figure 5.1. But such a large stencil is more appropriate for quadratic recovery.

To carry out quadratic recovery, for each node set we need to compute the coefficients in an expansion of the form

$$\pi_i(x) := a_{00} + a_{10}(x_1 - c_{i,1}) + a_{01}(x_2 - c_{i,2})$$
$$+ a_{11}(x_1 - c_{i,1})(x_2 - c_{i,2}) + \tfrac{1}{2}a_{20}(x_1 - c_{i,1})^2 + \tfrac{1}{2}a_{02}(x_2 - c_{i,2})^2$$

by matching its average over each triangle in the node set with that of U. Although the integrals occurring in the coefficient matrix of this system can be computed exactly it makes more sense to use an appropriate quadrature rule.

The real problem is the selection of the node sets. In Harten and Chakravarthy (1991) a sectorial search strategy is advocated in which the region outside the central triangle T_i is divided into sectors by continuing the triangle sides in both directions, giving three based on the triangle sides alternating with three based on its vertices; then these are treated in a way that corresponds to the two directions in the one-dimensional case, and typically will give 18 possible ways of adding the three triangles needed for the quadratic node set. Alternatively, as in Abgrall (1994a) we can argue as follows: referring to Figure 5.1 suppose that the chosen linear node set consisted of $\{T_i, T_{i_1}, T_{i_2}\}$; then there are three pairs of edges on the outer perimeter that correspond to a pair of triangles, such as $\{T_i, T_{i_1}\}$, and on each edge we consider the neighbouring triangle and its two neighbours, such as T_{i_3} with $T_{i_{3_1}}$ and $T_{i_{3_2}}$ on the outer edge of T_1 as shown in the figure; this will give us six choices of three triangles: T_{i_3} and its two neighbours, $T_{i_{1_1}}$ and its two neighbours, and both T_{i_3} and $T_{i_{1_1}}$ with one of their four neighbours. Again we have 18 choices!

Once a set of possible quadratic polynomials is computed, one has to be selected by some set of criteria. To ensure that the choice is the least oscillatory in some sense, we could use the criterion

$$W(\pi) := \sqrt{\sum_{\mu=1}^{2} \sum_{|\alpha|=\mu} a_\alpha^2} \qquad (5.4)$$

and choose the polynomial which gives a minimal value of W. However, there are many other possible criteria; and it is far from clear that the Taylor series used above is best suited for defining these criteria, or for computing the quadratic and higher-order approximations. We will take up these points in more detail in the next subsection.

5.3. Recovery on secondary grids

It is argued by Abgrall (1994b) that there are fewer node sets to consider for higher-order recovery algorithms in this case, and this will lead to important advantages of node-centred schemes over their cell-centre counterparts: we will consider them by reference to Figure 5.2, where the nodes are now the vertices of the triangular grid. In the first stage, for linear recovery, we consider all the triangles which share a given vertex, say i_0, and construct a linear function for each such that its average over each box centred on one of its vertices matches the corresponding average of U. Then we choose that with the smallest gradient: in the figure this is labelled $T_{\min}$. We will describe the quadratic recovery stage in more detail before we come back to this linear stage.

There are several very important developments presented in Abgrall (1994b) which elaborate on the advantages of the node-centred methods. The first has to do with the selection of successive node sets: as can be seen from Figure 5.2, the three further vertices needed for quadratic recovery can be obtained from the triangle (and its two further neighbours) that shares one of its sides with $T_{\min}$: this gives a choice of three node sets $K(B_{i_0})$ for this stage. Indeed, it is claimed in the paper that at each further stage only three possible node sets need to be considered.

An even more important aspect of defining a true generalization of the ENO process is to select a representation of the approximation at each

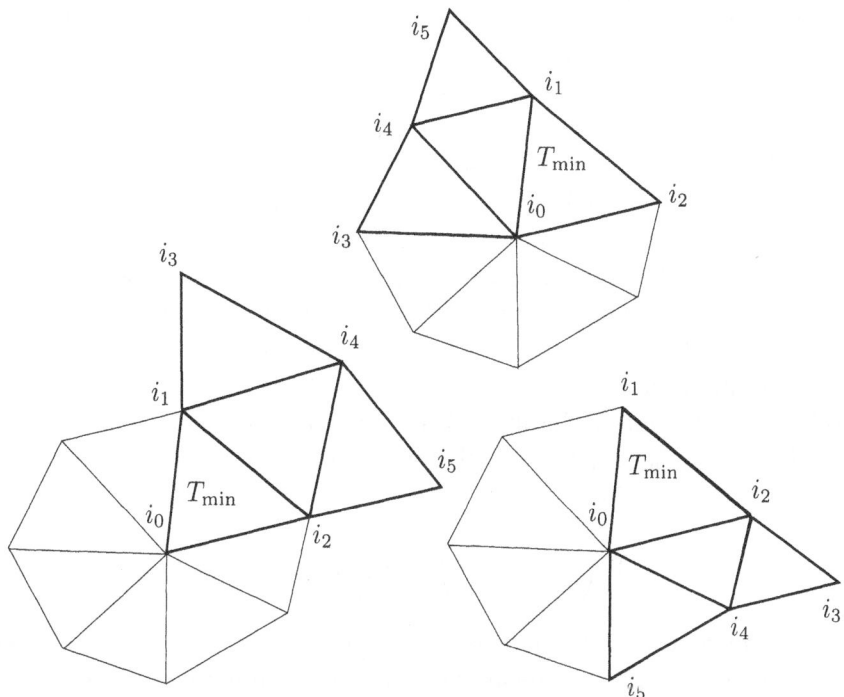

Figure 5.2. Three possible sets $K(B_{i_0})$ for quadratic recovery.

stage which has some of the important properties of the Newton divided differences used in the one-dimensional schemes. In Abgrall (1994b) the barycentric coordinates of $T_{\min}$ are used for the quadratic expansion; that is, with a cyclic ordering of $(1, 2, 3)$, and now with $\mathbf{x}_1, \mathbf{x}_2, \mathbf{x}_3$ the vertices of $T_{\min}$, we have the coordinates

$$\lambda_1(\mathbf{x}) = \frac{1}{2|T_{\min}|} \big[(x_{3,1} - x_{2,1})(x_2 - x_{2,1}) - (x_1 - x_{2,1})(x_{3,2} - x_{2,2})\big].$$

Then a quadratic polynomial for the box centred at i_0 can be written as

$$\pi_{i_0}(\mathbf{x}) = \sum_{m=1}^{3} \left(a_m \lambda_m(\mathbf{x}) + \sum_{n>m} A_{mn} \lambda_m(\mathbf{x}) \lambda_n(\mathbf{x}) \right);$$

and Abgrall could prove that the coefficient matrix of the system

$$\mathcal{A}(B_j)\pi_{i_0} = U_j, \qquad B_j \in K(B_{i_0}) \tag{5.5}$$

has condition number of order 1. This also holds for higher-order recovery, which is in sharp contrast to the poor conditioning obtained with the Taylor series expansion. In addition, by following the analysis of Ciarlet and Raviart (1972) for interpolation by finite element approximations, he was able to show that the derivatives of smooth functions were approximated similarly, with a similar dependence on the mesh quality. Thus the recovered functions should provide reliable indicators of the smoothness of the unknown solution.

For the quadratic recovery Abgrall also showed that the system can be factored into two subsystems of size 3×3, with one of them corresponding to the system needed for the linear recovery stage. So this captures another key feature of the one-dimensional algorithm. Subsequently, in Abgrall and Sonar (1997) it is shown that this property also holds at all orders by exploiting the generalization of Newton divided differences developed by Mühlbach (1978). We will not give any details here, but it is useful to outline the main ideas.

Mühlbach introduced the idea of a *complete Chebyshev system* of functions $(f_1, f_2, \ldots, f_k, \ldots, f_n)$, which for simplicity we can take to be real functions on $\mathbb{R}$, for which

$$V \begin{pmatrix} f_1, \ldots, f_k \\ x_1, \ldots, x_k \end{pmatrix} := \det f_j(x_i) \neq 0$$

is true for any distinct set of points $(x_1, \ldots, x_n)$ and $k = 2, 3, \ldots, n$. For such a system it is clear that a linear interpolatory formula can be constructed for another function $f(\cdot)$ which, with its error, he denotes as follows:

$$p_n f \equiv pf \begin{bmatrix} f_1, \ldots, f_n \\ x_1, \ldots, x_n \end{bmatrix}, \quad r_n f := f - p_n f.$$

Moreover, this can be expressed in a series whose terms are *generalized divided differences* of the function f, with that of order k given by

$$\begin{bmatrix} f_1, \ldots, f_k, \\ x_1, \ldots, x_k \end{bmatrix} f := \frac{V \begin{pmatrix} f_1, \ldots, f_{k-1}, f \\ x_1, \ldots, x_{k-1}, x_k \end{pmatrix}}{V \begin{pmatrix} f_1, \ldots, f_{k-1}, f_k \\ x_1, \ldots, x_{k-1}, x_k \end{pmatrix}};$$

the series can then be written in the form

$$p_n f \equiv pf \begin{bmatrix} f_1, \ldots, f_n \\ x_1, \ldots, x_n \end{bmatrix} = \sum_{k=1}^{n} \begin{bmatrix} f_1, \ldots, f_k, \\ x_1, \ldots, x_k \end{bmatrix} f \, g_k, \tag{5.6}$$

where

$$g_1 := f_1, \quad g_k := r_{k-1} f_k, \quad \text{for } k = 2, \ldots, n.$$

Finally, a recurrence relation for the divided differences

$$\begin{bmatrix} f_1, \ldots, f_k, \\ x_1, \ldots, x_k \end{bmatrix} f = \frac{\begin{bmatrix} f_1, \ldots, f_{k-1}, \\ x_2, \ldots, x_k \end{bmatrix} f - \begin{bmatrix} f_1, \ldots, f_{k-1}, \\ x_1, \ldots, x_{k-1} \end{bmatrix} f}{\begin{bmatrix} f_1, \ldots, f_{k-1}, \\ x_2, \ldots, x_k \end{bmatrix} f_k - \begin{bmatrix} f_1, \ldots, f_{k-1}, \\ x_1, \ldots, x_{k-1} \end{bmatrix} f_k}$$

was shown by Mühlbach to follow from a general Neville–Aitken recurrence formula.

What was shown in Abgrall and Sonar (1997) was that all of this could be generalized to the recovery problem in more than one space dimension and using linear functionals on the solution space, with the given functionals corresponding to the given function values and the unknown function to the sought-after linear functional. Then the Vandemonde determinants that occur in the definition of the generalized divided differences correspond to recovery equations such as (5.5) that have to be solved; and the recurrence relation for these divided differences expresses the fact that at each stage the system can be solved by solving similar systems corresponding to earlier stages. In particular, quadratic recovery can be implemented by twice solving the sort of 3×3 system needed for linear recovery.

Thus ENO schemes using quadratic and higher-order recovery become a very practical proposition. Moreover, so do the WENO schemes which require the calculation of more recovery approximations. On each box B_i we have to compute a set of recovery polynomials $\pi_i^{(k)}$ where k denotes the number of the stencil; then we compute a weighted sum

$$\pi_i := \sum_k \Omega_k \pi_i^{(k)}$$

where the Ω_k are weights with $\sum_k \Omega_k = 1$. An *oscillation indicator OI* is used to compute the weights: for example, we may use a Sobolev seminorm

$$OI(\pi_i^{(k)}) := \|\nabla \pi_i^{(k)}\|_{L^2(B_i)}$$

as an oscillation indicator; then the weights are computed from

$$\Omega_k := \frac{\omega(k)(\varepsilon + OI(\pi_i^{(k)}))^{-\beta}}{\sum_j \omega(j)(\varepsilon + OI(\pi_i^{(j)}))^{-\beta}}.$$

Here, $\omega(j), \omega(k)$ are weights which allow a different weighting of different stencils. The parameter ε is chosen to avoid the division by zero and β is a measure of sensitivity of the weights on the oscillation indicator. We set $\varepsilon = 10^{-16}$, $\beta := 8$ and $\omega(k) = 12$ for a central stencil while $\omega(j) = 1$ for a one-sided stencil. Such WENO schemes were developed by Friedrich (1998) and an example of their effectiveness is shown in Sonar (2002).

5.4. Splines and radial basis functions

In one dimension, splines are commonly introduced through a variational principle: the linear spline interpolant of a function at a given set of knots is that interpolant that minimizes the L^2-norm of its derivative: and a cubic spline interpolant is similarly an interpolant that minimizes the norm of the second derivative: see, for instance, de Boor (2001). More generally they can be characterized as centres of hypercircles in certain semi-Hilbert spaces, that is, a space with seminorm $|\cdot|_V$ for which there may exist nontrivial functions $w \in V$ for which $|w|_V = 0$ holds (*i.e.*, the seminorm has a 'hole': its kernel $\ker |\cdot|_V$ contains more than the null function). A spline in a semi-Hilbert space is then defined to be a function $\Phi \in V$ which minimizes the seminorm: that is, for the given information operator $\mathcal{I}$,

$$|\Phi|_V = \inf_{\substack{v \in V \\ \mathcal{I}v = \mathcal{I}u}} |v|_V$$

is to be satisfied.

Thus a one-dimensional cubic spline interpolant minimizes the seminorm given by the L^2-norm of the second derivative, which has a kernel consisting of linear polynomials, and these have to be specified by some side conditions; for example, the natural cubic spline is determined by setting to zero the second derivative at each boundary. In two dimensions the cubic spline generalizes to the *thin plate spline*, so-called because it is the solution of the biharmonic equation. In our setting of recovery from cell average data, Sonar (1996) has shown that it is given by

$$\Phi(\mathbf{x}) = \sum_{j=0}^{M-1} \alpha_j \mathcal{A}^{(\mathbf{y})}(\sigma_{i_j}) \big[|\mathbf{x} - \mathbf{y}|^2 \ln(|\mathbf{x} - \mathbf{y}|) \big] + a_{01}x_1 + a_{10}x_2 + a_{00}, \quad (5.7)$$

where $\mathcal{A}^{(\mathbf{y})}$ denotes application of the cell average operator with respect to the variable $\mathbf{y}$, and the additional linear polynomial is the contribution from the kernel of the seminorm. We now have to determine $M + 3$ coefficients $\alpha_0, \ldots, \alpha_{M-1}, a_{10}, a_{01}, a_{00}$ but we have only M conditions given by the information $\mathcal{I}u = \{\bar{u}\}$, the cell averages on the node set. If we require the condition

$$\forall q \in \ker |\cdot|_V : \quad \sum_{j=0}^{M-1} \alpha_j \mathcal{A}(\sigma_{i_j})q = 0$$

we get the remaining three conditions needed to determine all coefficients: we can think of this condition as 'fixing the hole' in the seminorm.

One can easily prove that the thin plate spline reproduces linear polynomials, so that recovering from three given cell averages just gives the linear polynomial which is constructed to fix the hole in the seminorm. Hence we need to have more than three cells in a node set. In applying this recovery to the cell T_i in Figure 5.1 we therefore use node sets comprised of four neighbouring triangles: there is one central node set $K_0(T_i) := T_i \cup T_{i_1} \cup T_{i_2} \cup T_{i_3}$; and the three one-sided node sets $K_1(T_i) := T_i \cup T_{i_1} \cup T_{i_{1_1}} \cup T_{i_{1_2}}$, $K_2(T_i) := T_i \cup T_{i_2} \cup T_{i_{2_1}} \cup T_{i_{2_2}}$, and $K_3(T_i) := T_i \cup T_{i_3} \cup T_{i_{3_1}} \cup T_{i_{3_2}}$. On each of the node sets we solve the linear system

$$\mathcal{A}(T)\Phi = \mathcal{A}(T)u =: \bar{u}_j, \quad T \in K_j(T_i), \quad j = 0, 1, 2, 3, \tag{5.8}$$

together with the conditions

$$\sum_{j=0}^{3} \alpha_j \mathcal{A}(T) = \sum_{j=0}^{3} \alpha_j \mathcal{A}(T) x_1 = \sum_{j=0}^{3} \alpha_j \mathcal{A}(T) x_2 = 0. \tag{5.9}$$

Denoting respectively by M and N the matrices in these two systems, we can write the equations for the seven unknowns in this thin plate recovery spline as

$$\begin{bmatrix} M & N^T \\ N & 0 \end{bmatrix} \begin{bmatrix} \alpha_0 \\ \vdots \\ \alpha_3 \\ a_{10} \\ a_{01} \\ a_{00} \end{bmatrix} = \begin{bmatrix} \bar{u}_0 \\ \vdots \\ \bar{u}_3 \\ 0 \\ 0 \\ 0 \end{bmatrix}. \tag{5.10}$$

Explicit expressions for the matrices M and N are relatively easy to calculate. Then, after computing the four recovery splines on the four node sets, that with smallest total variation over T_i is chosen to be the recovery spline on T_i.

It is shown in Iske and Sonar (1996) that the thin plate spline is just one example of a *radial basis function* that may be used in a cell average recovery algorithm. Conditions which have to be satisfied by any such function of the form $\psi(|\mathbf{x} - \mathbf{y}|)$ are given, as well as several other examples. They all require considerably more computation than the more conventional polynomial recovery. The thin plate spline has been experimented with most widely, but alternatives are easier to use and equally effective.

A simple model problem that can be used to compare some of the recovery procedures that we have discussed is the following: the initial data consists of a straight-sided cone, of unit height and with the radius of its base 0.15, whose centre is at (0.5,0); it is then convected in a circle about the origin, and we plot the results of various computations at a time corresponding to half a revolution. In Figure 5.3(a), the left plot shows the mesh (of 2500 nodes and 4802 triangles), and the right plot shows the initial data and contours of the solution without recovery after half a revolution, using the Engquist–Osher flux and a simple explicit time-stepping. Figure 5.3(b) shows the corresponding result obtained with the linear recovery procedure due to Durlofsky *et al.* (1992); and (c) is that obtained with a thin plate spline recovery step by Sonar (1996). The figure shows that the thin plate spline reproduces the cone much more accurately than the alternatives, both

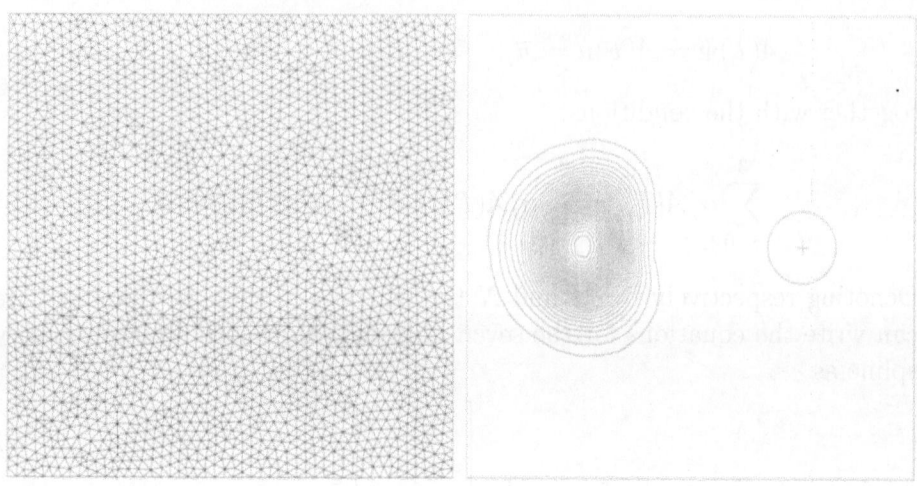

(a) Computational grid (*left*), base and centre of initial cone and solution without recovery

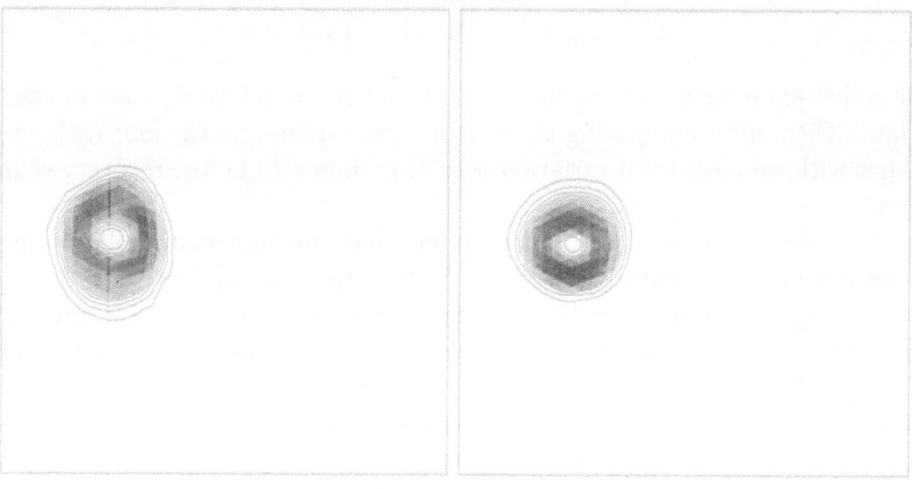

(b) Solution with linear recovery   (c) Solution with thin plate spline recovery

Figure 5.3. The rotating cone problem.

Table 5.1.

$\bar{u}$	basic	DEO	mingrad	quad1	quad2	TPS
min	0.0	0.0	-1.4×10^{-5}	-2.0×10^{-6}	0.0	0.0
max	0.382	0.635	0.753	1.04	0.764	0.974

as regards its compactness and its circular shape. Table 5.1, which shows minimum and maximum cell average heights, emphasizes the comparisons.

In the headings, *basic* denotes the basic unrecovered scheme, *DEO* linear recovery using the Durlofsky *et al.* (1992) algorithm, while *mingrad* denotes the simpler algorithm which chooses the linear recovery with smallest gradient; similarly *quad1* denotes quadratic recovery using the simple criterion (5.4), *quad2* is a more sophisticated quadratic recovery using a sector search algorithm and, finally, *TPS* denotes the thin plate spline recovery just described.

6. Grid adaptivity: *a posteriori* error control

Except in rather simple and special cases it is impractical to use any form of shock fitting to achieve sharp definition of shocks. The practical alternative is to use local mesh refinement. This is simplest with a triangular or tetrahedral mesh, and a large literature on both the practical and theoretical techniques has developed for application to elliptic problems: for a general introduction which leads towards our present CFD problems, see Eriksson, Estep, Hansbo and Johnson (1995) and the references therein. In order to build on this for our finite volume methods, it is best to use node-centred methods: then the cell averages over the boxes centred on each vertex are recovered to give local polynomial approximations on each box, as described in the previous section; restricting these to the vertices of the primary triangular mesh gives a continuous piecewise linear approximation on which to base criteria for mesh refinement or coarsening. This will form the basis of the methods described below.

There are, however, many differences in what is required for a compressible flow calculation from that for the approximation of a scalar elliptic problem. Estimating the *a posteriori* error is probably the major difference: which component or combination of components should be used to measure the error; what norm should be used; how to distinguish the measured error from its source; and then how best to do all of this when the solution may be changing rapidly with time? We shall pay less attention to this last aspect: we will consider problems of mesh refining and coarsening arising from shock movement; but we will not consider the estimation and use of variable time steps.

6.1. Mesh refinement and recoarsening

We will not give here all the details of any particular procedures, but it is important to outline the key ideas in order to understand the properties of the resultant approximations. At the end of each refinement or recoarsening we will ensure that we have a conforming triangulation; and we start with a conforming triangulation that is everywhere the coarsest that will be used. The algorithms summarized below follow the general strategies of Bank, Sherman and Weiser (1983); details can be found in Sonar (2002).

There are two main types of refinement: the so-called *red-refinement* of a triangle in which the mid-points of the sides are joined so that the triangle is divided into four similar triangles as in Figure 6.1; this will make the neighbouring triangles non-conforming, so that a *green-refinement* would be needed in which a mid-side is joined to the opposite node so as to divide the triangle into two as in the figure. The triangles resulting from a red-refinement are called *red triangles*, and are termed the *daughters* of the original *mother* triangle: similar terminology is used for the green-refinement.

These basic refinements of a single triangle can be used to define a refinement procedure for a conforming triangulation $\mathcal{T}$ in which a subset of its triangles have been marked for refinement:

Algorithm 1

1 Eliminate all green-refinements in $\mathcal{T}$ by restoring the mothers of green triangles. If a green triangle was specified for refinement the restored mother is also marked for refinement.
2 Red-refine all triangles which are marked for refinement.
3 While there exist triangles in $\mathcal{T}$ with more than one non-conforming node, they are red-refined.
4 Apply the green-refinement for all triangles which have exactly one non-conforming node.

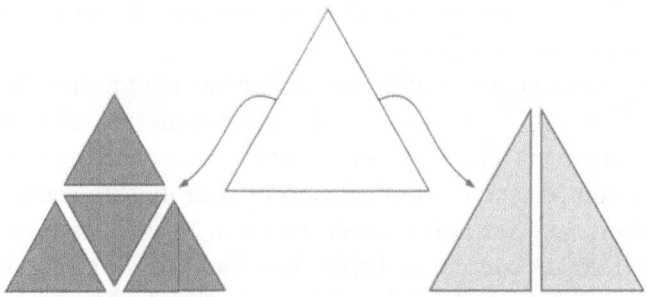

Figure 6.1. Red-refinement (*left*) and green-refinement (*right*) of a triangle.

The algorithm terminates with a conforming triangulation. Since all children of red-refinements are similar to their mothers and green triangles will be removed in the next refinement step, the refinement procedure is *stable*: that is, the inner angles of the triangles are bounded from below in any sequence of grid refinements.

In order to carry out a recoarsening procedure it is necessary to carry with each triangle a compact data structure called `History`. This will include whether the triangle is the result of a green-refinement, and if so the identifier of its sister; also it must hold the number of red-refinements that led to this triangle together with a data stack specifying its sisters and its antecedents. Then the following algorithm can be applied to a *resolvable patch*:

Algorithm 2

for all triangles $T \in \mathcal{T}$ which are specified for recoarsening
   for all three vertices P of T
      if the pair (P, T) spans a *resolvable patch* $\mathcal{P}$
      recoarsen this resolvable patch $\mathcal{P}$.

Figure 6.2 illustrates on the left-hand side some configurations for resolvable patches around a vertex, and on the right-hand side possible recoarsenings of the patch without producing hanging nodes. The bottom part of the figure illustrates a situation at the boundary of a triangulation.

The actual recoarsening of the triangles in a resolvable patch $\mathcal{P}$, spanned by triangle $T \in \mathcal{T}$ and one of its vertices P, is carried out as follows:

Algorithm 3

1. Remove all triangles in $\mathcal{P}$ and restore their mothers.
2. Remove all green triangles which produce hanging nodes in the mothers of $\mathcal{P}$, and restore the mothers of these green triangles.
3. Green-refine all triangles which have one non-conforming node.

The result of such recoarsening is a conforming triangulation that would be obtainable from the original triangulation by a sequence of refinement steps. Indeed, a sequence of such recoarsenings could lead back to the original triangulation.

6.2. Weighted L^2-norm error control

It was Johnson and his collaborators – see, *e.g.*, Hansbo and Johnson (1991) and Eriksson and Johnson (1993) – who introduced the idea of residual-based error indicators to CFD from their well-developed use with elliptic equations. We denote by $\mathbf{u}^h$ the continuous piecewise linear approximation constructed from a finite volume computation, and if $\mathcal{L}$ is a first-order differential operator, the corresponding *residual* for the problem $\mathcal{L}\mathbf{u} = \mathbf{0}$ can

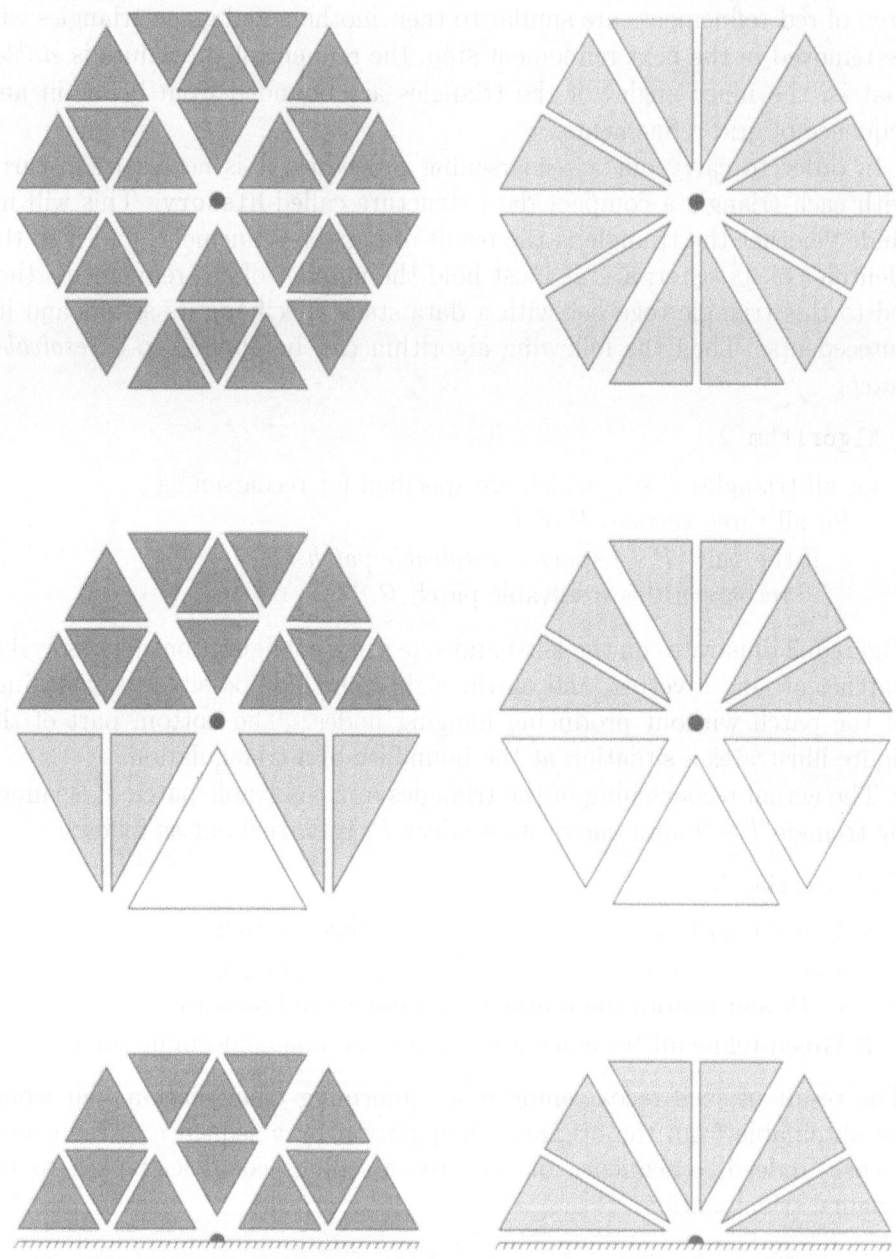

Figure 6.2. Resolvable patches (*left*) and recoarsened triangles (*right*).

be calculated as

$$\mathbf{r}^h := \mathcal{L}\mathbf{u}^h.$$

Our objective is local error control based on such a residual. In application to the Euler equations we have four components, and might first consider using the sum of the L^2-norms of each component over each triangle T, *i.e.*,

$$\|\mathbf{r}^h\|_{L^2(T)} := \sum_{i=1}^{4} \|r_i^h\|_{L^2(T)}$$

defined on the triangles of the primary grid, with the components of the residual corresponding to the continuity equation, the two momenta and energy equations.

Although our initial aim is to use this residual to guide the selection of triangles for refinement or recoarsening, the more ambitious target (which is achievable for the finite element approximation of elliptic problems) would be to define a residual which provides an efficient and reliable bound on the actual error in the approximation, $\mathbf{e}^h := \mathbf{u}^h - \mathbf{u}$: if the operator $\mathcal{L}$ has a bounded inverse from some space Y to a space X, we would like to establish bounds of the form

$$C_1\|\mathbf{r}^h\|_Y \leq \|\mathbf{e}^h\|_X \leq C_2\|\mathbf{r}^h\|_Y \tag{6.1}$$

for some computable constants C_1, C_2. Although we cannot expect to achieve this for the nonlinear Euler equations, it should be borne in mind as an eventual aim and thus give some guidance on how to measure and weight the contribution from each triangle. Moreover, we will below get quite close to realizing this aim for closely related PDE systems.

As a first step, let us consider using the unweighted L^2-norm of a shocked flow: in particular, suppose that the continuous piecewise linear numerical approximation, on a uniform mesh of size h, given by

$$u^h(x) = \begin{cases} 0 & 0 \leq x < x_i, \\ (x - x_i)/h & x_i \leq x < x_{i+1}, \\ 1 & x_{i+1} \leq x \leq 1, \end{cases}$$

approximates the function u which jumps from 0 to 1 at the mid-point of the interval $[x_i, x_{i+1}]$; and suppose also that the first-order differential operator $\mathcal{L}$ is just ∂_x. Then the L^2-norm of the residual on the interval $[x_i, x_{i+1}]$ is easily calculated to be

$$\|r^h\|_{L^2([x_i,x_{i+1}])} = \sqrt{\int_{x_i}^{x_{i+1}} |r^h|^2 \, \mathrm{d}x} = \sqrt{\int_{x_i}^{x_{i+1}} \frac{1}{h^2} \, \mathrm{d}x} = \frac{1}{\sqrt{h}},$$

and this quantity blows up at discontinuities as the grid is refined. Although in two dimensions on a square mesh this would be avoided for the norm on

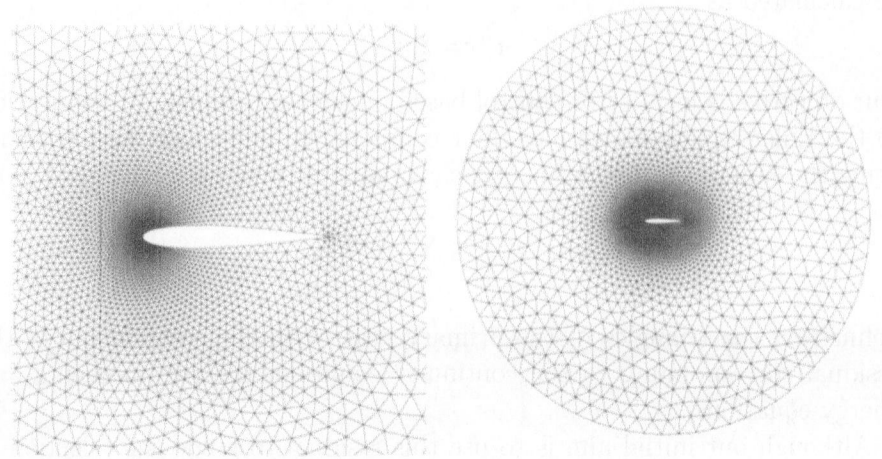

Figure 6.3. Initial grid for the NACA0012 aerofoil.

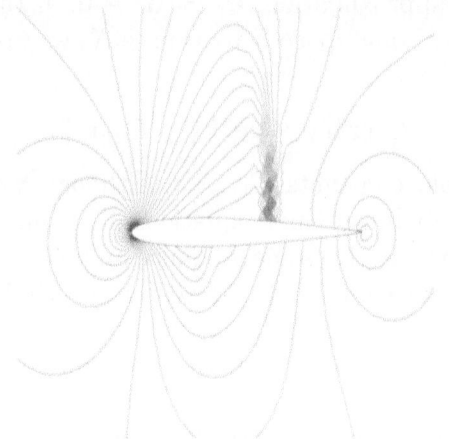

Figure 6.4. Pressure distribution on the initial grid.

an individual mesh square or triangle, for a shock extending a finite distance the norm over the region covering it would blow up in the same way. So the unweighted norm would give excessive refinement near such a shock.

On the other hand, in the finite element methods of Hansbo and Johnson (1991) it was found that the local triangle diameter should be used as a weight factor:

$$\|\mathbf{r}^h\|_{L^2_h(T)} := h_T \|\mathbf{r}^h\|_{L^2(T)}, \qquad (6.2)$$

where h_T denotes the length of the longest side of T. In Sonar (2002) its use as a refinement indicator for finite volume methods was compared with the use of other powers of h_T, and the use of more heuristic alternatives

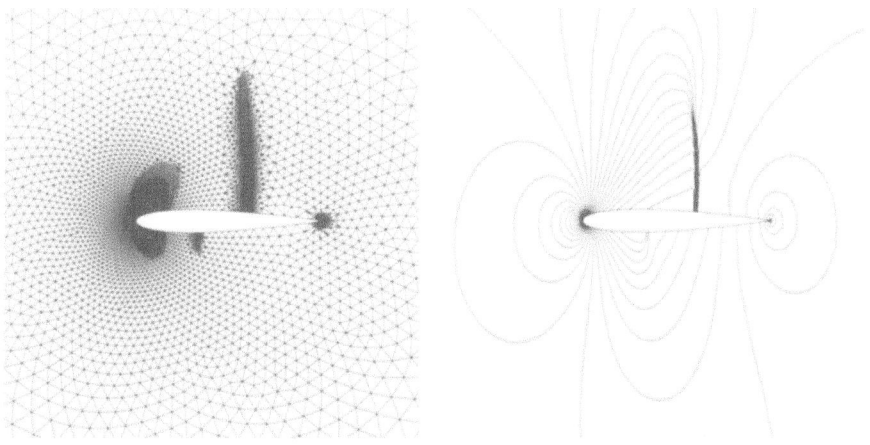

Figure 6.5. Grid after three refinement cycles with the unweighted norm indicator (*left*). Pressure distribution on this grid (*right*).

which had been advocated by other authors, for a number of flow problems. The simplest test problem was the very standard problem of the flow about the NACA0012 aerofoil, in which the Mach number of the incoming flow is Ma $= 0.8$ and the angle of attack is $\alpha = 1.25°$. The flow should contain a strong shock on the upper side of the aerofoil and a weak one on the lower side. The initial grid in the vicinity of the profile and also the whole grid are shown in Figure 6.3. Note that the leading edge region of the profile is already overly refined in this grid, which the present algorithms are unable to correct. Using a second-order box method (*i.e.*, a node-centred scheme with continuous piecewise linear recovery, as described in earlier sections) one obtains for the pressure distribution the results shown in Figure 6.4. Note that although both shocks are visible, the weak lower side shock is quite badly resolved.

Applying three refinement cycles using the unweighted L^2-norm as the refinement indicator results in the mesh shown in Figure 6.5. The corresponding pressure distribution is shown on the right-hand side. The results are undoubtedly much improved but there has been a lot of unnecessary refinement around the leading edge.

For comparison, the grid after three refinement cycles with the weighted L^2-norm indicator and the corresponding Mach number distribution are shown in Figure 6.6. There is clearly a much more appropriate refinement of the mesh, resulting in a much better defined solution.

However, it was the use of the dual problem in the *a posteriori* analysis of finite element approximations to elliptic equations that initially led to the special place of the L^2-norm. Developments for convection-diffusion

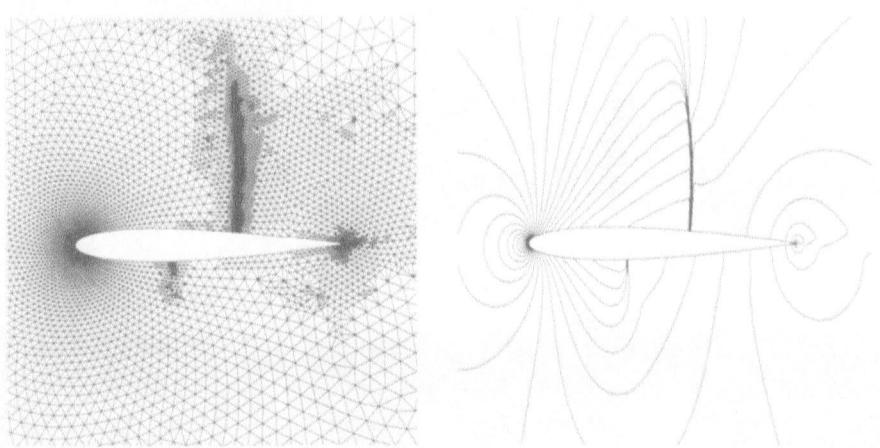

Figure 6.6. Grid after three refinement cycles with the weighted
norm indicator; and the calculated Mach number distribution.

problems in Eriksson and Johnson (1993) and earlier papers then led to the
weighted norm (6.2). But it was less clear whether this should be carried
over to hyperbolic problems approximated by finite volume methods. Sonar
(1993b) therefore experimented with various alternatives to this norm: in
particular, there are theoretical arguments for using the weak norm

$$\|\mathbf{r}^h\|_{H^{-1}(T)} := \sup_{\Phi \in H_0^1} \frac{|(\mathbf{r}^h, \Phi)_T|}{\|\Phi\|_{H_0^1(T)}},$$

where the supremum is taken over all H_0^1-functions on triangle T. To approximate this, each edge of each triangle was divided into four equal parts
and straight lines parallel to the edges drawn between them; their intersections define three interior points of the triangle and thence three hat
functions Φ_i, $i = 1, 2, 3$, which take the value 1 at the subdivision node i
and 0 elsewhere. Then the weak norm is approximated by taking the maximum over these choices, rather than the supremum over all $\Phi \in H_0^1(T)$.
Results obtained with this norm were never better than with the weighted
L^2-norm, were more sensitive to chosen tolerances and in the NACA0012
problem led to unwarranted mesh refinement well away from the profile.

More recently, Süli and Houston (1997) have given a very clear account
of the Johnson (1994) paradigm, and then adopted an alternative but related approach for general finite element approximations to hyperbolic equations, together with an application to the cell-vertex method. These results
together with the success of the weighted L^2-norm refinement indicator
have prompted further theoretical developments which we will describe next
before coming back to more extensive numerical tests. The analysis was first

developed for the symmetric positive PDEs studied by Friedrichs (1958), and now called *Friedrichs systems*; so we begin by putting the Euler equations in this form.

6.3. Symmetrizing the Euler equations

It is part of the general theory of hyperbolic systems that they may be symmetrized (*i.e.*, put in a form in which the flux Jacobian matrices are symmetric) by changing to *entropy variables* (Moch 1980), and this has been exploited for the Euler equations by Hughes, Franca and Mallet (1986). In the case of an ideal gas the entropy density is given by

$$\eta(\mathbf{u}) := -\rho s,$$

where $s := \ln(p\rho^{-\gamma})$ denotes the thermodynamic entropy. We then introduce new variables, the *entropy variables*, by means of the transformation

$$\mathbf{u} \longmapsto \mathbf{U}(\mathbf{u}) := \nabla_{\mathbf{u}}\eta(\mathbf{u}). \tag{6.3}$$

Explicitly, with the pressure p given by (2.13) and in terms of the primitive variables, this has the form

$$\mathbf{U}(\mathbf{u}) = \frac{\gamma - 1}{p} \begin{bmatrix} \frac{p}{\gamma-1}(\gamma + 1 - s) - \rho E \\ \rho v_1 \\ \rho v_2 \\ -\rho \end{bmatrix} =: \begin{bmatrix} U_1 \\ U_2 \\ U_3 \\ U_4 \end{bmatrix}, \tag{6.4}$$

and the inverse mapping $\mathbf{U} \longmapsto \mathbf{u}$ is given by

$$\mathbf{u}(\mathbf{U}) = \frac{p}{\gamma - 1} \begin{bmatrix} -U_4 \\ U_2 \\ U_3 \\ 1 - \frac{1}{2}\frac{U_2^2+U_3^2}{U_4} \end{bmatrix}, \tag{6.5}$$

where we now need to write $p = p(\mathbf{U})$ in terms of the entropy variables. Substituting for $\mathbf{u}$ in the Euler equations from (6.5), applying the chain rule and using the notation $A_0(\mathbf{U}) := \nabla_{\mathbf{U}}\mathbf{u}$, we can then write them as

$$A_0(\mathbf{U})\partial_t\mathbf{U} + \sum_{i=1}^{2}(\nabla_{\mathbf{u}}\mathbf{f}_i(\mathbf{u}(\mathbf{U})))A_0(\mathbf{U})\partial_{x_i}\mathbf{U} = 0, \tag{6.6}$$

which is in the form we are seeking.

Explicit expressions for A_0 and its inverse A_0^{-1} were derived by Hughes *et al.* (1986) and are given in Sonar and Süli (1998); they are quite complicated and it is simplest to consider them via the intermediate system of primitive variables, using the transformation (2.16). Fortunately, they are not needed in order to show that all the coefficient matrices in this new form of the equations are symmetric: that is, if we write the system in the compact notation

$$\sum_{i=0}^{2} A_i(\mathbf{U})\partial_{x_i}\mathbf{U} = \mathbf{0}, \tag{6.7}$$

with $x_0 := t$ and in which $A_i(\mathbf{U}) := \nabla_{\mathbf{u}}\mathbf{f}_i(\mathbf{u}(\mathbf{U}))A_0(\mathbf{U}) \equiv \nabla_{\mathbf{U}}\mathbf{f}_i$, $i = 1, 2$, these matrices and $A_0(\mathbf{U})$ are all symmetric 4×4 matrices. Following Sonar and Süli (1998), to show that this is so we introduce the scalar quantities

$$r(\mathbf{U}) = \mathbf{U}^T\mathbf{u} - \eta, \quad \text{and} \quad s_i(\mathbf{U}) = \mathbf{U}^T\mathbf{f}_i - q_i, \quad i = 1, 2,$$

where the q_i are the entropy fluxes. Then it is easily seen that

$$\nabla_{\mathbf{U}}r(\mathbf{U}) = \mathbf{u} + (\nabla_{\mathbf{U}}\mathbf{u})^T\mathbf{U} - (\nabla_{\mathbf{U}}\mathbf{u})^T\nabla_{\mathbf{u}}\eta = \mathbf{u};$$

and in the same way, by making use of the relation (2.5) satisfied by the entropy fluxes, we have

$$\nabla_{\mathbf{U}}s_i(\mathbf{U}) = \mathbf{f}_i + (\nabla_{\mathbf{U}}\mathbf{f}_i)^T\mathbf{U} - (\nabla_{\mathbf{U}}\mathbf{u})^T\nabla_{\mathbf{u}}q_i = \mathbf{f}_i.$$

It follows that $A_0(\mathbf{U}) \equiv \nabla_{\mathbf{U}}\mathbf{u}$ is the Hessian of the scalar r and is therefore symmetric: similarly, for $i = 1, 2$, $A_i(\mathbf{U}) \equiv \nabla_{\mathbf{U}}\mathbf{f}_i$ is the Hessian of the scalar s_i and is therefore symmetric.

In order to carry forward the error analysis developed for Friedrichs systems, we next need to carry out a local linearization. This is done about a constant mean state in each cell: that is, we assume the existence of a constant state $\mathbf{U}_c \in \mathbb{R}^4$ such that the decomposition

$$\mathbf{U} = \mathbf{U}_c + \mathbf{V}$$

holds for a small non-constant perturbation function $\mathbf{V}$. It follows that

$$\mathbf{u}(\mathbf{U}) = \mathbf{u}(\mathbf{U}_c + \mathbf{V}) = \mathbf{u}(\mathbf{U}_c) + \nabla_{\mathbf{U}}\mathbf{u}(\mathbf{U}_c)\mathbf{V} + \mathcal{O}(|\mathbf{V}|^2)$$
$$= \mathbf{u}(\mathbf{U}_c) + A_0(\mathbf{U}_c)\mathbf{V} + \mathcal{O}(|\mathbf{V}|)^2;$$

and in a similar way, with $\mathbf{F}_i(\mathbf{U}) := \mathbf{f}_i(\mathbf{u}(\mathbf{U}))$, we have

$$\mathbf{f}_i(\mathbf{u}(\mathbf{U})) = \mathbf{F}_i(\mathbf{U}_c) + \nabla_{\mathbf{u}}\mathbf{f}_i(\mathbf{u}(\mathbf{U}_c))A_0(\mathbf{U}_c)\mathbf{V} + \mathcal{O}(|\mathbf{V}|^2). \tag{6.8}$$

Writing $\mathbf{u}_c := \mathbf{u}(\mathbf{U}_c)$ and dropping the $\mathcal{O}(|\mathbf{V}|^2)$ terms, we then obtain the symmetric system

$$A_0(\mathbf{U}_c)\partial_t\mathbf{V} + \sum_{i=1}^{2} \nabla_{\mathbf{u}}\mathbf{f}_i(\mathbf{u}_c)A_0(\mathbf{U}_c)\partial_{x_i}\mathbf{V} = \mathbf{0}, \tag{6.9}$$

in which all matrix elements are constant. Finally, we write this in standard form as

$$L_E\mathbf{V} := \sum_{i=0}^{2} A_i(\mathbf{U}_c)\partial_{x_i}\mathbf{V} = \mathbf{0}, \tag{6.10}$$

where the A_i are given by

$$A_i(\mathbf{U}_c) := \nabla_{\mathbf{u}} \mathbf{f}_i(\mathbf{u}_c) A_0(\mathbf{U}_c), \quad i = 1, 2. \tag{6.11}$$

6.4. Dual graph-norm error indicators for Friedrichs systems

The *a posteriori* error analysis developed in Houston, Mackenzie, Süli and Warnecke (1999) and earlier papers was for a more general Friedrichs system than (6.10), as was that in Sonar and Süli (1998) where it was applied to the Euler equations. In summarizing these presentations, we therefore consider the system

$$L\mathbf{U} := \sum_{j=0}^{2} A_j(\mathbf{x}) \partial_{x_j} \mathbf{U} + C(\mathbf{x}) \mathbf{U} = \mathbf{0}, \tag{6.12}$$

where $\mathbf{x} = (x_0, x_1, x_2)^T := (t, x_1, x_2)^T$ is a space-time coordinate. Here the matrices A_j are symmetric with Lipschitz-continuous elements, the matrix C has continuous elements and we will assume that A_0 is positive definite. By assuming that it is symmetric positive definite in a region Ω we mean that there exists a positive constant $c_0 \equiv c_0(\Omega)$ such that

$$\tfrac{1}{2}\big(K(\mathbf{x}) + K^*(\mathbf{x})\big) \geq c_0 I, \tag{6.13}$$

for all $\mathbf{x} \in \overline{\Omega}$, where for some $\xi \in \mathbb{R}^3$, with $|\xi| = 1$, the matrix K is defined by

$$K := C - \tfrac{1}{2} \sum_{j=0}^{2} \partial_{x_j} A_j + \sum_{j=0}^{2} \xi_j A_j.$$

For the error analysis of such a system we need to distinguish the error generated within each cell and that transported from one cell to its neighbours. So for the unsteady problem integrated over one time step we consider a space-time prism $P_i^n := (n\Delta t, (n+1)\Delta t) \times T_i$ based on a triangle $T_i \in \mathcal{T}^h$, and introduce the matrix

$$B(\mathbf{x}) := \sum_{j=0}^{2} n_j A_j(\mathbf{x}),$$

where $\mathbf{n} = (n_0, n_1, n_2)^T$ denotes the unit outward normal vector to its boundary ∂P_i^n at a point $\mathbf{x}$. We suppose that B is non-singular at each such point, *i.e.*, ∂P_i^n is a non-characteristic hypersurface for the operator L. Now we split B into a negative semi-definite part B^- and a positive semi-definite part $B^+ = B - B^-$. We call $B^-\mathbf{U}$ the inflow part of the vector field $\mathbf{U}$ and $B^+\mathbf{U}$ its outflow part. Then it was shown in Friedrichs (1958) and Lax and Phillips (1960) that symmetric hyperbolic systems have unique strong solutions subject to a boundary condition that specifies $B^-\mathbf{U}$.

Now suppose that our numerical approximation $\mathbf{u}^h$ is converted into entropy variables to give $\mathbf{U}^h$. We consider the following boundary value problem in P_i^n:

$$L\hat{\mathbf{U}}^h = \mathbf{0} \quad \text{on} \quad P_i^n$$
$$B^-\hat{\mathbf{U}}^h|_{\partial P_i^n} = B^-\mathbf{U}^h|_{\partial P_i^n}.$$

We interpret the function $\hat{\mathbf{U}}^h$ as the exact solution of (6.12) in P_i^n with inflow boundary data contaminated by the *transported error* carried by $\mathbf{U}^h$. Hence we define the *cell error* by

$$\mathbf{e}_c \equiv \mathbf{e}_{P_i^n}^{\text{cell}} := \mathbf{U}^h - \hat{\mathbf{U}}^h, \tag{6.14}$$

being the error in the numerical solution which is produced on P_i^n; while the transported error is given by

$$\mathbf{e}_t \equiv \mathbf{e}_{P_i^n}^{\text{trans}} := \hat{\mathbf{U}}^h - \mathbf{U}. \tag{6.15}$$

The sum of these two is the total error $\mathbf{e}_{P_i^n} \equiv \mathbf{U}^h - \mathbf{U}$.

It is clear that the residual calculated in a given prism has no control over the transported error, which is just advected into the cell from upwind: while the cell error is governed directly by the residual via the relation

$$\mathbf{r}^h = L\mathbf{e}_{P_i^n} = L\mathbf{e}_{P_i^n}^{\text{cell}} \equiv L\mathbf{e}_c \quad \text{on} \quad P_i^n, \tag{6.16}$$

which is subject to a zero inflow boundary condition. Our next objective then is to obtain two-sided bounds of the form (6.1) for the cell error and cell residual. Note that this will only give some confidence in the overall accuracy of a computation if the cell errors dominate the transported errors; but in any case it should give a reliable indicator for local mesh refinement.

To develop such error bounds, it was shown in Houston *et al.* (1999) that one can define the following spaces and their associated norms: for certain weight functions w_i^n, the weighted graph-norm $\|\cdot\|_{D(L,P_i^n)}$ on

$$D_-(L, P_i^n) := \{\phi \in L^2(P_i^n) \mid L\phi \in L^2(P_i^n), \ B^-\phi = 0 \ \text{on} \ \partial P_i^n\},$$

is defined by

$$\|\phi\|_{D(L,P_i^n)} = \left[\|w_i^n\phi\|_{L^2(P_i^n)}^2 + \|w_i^n L\phi\|_{L^2(P_i^n)}^2\right]^{1/2},$$

and the associated dual graph-norm by

$$\|v\|_{D'(L,P_i^n)} := \sup_{\phi \in D_-(L,P_i^n)} \frac{|(v, \phi)_{P_i^n}|}{\|\phi\|_{D(L,P_i^n)}},$$

where $(\cdot, \cdot)_{P_i^n}$ denotes the usual L^2 inner product on P_i^n. Similarly, by introducing the formal adjoint

$$L^* \phi := -\sum_{j=0}^{2} \partial_{x_j}(A_j \phi) + C^* \phi,$$

with

$$\phi \in D_+(L^*, P_i^n) := \{\phi \in L^2(P_i^n) \,|\, L^* \phi \in L^2(P_i^n), \ B^+ \phi = 0 \ \text{on} \ \partial P_i^n\},$$

we can equip $D_+(L^*, P_i^n)$ with a corresponding graph-norm and associated dual graph-norm.

Establishing the necessary trace theorems in these spaces is a nontrivial part of the analysis which goes on to establish the following local *a posteriori* error bound.

Theorem 6.1. For the symmetric hyperbolic system (6.12), the cell error satisfies the following inequalities:

$$\left(\min_{P_i^n} w_i^n\right) \|\mathbf{r}^h\|_{D'(L^*, P_i^n)} \le \|\mathbf{e}_c\|_{L^2(P_i^n)} \tag{6.17}$$

$$\le \left(1 + \frac{1}{c_0^2}\right)^{1/2} \left(\max_{P_i^n} w_i^n\right) \|\mathbf{r}^h\|_{D'(L^*, P_i^n)},$$

where $c_0 \equiv c_0(P_i^n)$ is as defined in (6.13).

Proof. See Sonar (2002) or Sonar and Süli (1998).   □

In order to exploit this theorem by calculating the dual graph-norm of a residual on a space-time prism, we need first to consider the inflow and outflow boundary conditions for the Euler equations. It is more convenient to work with the unsymmetric form of the equations in the entropy variables obtained by premultiplying (6.9) by A_0^{-1}. Thus, denoting the resultant matrices by $\tilde{A}_j$, we have

$$\tilde{B}(P_i^n) = \sum_{j=0}^{2} n_j \tilde{A}_j \equiv n_0 I + \sum_{j=1}^{2} n_j A_0^{-1}(\mathbf{U}_{ci}) \nabla_{\mathbf{u}} \mathbf{f}_j(\mathbf{u}_{ci}) A_0(\mathbf{U}_{ci}), \tag{6.18}$$

where $\mathbf{U}_{ci}, \mathbf{u}_{ci}$ denote constant mean states of entropy and conservative variables, respectively, within the prism P_i^n. Then, as above, we have the inflow/outflow subdivision

$$\tilde{B}(P_i^n) = \tilde{B}^-(P_i^n) + \tilde{B}^+(P_i^n).$$

In particular, it is clear that the bottom of the prism is an inflow boundary while the top is an outflow boundary.

For the sides, we note that the matrices $\tilde{A}_j$ are similar to $\nabla_\mathbf{u}\mathbf{f}_j$, for $j = 1, 2$, and $\tilde{A}_0$ is similar to I; so each eigenvalue of $\tilde{B}(P_i^n)$ is given by

$$\mathrm{eig}(\tilde{B}(P_i^n)) = n_0 + \mathrm{eig}\left(\sum_{j=1}^2 n_j \nabla_\mathbf{u}\mathbf{f}_j(\mathbf{u}_{ci})\right),$$

in terms of the eigenvalues of $\nabla_\mathbf{u}\mathbf{f}_j$ which were given in Section 2.3. Thus we write

$$\Lambda(\mathbf{u}_{ci}, n_1, n_2) := \mathrm{diag}\{v_n, v_n, v_n + c_{ci}|(n_1, n_2)|, v_n - c_{ci}|(n_1, n_2)|\},$$

where $v_n = \sum_{j=1}^2 n_j v_{ci,j}$ is the flow speed in the normal direction and c_{ci} is the mean constant speed of sound in P_i^n. We split Λ into a matrix Λ^+ containing the positive eigenvalues, and Λ^- containing the negative eigenvalues. So we finally have a representation for the boundary matrix $\tilde{B}(P_i^n)$ in the form

$$\tilde{B}(P_i^n) = n_0 I + A_0^{-1}(\mathbf{U}_{ci})\left[P\Lambda^+ P^{-1}(\mathbf{u}_{ci}, n_1, n_2)\right]A_0(\mathbf{U}_{ci}) \qquad (6.19)$$

$$+ A_0^{-1}(\mathbf{U}_{ci})\left[P\Lambda^- P^{-1}(\mathbf{u}_{ci}, n_1, n_2)\right]A_0(\mathbf{U}_{ci}), \qquad (6.20)$$

where $P(\mathbf{u}_{ci}, n_1, n_2)$ is the matrix which diagonalizes $\sum_{j=1}^2 n_j \nabla_\mathbf{u}\mathbf{f}_j(\mathbf{u}_{ci})$. Note that this subdivision essentially corresponds to the flux vector splitting of Steger and Warming (1981).

There are several ways in which these formulae may be approximated to yield a practical refinement indicator. The simplest, whose use is reported on in Sonar (2002), is based on using an explicit time-stepping procedure so that divided differences are used instead of space-time basis functions in the prisms P_i^n. Then the graph-norm calculation is approximated by subdividing each triangle into 16 equal subtriangles, in the manner described above in connection with calculating the weak H^{-1}-norm. In this case we obtain three interior nodes and twelve boundary nodes for each triangle, giving 15 different test functions. Details of the calculation are given in Sonar (2002).

The adaptive procedure using this graph-norm refinement indicator makes use of two tolerances $\mathrm{TOL}_{\mathrm{refine}}$ and $\mathrm{TOL}_{\mathrm{coarse}}$ for the refinement and coarsening algorithms described above, and their choice in this case is quite critical. A typical mesh obtained in the case of the transonic NACA0012 flow problem described earlier is shown in Figure 6.7. The flow features are very well captured and the indicator has started to detect the supersonic region on the upper side. Another nice feature in comparison with the

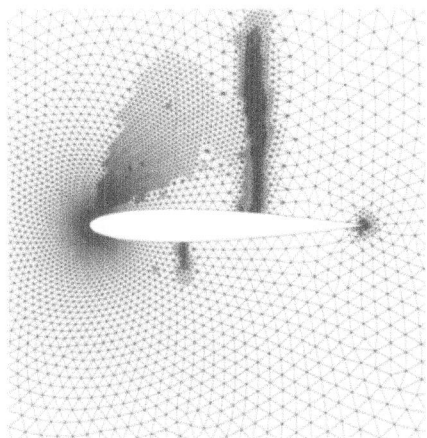

Figure 6.7. Grid after three refinement/coarsening cycles with the graph-norm indicator for transonic flow about the NACA0012 aerofoil.

H^{-1}-indicator described above is the absence of any noise spoiling the grid far away from the obstacle.

Although these and other results look quite promising there are several problems associated with the dual graph-norm refinement indicator. First of all, it is very expensive to compute in comparison with the weighted L^2-norm indicator. The second problem concerns its sensitivity: it turns out that in some calculations a small change in tolerance can influence the adapted grid enormously, which makes it hard to use in practice. This seems to arise from the inability of the indicator to detect contact discontinuities. Results which illustrate these points may be found in Sonar (2002).

6.5. Closing the loop and further tests

As claimed in Sonar and Süli (1998), the dual graph-norm error indicator which was described in the previous subsection seems to have been the first effective refinement indicator for the Euler equations with a sound mathematical foundation. On the other hand it has several practical disadvantages which were also indicated there. The more practical indicator would seem to be that based on the weighted L^2-norm, which gave the results in Figure 6.6.

It was therefore an important step towards resolving this dilemma when the following result was proved in Houston *et al.* (1999): for positive constants C and C' we have

$$C'h_0\|P_{h_0}\mathbf{r}^h\|_{L^2(P_i^n)} \leq \|\mathbf{e}_c\|_{L^2(P_i^n)} \leq Ch\|\mathbf{r}^h\|_{L^2(P_i^n)}, \qquad (6.21)$$

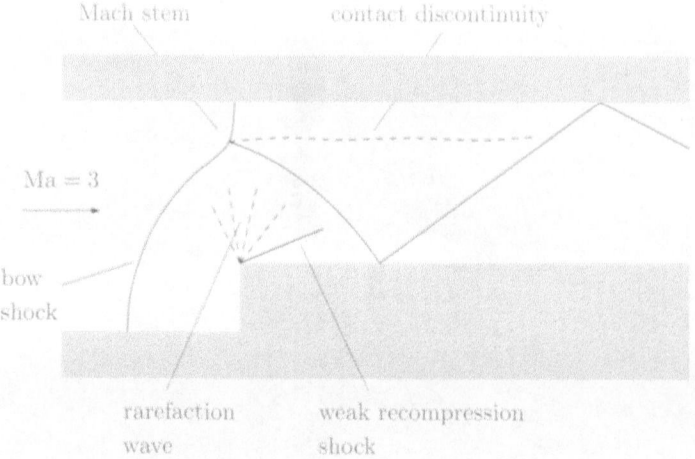

Figure 6.8. Flow phenomena in the
channel with forward facing step.

where h denotes the diameter of P_i^n, and P_{h_0} is the orthogonal projector
onto a finite element space on a micropartition of P_i^n of diameter h_0. This
micropartition corresponds to that used for approximating the graph-norms
of the bounds in (6.17).

With this result we finally have a practical refinement indicator which
rests on a firm theoretical base. So we conclude this survey by giving some
results for the test problem due to Woodward and Colella (1984), which
we used earlier. This is actually an unsteady flow problem in which the
forward-facing step is inserted into a steady, uniform Ma = 3.0 flow down
the channel. A complicated shock system develops whose steady state is
sketched in Figure 6.8: note too the contact discontinuity which emerges
from the point where the bow shock joins the Mach stem attached to the
upper boundary.

Computations for this problem were carried out with the unsteady DLR-
τ-code of Sonar (1993a), Sonar, Hannemann and Hempel (1994) and Meister
(1994), and reported in Sonar (2002). In Figure 6.9 we show the meshes
that were generated at various times using the weighted L_2-norm refinement
indicator. It is seen to have detected all the relevant flow phenomena: the
shock system is clearly visible, as is the contact discontinuity starting at
the Mach stem. Note that the corner point of the step is a true corner
singularity since it corresponds to the centre of a rarefaction wave: the
indicator has also detected phenomena associated with this special point
and refined the region in its vicinity. In Figure 6.10 we show the density
distributions obtained on these meshes.

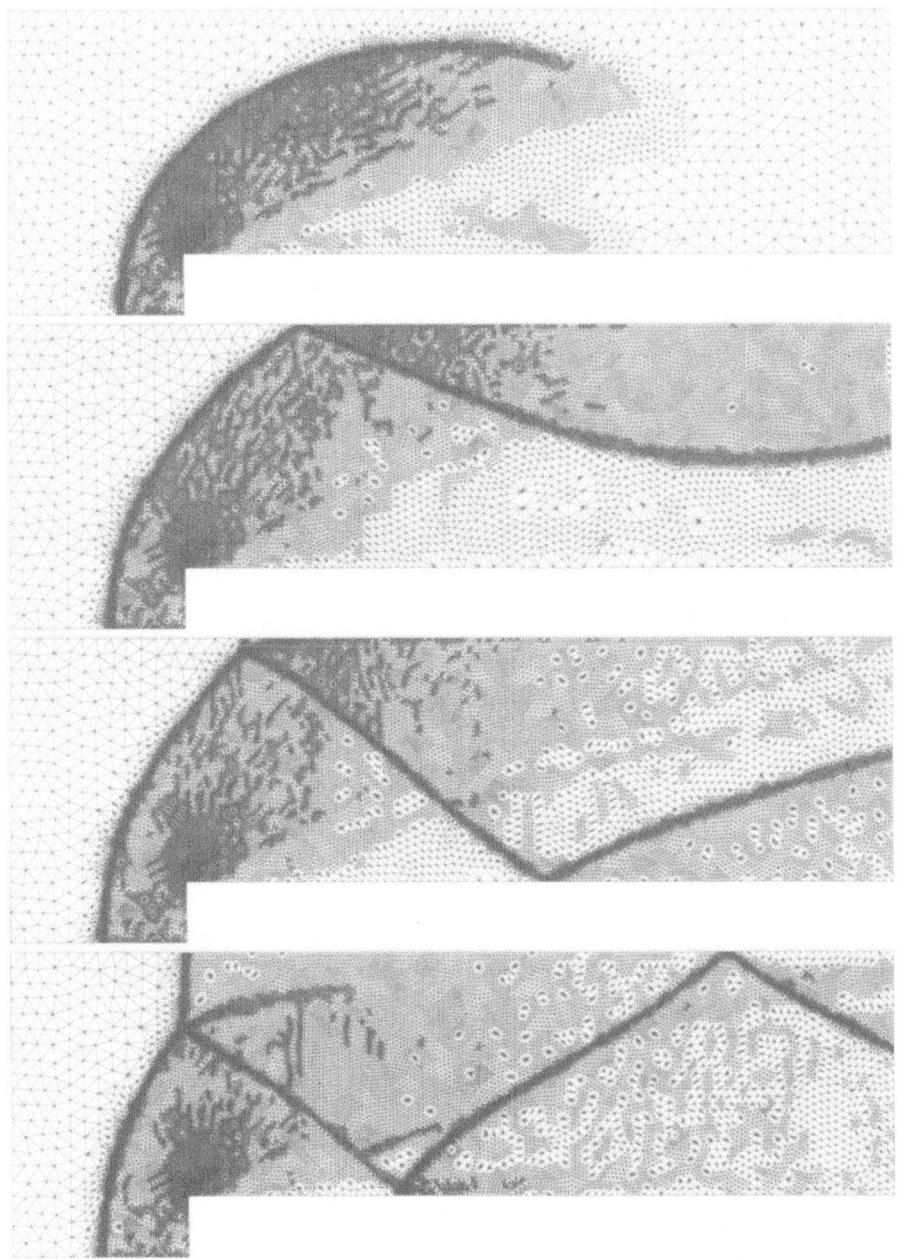

Figure 6.9. Adapted grids at different times
for the Woodward and Colella problem.

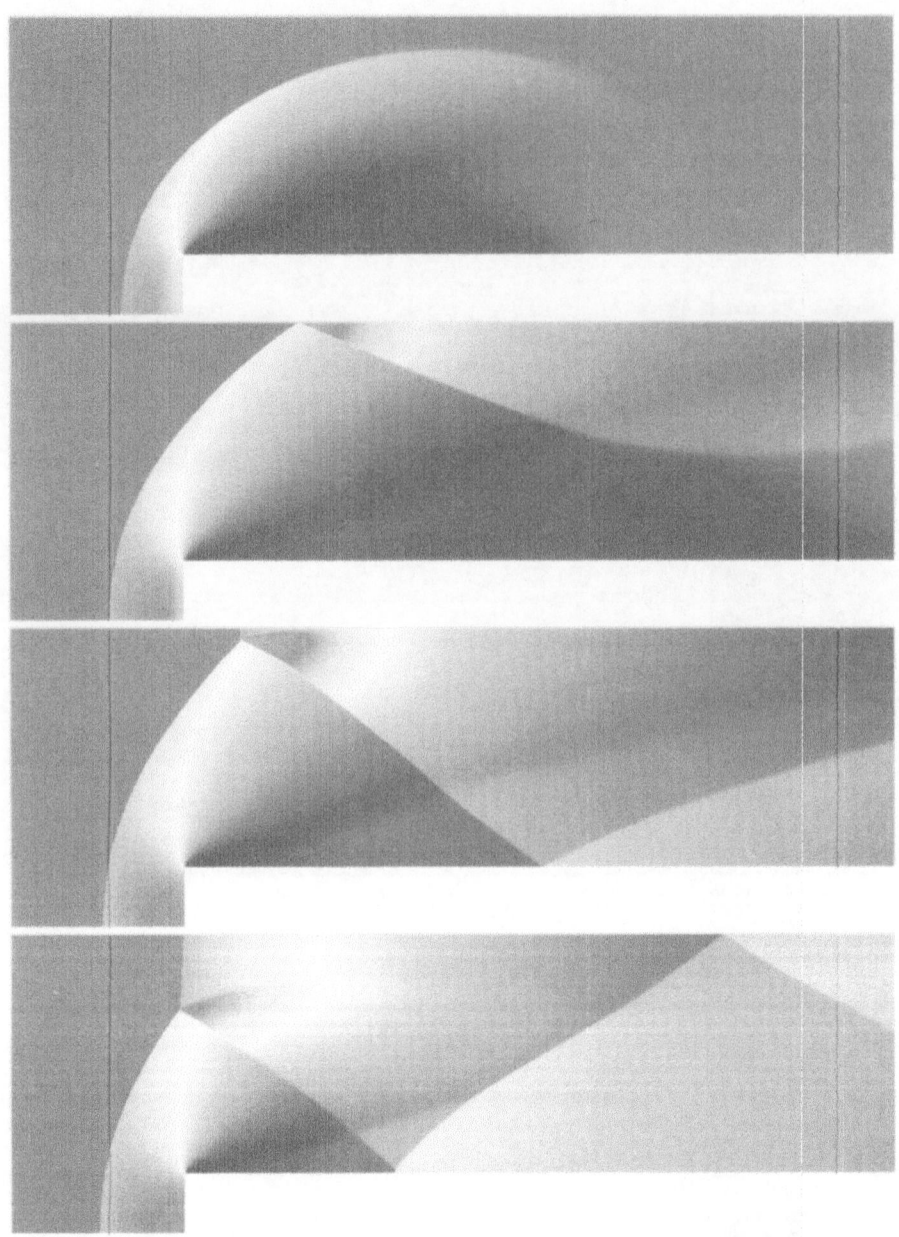

Figure 6.10. Density distributions
corresponding to the grids in Figure 6.9.

7. Concluding remarks

- Finite volume methods share with finite element methods the viewpoint that it is primarily the solution that is being approximated, rather than the equation or any operator in it.

- Their key guiding principle is exact satisfaction of the integral conservation laws; so they are at their most effective where solutions contain shocks or other discontinuities.

- Their advantage over evolution-Galerkin methods is that even explicit methods need a less good approximation to the evolution operator defined by the PDE as it is used only to calculate the fluxes.

- Limited as they are to using only piecewise constant functions as test functions, with the consequential heavy dependence on the recovery stage, they may well be superseded by discontinuous Galerkin methods, but only when these methods recognize finite volume methods as their proper antecedents and learn from them.

- The jury was out for a long time in the judgement between cell-centre and cell-vertex methods; but it now seems that node-centred schemes have acquired an edge over either. Cell-centre schemes still hold centre stage as regards practical codes, because of their more reliable representation of shocks as compared with cell-vertex methods. But node-centred methods have advantages over cell-centre methods in regard to generating hierarchies of recovery procedures.

- Although the ideal of a guaranteed error bound derived from an *a posteriori* residual is still not achievable for finite volume computations of nonlinear hyperbolic conservation laws, progress to that end during the past decade has been quite remarkable: in particular, soundly based practical mesh refinement indicators are now available.

- Further progress with these methods is likely to lie with implicit algorithms, exploiting the techniques developed in the optimization field for the rapid solution of large nonlinear systems of equations. This is consistent with the practical requirement in aerodynamics that flow calculations should be fully incorporated into the design cycle.

Acknowledgements

We are grateful to Phil Roe, Gil Strang and Endre Süli for their comments on a draft of the paper.

REFERENCES

R. Abgrall (1994a), 'An essentially non-oscillatory reconstruction procedure on finite-element type meshes: Application to compressible flows', *Comput. Methods Appl. Mech. Engrg.* **116**, 95–101.

R. Abgrall (1994b), 'On essentially non-oscillatory schemes on unstructured meshes: Analysis and implementation', *J. Comput. Phys.* **114**, 45–58.

R. Abgrall and T. Sonar (1997), 'On the use of Mühlbach expansions in the recovery step of ENO methods', *Numer. Math.* **76**, 1–27.

K. J. Badcock and B. E. Richards (1995), 'Implicit time stepping methods for the Navier–Stokes equations', *AIAA Journal* **34**, 555–559.

R. E. Bank, A. H. Sherman and A. Weiser (1983), Refinement algorithms and data structures for local regular mesh refinements, in *Scientific Computing* (R. Stepleman et al., eds), IMACS North-Holland, pp. 3–17.

J. W. Barrett, G. Moore and K. W. Morton (1988a), 'Optimal recovery in the finite element method, Part I: Recovery from weighted L^2 fits', *IMA J. Numer. Anal.* **8**, 149–184.

J. W. Barrett, G. Moore and K. W. Morton (1988b), 'Optimal recovery in the finite element method, Part II: Defect correction for ordinary differential equations', *IMA J. Numer. Anal.* **8**, 527–540.

M. Ben-Artzi and J. Falcovitz (1984), 'A second order Godunov-type scheme for compressible fluid dynamics', *J. Comput. Phys.* **55**, 1–32.

G. Birkhoff (1983), 'Numerical fluid dynamics', *SIAM Rev.* **25**, 1–34.

C. de Boor (2001), *A Practical Guide to Splines*, revised edn, Springer, New York.

A. Brandt (1977), 'Multilevel adaptive solutions to boundary-value problems', *Math. Comp.* **31**, 333–390.

Y. Brenier (1984), 'Average multi-valued solutions for scalar conservation laws', *SIAM J. Numer. Anal.* **21**, 1013–1037.

G. Bruhn (1985), 'Erhaltungssätze und schwache Lösungen in der Gasdynamik', *Math. Methods Appl. Sci.* **7**, 470–479.

D. S. Butler (1960), 'The numerical solution of hyperbolic systems of partial differential equations in three independent variables', *Proc. Roy. Soc.* **255A**, 233–252.

P. N. Childs and K. W. Morton (1990), 'Characteristic Galerkin methods for scalar conservation laws in one dimension', *SIAM J. Numer. Anal.* **27**, 553–594.

P. G. Ciarlet (1987), *The Finite Element Method for Elliptic Problems*, 2nd edn, North-Holland.

P. G. Ciarlet and P. R. Raviart (1972), 'General Lagrange and Hermite interpolation in R^n with applications to the finite element method', *Arch. Rat. Mech. Anal.* **46**, 177–199.

B. Cockburn, G. E. Karniadakis and C. W. Shu (2000), The development of discontinuous Galerkin methods, in *Proc. First International Symposium on Discontinuous Galerkin Methods*, Springer, New York, pp. 3–50.

P. Colella and P. R. Woodward (1984), 'The piecewise parabolic method (PPM) for gas-dynamical simulations', *J. Comput. Phys.* **54**, 174–201.

R. Courant, K. O. Friedrichs and H. Lewy (1928), 'Über die partiellen Differenzengleichungen der Physik', *Math. Ann.* **100**, 32–74.

P. I. Crumpton, J. A. Mackenzie and K. W. Morton (1993), 'Cell vertex algorithms for the compressible Navier–Stokes equations', *J. Comput. Phys.* **109**, 1–15.

J. A. Cunge, F. M. Holly and A. Verwey (1980), *Practical Aspects of Computational River Hydraulics*, Pitman, London.

H. Deconinck, P. L. Roe and R. Struijs (1993), 'A multidimensional generalization of Roe's flux difference splitter for the Euler equations', *Comput. Fluids* **22**, 215–222.

J. Douglas, Jr. and T. F. Russell (1982), 'Numerical methods for convection-dominated diffusion problems based on combining the method of characteristics with finite element or finite difference procedures', *SIAM J. Numer. Anal.* **19**, 321–352.

L. J. Durlofsky, B. Engquist and S. Osher (1992), 'Triangle based adaptive stencils for the solution of hyperbolic conservation laws', *J. Comput. Phys.* **98**, 64–73.

M. Eiermann and O. G. Ernst (2001), Geometric aspects of the theory of Krylov subspace methods, in *Acta Numerica*, Vol. 10 (A. Iserles, ed.), Cambridge University Press, pp. 251–312.

H. C. Elman, D. J. Silvester and A. J. Wathen (2005), *Finite Elements and Fast Iterative Solvers*, Oxford University Press.

B. Engquist and S. Osher (1981), 'One-sided difference approximations for non-linear conservation laws', *Math. Comp.* **36**, 321–352.

K. Eriksson and C. Johnson (1993), 'Adaptive streamline diffusion finite element methods for stationary convection-diffusion problems', *Math. Comp.* **60**, 167–188.

K. Eriksson, D. Estep, P. Hansbo and C. Johnson (1995), Introduction to adaptive methods for differential equations, in *Acta Numerica*, Vol. 4 (A. Iserles, ed.), Cambridge University Press, pp. 105–158.

R. P. Federenko (1964), 'The speed of convergence of one iterative process', *USSR Comp. Math. and Math. Phys.* **4**, 227–235.

L. Fezoui and B. Stoufflet (1989), 'A class of implicit upwind schemes for Euler simulations with unstructured meshes', *J. Comput. Phys.* **84**, 174–206.

M. A. Freitag and K. W. Morton (2007), 'The Preissmann box scheme and its modification for transcritical flows', *Internat. J. Numer. Methods Engrg.*, to appear.

O. Friedrich (1998), 'Weighted essentially non-oscillatory schemes for the interpolation of mean values on unstructured grids', *J. Comput. Phys.* **144**, 194–212.

K. O. Friedrichs (1958), 'Symmetric positive linear differential operators', *Comm. Pure Appl. Math.* **11**, 333–418.

E. Godlewski and P.-A. Raviart (1991), *Hyperbolic Systems of Conservation Laws*, Ellipses, Paris.

S. K. Godunov (1959), 'A finite difference method for the numerical computation of discontinuous solutions of the equations of fluid dynamics', *Mat. Sb.* **47**, 271–306.

M. Golomb and H. F. Weinberger (1959), Optimal approximation and error bounds, in *Symposium on Numerical Approximation* (R. E. Langer, ed.), University of Wisconsin Press, Madison, pp. 117–190.

A. Haar (1919), 'Über die Variation der Doppelintegrale', *J. Reine Angew. Math.* **149**, 1–18.

E. Hairer and G. Wanner (1996), *Solving Ordinary Differential Equations II: Stiff and Differential-Algebraic Problems*, Springer, Berlin/Heidelberg.

P. Hansbo and C. Johnson (1991), 'Adaptive streamline diffusion methods for compressible flow using conservation variables', *Comput. Methods Appl. Mech. Engrg.* **87**, 267–280.

A. Harten (1983), 'High resolution schemes for conservation laws', *J. Comput. Phys.* **49**, 357–393.

A. Harten (1984), 'On a class of high resolution total-variation-stable finite-difference schemes', *SIAM J. Numer. Anal.* **21**, 1–23.

A. Harten (1989), 'ENO schemes with subcell resolution', *J. Comput. Phys.* **83**, 148–184.

A. Harten and S. R. Chakravarthy (1991), Multi-dimensional ENO schemes for general geometries. Report no. 91-76, ICASE.

A. Harten and S. Osher (1987), 'Uniformly high-order nonoscillatory schemes I', *SIAM J. Numer. Anal.* **24**, 279–309.

A. Harten, B. Engquist, S. Osher and S. R. Chakravarthy (1987), 'Uniformly high order accurate essentially non-oscillatory schemes III', *J. Comput. Phys.* **71**, 231–303.

A. Harten, P. Lax and B. van Leer (1983), 'On upstream differencing and Godunov-type schemes for hyperbolic conservation laws', *SIAM Rev.* **25**, 35–61.

A. Harten, S. Osher, B. Engquist and S. R. Chakravarthy (1986), 'Some results on uniformly high-order accurate essentially nonoscillatory schemes', *Appl. Numer. Math.* **2**, 347–377.

M. R. Hestenes and E. Stiefel (1952), 'Methods of conjugate gradients for solving linear problems', *J. Res. Nat. Bur. Stand.* **49**, 409–436.

C. Hirsch (1988), *Numerical Computation of Internal and External Flows, Vol. 1: Fundamentals of Numerical Discretization*, Wiley.

C. Hirsch (1990), *Numerical Computation of Internal and External Flows, Vol. 2: Computational Methods for Inviscid and Viscous Flows*, Wiley.

P. Houston, J. M. Mackenzie, E. Süli and G. Warnecke (1999), '*A posteriori* error analysis for numerical approximation of Friedrichs systems', *Numer. Math.* **82**, 433–470.

T. J. R. Hughes, L. P. Franca and M. Mallet (1986), 'A new finite element formulation for compressible fluid dynamics I: Symmetric forms of the compressible Euler and Navier–Stokes equations and the second law of thermodynamics', *Comput. Methods Appl. Mech. Engrg.* **54**, 223–234.

A. Iske and T. Sonar (1996), 'On the structure of function spaces in optimal recovery of point functionals for ENO-schemes by radial basis functions', *Numer. Math.* **74**, 177–201.

A. M. Jaffe (2006), 'The Millennium Grand Challenge in Mathematics', *Notices Amer. Math. Soc.* **53**, 652–660.

A. Jameson (1979), Acceleration of transonic potential flow calculations on arbitrary meshes by the multiple grid method, in *AIAA 4th Computational Fluid Dynamics Conference*, Paper 79-1458.

A. Jameson, W. Schmidt and E. Turkel (1981), Numerical solution of the Euler equations by finite volume methods using Runge–Kutta time stepping schemes, in *AIAA 14th Fluid and Plasma Dynamics Conference*, Paper 81-1259.

C. Johnson (1994), A new paradigm for adaptive finite element methods, in *The Mathematics of Finite Element Methods and Applications: Highlights 1993* (J. R. Whiteman, ed.), Wiley, pp. 105–120.

R. Klötzler (1970), *Mehrdimensionale Variationsrechnung*, Birkhäuser.

S. Krŭzkov (1970), 'First-order quasilinear equations in several variables', *Math. USSR Sb.* **10**, 217–243.

P. D. Lax and R. S. Phillips (1960), 'Local boundary conditions for dissipative symmetric linear differential operators', *Comm. Pure Appl. Math.* **13**, 427–455.

P. D. Lax and B. Wendroff (1960), 'Systems of conservation laws', *Comm. Pure Appl. Math.* **13**, 217–237.

B. van Leer (1979), 'Towards the ultimate conservative difference scheme V: A second order sequel to Godunov's method', *J. Comput. Phys.* **32**, 101–136.

B. P. Leonard (1991), 'The ULTIMATE conservative difference scheme applied to unsteady one-dimensional advection', *Comput. Methods Appl. Mech. Engrg.* **88**, 17–74.

P. Lesaint (1977), Numerical solution of the equation of continuity, in *Topics in Numerical Analysis III* (J. J. H. Miller, ed.), Academic Press, pp. 199–222.

R. J. LeVeque (2002), *Finite Volume Methods for Hyperbolic Problems*, Cambridge University Press.

P. Lin, K. W. Morton and E. Süli (1993), 'Euler characteristic Galerkin scheme with recovery', *Math. Modelling Numer. Anal.* **27**, 863–894.

P. Lin, K. W. Morton and E. Süli (1997), 'Characteristic Galerkin schemes for scalar conservation laws in two and three space dimensions', *SIAM J. Numer. Anal.* **34**, 779–796.

M.-S. Liou and C. J. Steffen, Jr. (1993), 'A new flux splitting scheme', *J. Comput. Phys.* **107**, 23–39.

X.-D. Liu, S. Osher and T. Chan (1994), 'Weighted essentially non-oscillatory schemes', *J. Comput. Phys.* **115**, 200–212.

M. Lukáčová-Medviďová, K. W. Morton and G. Warnecke (2000), 'Evolution-Galerkin methods for hyperbolic systems in two space dimensions', *Math. Comp.* **69**, 1355–1384.

M. Lukáčová-Medviďová, K. W. Morton and G. Warnecke (2002), 'Finite volume evolution-Galerkin methods for Euler equations of gas dynamics', *Internat. J. Numer. Methods Fluids* **40**, 425–434.

M. Lukáčová-Medviďová, K. W. Morton and G. Warnecke (2004), 'Finite volume evolution-Galerkin methods for hyperbolic systems', *SIAM J. Sci. Comput.* **26**, 1–30.

P. W. McDonald (1971), The computation of transonic flow through two-dimensional gas turbine cascades, in *ASME Proc.*, Paper 71-GT-89, ASME, New York.

D. J. Mavriplis (1995), Multigrid techniques for unstructured meshes, in *VKI Lecture Series* VKI-LS 1995-02, von Karman Institute for Fluid Dynamics, Belgium.

D. J. Mavriplis (2002), 'An assessment of linear versus nonlinear multigrid methods for unstructured meshes', *J. Comput. Phys.* **175**, 302–325.

A. Meister (1994), Ein Beitrag zum DLR-τ-Code: Ein explizites und implizites Finite-Volume Verfahren zur Berechnung instationärer Strömungen auf unstructurierten Gittern. DLR Internal Report IB 223-94 A 36, Institute for Fluid Mechanics, DLR Göttingen.

A. Meister (1998), 'Comparison of different Krylov subspace methods embedded in an implicit finite volume scheme for the computation of viscous and inviscid flow fields on unstructured grids', *J. Comput. Phys.* **140**, 311–345.

A. Meister and M. Oevermann (1996), Computation of laminar and turbulent flow fields on unstructured grids with a finite volume scheme, in *Proc. 2nd Seminar on Euler and Navier–Stokes Equations, Prague*, pp. 61–62.

A. Meister and C. Vömel (2001), 'Efficient preconditioning of linear systems arising from the discretization of hyperbolic conservation laws', *Adv. Comput. Math.* **14**, 49–73.

C. A. Micchelli and T. J. Rivlin (1977), A survey of optimal recovery, in *Optimal Estimation in Approximation Theory* (C. A. Micchelli and T. J. Rivlin, eds), Plenum, pp. 1–54.

M. S. Moch (1980), 'Systems of conservation laws of mixed type', *J. Diff. Equations* **37**, 70–88.

C. B. Morrey (1960), 'Multiple integral problems in the calculus of variations and related topics', *Ann. Scuola Norm. Pisa* (III) **14**, 1–61.

K. W. Morton (1996), *Numerical Solution of Convection-Diffusion Problems*, Chapman and Hall.

K. W. Morton (1998), 'On the analysis of finite volume methods for evolutionary problems', *SIAM J. Numer. Anal.* **35**, 2195–2222.

K. W. Morton (2001), 'Discretization of unsteady hyperbolic conservation laws', *SIAM J. Numer. Anal.* **39**, 1556–1597.

K. W. Morton and D. F. Mayers (2005), *Numerical Solution of Partial Differential Equations*, 2nd edn, Cambridge University Press.

K. W. Morton and M. F. Paisley (1989), 'A finite volume scheme with shock fitting for the steady Euler equations', *J. Comput. Phys.* **80**, 168–203.

K. W. Morton and M. A. Rudgyard (1988), Shock recovery and the cell vertex scheme for the steady Euler equations, in *11th International Conference on Numerical Methods in Fluid Dynamics* (D. L. Dwoyer, M. Y. Hussaini and R. G. Voigt, eds), Springer, pp. 424–428.

K. W. Morton and S. M. Stringer (1998), Artificial dissipation as a feedback control with application to the cell vertex method, in *Computational Fluid Dynamics Review 1998*, Vol. 1 (M. Hafez and K. Oshima, eds), World Scientific, pp. 262–279.

G. Mühlbach (1978), 'The general Neville–Aitken algorithm and some applications', *Numer. Math.* **31**, 97–110.

C. Müller (1957), *Grundprobleme der Mathematischen Theorie Elektromagnetischer Schwingungen*, Springer.

E. M. Murman and J. D. Cole (1971), 'Calculation of plane steady transonic flows', *AIAA Journal* **9**, 114–121.

R. N. Ni (1982), 'A multiple grid system for solving the Euler equations', *AIAA Journal* **20**, 1565–1571.

E. J. Nielsen, W. K. Anderson, R. W. Walters and D. E. Kayes (1995), Application of Newton–Krylov methodology to a three-dimensional Euler code, in *Proc. 12th IAAA CFD Conf.*, Paper 95-1733-CP.

O. Oleinik (1957), 'Discontinuous solutions of nonlinear differential equations', *Usp. Mat. Nauk.* (NS) **12**, 3–73.

S. Osher and F. Solomon (1982), 'Upwind difference schemes for hyperbolic systems of conservation laws', *Math. Comp.* **38**, 339–374.

S. Ostkamp (1997), 'Multidimensional characteristic Galerkin schemes and evolution operators for hyperbolic systems', *Math. Methods Appl. Sci.* **20**, 1111–1125.

O. Pironneau (1982), 'On the transport-diffusion algorithm and its application to the Navier–Stokes equations', *Numer. Math.* **38**, 309–332.

A. Preissmann (1961), Propagation des intumescences dans les canaux et rivières, in *1st Congr. de l'Assoc. Française de Calcul*, Association Française de Calcul, Grenoble, France, pp. 433–442.

A. Quarteroni and A. Valli (1994), *Numerical Approximation of Partial Differential Equations*, Springer, Heidelberg.

M. Ricciutto, A. Csik and H. Deconinck (2005), 'Residual distribution for general time-dependent conservation laws', *J. Comput. Phys.* **209**, 249–289.

R. D. Richtmyer and K. W. Morton (1967), *Difference Methods for Initial-Value Problems*, Interscience, New York.

A. W. Rizzi and M. Inouye (1973), 'Time split finite volume method for three dimensional blunt-body flows', *AAIA Journal* **11**, 1478–1485.

P. L. Roe (1981), 'Approximate Riemann solvers, parameter vectors and difference schemes', *J. Comput. Phys.* **43**, 357–372.

P. L. Roe (1982), Fluctuations and signals: A framework for numerical evolution problems, in *Numerical Methods for Fluid Dynamics* (K. W. Morton and M. J. Baines, eds), Academic Press, pp. 219–257.

P. L. Roe (2001), Chapter 6, Numerical Methods, in *Handbook of Shockwaves*, Vol. 1 (G. Ben-dor, O. Igra and T. Elperin, eds), Academic Press, pp. 788–876.

Y. Saad and M. H. Schultz (1986), 'A generalized minimal residual algorithm for solving nonsymmetric linear systems', *J. Sci. Statist. Comput.* **7**, 856–869.

C. Shu and S. Osher (1988), 'Efficient implementation of Essentially Non-Oscillatory shock-capturing schemes', *J. Comput. Phys.* **77**, 439–471.

J. Smoller (1983), *Shock Waves and Reaction-Diffusion Equations*, Springer, New York.

G. A. Sod (1978), 'A survey of several finite difference methods for systems of nonlinear hyperbolic conservation laws', *J. Comput. Phys.* **27**, 1–31.

T. Sonar (1993a), 'On the design of an upwind scheme for compressible flow on general triangulations', *Numer. Algorithms* **4**, 135–149.

T. Sonar (1993b), 'Strong and weak norm refinement indicators based on the finite element residual for compressible flow computation', *IMPACT of Comp. in Science and Eng.* **5**, 111–127.

T. Sonar (1996), 'Optimal recovery using thin plate splines in finite volume methods for the numerical solution of hyperbolic conservation laws', *IMA J. Numer. Anal.* **16**, 549–581.

T. Sonar (2002), Chapter 3, 'Methods on unstructured grids, WENO and ENO recovery techniques', in *Hyperbolic Partial Differential Equations: Theory, Numerics and Applications* (A. Meister and J. Struckmeier, eds), Vieweg, pp. 115–232.

T. Sonar and E. Süli (1998), 'A dual graph-norm refinement indicator for finite volume approximations of the Euler equations', *Numer. Math.* **78**, 619–658.

T. Sonar, V. Hannemann and D. Hempel (1994), 'Dynamic adaptivity and residual control in unsteady compressible flow computation', *Math. Comput. Modelling* **20**, 201–213.

A. Staniforth and J. Côté (1991), 'Semi-Lagrangian integration schemes and their application to environmental flows', *Mon. Weather Rev.* **119**, 2206–2223.

J. L. Steger and R. F. Warming (1981), 'Flux vector splitting of the inviscid gas-dynamic equations with applications to finite difference methods', *J. Comput. Phys.* **40**, 263–293.

E. Süli and P. Houston (1997), Finite element methods for hyperbolic problems: *a posteriori* error analysis and adaptivity, in *The State of the Art in Numerical Analysis* (I. S. Duff and G. A. Watson, eds), Clarendon Press, Oxford, pp. 441–471.

J. L. Synge (1957), *The Hypercircle in Mathematical Physics*, Cambridge University Press.

R. S. Varga (1962), *Matrix Iterative Analysis*, Prentice-Hall International, London.

H. A. van der Vorst (2003), *Iterative Krylov Methods for Large Linear Systems*, Cambridge University Press.

Y. Wada and M.-S. Liou (1994), A flux splitting scheme with high resolution and robustness for discontinuities. AIAA Paper 94-0083.

P. Wesseling (1992), *An Introduction to Multigrid Methods*, Wiley, Chichester.

K. H. Winters, J. Rae, C. P. Jackson and K. A. Cliffe (1981), 'The finite element method for laminar flow with chemical reaction', *Internat. J. Numer. Methods Engrg.* **17**, 239–253.

P. Woodward and P. Colella (1984), 'The numerical solution of two-dimensional fluid flow with strong shocks', *J. Comput. Phys.* **54**, 115–173.

Acta Numerica (2007), pp. 239–303
doi: 10.1017/S096249290631001X

© Cambridge University Press, 2007

Semi-analytic geometry
with R-functions

Vadim Shapiro

Mechanical Engineering & Computer Sciences,
University of Wisconsin, Madison, 1513 University Avenue,
Madison, Wisconsin 53706, USA
E-mail: vshapiro@engr.wisc.edu

Dedicated to V. L. Rvachev, 1926–2005

V. L. Rvachev called R-functions 'logically charged functions' because they encode complete logical information within the standard setting of real analysis. He invented them in the 1960s as a means for unifying logic, geometry, and analysis within a common computational framework – in an effort to develop a new computationally effective language for modelling and solving boundary value problems. Over the last forty years, R-functions have been accepted as a valuable tool in computer graphics, geometric modelling, computational physics, and in many areas of engineering design, analysis, and optimization. Yet, many elements of the theory of R-functions continue to be rediscovered in different application areas and special situations. The purpose of this survey is to expose the key ideas and concepts behind the theory of R-functions, explain the utility of R-functions in a broad range of applications, and to discuss selected algorithmic issues arising in connection with their use.

CONTENTS

1. From Descartes to Rvachev

Descartes (1637) is usually credited with conceiving the coordinate method that allows *investigating* geometric objects by algebraic means. Centuries of remarkable progress in understanding and classifying local and global properties of analytic and algebraic representations followed, but the need for systematic *construction* of such representations did not materialize until the middle of the twentieth century. The arrival of the computer created the need to represent and manipulate sets of points, particularly (but not exclusively) in Euclidean three-dimensional space, for the purposes of visualization, animation, geometric design, analysis, simulation, and so on. How does one represent a set of points on a computer using their coordinates? There are really only two methods: by providing a rule for *generating* points in the set, or by designing a method for *testing* a point $\mathbf{p} \in R^n$ with known coordinates against some predicate that distinguishes the points in the set from other points. The first method assumes an ability to *parametrize* the point set (for example, as a spline, or a triangulation, subdivision scheme, or some other procedural definition); see Farin, Hoschek and Kim (2002) for a recent update on progress in parametric modelling. The second method implies an ability to represent a geometric object Ω *implicitly* by a predicate $S(\mathbf{p})$, as

$$\Omega = \{\mathbf{p} : S(\mathbf{p}) \text{ is } true\}. \tag{1.1}$$

Parametrizations and predicates are well known for simple objects, such as lines, conic sections, and quadric surfaces, but not for the majority of useful geometric objects arising in the physical world and non-trivial modelling situations. Such objects require constructions to be carried out in a *piecewise* fashion, leading to non-trivial data structures and algorithms in geometric modelling and computational geometry. In this survey, we will not deal with parametric representations, but focus instead on the construction of implicit representations. The primitive form of such a representation is an equation $f(\mathbf{p}) = 0$ or an inequality $f(\mathbf{p}) \geq 0$. In this case, the predicate $S(\mathbf{p})$ is defined by the sign of $f(\mathbf{p})$, for example, to be true if $f(\mathbf{p}) = 0$ or $f(\mathbf{p}) \geq 0$, and false otherwise. For convenience of notation, we will use the predicates $(f(\mathbf{p}) = 0)$ or $(f(\mathbf{p}) \geq 0)$, respectively, to denote the set Ω of points for which the predicate $S(\mathbf{p})$ is true.

But how does one write an equation for a rectangle? V. L. Rvachev struggled with this question in the 1960s while using the method of Kantorovich (Kantorovich and Krylov 1958) to solve a boundary value problem of contact mechanics on a square domain. He finally came up with the equation for the boundary of a rectangle with sides $2a$ and $2b$, respectively, as:

$$a^2 + b^2 - x^2 - y^2 - \sqrt{(a^2 - x^2)^2 + (b^2 - y^2)^2} = 0. \tag{1.2}$$

However, he could not explain his own constructions using the classical

methods of analytic and algebraic geometry that focus on *direct* problems of
investigating given equations and inequalities. In contrast, Rvachev wanted
to devise a methodology for solving what he termed the *inverse problem of
analytic geometry*: constructing equations and inequalities for given geometric objects. This quest resulted in his seminal publication, Rvachev (1963),
followed by the comprehensive *theory of R-functions* that has been developed over the last forty-plus years.[1] In a nutshell, R-functions operate on
real-valued inequalities as differentiable logic operations; the resulting theory solves the inverse problem of analytic geometry and has a wide range of
applications, with particular emphasis on solutions of boundary value problems. As of 2001, the bibliography on the theory of R-functions included
more than fifteen monographs and over five hundred technical articles co-
authored by Rvachev and his followers (Matsevity 2001). R-functions were
introduced into the Western literature by the author (Shapiro 1988), and
are now widely used in geometric modelling, computer graphics, robotics,
engineering analysis, and other computational applications.

The goal of this paper is to expose the key concepts in the theory of
R-functions, without trying to be comprehensive. This subject will take us
up to Section 4. Additional references in English are now accessible, notably the review by Rvachev and Sheiko (1995). Among the references in
Russian, the monograph by Rvachev (1982) continues to serve as an ency-
clopedic source of many key ideas and results. The utility of R-functions
cannot be fully appreciated without some discussion of how they may be
used algorithmically to solve inverse problem of analytic geometry (Section 5). Sections 6 and 7 are devoted to applications of R-functions and
derived constructions to computational tasks in geometric modelling and
boundary value problems, respectively.

2. Functions for shapes with corners

2.1. Many equations of a rectangle

How *does* one write an equation for a rectangle? Serendipitously, it is not
that difficult to come up with several methods, though none of them would
directly yield expression (1.2). For example, recalling the definition of the
L_p-norm, we know that the equation $1 - |x|^n - |y|^n = 0$ describes a family
of shapes for $n = 2, 3, \ldots$, that vary from the circle, when $n = 2$, to the

---

[1] The origin of the term R-function is not entirely clear. The use of a Roman (not
   Cyrillic!) symbol R suggests that it is a mathematical symbol corresponding to the set
   of reals. A conflicting personal account by Rvachev (1996) suggests that R does stand
   for Rvachev, but that it was coined by his sister, also a noted Ukrainian mathematician,
   supposedly in memory of their father.

unit square, as $n \to \infty$. Scaling x by $\alpha = 1/a$ and y by $\beta = 1/b$, we obtain the widely used equation of a superellipse:

$$1 - |\alpha x|^n - |\beta y|^n = 0. \tag{2.1}$$

For even n, this is a polynomial equation since we can then drop the absolute value sign. Of course, this is only an approximation of the rectangle's boundary, since we can never quite get into its corners. Alternatively, if we take the limit, we end up with the L_∞-norm and the corresponding equation

$$1 - \max(|\alpha x|, |\beta y|) = 0, \tag{2.2}$$

which defines the exact desired boundary of the rectangle, even if we no longer have the nice analytic properties of a polynomial.

But what if we want equations for other shapes, say a polygon? A more general method for constructing such equations and inequalities would be needed, and it should also work for the rectangle of course. So let us consider the rectangle once more, and try to compose its equation from simpler primitive equations. Even for the rectangle, there are at least two distinct ways to do this.

First observe that the rectangle is the *intersection* of the vertical stripe $\Omega_1 = (a^2 - x^2 \geq 0)$ and horizontal stripe $\Omega_2 = (b^2 - y^2 \geq 0)$ defined by two inequalities. The corresponding equations define their respective boundaries (pairs of vertical or horizontal lines). Then the rectangle is defined by

$$(a^2 - x^2 \geq 0) \cap (b^2 - y^2 \geq 0), \quad \text{or} \quad (f = \min(a^2 - x^2, \ b^2 - y^2) \geq 0). \tag{2.3}$$

Furthermore, the equality $f = 0$ defines the boundary of the rectangle. Once again, the equation is not polynomial and we may not like the differential properties of the min function, but the construction itself is perfectly valid and generalizes to other shapes that can be defined by intersections of simpler shapes. In fact, we are only one step away from obtaining the expression in (1.2).

Alternatively, we observe that the *boundary* of the rectangle is the *union* of four line segments. *If* we could write an equation $f_i(x, y) = 0$, $i = 1, 2, 3, 4$, for each of the line segments, then the equation of the rectangle is obtained by a simple product as

$$f_1 f_2 f_3 f_4 = 0. \tag{2.4}$$

This construction generalizes to arbitrary polygonal boundaries, and requires only multiplication – provided, of course, that we find a method for writing an equation for each line segment. Perhaps we could try to represent each line segment as the *intersection* of a line and a circular disk, but this would require using again the min operation, and so on.

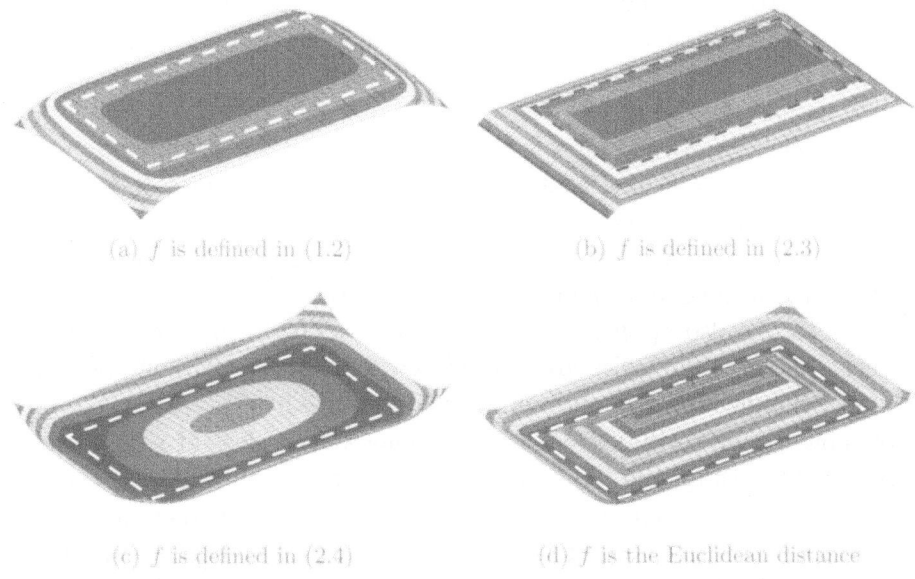

(a) f is defined in (1.2) (b) f is defined in (2.3)

(c) f is defined in (2.4) (d) f is the Euclidean distance

Figure 2.1. Four different implicit representations $(f = 0)$ of the rectangle.

Clearly there are many different ways to construct an equation $f(x, y) = 0$ for the rectangle's boundary. Four such functions corresponding to some of the above constructions are plotted in Figure 2.1. The theory of R-functions explains, systematizes, and expands the above constructions to a virtually unlimited variety of shapes and functions. But for now, the above observations naturally raise two interrelated questions.

- What other *useful* point sets (shapes) admit similar solutions to the inverse problem of analytic geometry?
- Which types of functions are *possible* and *preferable* for such shapes?

The answers to these questions depend on what is meant by 'useful' and 'preferable', as we discuss next.

2.2. Useful shapes

If one is mostly interested in describing domains of boundary value problems (as Rvachev was), then it is reasonable to assume that the point sets in question should include their boundaries, *i.e.*, be closed subsets of $\mathbb{R}^d$. For any such closed set Ω, there exists a suitable continuous function, namely the Euclidean distance function

$$d(\mathbf{p}) \equiv \inf_{\mathbf{x} \in \partial\Omega} \|\mathbf{p} - \mathbf{x}\| \qquad (2.5)$$

and, conversely, for every continuous function f, the equality $f(\mathbf{p}) = 0$ represents a closed subset of $\mathbb{R}^d$. For the rectangle, this function is plotted in Figure 2.1(d). Furthermore, according to Whitney (1934), whenever it is possible to construct such a continuous function f, it is also possible to construct a C^∞-function that vanishes on the set of points. Specific constructions may be found in Rvachev and Rvachev (1979), but this would take us well beyond the scope of the present survey; however, we shall see that constructing C^n-functions is fairly straightforward with R-functions. We should not expect to do better than that in general, because arguments based on Taylor series expansion easily show that no such function can be analytic in the neighbourhood of a corner of the square (Rvachev 1982). This limitation also applies to a wide variety of shapes in science and engineering whose boundaries are only piecewise smooth, *i.e.*, composed in a piecewise manner from smooth curves and surfaces.

We now recognize that such shapes belong to the class of *semi-analytic sets*, defined as those sets that can be constructed as a finite Boolean combination (*i.e.*, a finite composition of unions, intersections, and complements) of sets $(f_i \geq 0)$, where f_i are real analytic functions. Originally proposed by Lojasiewicz (1964) as a natural generalization of semi-algebraic sets, semi-analytic sets are now widely adopted as a proper setting for most geometric modelling applications (Requicha 1977, 1980, Shapiro 1994, 2002), though semi-algebraic sets continue to dominate most practical implementations. Semi-analytic sets clearly include all algebraic and analytic varieties of the form $(f = 0)$, but our goal is to be able to construct functional representations for any closed semi-analytic set, without restrictions on its dimension, codimension, or other topological properties. Furthermore, the distinction between sets defined by inequalities $(f \geq 0)$ and equalities $(f = 0)$ is artificial. A point set Ω that is represented as $(f = 0)$ is also represented by $(-f^2 \geq 0)$; and whenever Ω is represented by $(g \geq 0)$, it is also represented by $(g - |g| = 0)$.

The characterization of useful shapes as semi-analytic sets points the way towards the constructive solution of the problem of inverse analytic geometry. Given any shape Ω, we may subdivide it into *primitive* 'pieces' for which the inverse problem is either trivial or has already been solved. The corners usually provide a good hint on where the shape should be subdivided. We can then combine these primitive solutions using the logical operations of $\wedge, \vee, \neg$ into a single predicate $S(\mathbf{p})$ that represents Ω. Three of the rectangle's equations were constructed using this idea: using intersection of vertical and horizontal stripes, using union of the line segments, and the very first equation (1.2), though the latter may not yet be obvious. On the other hand, introduction of logical (or equivalently set-theoretic) operations also leads to a conceptual difficulty. We no longer have a single real-valued inequality $f \geq 0$, but a logical predicate defining the set of points in Ω. In writing the equations for the square, we carefully translated the logical predicates into the corresponding real-valued function f, but to what end?

2.3. Preferable functions

If the function f is used only as a characteristic function that distinguishes points in the set Ω from all other points, then it really does not matter what f is, as long as it can be evaluated in a reasonably efficient manner. What else would we use the function f for?

Rvachev's original goal was to extend the method of Kantorovich for solving boundary value problems. Briefly, the method is a technique for constructing the coordinate basis functions satisfying the Dirichlet boundary conditions $u_{|\partial\Omega} = \varphi$. The idea of the method is based on the observation that in this case, the solution of a differential equation can be represented in the form

$$u = \varphi + \omega\Psi, \tag{2.6}$$

where $\omega : \mathbb{R}^n \to \mathbb{R}$ is a known function that takes on zero value on the boundary $\partial\Omega$ of the domain and is positive in the interior of Ω; Ψ is some function to be determined. Representing $\Psi = \sum_{i=1}^{n} C_i\chi_i$, as a linear combination of basis functions from some sufficiently complete space (polynomial, splines, *etc.*) reduces the original boundary value problem to that of determining the coefficients C_i that solve the corresponding variational problem. But it is not obvious how this method may be extended to solve boundary value problems with other types of boundary conditions or what the function ω should be in general. Rvachev recognized expression (2.6) as the beginning of a Taylor series expansion. In one dimension, expression (2.6) becomes

$$u = u(x_0) + (x - x_0)\Psi. \tag{2.7}$$

In other words, the function ω appears to play the role of the *distance* to the boundary, at least in the vicinity of the boundary $\partial\Omega$, where the boundary conditions are prescribed. Thus, the Kantorovich method may be viewed as a power series expansion of a field function u in powers of the distance ω to the boundary $\partial\Omega$; higher-order boundary conditions should be associated with powers of ω, and ω should be sufficiently smooth (Rvachev, Sheiko, Shapiro and Tsukanov 2000).

But there is another problem. If different boundary conditions are prescribed on different portions $\partial\Omega_i$ of the boundary, they must be somehow interpolated. In one dimension, the interpolation problem is well understood. If function values are specified at n points $x_1, \ldots, x_n$, the key to interpolation is to construct weight functions W_i that take on the value of 1 at x_i and are 0 at all other points $x_j, j \neq i$. For example, W_i can be taken as the Lagrange basis polynomial L_i written in a general form as

$$L_i(x) = k_i \prod_{j \neq i} (x - x_j), \tag{2.8}$$

where k_i is chosen so that $L_i(x_i) = 1$ (Lancaster and Salkauskas 1986).

Once again, we see that the key to constructing the weight appears to be the *distances* to the points x_j where the data is prescribed. It takes a bit of imagination to realize that the same technique should work in higher dimensions to *transfinitely* interpolate values prescribed on boundaries $\partial\Omega_i$, if every term $(x-x_j)$ is replaced by a function ω_j that measures the distance to the boundary portion $\partial\Omega_j$ (Rvachev, Sheiko, Shapiro and Tsukanov 2001).

Implicit representations and distance functions are also used widely for computer shape and solid modelling (Shapiro 1994, Pasko, Adzhiev, Sourin and Savchenko 1995, Bloomenthal 1997). In these applications, it is often assumed that $(\omega \geq 0)$ is a solid shape Ω, whose boundary $\partial\Omega$ is $(\omega = 0)$, and whose interior $\mathrm{int}\,\Omega$ is $(\omega > 0)$. The functions used in equations (1.2), (2.2), and (2.3) for the rectangle exhibit this property, but it clearly does not hold in general. For example, the function in equation (2.4) is strictly positive everywhere except the rectangle's boundary and does not distinguish between the rectangle's interior and exterior. Furthermore, if $(\omega \geq 0)$ is a point set, then $(\omega g \geq 0)$ is the same set for any function g such that $(g \geq 0) \subset (f \geq 0)$, but $(\omega g = 0)$ is not necessarily the set's boundary. These simple examples illustrate that topological properties of implicitly defined point sets vary widely, depending on application, construction procedures, and type of functions f. We will discuss limited results related to solid modelling in Section 5, but we also note that topological properties of semi-analytic and semi-algebraic sets have been studied extensively, for example, in Whitney (1957, 1965), Andradas, Bröcker and Ruiz (1996) and Bochnak, Coste and Roy (1998).

To summarize, the useful properties of the functions sought include (1) the identification of the function's sign with the membership in the set, (2) some degree of smoothness, and (3) distance-like properties. When the geometry of the point set is encoded using such a function, many otherwise difficult problems involving geometry of the point set become amenable to standard techniques from classical one-dimensional functional and numerical analysis. For boundary value problems, this implies an ability to construct bases of coordinate functions satisfying any and all types of boundary conditions. In retrospect, smooth distance-like functions may be constructed for most shapes by a variety of approximate techniques (Freytag, Shapiro and Tsukanov 2006), but to construct them *exactly* everywhere, including the corners, we need R-functions.

3. R-functions

The main utility of the theory of R-functions is to replace the logical and set-theoretic constructions with the corresponding real-valued functions, yielding an implicit representation $\omega(\mathbf{p}) \geq 0$ for any semi-analytic set Ω. Based on the above discussion, this task would be impossible with algebraic or analytic functions ω. So we must look for additional operations, and it turns out that all we really need is the square root. The material in this section follows roughly Shapiro (1988), but is essentially an annotated and distilled version of thorough expositions by Rvachev (1967, 1974, 1982).

3.1. Logically charged real functions

Logically charged real functions is the name given by Rvachev to real-valued functions of real variables having the property that their signs are completely determined by the signs of their arguments, and are independent of the magnitude of the arguments. For example, consider the following functions:

$$W_1 = xyz,$$

$$W_2 = x + y + \sqrt{xy + x^2 + y^2},$$

$$W_3 = 2 + x^2 + y^2 + z^2,$$

$$W_4 = x + y + z - \sqrt{x^2 + y^2} - \sqrt{x^2 + z^2} - \sqrt{y^2 + z^2} + \sqrt{x^2 + y^2 + z^2},$$

$$W_5 = xy + z + |z - yx|.$$

Table 3.1 shows how the signs of these functions depend on the signs of their arguments. In contrast, here are some functions whose sign depends not only on the sign of the arguments but also on their magnitude:

$$W_6 = xyz + 1, \qquad W_7 = \sin xy, \qquad W_8 = x + y + z - \sqrt{x^2 + y^2},$$

and so on. Specifying distributions of signs for the arguments of functions $W_1, \ldots, W_5$ completely determines the corresponding sign distribution of the functions; functions W_6, W_7 and W_8 do not behave in this fashion.

These simple examples illustrate the key idea. We view the set of reals $\mathbb{R}$ as consisting of two subsets: $\Delta = \{(-\infty, 0], [0, +\infty)\}$, and then seek

Table 3.1. The signs of real functions $W_1, \ldots, W_5$ depend only on the signs of their arguments x, y, and z

x	y	z	W_1	W_2	W_3	W_4	W_5
−	−	−	−	−	+	−	+
−	−	+	+	−	+	−	+
−	+	−	+	+	+	−	−
−	+	+	−	+	+	−	+
+	−	−	+	+	+	−	−
+	−	+	−	+	+	−	+
+	+	−	−	+	+	−	+
+	+	+	+	+	+	+	+

the set $R(\Delta)$ of those functions $f : \mathbb{R}^n \to \mathbb{R}$ that predictably inherit the membership in each of the two subsets of Δ. For the time being, we assume that zero is always signed: $+0$, or -0; this allows us to determine whether it belongs to the set of positive or negative numbers.[2] Summarizing the inheritance properties of such functions in a sign table, such as Table 3.1, immediately suggests the connection between functions in $R(\Delta)$ and the binary logic functions. The connection is made precise by using the (Heaviside) characteristic function $S_2 : \mathbb{R} \to \mathbb{B} \equiv \{0, 1\}$ of the interval $[0+, \infty)$:

$$S_2(x) = \begin{cases} 0 & \text{if } x \leq -0, \\ 1 & \text{if } x \geq +0. \end{cases}$$

Definition 1. A function $f_\Phi : \mathbb{R}^n \to \mathbb{R}$ is an *R-function* if there exists a (binary) logic function $\Phi : \mathbb{B}^n \to \mathbb{B}$ satisfying the commutative diagram:

$$
\begin{array}{ccc}
\mathbb{R}^n & \xrightarrow{\ f_\Phi\ } & \mathbb{R} \\
{\scriptstyle S_2^n}\big\downarrow & & \big\downarrow{\scriptstyle S_2} \\
\mathbb{B}^n & \xrightarrow[\ \Phi\ ]{} & \mathbb{B}
\end{array}
\qquad (3.1)
$$

It is well known that the logic functions form a Boolean algebra with truth value 1 and false value 0. Such functions are usually defined using logic operations $\wedge$ (and), $\vee$ (or), and $\neg$ (negation) on n logic variables. The logic function Φ in the above definition is called the *companion* function of the R-function f_Φ. The commutative diagram implies that

$$S_2(f_\Phi(x_1, x_2, \ldots, x_n)) = \Phi(S_2(x_1), S_2(x_2), \ldots, S_2(x_n)). \qquad (3.2)$$

Informally, a real function f_Φ is an R-function if it can change its property (sign) only when some of its arguments change the same property (sign). We will adopt the above definition of R-functions for the purposes of this survey. But in fact, the notion of R-functions is a special case of a more general concept of R-mapping that is associated with qualitative k-partitions of arbitrary domains and multi-valued logic functions (Rvachev 1982). We will touch briefly on this subject in Section 8.

It follows that every logic function Φ is a companion to infinitely many R-functions. For example, the companion logic function for the R-function $w_1 = xy$ is $X \Leftrightarrow Y$ (X is equivalent to Y); we just check that

$$S_2(xy) = (S_2(x) \Leftrightarrow S_2(y)).$$

But the logical equivalence is also a companion function for R-functions

---

[2] Including zero in both intervals may seem strange, and we will revisit this issue in Section 3.7.

such as

$$w_2 = xy(1 + x^2 + y^2), \qquad w_3 = (1 - 2^{-x})(3^y - 1),$$

and so on. The set of all R-functions that have the same logic companion function is called a *branch* of the set of R-functions. Since the number of distinct logic functions of n arguments is 2^{2^n}, there are also 2^{2^n} distinct branches of R-functions of n arguments.

3.2. Branches of R-functions

The set of R-functions is infinite. However, for applications, it is not necessary to know all R-functions; we only need to be able to construct R-functions that belong to a specified branch. The recipes for such constructions are implied by the general properties of R-functions that follow almost immediately from their definition. Complete proofs, as well as many additional properties, can be found in the references, notably in Rvachev (1967) and (1982).

(1) The set of R-functions is closed under composition. In other words, any composition of R-functions is also an R-function.

(2) If a continuous function $f(x_1, \ldots, x_n)$ has zeros only on coordinate hyperplanes (*i.e.*, $f = 0$ implies that one or more $x_j = 0, j = 1, 2, \ldots, n$), then f is an R-function.

(3) The product of R-functions is an R-function (because the logical companion of the product is equivalence). If the R-function $f(x_1, \ldots, x_n)$ belongs to some branch, and $g(x_1, \ldots, x_n) > 0$ is an arbitrary function, then the function fg also belongs to the same branch.

(4) If f_1 and f_2 are R-functions from the same branch, then the sum $f_1 + f_2$ is an R-function belonging to the same branch.

(5) If f_Φ is an R-function whose logic companion function is Φ, and C is some constant, then Cf_Φ is also an R-function. The logic companion function of Cf is Φ if $C > 0$, or $\neg\Phi$ if $C < 0$.

(6) If $f_\Phi(x_1, \ldots, x_n)$ is an R-function whose logic companion function is $\Phi(X_1, \ldots, X_n)$ and f can be integrated with respect to x_i, then the function $\int_0^{x_i} f(x_1, \ldots, x_n) \, dx_i$ is an R-function whose logic companion function is $X_i \Leftrightarrow \Phi(X_1, \ldots, X_n)$.

The above list of properties is not exhaustive, but it is enough to suggest that more complex R-functions may be constructed from simpler functions. In particular, the closure under composition leads to the notion of *sufficiently complete* systems of R-functions, *i.e.*, collections of R-functions that can be composed in order to obtain an R-function from any branch.

Theorem 1. (Rvachev 1967) Let H be some system of R-functions, and let G be the corresponding system of the logic companion functions. The system H is sufficiently complete if the system G is complete.

The criteria for completeness of a system of Boolean logic functions are well understood. For example, take $G = \{1, \neg X, X_1 \wedge X_2, X_1 \vee X_2\}$. It is well known that all logic functions can be constructed using these basic functions; in other words, G is complete. It is neither unique nor minimal, since the same functions can also be constructed using only conjunction and negation, or disjunction and negation. Furthermore, all logic functions can be constructed using only one operation, the so-called Sheffer's stroke (Sheffer 1913), popularly known as the *nand* ('not and') operation. For geometric applications, the logical operations $\vee$ and $\wedge$ are both convenient and intuitive, because they define the set operations of union and intersection respectively. Thus, we adopt the system G as the primary system of companion functions and, following Theorem 1, seek the R-functions from the corresponding branches. We will refer to these functions, respectively, as R-negation, R-disjunction, and R-conjunction.

3.3. Sufficiently complete systems of R-functions

It is fairly easy to come up with any number of sufficiently complete systems of R-functions. For example, it is easy to check that the following functions are R-functions (their logic companion function in parentheses):

$$
\begin{array}{lll}
C & \equiv \text{const} & (\text{logical } 1), \\
\overline{x} & \equiv -x & (\text{logical negation } \neg), \\
x_1 \wedge_1 x_2 & \equiv \min(x_1, x_2) & (\text{logical conjunction } \wedge), \\
x_1 \vee_1 x_2 & \equiv \max(x_1, x_2) & (\text{logical disjunction } \vee).
\end{array}
$$

Theorem 1 states that an R-function from any branch can be defined using a composition of these four functions. We shall see that this system of R-functions, which we will call $R_1(\Delta)$, has a number of attractive properties, but the resulting R-functions are not differentiable. For applications where differentiability is important, for example in solutions of boundary value problems, we need another system. Below we compare several such systems in terms of simplicity, differential properties, and convenience of use. We will adopt a constant function and R-negation as above for all systems of R-functions, so that the differences between various systems amount to choosing only two operations: R-conjunction and R-disjunction.

A particularly elegant method for deriving a simple but powerful sufficiently complete system of R-functions relies on the triangle inequality. Suppose that we want to construct an R-conjunction. We are looking for a function f of two arguments x_1 and x_2, whose sign is positive if and only if both x_1 and x_2 are positive. Consider a triangle with two sides of length x_1

and x_2. The square of the third side is determined by the law of cosines as $x_1^2 + x_2^2 - 2\alpha x_1 x_2$, where α is the cosine of the angle between the two sides. It is easy to see that the function

$$f = x_1 + x_2 - \sqrt{x_1^2 + x_2^2 - 2\alpha x_1 x_2}$$

satisfies the desired properties. When both x_1 and x_2 are positive, their sum must exceed the length of the third side of the triangle, and therefore $f > 0$. If either x_1 or x_2 are negative, then by the same argument $f < 0$. In other words, f is the R-conjunction for any value of $-1 < \alpha < 1$. A similar argument leads immediately to the conclusion that the function

$$g = x_1 + x_2 + \sqrt{x_1^2 + x_2^2 - 2\alpha x_1 x_2}$$

is the corresponding R-disjunction of the two real variables x_1, x_2. Together with R-negation, these R-functions constitute a sufficiently complete system, which allows construction of all other R-functions by composition. In fact, we are only one step away from what is considered to be the principal system of R-functions.

Based on our observations above, we define a system of R-functions by

$$R_\alpha(\Delta): \quad \frac{1}{1+\alpha}\left(x_1 + x_2 \pm \sqrt{x_1^2 + x_2^2 - 2\alpha x_1 x_2}\right), \tag{3.3}$$

with $(+)$ defining R-disjunction $x_1 \vee_\alpha x_2$ and $(-)$ defining R-conjunction $x_1 \wedge_\alpha x_2$, respectively. The scalar factor $\frac{1}{1+\alpha}$ remains positive and does not affect our derivation above. It will prove useful for other distance-related properties of R-functions. But is the system R_α better than the system R_1?

It may seem that we have not improved all that much, because the two systems are closely related. Observe that $\min(x_1, x_2)$ and $\max(x_1, x_2)$ are the smallest and the largest root, respectively, of the equation

$$z^2 - (x_1 + x_2)z + x_1 x_2 = 0,$$

because $\min(x_1, x_2) + \max(x_1, x_2) = x_1 + x_2$, and $\min(x_1, x_2)\max(x_1, x_2) = x_1 x_2$. Solving this equation for z, we get two roots,

$$\frac{1}{2}\left[x_1 + x_2 \pm \sqrt{(x_1 - x_2)^2}\right],$$

which are simply $\max(x_1, x_2)$ and $\min(x_1, x_2)$, depending on whether we choose $(+)$ or $(-)$, respectively. Apparently, the system of R-functions $R_1(\Delta)$ is indeed the system $R_\alpha(\Delta)$ with $\alpha = 1$, which we previously excluded because it corresponds to a singular triangle with a zero angle. Whenever the expression under the square root vanishes, the resulting R_α-functions become non-differentiable; thus, R_1-functions are not differentiable when

$x_1 = x_2$. But let us choose $\alpha = 0$. The system $R_\alpha(\Delta)$ becomes

$$R_0(\Delta): \quad x_1 + x_2 \pm \sqrt{x_1^2 + x_2^2}. \tag{3.4}$$

If we think of x_1, x_2 as sides of a triangle, then $\alpha = 0$ implies that this triangle is right, and the R_0-functions are based on the Pythagoras theorem! A remarkable property of these R_0-functions is that they are analytic everywhere, except at the origin where $x_1 = x_2 = 0$.

We can improve further, at least theoretically, by upgrading R_0-functions to the class of C^m-functions defined as:

$$R_0^m(\Delta): \quad \left(x_1 + x_2 \pm \sqrt{x_1^2 + x_2^2}\right)(x_1^2 + x_2^2)^{\frac{m}{2}}. \tag{3.5}$$

The additional factor of $(x_1^2 + x_2^2)^{\frac{m}{2}}$ makes these functions differentiable at the origin as as well, but with vanishing derivatives. At all other points, this factor stays positive, and hence does not affect the logical properties of the R_0-function. Another useful generalization of the R_0-system comes from restating the triangle inequality with the L_p-norm. The resulting system of $R_p(\Delta)$-functions becomes[3]

$$R_p(\Delta): \quad x_1 + x_2 \pm (x_1^p + x_2^p)^{\frac{1}{p}}, \tag{3.6}$$

for any even positive integer p.

One may wonder whether the above R-functions are as simple as possible. For example, can we find a sufficiently complete system of R-functions among polynomials? The answer is no. It was shown in Rvachev (1967) that a sufficiently complete system of R-functions cannot be constructed using addition and multiplication alone. On the other hand, a sufficiently complete system does not have to use the root operation, and other systems of R-functions, including those constructed in a piecewise manner, are discussed in Rvachev (1982). We will use the notation $R^*(\Delta)$ to refer to a generic sufficiently complete system of R-functions, and the corresponding R^*-functions as $\wedge^*, \vee^*$, etc. It is convenient to compare the different systems of R-functions by plotting their level sets on the $x_1 x_2$ plane (see Figure 3.1). From the definition, all R-functions from the same branch have identical signs in every quadrant. Thus, all R-conjunctions are positive in the first quadrant and negative in the other three. Similarly, all R-disjunctions are positive in all quadrants except the third quadrant, where both x_1 and x_2 are negative. It is also clear that similarities between different systems of R-functions end somewhere in the neighbourhoods of the coordinate axes, and the differences may become more pronounced as we go away from the coordinate axes.

---

[3] An unfortunate consequence of this notation, preserved from Rvachev (1982), is that $R_p(\Delta)$ with $p = 2$ is identical to $R_d(\Delta)$ with $d = 0$.

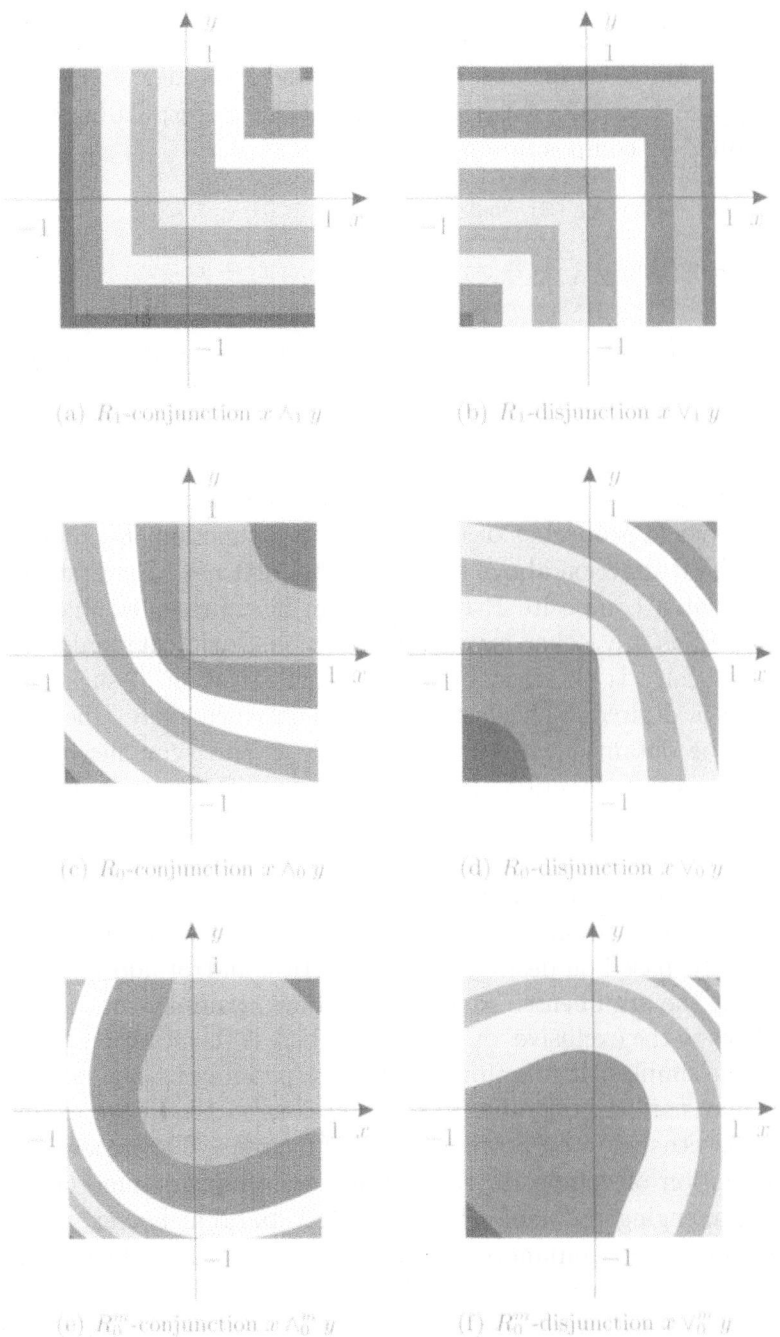

(a) R_1-conjunction $x \wedge_1 y$

(b) R_1-disjunction $x \vee_1 y$

(c) R_0-conjunction $x \wedge_0 y$

(d) R_0-disjunction $x \vee_0 y$

(e) R_0^m-conjunction $x \wedge_0^m y$

(f) R_0^m-disjunction $x \vee_0^m y$

Figure 3.1. R-conjunctions (*left*) and R-disjunctions (*right*) for the three popular systems of R-functions: $R_1(\Delta)$, $R_0(\Delta)$, and $R_0^m(\Delta)$.

3.4. Composite and direct R-functions

Given any such sufficiently complete system, composition may be used to construct R-functions for any branch specified by a logic companion function. Suppose $\Phi = (\neg X_1 \wedge X_2) \vee (X_1 \wedge \neg X_2)$. The corresponding R_0-function f is obtained by composition as

$$f = (\overline{x}_1 \wedge_0 x_2) \vee_0 (x_1 \wedge_0 \overline{x}_2)$$

$$= -x_1 + x_2 - \sqrt{x_1^2 + x_2^2} + x_1 - x_2 - \sqrt{x_1^2 + x_2^2}$$

$$+ \left(\left(-x_1 + x_2 - \sqrt{x_1^2 + x_2^2} \right)^2 + \left(x_1 - x_2 - \sqrt{x_1^2 + x_2^2} \right)^2 \right)^{\frac{1}{2}}$$

$$= 2 \left(\sqrt{x_1^2 + x_2^2 - x_1 x_2} - \sqrt{x_1^2 + x_2^2} \right),$$

and, of course, the factor of 2 may be dropped as well, because it does not affect the logical properties of the composite R-function. Each of the three forms of the R-function above suggests a different use. The first expression suggests that a composite R-function may be represented and evaluated procedurally as any other expression, given its logic companion function. The second form is obtained by syntactic substitution, if such an explicit expression is desired. The last expression is much more efficient, but it could not be obtained without direct analysis and symbolic optimization of the composite R-function. From now on we will use R-conjunctions and R-disjunctions as elementary functions, i.e., we will just write $x_1 \wedge_\alpha^m x_2$, $x_1 \vee_1 x_2$, etc. We know how to evaluate and differentiate these functions, and the notation becomes much simpler.

In many special situations, it may be advantageous to construct R-functions directly, based on desired logic properties, and/or additional assumptions about the arguments. Thus, in the above example, the logic function Φ simplifies to the exclusive 'or' function, which is the negation of the equivalence. The simplest R-function in the corresponding branch is $-xy$, which may or may not be preferable to the functions in the above example, depending on other desired properties of R-functions. The general questions of optimization of composite R-functions according to some criteria may lead to challenging problems (Rvachev 1982, p. 127). For example, direct constructions of R_1-conjunction and R_1-disjunction have been generalized to n-ary logical operations, but similar generalizations of R_0-functions, that are analytic almost everywhere, have been established only up to 5 arguments.

We will discuss several other direct constructions in the context of applications in Sections 6 and 7. A particularly useful concept is that of a function $f : \mathbb{R}^n \to \mathbb{R}$ which is an R-function only on some subdomain $G \subset \mathbb{R}^n$. Such a function f is called a *conditional* R-function in Rvachev

(1982). For example, consider the function $f : \mathbb{R}^4 \to \mathbb{R}$,

$$f = x_1 + x_2 + s_1\sqrt{x_1^2 + x_2^3} + s_2, \qquad (3.7)$$

which is generally not an R-function. However when $s_2 = 0$ and $s_1 \in \{-1, 1\}$, f is an R-function of x_1, x_2: R_0-disjunction if $s_1 = 1$ and R_0-conjunction if $s_1 = -1$. Conditional R-functions may also be defined over finite intervals of its arguments, for example over the interval $x_i \in [-1, 1]$.

3.5. Logic properties of R-functions

Since R-functions mimic the corresponding companion logic functions, one might expect that they should also inherit some properties of the Boolean logic algebra. In particular, we should be able to rely on the laws of Boolean algebra to transform R-functions without affecting their logical properties. For example, from the definition, the R-functions

$$x_1 \wedge^* (x_2 \wedge^* x_3) \quad \text{and} \quad (x_1 \wedge^* x_2) \wedge^* x_3$$

belong to the same branch, by the associative law, as do the R-functions

$$x_1 \wedge^* (x_2 \vee^* x_3) \quad \text{and} \quad (x_1 \wedge^* x_2) \vee^* (x_1 \wedge^* x_3)$$

by the distributive law, and so on.

Unfortunately, we made one important assumption that directly contradicts the properties of the Boolean algebra. Because zero is included with both positive and negative numbers, the range of $x \wedge^* \bar{x}$ includes both negative numbers and 0. This statement is at odds with the requirement of Boolean algebra that $X \wedge \neg X = \emptyset$. Thus, strictly speaking, the logical properties of R-functions are driven by the properties of a distributive lattice, and not those of the Boolean algebra. As a consequence, properties of R-functions that involve the negation operation have to be considered case by case. For example, it is easy to show that R-functions do satisfy the usual De Morgan's laws. We shall also see in Section 5 that inclusion of zero in both positive and negative numbers causes non-trivial complications in solving the inverse problem of analytic geometry. However, observe that the same assumption about the partition Δ is crucial for the following result.

Theorem 2. (Rvachev 1974, p. 62) Every branch of R-functions contains at least one continuous R-function.

Since the R-conjunction and the R-negation as defined above are both continuous functions, this result follows directly from Theorem 1. If instead the real axis were subdivided into two intervals $(-\infty, 0)$ and $[0, +\infty)$, and 0 were considered a positive number, then Theorem 2 would not be true. To see this, observe that any R-negation would have to be discontinuous at $x = 0$ (Rvachev 1974, p. 58).

As *real-valued* functions, R-functions may possess a number of additional logic-related properties. For example, it is easy to check that

$$\bar{\bar{x}} = x, \qquad \overline{x_1 \wedge^* x_2} = \bar{x}_1 \vee^* \bar{x}_2, \qquad \overline{x_1 \vee^* x_2} = \bar{x}_1 \wedge^* \bar{x}_2.$$

Since R-disjunctions and R-conjunctions are symmetric with respect to the two arguments x_1, x_2, we have the usual commutative laws:

$$x_1 \wedge^* x_2 = x_2 \wedge^* x_1, \qquad x_1 \vee^* x_2 = x_2 \vee^* x_1.$$

Other properties may be derived for specific systems of R-functions and used in their construction and simplification. However, most systems of R-functions do *not* obey the associative or distributive laws. A notable exception to this rule is the system $R_1(\Delta)$, where min and max operations are clearly associative and distribute over each other. These properties and other computational considerations make R_1-functions preferable to other systems whenever differentiability of the functions constructed is not required.

3.6. *Differential properties of the elementary R-functions*

Differential properties of R-functions are determined by the properties of the chosen system of R-functions and vary considerably. The smoothness of functions and magnitudes of their derivatives are important in geometric applications described in Section 6, and are critical for correctness of solutions to boundary value problems discussed in Section 7.

Directly differentiating R-functions in the system $R_p(\Delta)$ yields

$$\frac{\partial f}{\partial x_i} = 1 \pm \frac{x_i^{p-1}}{\left(x_1^p + x_2^p\right)^{\frac{p-1}{p}}}, \qquad i = 1, 2, \tag{3.8}$$

where f is either an R-conjunction or an R-disjunction, depending on the choice of the sign. We observe that these R-functions are analytic everywhere, except at $x_1 = x_2 = 0$ where the derivative values change with the direction of approach but remain bounded. On the coordinate axes where one of the variables x_i is zero and the R-function changes its sign, the derivative with respect to this variable is 1, and the derivative with respect to the other variable is 0. At the same points, all higher-order derivatives up to order $p - 1$ vanish. In other words, the R_p-functions behave as the $(p-1)$th-order approximation to the distance functions in the vicinity of the coordinate axes where the R-functions change their sign. This behaviour is clearly visible in the plots of R-functions in Figures 3.1(c) and (d). Recall that the R_0-system is a special case of the R_p-system with $p = 2$.

A similar analysis of R_1-functions reveals that min and max behave as exact distances near the same coordinate axes, but these functions are also not differentiable along the line $x_1 = x_2$. Both of these facts are clearly visible in Figures 3.1(a) and (b). Finally, the R_0^m-functions are m times differentiable everywhere, including at the origin where the first m derivatives are 0. The significant drawback of these R-functions is that they no longer possess the distance properties of the other systems (see Figures 3.1(e) and (f)). As we shall see in Section 6, this severely limits their usefulness in most applications. For additional detailed analysis of differential properties of the popular systems of R-functions, the reader is referred to Shapiro and Tsukanov (1999a).

3.7. Other partitions of the real axis

We have chosen the sign of a real number as the criterion for partitioning the real axis $\mathbb{R}$, but it is not obvious that choice of the partition Δ was 'correct'. For example, we could also choose other partitions, such as:

$$\Delta_2 = \{(-\infty, 0), \ [0, +\infty)\}, \qquad \Delta_3 = \{(-\infty, 0), \ 0, \ (0, +\infty)\}.$$

All three partitions distinguish between the positive and the negative real numbers. Note that the Δ_3 partitions the real axis into three intervals, not two. This forces us to redefine the notion of R-function in terms of companion functions of 3-valued logic, as opposed to the Boolean functions, and generalizations to multi-valued logic are briefly discussed in Section 8. The three partitions are different in the handling of zero, and we have already observed the importance of choosing the Δ partition as opposed to Δ_2 in ensuring continuity of the associated R-functions. Observe that the sets $R(\Delta_2)$, $R(\Delta)$ and $R(\Delta_3)$ intersect (Rvachev 1974, p. 57). For example, the function $x_1 + x_2 - |x_1 - x_2|$ is an R-function for each of the above partitions. At the same time, the function $x_1 x_2$ is in $R(\Delta_3)$ and $R(\Delta)$ but not in $R(\Delta_2)$, and function $x_1^2 x_2^2 (1 - x_1)^2$ is in $R(\Delta_2)$ but not in $R(\Delta_3)$, and so on.

One may wonder why we would bother with Δ_3 to begin with. Such a partition allows one to distinguish the zero value from all other values, which is not possible with either Δ_2 or Δ and may be important for some applications. On the other hand, $R(\Delta_3)$ contains some R-functions with 'undesirable' properties and 3-valued logic brings complications of its own (Rvachev 1982). To cut a long story short, it turns out that all *continuous* functions in $R(\Delta_3)$ are also in $R(\Delta)$. Thus, we rely on Boolean algebra and use only $R(\Delta)$-functions, but occasionally treat them as $R(\Delta_3)$-functions. This allows us to use 3-valued logic in order to identify and *rule out* any situations where zero values may cause anomalies or ambiguities.

These and and other partitions are formally studied in Rvachev (1982), Rvachev and Rvachev (1979), and Rvachev (1974). The partition Δ was used originally in Rvachev (1967), while Δ_2 was employed in Rvachev, Kurpa, Sklepus and Uchishvili (1973) and Rvachev and Slesarenko (1976) which are more concerned with applications.

4. From inequalities to normalized functions

In this section, we explain what is considered the main result of the theory of R-functions. R-functions allow us to construct a smooth distance-like real-valued function for any point set described by a logical predicate on a collection of inequalities. The construction is essentially a syntactic substitution.

4.1. Inequalities from logical predicates

A composition of R-functions involves applying an R-function to other R-functions. But consider what happens if the arguments of an R-function are some other arbitrary functions (that are not necessarily R-functions). Consider a function $f \equiv \phi_1 \wedge^* \phi_2$, where ϕ_1, ϕ_2 are any real-valued functions. By definition, the composite function f is positive if and only if both ϕ_1 and ϕ_2 are positive. In other words,

$$(\phi_1 \wedge^* \phi_2) \geq 0 \iff (\phi_1 \geq 0) \wedge (\phi_2 \geq 0), \tag{4.1}$$

which means that a logical conjunction of two inequalities on the right-hand side can be replaced by an equivalent single inequality on the left-hand side. For example, recall that a rectangle is an intersection of two primitive sets $(a^2 - x^2 \geq 0)$ and $(b^2 - y^2 \geq 0)$. Substituting these into (4.1), and using R_0-conjunction, we have

$$(a^2 - x^2) \wedge_0 (b^2 - y^2) \geq 0 \iff (a^2 - x^2 \geq 0) \wedge (b^2 - y^2 \geq 0).$$

The function in the inequality on the left-hand side is identical to the function in equation (1.2). By construction, the function is zero only on the points of the rectangle's boundary, and positive inside; furthermore the constructed function is analytic everywhere except at the corners of the rectangle, where both primitive functions are zero.

The above reasoning generalizes in a straightforward fashion to arbitrary predicates on sets. Let real-function inequalities $\omega_i(x_1, \ldots, x_n) \geq 0$, $i = 1, \ldots, m$ define the primitive geometric point sets $\Omega_i \subseteq \mathbb{E}^n$, and let $\Phi : \mathbb{B}^m \to \mathbb{B}$ be a predicate constructed using logical functions $\wedge, \vee, \neg$. Then the statement

$$\Phi(S_2(\omega_1), \ldots, S_2(\omega_m)) = 1 \tag{4.2}$$

represents a set $\Omega \subseteq \mathbb{E}^n$ of points where the predicate is true. The logic function Φ defines the corresponding set-valued function $\mathbf{\Phi} : \mathbb{E}^{nm} \to \mathbb{E}^n$, constructed with set operations $\cap, \cup, -$, respectively, so that

$$\Omega = \mathbf{\Phi}(\Omega_1, \ldots, \Omega_m). \tag{4.3}$$

We seek a single real-function inequality $f(x_1, \ldots, x_n) \geq 0$ that defines the composite object Ω, which is readily obtained following the general result in Rvachev (1974).

Theorem 3. Suppose the logic function $\Phi(X_1, \ldots, X_m)$ is the companion of a continuous R-function $f_\Phi(x_1, \ldots, x_m)$ and the corresponding set function Φ maps closed sets into closed sets. If the closed set $\Omega \subset \mathbb{E}^n$ is represented as in (4.2), then it is also represented by the inequality

$$f_\Phi(\omega_1, \ldots, \omega_m) \geq 0. \tag{4.4}$$

In other words, to obtain a real function inequality $f \geq 0$ defining the set Ω constructed from primitive sets $(\omega_i \geq 0)$, it suffices to construct an appropriate R-function and substitute for its arguments the real functions ω_i defining the primitive sets Ω_i. The proof of the theorem follows immediately from the definition of R-functions, as expressed by equation (3.2), where membership of a point in a set is identified by the non-negative sign of the corresponding defining function evaluated at this point.

The restriction to closed sets in Theorem 3 is awkward. On one hand, it would make sense to restrict our attention to the lattice of the closed sets with operations of $\cap$, $\cup$. On the other hand, we do want the complement operation, partly for convenience, but also because it is the companion to the R-negation operation that behaves more like the *pseudo-complement* (defined as the closure of the complement) than the usual complement. For example, technically speaking, the theorem cannot be used with set difference operation $\Omega_1 \setminus \Omega_2$, because

$$(\omega_1 \wedge^* \bar{\omega}_2) \geq 0 \; \nLeftrightarrow \; (\omega_1 \geq 0) \wedge \neg(\omega_2 \geq 0). \tag{4.5}$$

We can get around this difficulty whenever the *closure* of $\neg(\omega_2 \geq 0)$ is $(-\omega_2 \geq 0)$, since R-negation is defined as $\bar{\omega}_2 = -\omega_2$. So, in this particular case, we can obtain a *pseudo-difference* operation by using $-\omega_2$ on both sides of the equivalence statement (4.5) in place of R-negation on the left, and logical $\neg$ on the right-hand side. For the time being, we will rely on such case-by-case analysis to construct equations and inequalities for curves, surfaces, and regions in Euclidean space, but such difficulties need to be accounted for when discussing the algorithmic construction of the Boolean companion functions in Section 5.

4.2. Examples

Example 1. The three-dimensional model of a chess pawn, shown in Figure 4.1(a), can be constructed as a set expression adopted from Rvachev (1967):

$$\Omega = (\Omega_1 \cap \Omega_2 \cap \Omega_3) \cup \Omega_4 \cup \Omega_5,$$

where the primitive regions $\Omega_i = (\omega_i(x, y, z) \geq 0)$ are halfspaces defined in Table 4.1. Following Theorem 3, a single inequality $(\omega \geq 0)$ defining the same point set Ω is obtained by syntactic substitution:

$$\omega = (\omega_1 \wedge^* \omega_2 \wedge^* \omega_3) \vee^* \omega_4 \vee^* \omega_5 \geq 0. \tag{4.6}$$

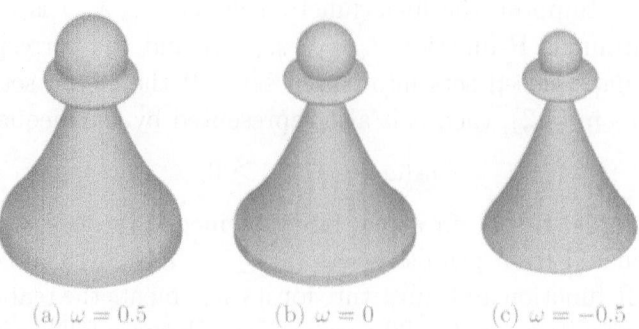

(a) $\omega = 0.5$ (b) $\omega = 0$ (c) $\omega = -0.5$

Figure 4.1. Isosurfaces $\omega = c$ of function ω. The pawn is
defined implicitly by ($\omega \geq 0$) and its boundary by ($\omega = 0$).

Table 4.1.

Primitive Ω_i	Function ω_i
Ω_1 solid of revolution	$\omega_1 = -z + \frac{7}{16}\left(\sqrt{x^2 + y^2} - 4.0\right)^2$
Ω_2 cylinder	$\omega_2 = 9.0 - x^2 - y^2$
Ω_3 horizontal slab	$\omega_3 = z(7 - z)$
Ω_4 sphere	$\omega_4 = 1 - x^2 - y^2 - (7 - z)^2$
Ω_5 ellipsoid	$\omega_5 = 2 - x^2 - y^2 - 9(6 - z)^2$

Figure 4.1(b) shows the isosurface $\omega = 0$ constructed with R_0-functions and
computed by polygonization, while Figures 4.1(a) and (c) show isosurfaces
$\omega = 0.5$ and $\omega = -0.5$, respectively.

Example 2. In this case, the goal is to construct a function ω that vanishes on the dotted line boundary of a 'flag' shape shown in Figure 4.2.
The boundary of the flag itself consists of four segments: circular on the
right, sinusoidal on the top, linear on the left and the bottom of the flag.
In addition, the flag's 'handle' is the dangling line segment that does not
bound any interior. Define four primitive halfspaces $\Omega_i = (\omega_i \geq 0)$ using the
functions in Table 4.2. It is easy to see that the bounded flag itself is simply
$\bigcap_{i=1}^{4} \Omega_i$, but the handle is more problematic. First of all it is a *segment* of
the *line*, not a halfspace. But we can always write the line as a halfspace
$(-|\omega_3| \geq 0)$, and intersect it with some region Ω_5, say a unit circular disk
$\Omega_5 = (1 - (x - 2)^2 - (y - 2)^2 \geq 0)$, to select the required segment. The union

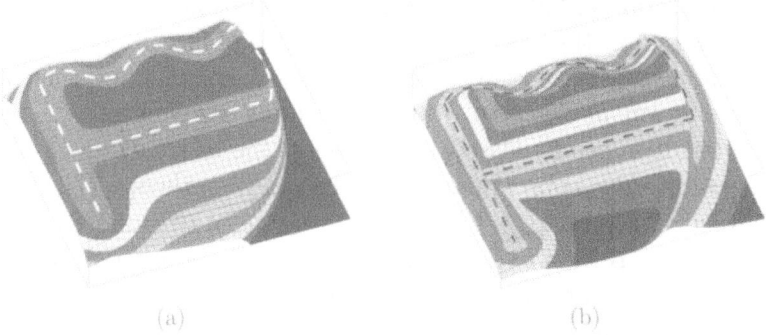

(a)   (b)

Figure 4.2. Implicit representation ($\omega = 0$) of the one-dimensional planar shape shown as a dotted line. (a) Plot of the function ω defined in (4.7). (b) Normalized function ω_1 constructed from ω using (4.9).

Table 4.2.

Primitive Ω_i	Function ω_i
Ω_1 circular	$\omega_1 = 4.5^2 - (x+2)^2 - y^2$
Ω_2 sinusoidal	$\omega_2 = 1 + 0.25\sin(\pi x) - y$
Ω_3 vertical linear	$\omega_3 = x + 2$
Ω_4 horizontal linear	$\omega_4 = y + 1$

of the flag and the handle gives

$$\Omega = (\Omega_1 \cap \Omega_2 \cap \Omega_3 \cap \Omega_4) \cup ((-|\omega_3| \geq 0) \cap \Omega_5).$$

Applying Theorem 3 yields a single inequality constructed with R-functions:

$$\omega = (\omega_1 \wedge^* \omega_2 \wedge^* \omega_3 \wedge^* \omega_4) \vee^* (-|\omega_3| \wedge^* \omega_5) \geq 0. \qquad (4.7)$$

The function ω constructed with R_0-functions is plotted in Figure 4.2(a).

4.3. Distance and normalized functions

Given a closed point set Ω, the Euclidean distance function $d : E^n \to \mathbb{R}$ defined by (2.5) gives for every point of space the shortest distance to the boundary $\partial\Omega$. As we discussed in Section 2, distances play an important role in many applied computational problems. Traditionally, the distance functions are written in a closed form for simple geometric shapes, for example a line, a plane, a circle, a sphere, but more complex shapes usually

involve procedural definitions, numerical computations, and/or approximations. Furthermore, the distance function is not differentiable at all points
that are equidistant[4] from two or more points of $\partial\Omega$, making them unsuitable for use in any application where smoothness is required.

Both difficulties are bypassed by replacing the exact distance functions
with their mth-order approximations in the neighbourhood of $\partial\Omega$. Let ν be
the unit vector at point $\mathbf{p} \in \partial\Omega$ pointing away from $\partial\Omega$ towards the points
that are closer to $\mathbf{p}$ than to any other point in $\partial\Omega$. In other words, ν is the
unit normal on regular (smooth) points of the boundary $\partial\Omega$, but it is also
well defined in neighbourhoods of all other points, including sharp corners.
A suitable mth-order approximation of the distance function d is a function
ω whose derivatives agree with d up to the order m in all normal directions
ν. In other words, we say that the function ω is *normalized* up to the mth
order if its directional derivatives D_ν^k in the direction ν near $\partial\Omega$ satisfy

$$D_\nu \omega = 1, \qquad D_\nu^k \omega = 0, \ k = 2, 3, \ldots, m. \tag{4.8}$$

The exact distance function d is normalized to any order, and equations of
the form $d = 0$ are often called normal.

Without additional assumptions, there is no reason to expect that a
function f constructed using the theory of R-functions should possess any
distance-related properties. But suppose that we constructed a function
$\omega \in C^m$ such that $\omega(\mathbf{p}) = 0$ and $D_\nu^1 \omega(\mathbf{p}) \neq 0$ on all points $\mathbf{p} \in \partial\Omega$. Then
the scaled version of this function

$$\omega_1 \equiv f(f^2 + \|\nabla f\|^2)^{-\frac{1}{2}} \in C^{m-1} \tag{4.9}$$

is normalized to the first order. Straightforward differentiation confirms
that $D_\nu \omega_1 = \|\nabla \omega_1\| = 1$ on all regular points of $\partial\Omega$. Figure 4.2(b) shows
the plot of the function ω_1, normalized to the first order by applying (4.9) to
the function ω defined in (4.7) and plotted in Figure 4.2(a). Furthermore,
if ω_1 is normalized to the first order, then the function ω_m normalized to
the mth order may be constructed by recursively subtracting the non-zero
contribution of the higher-order terms:

$$\omega_m = \omega_{m-1} - \frac{1}{m!}\omega_1^m D_\nu^m \omega_{m-1}. \tag{4.10}$$

This method of normalization is particularly effective as an analytical tool,
or when the initial function ω is relatively simple. However, if the function $\omega = f(\omega_1, \ldots, \omega_m)$ is constructed as an R-function on a large number
of primitive functions ω_i in (4.4), it is not likely to satisfy the required
smoothness conditions, and the method becomes impractical.

---

[4] This includes all points on the medial axis, or on the Voronoi diagram of $\partial\Omega$.

A more constructive approach to normalization is to start with analytic or sufficiently smooth primitive functions ω_i that are already normalized, and then choose the R-function f so that it preserves the normalization at all regular points of the boundary $\partial\Omega$. In particular, we observed in Section 3.6 that many of the R-functions themselves are normalized near their zero level sets, as is evident from expression (3.8). This is all that is required to establish the following result.

Theorem 4. Suppose that the argument x_i appears in the R_p-function $f(x_1, x_2, \ldots, x_n)$ only once and has an inversion degree[5] of r. Let the functions $\omega_1, \omega_2, \ldots, \omega_n$ be in C^s and the boundary $\partial\Omega = (f(\omega_1, \omega_2, \ldots, \omega_n) = 0)$. Then, at every regular point $\mathbf{p} \in \partial\Omega$ where

$$\omega_i(\mathbf{p}) = 0, \qquad \omega_j(\mathbf{p}) \neq 0, j = 1, \ldots, n, j \neq i,$$

for every direction μ and $k \leq s < p$,

$$D^k_\mu f(\omega_1, \omega_2, \ldots, \omega_n) = (-1)^r D^k_\mu \omega_i.$$

Thus, the normalization of functions constructed with R_p-functions, following Theorem 3, comes at no extra cost, provided that the primitive functions ω_i are themselves normalized to the required order. Furthermore, if a point $\mathbf{p} \in \partial\Omega$ belongs to the boundary of exactly one primitive $(\omega_i(\mathbf{p}) = 0)$, then all differential properties of the composite function f at $\mathbf{p}$ are completely determined by the differential properties of $\omega_i(\mathbf{p})$. Rvachev (1982) derives sufficient conditions for Theorem 4 to hold with any system of R-functions, and proves specific results for other popular choices of R-functions.

5. The inverse problem of analytic geometry

5.1. The general problem

We now appear to have all the ingredients needed to solve the general problem of inverse analytic geometry. For any closed semi-analytic set Ω, the problem is solved in two steps: first represent Ω by a logical predicate Φ on analytic primitives $(\omega_i \geq 0)$, then translate this logical predicate into the corresponding inequality $(\omega_\Phi \geq 0)$ by syntactic substitution, as prescribed by Theorem 3. If this inequality represents a set Ω, then every point $\mathbf{p}$ on the boundary $\partial\Omega$ has the property that $\omega_\Phi(\mathbf{p}) = 0$. If there are no other points $\mathbf{p} \notin \partial\Omega$ where ω_Φ vanishes, this translation solves the inverse problem of analytic geometry. We also saw that R-functions may be chosen to

---

[5] Inversion degree of the argument x is the number of times subexpressions with x are negated during evaluation of f. For example, in $x_1 \wedge_p \overline{(x_2 \vee_p \overline{x_3})}$ the inversion degree of x_1 is 1, the inversion degree of x_2 is 2, and the inversion degree of x_3 is 3.

preserve the distance properties of the primitive functions ω_i at the regular points of the boundary $\partial\Omega$ to any desired order.

The above observations may be put to immediate practical use with a new generation of geometric languages that describe complex shapes by recursively combining simpler shapes using various set operations. An example of such a language is Constructive Solid Geometry (CSG) representation (Requicha and Voelcker 1977) that was particularly popular in the early days of solid modelling. Rvachev and Manko (1983) developed a similar language for describing domains of boundary value problems by overloading the usual logic operations with R-functions that combine basic primitives ($\omega_i \geq 0$) of several common types. The R-function constructions now appear in the core of modern computer graphics languages that produce implicit representations $\omega = 0$ of shapes and scenes, for example, in Pasko *et al.* (1995) and Wyvill, Guy and Galin (1999).

However, in many practical situations, point sets are given not in the required predicate form, but are more naturally described by their *boundaries*. Engineers and scientists tend to sketch or sculpt the shapes of interest, and digitally acquired shapes are often defined by reconstructed boundaries. Manually constructing predicates Φ or, equivalently, set-valued expressions Φ for such shapes is often a non-trivial proposition. We do not usually think of a rectangle as the intersection of two unbounded strips, and even the simple predicate expressions for the point sets in the last section are not obvious or unique. Thus, the general problem of the inverse analytic geometry may be formulated as follows.

P1: Given a piecewise (semi-)analytic boundary $\Gamma = \bigcup \Gamma_i$ of a set Ω, where $\Gamma_i \subseteq (\gamma_i = 0)$, construct a function ω such that $\Gamma = (\omega = 0)$ and ω is normalized to some order p on all regular points of the boundary Γ.

Both the boundary Γ and the set Ω are closed semi-analytic sets and, as such, can be represented by set expressions as required by Theorem 3. But these set expressions are neither known nor unique. Accordingly, there are two generic approaches to solving this problem using the theory of R-functions described below: the first one focuses on set representation of Γ and the second on set representation of the set Ω itself. Both approaches are based on the well-known fact that every Boolean set expression may be represented in a disjunctive canonical form as a union of intersection terms.

5.2. Normalized functions from boundaries

In many applications, it is understood that $\Gamma = \partial\Omega$ is the boundary of some domain Ω with non-empty interior, and we will consider such situations in Section 5.3. But in general this need not be so, and when Γ does not bound any interior, we have $\Gamma = \Omega$. The problem of constructing the normalized function ω such that $\omega = 0|_\Gamma$ is solved by constructing normalized functions ω_i for portions Γ_i of the boundary and then combining them, using R-functions in both steps of the procedure. In fact, we used this method in constructing the function ω in Example 2. The key steps are described in Rvachev (1982), but for detailed analysis, extensions, and experimental results the reader is referred to Shapiro and Tsukanov (1999a) and Biswas and Shapiro (2004). We will refer to the two steps in this construction procedure as *trimming* and *joining*, respectively.

Trimming

The trimming step of the construction procedure assumes that a portion Γ_i of the boundary Γ may be represented as $(\gamma_i = 0) \cap \Lambda$, where γ_i is normalized to some order and represents an unbounded curve, surface, or hypersurface $(\gamma_i = 0) \supset \Gamma_i$, and $\Lambda \subset \mathbb{E}^d$ is a full-dimensional region that contains its portion Γ_i. If Λ is defined implicitly by $(\lambda \geq 0)$ and λ is also normalized, then this construction translates directly into $\omega_i = -(-|\gamma_i| \wedge^* \lambda)$, with $(\omega_i = 0)$ defining the trimmed portion of Γ_i. This achieves the desired result, unless one is concerned with differential properties of ω_i. It is expected that ω_i is not differentiable on the points of the *trimmed* boundary Γ_i, but the constructed function ω_i is not differentiable on all points where $\gamma_i = 0$. It is also easy to check that ω_i is normalized on all regular points of Γ_i, but not near its end points where $\gamma_i = \lambda = 0$. Several improved alternative approaches to trimming are known. For example, it can be shown (Sheiko 1982) that

$$\omega_i = \sqrt{\gamma_i^2 + \frac{\left(\sqrt{\lambda^2 + \gamma_i^4} - \lambda\right)^2}{4}} \tag{5.1}$$

is normalized on all regular points of Γ, twice differentiable on the boundary of Λ, and analytic on all other points $\mathbf{p} \notin \Gamma_i$. Normalization at the end points of Γ_i depends on the local geometry of the intersection between the sets $(\gamma_i = 0)$ and $(\lambda_i = 0)$ and can be guaranteed via suitable coordinate transformation. (See examples in Shapiro and Tsukanov (1999a).)

The above approach to trimming works well for curve segments, because both planar and space curves can be trimmed by relatively simple trim regions, such as spheres, boxes, *etc.*, for which normalized implicit representation $(\lambda = 0)$ are easily constructed. Two difficulties arise when $\Gamma_i \subset \mathbb{E}^3$ is a trimmed surface. In this case, the trim region Λ may be described by a complex set-theoretic expression, closely related to Constructive Solid Geometry representations (Rossignac 1996). We will discuss how such expressions may be constructed automatically in Section 5.3.

The second issue relates to the differential properties of the function λ. Suppose that a trim region $\Lambda = \Phi(\Lambda_1, \ldots, \Lambda_n)$ is described by a set-theoretic

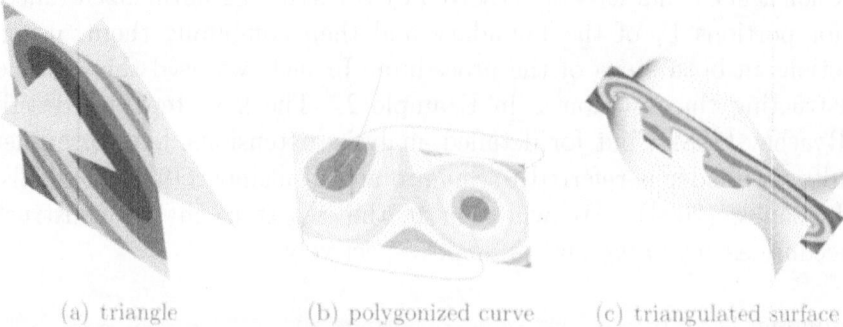

(a) triangle   (b) polygonized curve   (c) triangulated surface

Figure 5.1. Normalized functions, constructed for various sets Γ by the trim-and-join method, are differentiable on all points not in the set Γ.

expression Φ, where Λ_j are primitive halfspaces ($\lambda_j \geq 0$). Following Theorem 3, we immediately obtain the implicit representation for $\Lambda = (\lambda \geq 0)$, where $\lambda = f_\Phi(\lambda_1, \ldots, \lambda_n)$ is an R-function corresponding to Φ. As we discussed earlier, this function may not be differentiable at certain 'corner points' where more than one λ_j vanish. These points typically do *not* lie on the hypersurface ($\gamma_i = 0$), but the singularities will be inherited by ω_i in equation (5.1). The difficulty is resolved by noticing that the set-theoretic representation of the trim volume $\Phi(\Lambda_1, \ldots, \Lambda_n)$ is needed only on the surface ($\gamma_i = 0$) being trimmed. Then constructing $\lambda = f_\Phi$ using the *conditional* R_0-functions (3.7) with $s_2 = a\gamma_i^k$ guarantees that the function λ will be analytic everywhere in $\mathbb{E}^3$ except at the edges of the trimmed surface Γ_i. Parameters a, k can be used to control the overall shape of the trim region away from the surface. Figure 5.1(a) shows a normalized function for a triangle from Biswas and Shapiro (2004), where this method was proposed. In this case, Λ is the unbounded triangular prism perpendicular to the plane of the triangle. The function λ was constructed using the conditional R-conjunction on the three linear halfspaces bounding the prism. The function shown is twice differentiable on all points away from the triangle.

Joining

Suppose that $\Gamma = \bigcup \Gamma_i$ and we used the trimming operations described above to construct normalized functions ω_i such that $\Gamma_i = (\omega_i = 0)$. We now want to construct a single function ω such that $\Gamma = (\omega = 0)$. By the above construction, ω_i is strictly positive on all points $\mathbf{p} \notin \Gamma_i$. Applying Theorem 3 and De Morgan's law to the union $\bigcup_i \Gamma_i$ gives

$$\bigcup_i (\omega_i = 0) = \bigcup_i (-\omega_i \geq 0) = \left(\overset{*}{\bigvee_i} (-\omega_i) \geq 0 \right) = \left(\overset{*}{\bigwedge_i} \omega_i \geq 0 \right). \quad (5.2)$$

In particular, applying the $\wedge_p$ operations to normalized ω_i ensures the normalization of the resulting function ω to the pth order on all regular points of Γ.

Notice, however, that the functions ω_i are non-negative *everywhere*, and we are only interested in the zero set of the constructed function ω. This implies that the union of sets Γ_i is represented by the product of the respective functions ω_i. Since we also want to guarantee that ω is normalized, we replace multiplication with the R_p-equivalence operation defined as

$$ x \sim_p y = xy(|x|^p + |y|^p)^{-1/p}, \tag{5.3} $$

and construct the required function ω as

$$ \omega = \omega_1 \sim_p \omega_2 \sim_p \cdots \sim_p \omega_n. $$

Once again, the function ω is normalized if each ω_i is normalized. Furthermore, the R_p-equivalence operation is associative, whereas the R_p-conjunction used in equation (5.2) is not. As should be expected, neither joining operation maintains normalization at the corner points where $\Gamma_i \cap \Gamma_j \neq \emptyset$. The reader is referred to Shapiro and Tsukanov (1999a) and Biswas and Shapiro (2004) for additional discussion of differential properties of ω in the neighbourhoods of the corners, as well as possible means to control them. See also Section 7.3 for a related discussion.

The described trim-and-join technique can be used to construct an implicit representation ($\omega = 0$) for a variety of point sets Γ, including space curves, polygonized surfaces, polyhedra, piecewise-algebraic boundary representations, dimensionally heterogeneous complexes, *etc.* Figures 5.1(b) and (c) show examples of normalized functions from Biswas and Shapiro (2004) constructed by the described method.

One disadvantage of the above approach is that the constructed function ω is strictly positive everywhere away from the boundary Γ, and thus does not distinguish any interior points of Ω even when they are bounded by Γ. For example, the Jordan–Brouwer separation theorem (which subsumes the Jordan Curve theorem) guarantees that the bounded interior of $\Omega \subset \mathbb{E}^3$ is determined unambiguously whenever Γ is a compact two-dimensional C^0-manifold surface in $\mathbb{E}^3$. In this case, the *signed* function ω may be constructed for Ω, such that $\omega > 0$ for all points in the interior $\mathrm{int}\,\Omega$, and $\omega < 0$ for all points $\mathbf{p} \notin \Omega$. This can be achieved by multiplying ω by the characteristic function $\xi(\mathbf{p}, \Omega)$ defined to be 1 when $\mathbf{p} \in \Omega$ and -1 otherwise, but algorithms for computing ξ usually require non-trivial data structures to represent $\partial\Omega$ and numerically sensitive algorithms, for example to compute the winding number and/or mod 2 intersection computations (Shapiro 2002, O'Rourke 1998). We now consider an alternative approach that relies on constructing set-theoretic representation for the Ω (and not just its boundary $\partial\Omega$).

5.3. *Signed functions via set expressions*

When $\partial\Omega$ bounds a non-empty bounded interior $\operatorname{int}\Omega$, we know that the closed semi-analytic set Ω may be represented explicitly by a set-theoretic expression Φ on some set of primitive analytic halfspaces. We also wish for ω_Φ to be a *signed* function for Ω, that is, $\partial\Omega = (\omega_\phi = 0)$ and $\operatorname{int}\Omega = (\omega_\phi > 0)$, but we shall see that this may not always be the case. The general problem P1 may be restated as follows.

P2: Given the boundary $\partial\Omega = \bigcup \partial\Omega_i$ of the set Ω, let $\mathcal{H}$ be the set of bounding halfspaces $\Omega_i^+ = (\omega_i \geq 0)$ and $\Omega_i^- = (-\omega_i \geq 0)$ induced from boundary portions $\partial\Omega_i$. Construct a set expression $\Phi(\Omega_1^\pm, \ldots, \Omega_m^\pm)$ such that $\Omega = \Phi$, and ω_Φ is a signed function for Ω.

When $\partial\Omega$ bounds some non-empty interior, the halfspaces $\Omega_i^+ = (\omega_i \geq 0)$ may be chosen to include some points in the interior $\operatorname{int}\Omega$. Below we consider this problem for three different classes of sets Ω: simple two dimensional polygons, general semi-analytic sets, and manifold solids.

Simple polygons and extensions

If $\Omega \subset \mathbb{E}^2$ is a simple polygon, its boundary $\partial\Omega$ is a union of n line segments $\partial\Omega_i$. Thus, $(\omega_i = 0)$ is a line, Ω_i^+ is an induced closed linear halfspace, and Ω_i^- is its pseudo-complement. We will assume that the positive side of ω_i coincides with the interior of the polygon Ω. It is easy to see that Ω may be represented by a set expression using only halfspaces induced from the polygon's edges. Let $\mathcal{A}(L)$ be the linear arrangement of the collection of lines $L = \{(\omega_i = 0)\}$. It consists of all k-cells σ^k formed by nonempty intersections of the induced halfspaces and their pseudo-complements (Edelsbrunner 1987). It should be obvious that Ω is the union of all two-dimensional $\sigma_j^2 \subseteq \Omega$. The resulting set expression has the unique canonical disjunctive form

$$\Phi = \bigcup_j \bigcap_i S_i, \tag{5.4}$$

where $S_i \in \{\Omega_i^+, \Omega_i^-\}$. Expression (5.4) is inefficient, but can be optimized in a number of ways using Boolean optimization techniques.

Alternatively, the optimal representation of a polygon, with every primitive Ω_i^+ appearing exactly once, may be constructed efficiently using the algorithm described in Dobkin, Guibas, Hershberger and Snoeynik (1988). Any polygon can be represented as the intersection of two or more polygonal semi-infinite chains as illustrated in Figure 5.2(b). The chains intersect at the vertices of the polygon's convex hull. Each of the chains can be split recursively into smaller subchains. If the split occurs at a concave vertex of the *original* polygon, then the subchains are combined using set union;

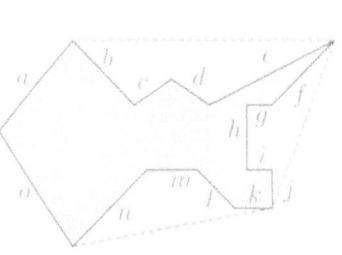

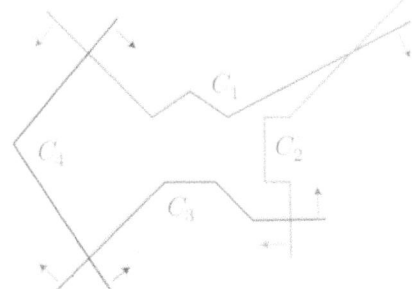

(a) a polygon and its convex hull

(b) a polygon can be represented
as the intersection of four
polygonal chains

Figure 5.2. A set representation for any polygon can be constructed using union and intersection on polygonal chains associated with polygon's edges.

the subchains are combined using intersection at the convex vertices of the original polygon.

Every subexpression of the resulting set expression corresponds to some polygonal chain. For the polygon in Figure 5.2, the resulting expression is

$$(b \cup (c \cap d) \cup e)((f \cap g) \cup h \cup (i \cap j))((k \cap l) \cup m \cup n) \cap (o \cap a),$$

where the literals correspond to the linear halfspaces Ω_i^+ associated with the polygon edges. The expression is the intersection of four subexpressions corresponding to the four chains shown in Figure 5.2. Each chain is either the union or the intersection of its subchains. For example, the chain C_1 is formed by three subchains, b, $c \cap d$, and e, meeting at the vertices of the convex hull of C_1; since these vertices are concave, C_1 is represented as a union of the three subchains, $b \cup (c \cap d) \cup e$, and so on.

A similar algorithm was articulated much earlier by Rvachev et al. (1973), who proposed recursively splitting the polygonal chains at the vertices of the convex hulls of the *bounded* (trimmed) polygonal chains, connected by the dashed lines in Figure 5.2(a). A counter-example in Peterson (1986) shows that such an algorithm does not always result in the correct expression. Other related algorithms to construct set expressions for polygons are described by Tor and Middleditch (1984) and Woodwark and Wallis (1982).

The construction algorithms for simple polygons based on the convex hull may be extended to some other point sets. Non-simple polygons are easily represented as a set combination of simple polygons. The approach may also be generalized to a large class of curved polygons (Shapiro 2001). Two-dimensional set representations are often used to construct representations for three-dimensional solids by translational or rotational extrusion of a planar shape in the direction normal to the plane (Woodwark and Wallis 1982, Peterson 1986, Shapiro and Vossler 1991b). In this case, linear halfplanes become either linear or quadric halfspaces in $\mathbb{E}^3$. Algorithms based on the convex hull can also be applied directly to some, but not all, three-dimensional polyhedra (Woo 1982, Kim and Wilde 1992).

General semi-analytic sets

General theoretical algorithms for constructing set expressions for semi-algebraic (but not semi-analytic) sets are known; for example, see Basu, Pollack and Roy (2003, 2005) and references therein. At the time of writing, none of these algorithms is practical enough to deal with realistic engineering problems, even when the space dimension and the number and degree of all polynomials are fixed and imply polynomial complexity. We seek a more intuitive, geometric characterization of the construction problem that can be used in restricted practical situations.

Conceptually, the approach to constructing set expressions for simple polygons, using the arrangement of primitives, generalizes to arbitrary semi-algebraic and semi-analytic sets (Shapiro 1991, 1997). It is based on the observation that a set of $2n$ (semi-)analytic primitives $(\pm\omega_i \geq 0) \subset \mathbb{R}^d$ generates a finite distributive lattice $\mathcal{L}$ of closed subsets of $\mathbb{R}^d$ under operations of $\cap$, $\cup$. This implies that Ω is an element of $\mathcal{L}$ *if and only if* it can be represented in the disjunctive canonical form (5.4). The key difference between the linear arrangement and the general case is in the intersection terms $J_k = \bigcap_i S_i$, $S_i \in \{\Omega_i^+, \Omega_i^-, (\omega_i = 0)\}$. They are no longer convex sets but can be heterogeneous, possibly disconnected, sets of arbitrary dimension. A non-empty intersection term J_k is called a *join-irreducible* element of the lattice $\mathcal{L}$ if

$$A \cup B = J_k \implies (A = J_k) \text{ or } (B = J_k),$$

for any sets $A, B \in \mathcal{L}$. In geometric terms, J_k is either a set that does not contain any other sets of the lattice, or J_k contains other elements as *proper* subsets. In the arrangement $\mathcal{A}\{(\omega_i = 0)\}$ of analytic primitives, the sets J_k play the same role as the k-cells σ^k play in the linear arrangement of lines. It follows that, to construct a set expression $\Phi(\Omega_1^{\pm}, \ldots, \Omega_2^{\pm})$ for a semi-analytic set Ω, we need to compute the decomposition of $\mathbb{R}^d$ into join-irreducible sets J_k of $\mathcal{L}$; the union of $J_k \subseteq \Omega$ yields the disjunctive canonical form (5.4) for Ω.

A potential problem with this approach is that Ω may not be an element of the lattice $\mathcal{L}$. Consider the shape Ω in the flag example (Example 2). Four primitive halfspaces ω_i, $i = 1, 2, 3, 4$ were induced from the given piecewise description of $\partial\Omega$. But the flag's handle is a trimmed portion of the line $(\omega_3 = 0)$ and cannot be represented without introducing an additional (non-unique) halfspace $(\omega_5 \geq 0)$. This happened because the

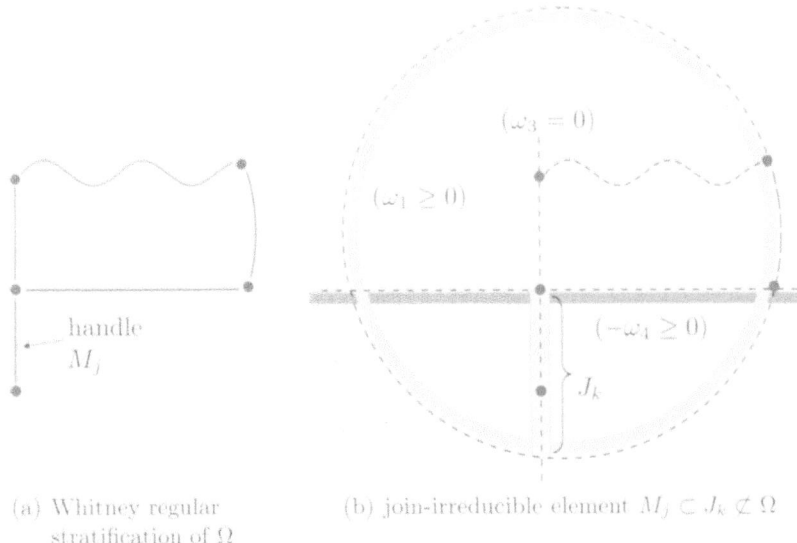

(a) Whitney regular   (b) join-irreducible element $M_j \subset J_k \not\subset \Omega$
    stratification of Ω

Figure 5.3. The flag shape Ω is not describable
by the halfspaces induced from $\partial\Omega$.

join-irreducible element

$$J_k = (\omega_1 \geq 0) \cap (\omega_3 = 0) \cap (-\omega_4 \geq 0)$$

contains the flag's handle but $J_k \not\subseteq \Omega$. See Figure 5.3. This example
illustrates the key difficulty in constructing set representations. The set
of halfspaces $\mathcal{H}$ induced from the boundary $\partial\Omega_i$ may not be sufficient for
representing Ω, because the decomposition of space into the join-irreducible
elements may not be fine enough. In this case, we say that the set Ω is *not
describable* by the primitives in $\mathcal{H}$.

Apparently, describability of Ω has something to do with the ability to
represent some subsets of Ω, such as in the example above. What are these
subsets and can they be enumerated? Any semi-analytic d-dimensional
shape Ω can be stratified into k-manifold cells M_j, for $k = 0, \ldots, d$, such
that all points in M_j have the same signs with respect to all primitives in $\mathcal{H}$.
Because we are concerned with closed sets, we require that the closure of cells
satisfy cl $M_j \subseteq \Omega$, which in turn requires that the closure of every cell M_j is
a union of other cells in the stratification. The coarsest (and therefore minimal) stratification satisfying these conditions is the sign-invariant Whitney
regular stratification into connected strata (Whitney 1965, Shapiro 1997).

Theorem 5. Let $\mathcal{L}$ be a finite distributive lattice generated by a set $\mathcal{H}$ of
halfspaces $(\pm\omega_i \geq 0), i = 1, \ldots, n$. Set $\Omega \in \mathcal{L}$ if and only if, for every k-cell
$M_j \subset \Omega$ in the connected sign-invariant Whitney regular stratification of Ω,
cl $M_j \subseteq J_k \subseteq \Omega$, for some join-irreducible element $J_k \in \mathcal{L}$.

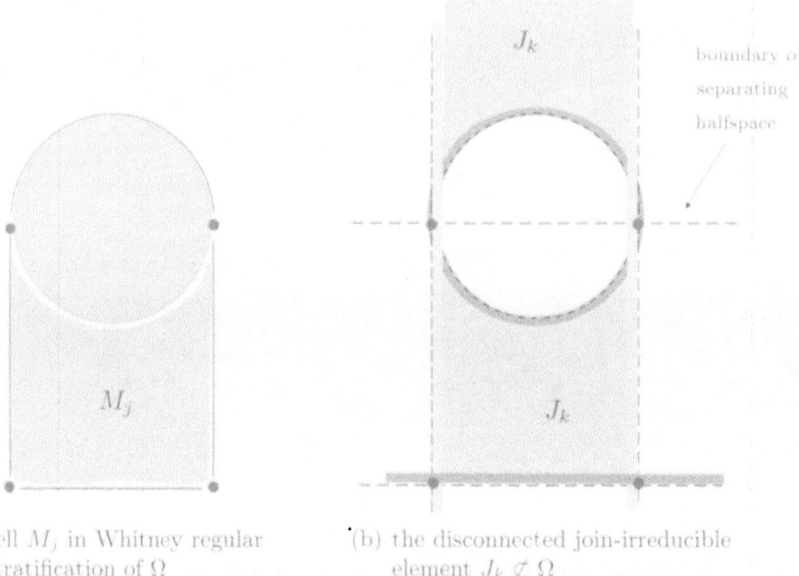

(a) cell M_j in Whitney regular
    stratification of Ω

(b) the disconnected join-irreducible
    element $J_k \not\subset \Omega$

Figure 5.4. The shape Ω in (a) is not describable by the halfspaces
induced from $\partial\Omega$, shown in (b), because the join-irreducible element J_k
containing the cell M_j is separated by the boundary $\partial\Omega$ into two
components. Additional separating halfspaces are needed to represent Ω.

The proof follows from results in Shapiro (1991). The theorem reaffirms
that the set Ω is not describable if it cannot be represented as a union of
join-irreducible elements, but it also explains why this happens. In general,
when the set $\mathcal{H}$ of primitive halfspaces is induced from the boundary $\partial\Omega$,
there is no guarantee that the intersection terms defining the join-irreducible
elements J_k satisfy the conditions of the theorem. This could happen when
$\mathrm{cl}(\omega_i > 0) \neq (\omega_i \geq 0)$, because the latter could contain some additional
points. Another common situation is that M_k are connected sets, but the
set $J_k \setminus \partial\Omega$ may contain several connected components. See the example
in Figure 5.4. In all such cases, the set Ω is not describable by the set
of halfspaces $\mathcal{H}$, and additional *separating* halfspaces must be used in constructing any set expression for Ω. The purpose of the separating halfspaces
is to break up the problematic join-irreducible elements into smaller join-
irreducible elements satisfying the conditions of Theorem 5. See Shapiro
(1997) for more details.

The above observations suggest that, in order to construct a set expression
for Ω, we may need to construct the Whitney regular stratification of Ω,
compute the relevant join-irreducible intersection terms J_k, add separating
halfspaces as needed, and finally construct the disjunctive canonical form.

Theoretically, all these steps are feasible, at least for semi-algebraic sets, but no practical algorithms are available, and this situation is not likely to change in the foreseeable future. The next section outlines a pragmatic and fully implemented approach to solving the describability problem for a limited but useful class of three-dimensional solid shapes.

Set expressions for solids

For a recent survey on solid modelling and many additional references, the reader is referred to Shapiro (2002). A closed bounded semi-analytic set $\Omega \subseteq \mathbb{E}^d$ is called a *solid* if it is *closed regular*, that is, $\mathrm{cl}(\mathrm{int}\,\Omega) = \Omega$ (Requicha 1980). A solid can be represented on a computer using one of many representation schemes, but the most common way to represent a solid model on a computer is by its boundary $\partial\Omega$ stored as a union of faces $\partial\Omega = \bigcup_i \partial\Omega_i$. It is also common to assume that $\partial\Omega$ is a C^0 orientable $(d-1)$-dimensional manifold, and every face $\partial\Omega_i$ is a subset of an analytic or algebraic hypersurface ($\omega_i = 0$). Two-dimensional polygons and three-dimensional polyhedra (curved or linear) are widely recognized examples of solids. We shall assume that ω_i are known, which is the case for polyhedral solids or solids bounded by second-degree surfaces (but may not be true for more general solids bounded by parametric surfaces).

If a solid Ω is not describable by halfspaces in $\mathcal{H}$, there is at least one join-irreducible element J_k that intersects both the interior $\mathrm{int}\,\Omega$ and the exterior $\mathrm{e}\Omega$. This means that for some points $\mathbf{p} \in \mathrm{int}\,\Omega$ and $\mathbf{q} \in \mathrm{e}\Omega$, $\omega_i(\mathbf{p})$ and $\omega_i(\mathbf{q})$ have the same sign for *all* primitive functions ω_i used to define the halfspaces in $\mathcal{H}$. On the other hand, because Ω is a solid, such points $\mathbf{p}$ and $\mathbf{q}$ must be separated by the boundary $\partial\Omega$. Figure 5.4 demonstrates the situation for a simple two-dimensional solid shape. These observations suggest that the construction of the additional separating halfspaces $\mathcal{G}$ can be guided by the faces $\partial\Omega_i$ in the boundary representation. For example, the following result is proved in Shapiro (1991).

Theorem 6. Let Ω be a solid, $\{\partial\Omega_i\}$ a set of faces in the boundary representation, and $\mathcal{H}$ a set of halfspaces induced from the faces. Suppose that the interior of every face $\partial\Omega_i$ is separated from the rest of the surface $(\omega_i = 0) \setminus \partial\Omega_i$ by a family $\mathcal{G}$ of linear halfspaces ($g_k \geq 0$). Then Ω is describable by $\mathcal{H} \cup \mathcal{G}$.

If Ω is a curved polygon with edges that do not change their sign of curvature, Theorem 6 implies that the polygon is describable by the halfplanes $\mathcal{H}$ induced from the polygon's edges and the linear halfplanes $\mathcal{G}$ associated with polygon's chords (Shapiro and Vossler 1991b). Figure 5.5 illustrates this result on a simple two-dimensional solid. Notice that in this case, the additional halfspaces are not necessary, because Ω is the union of three circular halfspaces. For three-dimensional solids, the theorem requires all

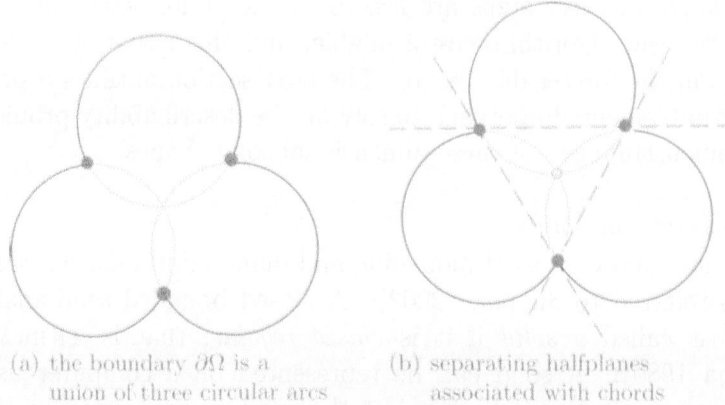

(a) the boundary $\partial\Omega$ is a   (b) separating halfplanes
union of three circular arcs   associated with chords

Figure 5.5. Illustration of Theorem 6. A set of halfspaces associated with the chords is sufficient but not always necessary for describability of Ω and its pseudo-complement $\mathrm{cl}(-\Omega)$.

faces to be bounded by planar curves, but the construction of the linear halfspaces $\mathcal{G}$ may be fully automated in many cases. Linear separators may also suffice when a solid's boundary contains non-planar edges, but in general this is not true. For example, when the functions ω_i are polynomials of degree k, the degree of separating primitives may be $\geq k/2$. Shapiro and Vossler (1993) used the above results to design and implement a fully automated procedure for constructing set representations for solids bounded by second-degree surfaces.

Optimization and signed functions
Once Ω is known to be describable by halfspaces in $\mathcal{H} \cup \mathcal{G}$, it could be represented in the canonical disjunctive form (5.4), but that is probably the most inefficient way to represent Ω. There are at least two different possibilities for optimization. First, the constructed set $\mathcal{G}$ is sufficient, but not all halfspaces in $\mathcal{G}$ are usually necessary for representing Ω. A (non-unique) minimal set $\mathcal{G}$ may be determined by incrementally removing halfspaces from $\mathcal{G}$ until the conditions of Theorem 5 are violated. In the example of Figure 5.5, all chordal halfspaces would be removed. Secondly, standard Boolean optimization techniques (Lawler 1964) may be used to optimize the constructed function Φ.

Note that one can write down 2^n distinct intersection terms, but there is only a polynomial number of non-empty join-irreducible elements J_k in the arrangement $\mathcal{A}\{(\omega_i = 0)\}$. Thus, the canonical form (5.4) may be computed as a decomposition of Ω and represented by a set of characteristic points, one point from each J_k. The intersection terms may be optimized by dropping all those halfspaces that do not bound the particular J_k. Containment relationships between the intersection terms may be exploited to obtain a nearly optimal union of intersection expressions, which intuitively correspond to computing a minimal convex cover of Ω (O'Rourke 1982). These observations were used to design practical algorithms for optimizing set representations for polygons and solids in Shapiro and Vossler (1991a, 1991b).

A more serious problem is that the function ω_Φ obtained from the optimized expression Φ via Theorem 3 may not be properly signed. It is true that $\partial\Omega \subseteq (\omega_\Phi = 0)$, but ω_Φ may *also* be zero at some points in the interior int Ω. In Figure 5.5, if Φ is the union of three circular disks, then $\omega_\Phi = 0$ at the point where all three circles intersect. Recall from Section 3 that the root cause of this problem lies in the adopted definition of R-functions that does not distinguish between zero and the positive numbers. On the other hand, notice that ω_ϕ is strictly negative at all points $\mathbf{p} \notin \Omega$. Similarly, if Υ is a set representation for the pseudo-complement of Ω defined as cl$(-\Omega)$, then the function $-\omega_\Upsilon > 0$ on all points $\mathbf{p} \in$ int Ω. It follows that the function $\omega_\Phi - \omega_\Upsilon$ is the signed normalized function for $\partial\Omega$ solving the problem P2 (Shapiro 1994). Notice that Theorem 6 applies to both Ω and its pseudo-complement. For example, the chordal halfplanes in Figure 5.5(b) are also sufficient for representing cl$(-\Omega)$ and, in this case, they are also necessary.

The above solution is general, but it is inelegant and expensive because it requires constructing set representations twice: once for the set Ω and again for its pseudo-complement cl$(-\Omega)$. It is reasonable to ask if this double effort may be avoided and, if so, under what conditions. For example, the function $-\omega_\Upsilon$ may not be properly signed outside of Ω, but it has all the properties required by the Kantorovich's method for solving boundary value problems and its generalizations, as described in Section 7. For solid models, Shapiro (1999) showed that the function ω_Φ is properly signed if and only if the set expression Φ and its dual[6] represent closed regular sets. Examples of such set expressions include set expressions for simple polygons computed by the recursive decomposition algorithm in Section 5.3, and monotone set expressions using primitives from a simple arrangement.

6. Geometric modelling

The solution to the inverse problem of analytic geometry afforded by R-functions encodes the geometric information in terms of sufficiently smooth and normalized real-valued functions. This, in turn, allows reformulation of many geometry-intensive computational problems in terms of simpler problems that can be solved using classical tools and algorithms for dealing with

---

[6] A dual of a set expression Φ is obtained from Φ by complementing all primitives and changing every $\cap$ to $\cup$ and *vice versa*.

such functions. This section briefly surveys several such applications, with particular focus on those areas where significant advances depend specifically on properties of R-functions.

The theory of R-functions remained largely unknown to Western researchers until the late 1980s, and even after its initial exposure many similar concepts were developed independently in the former Soviet Union and in the West. This brief survey does not attempt to be comprehensive or to establish priority among the proposed ideas. Over the last thirty years, geometric modelling has blossomed as a discipline, with implicit shape representations of the form $\omega = 0$ becoming increasingly popular and now used widely. Many construction methods for such representations are available, as described, for example, by Bloomenthal (1997), Velho, Gomes and de Figueiredo (2002), and others.

The popularity of implicit representations can be attributed to several factors. It is conceptually a very simple representation of a shape that determines point membership via the sign of the defining function ω. It puts no restrictions on the topological properties of the represented sets, but a variety of computational techniques have been developed to parametrize and render the set boundaries when they are manifold. These include the marching cube algorithm (Lorensen and Cline 1987), polygonization (Bloomenthal 1988), and other numerical continuation methods for piecewise-linear approximation of manifolds (Allgower and Georg 1990). Volumetric scan-conversion of such shapes is achieved via ray casting (Roth 1982), computed by intersecting (a grid of) lines with the implicit representation ($\omega = 0$).

6.1. Point membership classification

Computer modelling of complex shapes (point sets) as set combinations of simpler point sets was called 'constructive geometry' by Ricci (1973), who suggested that a three-dimensional point set may be represented as $X = f^{-1}(0, 1]$ where f is a non-negative real-valued function. He then observed that the set (pseudo-)complement is defined by $1/f$, the set operations $\cap$ and $\cup$ can be encoded in terms of max and min functions, respectively, and proposed their L_p-norm and polynomial approximations for construction of increasingly complex shapes and images. Ricci's representations have remained popular in many geometric modelling application: for example, see Storti, Ganter and Nevrinceanu (1992) and Blechschmidt and Nagasuru (1990). More generally, the wide use of Constructive Solid Geometry (CSG) (Requicha and Voelcker 1977) also promoted the use of min/max operators, for example for scan-converting CSG representations into a volumetrically defined Euclidean distance map (Breen, Mauch and Whitaker 1998), or for reformulating the boundary evaluation problem as a level-set marching method (Sethian 1996).

Translation of the theory of R-functions into English (Shapiro 1988) led to a widespread adoption of other systems of R-functions for computer graphics and shape modelling applications. Notably, Pasko *et al.* (1995) developed a powerful geometric language for computer graphics based on R-functions, which inspired others to add R-functions as basic geometric operations for combining functionally (implicitly) defined point sets: for example, Wyvill *et al.* (1999) and Fougerolle, Gribok, Foufou, Truchetet and Abidi (2005). Use of R-functions in solid modelling remains limited, partly because associating R-functions with *regularized* set operations is technically incorrect. The resulting sets ($\omega = 0$) may not be closed regular and may include points in the interior of the solid, as we discovered in Section 5.3. On the other hand, such non-regular and interior points may be identified from extremal properties of the constructed function ω, and every solid shape may be represented by a sufficiently smooth signed function ω using methods described in Section 5.3 (Shapiro 1994).

6.2. Blending

Unions and intersections of smooth shapes create sharp edges and corners, which may be undesirable in many applications. Such sharp features are the source of stress concentrations and other singular behaviours, cannot be manufactured by many manufacturing processes (for example, metal casting, stamping, *etc.*), and may not be aesthetically pleasing. The procedure for smoothing the sharp edges and corners is commonly known as *blending* (see Woodwark (1988) for an introductory survey of blending techniques). Ricci's operations and other polynomial approximations of R-functions are examples of *global* blending because the smoothing affects all points of the shape's boundary. In most practical situations, it is desirable to blend sharp features *locally* and in range-controlled fashion, for example with the desired radius as a function of distance from the feature, so that the points some distance away are not affected by the blends.

A general formulation for blending was proposed by Rockwood and Owen (1987). Suppose we want to blend the intersection of two implicitly defined primitive shapes ($\omega_i \geq 0$), $i = 1, 2$. Construct a binary blend function B_{12} : $\mathbb{R}^2 \to \mathbb{R}$ that blends the intersection of two linear halfspaces, say ($x \geq 0$) and ($y \geq 0$). Then the blend of the desired intersection is simply $B_{12}(\omega_1, \omega_2)$. A variety of blending functions B_{12} can be used, including superelliptic, circular, variable-radius, and others. It is also observed that n-primitive blending functions $B_{12\ldots n}$ can be constructed directly, or if a solid is defined constructively using set operations on implicitly defined primitives, then simply substituting B_{12} for the intersection and $-B_{12}(-\omega_1, -\omega_2)$ for the union blends all sharp features in the resulting shape. Rockwood (1989) explains (and resolves) several difficulties with this approach, including the

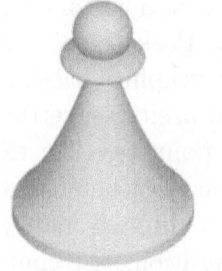 

(a) the original model from   (b) the normalized and
Figure 4.1(b)   blended model

Figure 6.1. The isosurface of the pawn in Figure 4.1(b)
after all primitives are normalized and all R-functions are
replaced by the conditional R-functions B_{12} with $\rho = 0.25$.

dependence of the blend on the metric properties of the defining functions
and how the blend itself is constructed in different regions of $\mathbb{R}^2$.

The concept of blending may seem to be at odds with R-functions, because
the sharp features (corners) is why we needed R-functions in the first place.
But in fact, the blending functions B_{12} are simply R-functions modified in
the neighbourhood of their zero set. This view was made explicit by Pasko,
Pasko, Ikeda and Kunii (2002) who proposed modifying R-functions as

$$B_{12} = x_1 \odot x_2 + d(x_1, x_2), \tag{6.1}$$

where $\odot$ is any binary R-function, and $d(x_1, x_2)$ is a 'displacement' function
whose behaviour defines the actual blend. If d is a non-negative function,
it affects the shape globally. If the blend is to be restricted to some neighbourhood of the corner, the displacement function $d(x_1, x_2)$ must vanish
outside the neighbourhood of the origin. Such displacement functions have
been proposed in Pasko *et al.* (2002) and more complex transition functions
that generalize the notion of blending are proposed in Barthe, Wyvill and
De Groot (2004) and Fayolle, Pasko, Schmitt and Mirenkov (2006).

Viewing blending and transition operations in terms of R-functions is particularly appealing because R_p-functions behave as approximate distances
that are smooth everywhere except at the origin, which corresponds exactly
to the sharp feature to be blended. Following Theorem 4, composition of
such blends preserves normalization properties of the primitive functions ω_i,
providing a convenient mechanism for controlling the shape of the composite
blends in the vicinity of sharp features. Furthermore, it is easy to see that
(6.1) is an instance of a *conditional* R-function, introduced in Section 3.4,
under the condition that $d(x_1, x_2) = 0$. Other conditional R-functions can

be used to define powerful blending techniques. For example, consider a
ρ-blending operation corrected from Sheiko (1982):

$$B_{12} = x_1 + x_2 + s\sqrt{x_1^2 + x_2^2 + \frac{1}{8\rho^2}s_\rho(s_\rho - |s_\rho|)},\qquad(6.2)$$

with $s_\rho = x_1^2 + x_2^2 - \rho^2$. Comparing (6.2) with (3.7), it should be clear that in
this case B_{12} is a conditional R_0-function. Everywhere outside the circular
region of radius ρ centred at the origin, the term $s_\rho - |s_\rho|$ vanishes, making
B_{12} a disjunction when $s = 1$ and a conjunction when $s = -1$. Figure 6.1
shows the result of normalizing all primitives and replacing all R-functions
in Example 1 by the conditional R-function B_{12} from (6.2) with $\rho = 0.25$.
Notice that the overall shape of the pawn is virtually unchanged, but all
the sharp edges are now replaced with smooth blends.

6.3. Envelopes and projections

Beyond the obvious applications in three-dimensional Euclidean space,
R-functions can and have been used in more abstract settings. For example,
Pasko et al. (1995) describe a general multi-dimensional framework for geometric modelling that includes a number of advanced operations relying on
R-functions. If $\Omega_1 \subset \mathbb{R}^k$ is defined by $f_1 : \mathbb{R}^k \to \mathbb{R}$ and $\Omega_2 \subset \mathbb{R}^m$ is defined
by $f_2 : \mathbb{R}^m \to \mathbb{R}$, then the Cartesian product $\Omega_3 \subset \mathbb{R}^{k+m}$ is immediately
given by an R-conjunction operation:

$$\Omega_1 \times \Omega_2 = f_1 \wedge^* f_2.$$

If $F : \mathbb{R}^n \to \mathbb{R}$, and $\Omega = (F \geq 0) \subset \mathbb{R}^n$, then a *section* of Ω is obtained by assigning a fixed value K to the ith variable x_i. For m values
of $K_{j+1} = K_j + \Delta x_i$ that are spaced apart by some Δx_i, we end up with
a stack of sections C_{ij}, $j = 1, \ldots, m$. As $\Delta x_i \to 0$, R-disjunction $\bigvee_j^* C_{ij}$
converges to the *projection* of Ω on the bounded interval of $\mathbb{R}^{n-1}$. Instead
of taking the limit, the R-disjunction of m sections may be blended together using one of the blending operations described above. These and
other derived functional operators provide a powerful arsenal for modelling
point sets in a multi-dimensional setting. For example, the *sweep* of a shape
Ω that is being transformed by a one-parameter affine transformation M_t,
$t \in [t_0, t_1]$ is defined as $\bigcup_{q \in M_t} \Omega_q$, where Ω_q is a transformed instance of
Ω by transformation q. Sweep is one of the fundamental operations in
solid modelling, with many applications in mechanical design and manufacturing (Abdel-Malek, Blackmore and Joy 2006). Sourin and Pasko (1995)
show how the general sweep may be formulated and computed using R-
functions. If $\Omega = (F(p) \geq 0) \subset \mathbb{R}^n$ is a static shape, then the dynamic
shape $(F(p,t) \geq 0) \subset \mathbb{R}^{n+1}$ is obtained by the coordinate substitution
$p \mapsto M_t^{-1}(p)$. Computing the sweep amounts to representing the projection

on $\mathbb{R}^n$ of the $(n+1)$-dimensional point set defined using R-functions as

$$F(p,t) \wedge^* (t - t_0) \wedge^* (t_1 - t),$$

and evaluating its boundary numerically, based on the requirement that $\partial F/\partial t = 0$. This technique can also be used to compute the boundary of the dual infinite intersection $\bigcap_{q \in M_t} \Omega_q$ operation described by Ilies and Shapiro (1999).

6.4. Symmetric and periodic coordinate transformations

Coordinate transformations, a standard tool in analytic geometry, can be used effectively with R-functions. If $\omega(x_1, \ldots, x_n)$ is a normalized function defining Ω, then $\omega(\mu_1, \ldots, \mu_n)$ is constructed by applying the coordinate transformations $\mu_i : \mathbb{R}^k \to \mathbb{R}$ to some or all variables x_i. The notion of the dynamic shape $(F(p,t) \geq 0)$ in the sweep operation described above relied on such a coordinate transformation, but a more familiar form involves sweeping a two-dimensional shape $(\omega(x_1, x_2) \geq 0) \subset \mathbb{R}^2$ in $\mathbb{R}^3$ by coordinate substitution $x_i \mapsto \mu_i(x_1, x_2, x_3)$. When $\mu_2 = \sqrt{x_2^2 + x_3^2}$, the inequality $(\omega(x_1, \mu_2) \geq 0)$ defines a body of revolution; when $\mu_i = x_i - \beta_i x_3$, $(\omega(\mu_1, \mu_2) \geq 0)$ is the prismatic body swept in the direction $(\beta_1, \beta_2, 1)$, and a general screw sweep when

$$\mu_1 = x \cos \phi(x_3) + y \sin \phi(x_3) + c_1(x_3),$$
$$\mu_2 = x \cos \phi(x_3) - y \sin \phi(x_3) + c_2(x_3),$$

(6.3)

where $\phi(x_3)$ and $c_i(x_3)$ are the parameters of the screw motion around the x_3-axis. Rvachev (1982) studied these and other coordinate transformations that construct three-dimensional shapes from two-dimensional sections, modifying them as needed to ensure differential and normalization properties of the transformed function $\omega(\mu_1, \ldots, \mu_n)$.

When the μ_is specify an isometry, they can be used to make copies of Ω that can be combined with Ω using R-functions. If $\mu_i(x_i)$ are periodic functions, then the shape $(\omega(\mu_1, \ldots, \mu_n) \geq 0)$ inherits the discrete symmetries of μ_i. For example, suppose that Ω is symmetric with respect to the coordinate axis and fits inside a coordinate box of width $2a$. Define the ith coordinate transformation $\mu_i(x_i) = x_i$ on the interval $-a \leq x_i \leq a$ and require it to be a periodic function with period greater than $2a$. Then the set $(\omega(x_1, \ldots, \mu_i(x_i), \ldots, x_n) \geq 0)$ has translational symmetry along the x_i-axis, reproducing the base shape Ω at regular intervals. The simplest function μ satisfying this requirement is the $45°$ saw-tooth pattern, but smooth approximations may be constructed using Fourier series, or by methods described by Rvachev (1974).

Rvachev, Sheiko and Shapiro (1999) observed that more complex coordinate transformation functions μ_i may be constructed as semi-analytic compositions of the above primitive coordinate transformations and R-functions. Such coordinate transformations may prescribe periodic, symmetric, and random properties that explicitly depend on the values of the coordinates x_i and additional external parameters. Assuming the typical situation where the complexity of the constructed normalized function ω significantly exceeds that of the coordinate transformations, this approach leads to constructions that are not only more intuitive, but also more efficient. Maksimenko-Sheiko, Matsevityi and Sheiko (2005) also used this technique to construct piecewise functions $\phi(x_3)$ in the coordinate transformations (6.3).

Example 3. The two-dimensional portion of the flag in Example 2 is defined by $(\omega(\mathbf{p}) = 0)$, where $\omega = \omega_1 \wedge_0 \omega_2 \wedge_0 \omega_3 \wedge_0 \omega_4$, and all functions ω_i have been normalized to the first order. Figure 6.2(b) shows the plot of the normalized function $\omega(M(\mathbf{p}))$, where $M = M_2 \circ M_1$ is a composition of two coordinate transformations:

$$M_2 = (x - 2h) \wedge_1 \mu_2(x) \wedge_1 (l - x), \qquad M_1 = Rot(\mu_1(\theta)).$$

The functions μ_1 and μ_2 are Fourier series approximations of the function μ shown in Figure 6.2(a). In this particular example, the function μ_1 is shifted by $\pi/7$, has a period of $4\pi/7$, and is a composition of rotation and reflection as seen in Figure 6.2(b). Transformation M_2 uses the R_1-conjunction to truncate the periodic transformation $\mu_2 = \mu$ as shown in Figure 6.2(a).

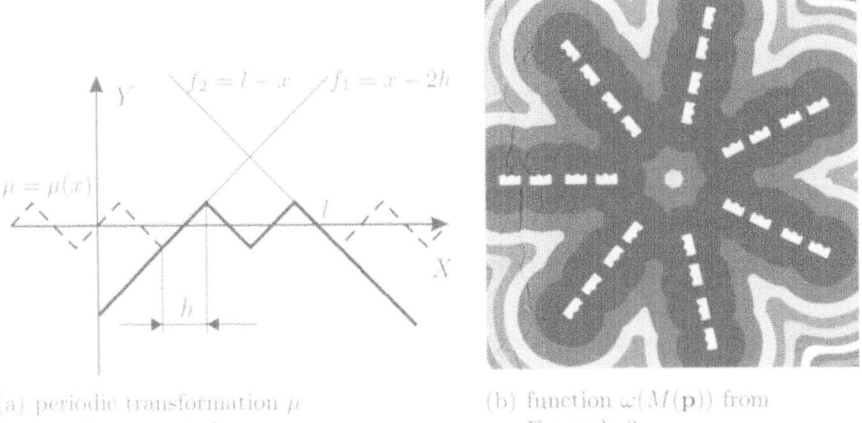

(a) periodic transformation μ may be truncated

(b) function $\omega(M(\mathbf{p}))$ from Example 3

Figure 6.2. Coordinate transformations may be constructed using compositions of symmetric and periodic functions with R-functions.

6.5. Planning and design

R-functions and derived constructions, such as those described above, and others, have been used extensively in a variety of shape modelling applications, for example in mechanical design (Kutsenko 1990, Ensz, Storti and Ganter 1998), robot motion planning (Shkel 1997, Rimon and Koditschek 1990), hair modelling (Sourin, Pasko and Savchenko 1996), Monte Carlo models of transport phenomena (Altiparmakov and Belicev 1990), or stochastic optimization techniques (Komkov 1989) that require repeated sampling of the domain and/or boundary. In addition to the usual advantages of implicit representations, parametric and differential properties of R-functions significantly expand both the range and the possibilities of implicit representations. Below, we consider two (not mutually exclusive) situations where modelling a region Ω with R-functions is useful: when Ω represents a constraint in physical or abstract space, and when Ω itself is the object of design and optimization.

A typical mathematical programming problem is to find a point $\mathbf{x}^0 \in \Omega \subset \mathbb{R}^n$ where some objective function $f(\mathbf{x})$ attains a maximum or a minimum value. R-functions allow us to represent virtually any constraint region Ω by a single inequality $(\omega \geq 0)$. In geometric modelling applications, the constraint $(\omega \geq 0)$ often describes the subset Ω of physical space that must be avoided or included for design or planning purposes. In robot motion planning, Ω usually corresponds to either the obstacle space (to be avoided by the robot) or to the free space (where the robot can move). For example, Rimon and Koditschek (1992) use R-functions to construct the artificial potential function ω representing free space Ω; the differential properties of ω are critical because the gradient information is used to navigate through Ω.

In more general situations, the constraint region Ω may be specified as a union of N systems of inequalities $\sigma_{ki}(\mathbf{x}) \geq 0$, $i = 1, \ldots, M$, where each $(\sigma_{ki} \geq 0)$ is a region Σ_{ik}, and $k = 1, \ldots, N$. In other words, the region of interest is given in disjunctive normal form as

$$\Omega = \bigcup_k \left(\bigcap_i \Sigma_{ki} \right)$$

or, using R-functions, by the inequality

$$\overset{*}{\bigvee_k} \left(\overset{*}{\bigwedge_i} \sigma_{ki} \right) \geq 0.$$

For example, when a rigid shape $\Omega_i \subset \mathbb{E}^3$ is given by $(\omega_i \geq 0)$ and is free to move in space, it can be represented in general position by $(\omega_i(\mathbf{x}, \mathbf{p_i}) \geq 0)$, where $\mathbf{x} \in \mathbb{E}^3$, and $\mathbf{p_i}$ is a vector of its location (position and orientation) parameters. The intersection of two such shapes is given by $\sigma_{ij}(\mathbf{x}, \mathbf{p_i}, \mathbf{p_j}) = (\omega_i(\mathbf{x}, \mathbf{p_i}) \wedge^* \omega_j(\mathbf{x}, \mathbf{p_j})) \geq 0$. It follows that the pairwise non-interference condition for the two objects can be formulated as the requirement that $-\max_{\mathbf{x}} \sigma_{ij}(\mathbf{x}, \mathbf{p_i}, \mathbf{p_j}) \geq 0$. The non-interference conditions generalize in an obvious fashion to collections of n rigid shapes and to the unions of rigid shapes, using R-conjunctions and R-disjunctions on pairwise condition functions σ_{ij}. These ideas were first discussed in Rvachev (1967), and further developed in Stoian (1975) and Stoian and Panasenko (1978), with many applications to problems of optimal placement, blank nesting, and packing.

When it is possible to estimate that the optimal value of the objective function $f(x)$ is $z_0 \in [z_{\min}, z_{\max}]$, R-functions can be used to transform the original constrained problem into a sequence of unconstrained optimization problems. Suppose we are looking for the minimum, and consider the region $Q(z) = \Omega \cap (z - f(\mathbf{x}) \geq 0)$, also defined by the inequality

$$q(\mathbf{x}, z) = \omega(\mathbf{x}) \wedge_\alpha (z - f(\mathbf{x})) \geq 0. \tag{6.4}$$

For a fixed value $z > z_0$, the region $Q(z)$ has interior points, and if $z < z_0$, then $Q(z)$ is an empty set $\emptyset$. Assuming that $Q(z)$ is bounded, and ω, f are continuous functions, it follows that the maximum value of the objective function $f(\mathbf{x})$ is achieved when $q_0(z) = \max_{\mathbf{x}} q(\mathbf{x}, z) = 0$. Note that $q_0(z)$ is monotonc in z. Thus, thc original problcm is transformed into a sequence of optimization problems (each problem is defined by a fixed value of $z \in [z_{\min}, z_{\max}]$) of the function $q(\mathbf{x}, z)$ without any constraints on $\mathbf{x}$. Additional details can be found in Rvachev (1967, 1982).

6.6. Shape optimization

In shape design and optimization problems, Ω is not a constraint, but is the object of study. In this case, the shape is parametrized as $\Omega(\mathbf{b})$, and the challenge is to determine the values of parameters $\mathbf{b} = \{b_1, \ldots, b_k\}$ that will optimize some objective function $F(\Omega(\mathbf{b}))$, for example, volume, energy, stress, *etc.* Shape optimization involves three tasks: computation of sensitivity dF/db_i, updating the model $\Omega(\mathbf{b})$, and (re)evaluating the objective function F; each of the tasks is simplified using R-functions. Consider a generic case, when the objective function F is defined on Ω as $F = \int_\Omega f \, d\Omega$. Then, it can be shown (Chen, Freytag and Shapiro 2007a) that computation of sensitivity requires calculation of

$$\int_{\partial\Omega} \frac{f}{|\nabla\omega|} \frac{\partial\omega}{\partial b_i} \, d\Gamma = \sum_{k=1}^{N} \int_{\partial\Omega_k} \frac{f}{|\nabla\omega_k|} \frac{\partial\omega_k}{\partial b_i} \, d\Gamma, \tag{6.5}$$

where $\partial\Omega = (\omega = 0)$, and $\partial\Omega_k = (\omega_k = 0)$ is the portion of the boundary that belongs to the kth primitive used to construct the domain Ω and is affected by the parameter b_i. This result is implied by the mere *existence* of the R_p-construction for ω in terms of the primitive functions ω_k, and Theorem 4 – even if this construction may not be known explicitly. On the other hand, if the domain is indeed represented by an R-function on n primitives,

$$\omega(\omega_1, \ldots, \omega_k(b_i), \ldots, \omega_n) \geq 0, \tag{6.6}$$

then updating the geometric model $\Omega(\mathbf{b})$ is simply a matter of syntactically updating the parameter b_i. Furthermore, when the objective function $F(\Omega(\mathbf{b}))$ can be computed directly from the implicit representation $\omega = 0$, it automatically inherits the same parametrization and may be (re)evaluated for any value of b_i. A great advantage of this approach to shape optimization is that it places no artificial constraints on the topology of Ω, which is free to change during the optimization process.

The above techniques work well when the objective function F is volume, mass, surface, and moments of inertia, but they also apply to more general situations where F may depend on the solution of a boundary value problem defined over Ω. Suppose that boundary conditions ϕ_k are prescribed on the portion of boundary Ω_k. In this case, Shapiro and Tsukanov (1999b) observed that parametrization of $\Omega(\mathbf{b})$ as above also induces a parametrization of the solution structure to the boundary value problem $u = B(\omega, \omega_k, \phi_k)[\Psi]$, where B can be viewed as an operator and Ψ is a suitable set of basis functions.[7] Such a parametrization supports fully automated (re)evaluation of boundary value problems, which is particularly effective for problems with deforming domains and moving boundary conditions. Chen, Shapiro, Suresh and Tsukanov (2007b) show that the computation of sensitivity for a large class of boundary value problems also reduces to boundary integration over primitive boundaries ($\omega_k = 0$). They also showed that the representation of Ω by (6.6) can include primitive halfspaces ($\omega_k \geq 0$) that define free-form boundaries and/or spatial constraints, leading to fully automated procedure for shape and topology optimization with parametric control.

7. Boundary value problems

Perhaps the most significant application of R-functions has been in the area of boundary value problems – the area that motivated Rvachev to invent the concept of R-functions and that he himself has always included under the general auspices of the 'theory of R-functions'. We summarize the key ideas in this section.

7.1. Generalized Taylor series

The classical Taylor formula approximates a function in the neighbourhood of a given point x_0 by a polynomial in $(x - x_0)$. The neighbourhood itself is described by the term $x - x_0$, which can be thought of as a one-dimensional

---

[7] The concept of solution structure is explained in Section 7.

distance function that vanishes at the point x_0. The generalized Taylor series expansion introduced by Rvachev (1974) represents a function u in the neighbourhood of the boundary $\partial\Omega$, as a polynomial in (powers of) distance to the boundary $\partial\Omega$. Suppose that $\partial\Omega$ is described by some function ω that vanishes on $\partial\Omega$ and is normalized to order m. Then, for a *known* function u,

$$u = u(0) + \sum_{k=1}^{m} \frac{1}{k!} u_k(0) \omega^k + O(\omega^{m+1}), \qquad (7.1)$$

where $u_k = \frac{\partial^k u}{\partial \omega^k} = D_\nu^k u$ (since ω is normalized) are evaluated at the boundary $\partial\Omega$, in the direction ν normal to $\partial\Omega$.

In most applications, the function u is not known, but must be reconstructed from its values and/or derivatives f_k specified at the boundary $\partial\Omega$. In order to be used as coefficients in the generalized Taylor series (7.1), not only must the functions f_k be defined everywhere, but they must also behave as constants in the direction normal to the boundary. This is achieved by conditioning the specified functions f_k through the coordinate transformation

$$f^*(\mathbf{x}) = f(\mathbf{x} - \omega \nabla \omega). \qquad (7.2)$$

Since ω is normalized, the result of this coordinate transformation is that the value of f_k^* at any point near $\partial\Omega$ is determined by the closest point on $\partial\Omega$ and

$$f^*(\mathbf{x})_{|\partial\Omega} = f(\mathbf{x})_{|\partial\Omega}, \qquad \frac{\partial^k f^*}{\partial \nu^k}_{|\partial\Omega} = 0, \qquad \text{for } k = 1, 2, \ldots, m. \qquad (7.3)$$

The functions f_k^* are called *normalizers* of functions f_k by ω in Rvachev (1974), and are used as coefficients in the generalized Taylor series expansion. Another technique to construct normalizers is described by Rvachev and Sheiko (1995).

Theorem 7. (Rvachev 1982) If the function $\omega(\mathbf{x})$ is normalized up to the mth order and a function $u(\mathbf{x})$ satisfies conditions

$$u(\mathbf{x})_{|\partial\Omega} = f_0(\mathbf{x}), \qquad \frac{\partial^k u}{\partial \nu^k}_{|\partial\Omega} = f_k(\mathbf{x}), \qquad \text{for } k = 1, 2, \ldots, m, \qquad (7.4)$$

then u can be represented in the neighbourhood of the boundary $\partial\Omega$ in the form

$$u = f_0^* + \sum_{k=1}^{m} \frac{1}{k!} f_k^* \omega^k + O(\omega^{m+1}), \qquad (7.5)$$

where $f_k^*(\mathbf{x})$, $k = 0, 1, \ldots, m$ are normalizers of the functions $f_k(\mathbf{x})$, $k = 0, 1, \ldots, m$ with respect to $\omega(\mathbf{x})$.

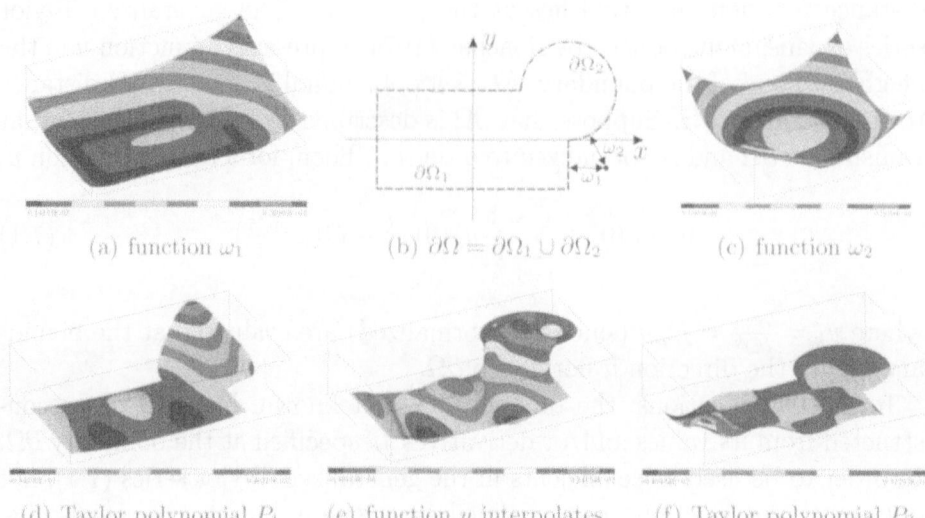

(a) function ω_1 (b) $\partial\Omega = \partial\Omega_1 \cup \partial\Omega_2$ (c) function ω_2

(d) Taylor polynomial P_1 (e) function u interpolates   (f) Taylor polynomial P_2
                                    P_1 and P_2

Figure 7.1. When portions of the boundary $\partial\Omega_1$ and $\partial\Omega_2$ are represented implicitly by normalized functions $(\omega_1 = 0)$ and $(\omega_2 = 0)$, respectively, interpolating boundary conditions is a matter of syntactic substitution.

Example 4. The boundary of the domain in Figure 7.1 is $\partial\Omega = \partial\Omega_1 \cup \partial\Omega_2$, where $\partial\Omega_2$ is the portion of the circle. The following boundary conditions are prescribed:

$$f|_{\partial\Omega_1} = \underbrace{1 + x^2}_{g_0}, \qquad \frac{\partial f}{\partial\nu}\Big|_{\partial\Omega_1} = \underbrace{10(\cos(\pi x) + y)}_{g_1}; \qquad (7.6)$$

$$f|_{\partial\Omega_2} = \underbrace{10 + (y - 2)^2}_{h_0}, \qquad \frac{\partial f}{\partial\nu}\Big|_{\partial\Omega_2} = \underbrace{10\sin(\pi x)\cos(\pi y)}_{h_1}. \qquad (7.7)$$

If the boundaries $\partial\Omega_i$ are represented by the respective normalized functions $(\omega_i = 0)$, extending these boundary conditions into the domain is a matter of syntactic substitution into the truncated Taylor series from (7.5). For $\partial\Omega_1$, we have

$$P_1 = g_0^* + g_1\omega_1 = 1 + \left(x - \frac{\partial\omega_1}{\partial x}\omega_1\right)^2 + 10(\cos(\pi x) + y)\omega_1,$$

$$P_2 = h_0^* + h_1\omega_2 = 10 + \left(y - 2 - \frac{\partial\omega_1}{\partial y}\omega_2\right)^2 + 10\sin(\pi x)\cos(\pi y)\omega_2.$$

The two functions are plotted over the domain Ω in Figures 7.1(d) and (f), respectively.

The generalization suggests that many techniques relying on Taylor polynomials in the classical univariate setting may also be applicable in the general multi-dimensional setting, once the normalized function ω is constructed for the boundary $\partial\Omega$ of interest. In particular, Rvachev (1967) recognized that Kantorovich's method for representing the solutions to boundary value problems with Dirichlet boundary conditions via (2.6) is a special case of (7.5) with $k = 1$. More generally, in the context of boundary value problems, the prescribed functions f_k correspond to the boundary conditions, while the remainder term $O(\omega^{m+1})$ must be determined to satisfy some additional constraint, e.g., a differential equation. Formally, the approach is justified following a generalization of the classical Weierstrass Approximation Theorem, modified from Kharrik (1963).[8]

Theorem 8.   Suppose that Ω is a bounded region of m-dimensional space with boundary $\partial\Omega$. Let $\omega(\mathbf{x}) \in C^s$ be a function defined in an open region that contains Ω and satisfies the following conditions:

(1) $\omega(\mathbf{x}) = 0 \iff \mathbf{x} \in \partial\Omega,$
(2) derivatives of ω up to order s satisfy the Lipschitz condition,
(3) $\nabla\omega(\mathbf{x})_{|\mathbf{x}\in\partial\Omega} \neq 0.$

If a function $\gamma \in C^s(\Omega)$ and vanishes on the boundary $\partial\Omega$ together with its derivatives up to order $k < s$, then for any positive ε there exists a polynomial Ψ such that

$$\|\gamma - \omega^{k+1}\Psi\|_{H^s(\Omega)} < \varepsilon. \qquad (7.8)$$

We can now rewrite expression (7.5) as

$$u = P + \omega^{k+1}\Psi, \qquad (7.9)$$

where P satisfies the boundary conditions $f_i, i = 0, 1, \ldots, k$. If f is the solution to the boundary value problem, the function $\gamma = f - P$ satisfies the conditions of Theorem 8, and approximating the solution to the boundary value problem amounts to choosing the undetermined polynomial Ψ. But before we discuss this task, we deal with yet another challenge. In most practical situations, different boundary conditions are prescribed at different portions $\partial\Omega_i$ of the boundary $\partial\Omega$. These boundary conditions must be interpolated in some way, and once again, the task is greatly simplified using normalized functions ω.

7.2. Transfinite interpolation

For every type of boundary condition specified on a portion of the boundary $\partial\Omega_i$, we can construct a generalized Taylor polynomial P_i of the form (7.5).

---

[8] Kharrik's original theorem also deals with issues related to the order of approximation and convergence that are outside the scope of this survey.

We would now like to interpolate all of them by a single function $\sum_{i=1}^{n} P_i W_i$. Rvachev (1967) proposed choosing the weights W_i as

$$W_i(\mathbf{x}) = \frac{\omega_i^{-\mu_i}(\mathbf{x})}{\sum_{j=1}^{n} \omega_j^{-\mu_j}(\mathbf{x})} = \frac{\prod_{j=1;j\neq i}^{n} \omega_j^{\mu_j}(\mathbf{x})}{\sum_{k=1}^{n} \prod_{j=1;j\neq k}^{n} \omega_j^{\mu_j}(\mathbf{x})}, \qquad (7.10)$$

which can be seen as a form of inverse distance interpolation, where the weights W_i form a partition of unity and are chosen to be inversely proportional to the power μ_i of the approximate distance ω_i from the locus $\partial\Omega_j$. When $\mu_i = 1$, the expression on the right appears to be a form of Lagrange interpolation (2.8). Inverse distance interpolation was used by Shepard (1968) to interpolate scattered point data, but Watson (1992) cites applications of this technique dating back to 1920s. It is well known (Hoschek and Lasser 1993) that the exponents μ_i control the behaviour of the interpolating function at the *loci* $\partial\Omega_i$: when $0 < \mu_i \leq 1$, the interpolant is not differentiable; values of $\mu_1 > 1$ ensure that the interpolant is differentiable $\mu_i - 1$ times at $\partial\Omega_i$.

Rvachev *et al.* (2001) study the above approach to interpolation and demonstrate its advantages in several applications. Note that (7.10) depends only on the knowledge of the normalized functions ω_i, and places no constraints on the differential or topological properties of the sets $\partial\Omega_i$. Thus, the method may be used without modification to interpolate functions and their derivatives that are specified over arbitrary points, curves, surfaces, or regions – with or without sharp corners – provided they are represented implicitly by normalized functions ω_i. Applications of this technique are found in many areas ranging from geographic information systems to modelling of material properties (Biswas, Shapiro and Tsukanov 2004).

If P_i are generalized Taylor polynomials in the form of (7.5) satisfying the boundary conditions of order k_i on the boundary $\partial\Omega_i$, then they can be interpolated into a single expression using weights (7.10). For example, the two Taylor polynomials P_1 and P_2 in the Example 4 are interpolated by

$$u_0 = \frac{P_1\omega_2^2 + P_2\omega_1^2}{\omega_1^2 + \omega_2^2},$$

which is shown in Figure 7.1(e). Theorem 8 may no longer apply because derivatives of different order may be indicated on different portions $\partial\Omega_i$. However, an extension of this result is proved by Rvachev *et al.* (2000).

Theorem 9. Let Ω be a closed region and $f \in C^s(\Omega)$ be defined in the interior of Ω. Values of the function f and its partial derivatives up to order $k_i < s$ are prescribed on boundaries $\partial\Omega_i \subset \partial\Omega$. Then, for any small ε there exists a polynomial Ψ such that the inequality

$$\|\gamma - \lambda\Psi\|_{H^s(\Omega)} < \varepsilon \tag{7.11}$$

is satisfied, where $\gamma = f - u_0$ is a function that vanishes on $\partial\Omega_i$ together with its partial derivatives up to order k_i, and $\lambda = \prod_{i=1}^{N} \omega_i^{k_i+1}$.

In other words, a general solution to a boundary value problem may be written as

$$u = u_0 + R = \frac{\sum\limits_{i=1}^{n} P_i \omega_i^{-k_i}}{\sum\limits_{i=1}^{n} \omega_i^{-k_i}} + \Psi \prod_{j=1}^{n} \omega_j^{k_j+1}. \tag{7.12}$$

The first term u_0 in the expression satisfies all imposed boundary conditions exactly, while the second term is simply a product of the remainder terms for each individual boundary condition. The power $k_j + 1$ of ω_j indicates that derivatives up to order k_j have been specified on $\partial\Omega_j$. Finding a solution to the boundary value problem reduces to constructing the polynomial Ψ.

7.3. Solution structures of boundary value problems

Theorem 9 suggests a non-traditional approach to solving boundary value problems that generalizes the method of Kantorovich. Given any representation of boundary $\partial\Omega = \bigcup \partial\Omega_i$ and associated boundary conditions, we first construct a normalized function ω_i for each portion of the boundary $\partial\Omega_i$, for example using R-functions and one of the methods described in Section 5. Syntactic substitution into expression (7.5) yields a Taylor polynomial corresponding to the boundary condition on $\partial\Omega_i$. Expression 7.12 interpolates all specified boundary conditions and defines the *solution structure* of the boundary value problem – a space of functions that satisfy all given boundary conditions and differ from each other only in the choice of the undetermined polynomial Ψ. Solving the boundary value problem amounts to choosing Ψ from a sufficiently complete space (multivariate polynomials, B-splines, finite elements, *etc.*) to approximate the differential equation using least squares, Ritz, or another variational method. For example, the expression

$$u = \underbrace{(\Psi_1 - \omega D_\nu(\Psi_1))}_{f_0^*(0)} + \underbrace{\varphi\omega}_{f_1(0)\omega} + \underbrace{\omega^2\Psi_2}_{O(\omega^2)} \tag{7.13}$$

defines a family of functions that satisfy the Neumann boundary conditions $\frac{\partial u}{\partial \nu}|_{\partial\Omega} = \varphi_0$, where φ is an extension of φ_0 into the interior of domain Ω. The first term f_0 in expression (7.13) represents a value of the function prescribed on the boundary $\partial\Omega$. Since the Neumann boundary condition does not explicitly prescribe the value of u on the boundary, f_0 is represented by a linear combination Ψ_1 of basis functions with coefficients that will be determined by the numerical solution procedure. Subtraction of $\omega D_\nu(\Psi_1)$ from Ψ_1 ensures that the first normal derivative of u_0 vanishes on the zero set of ω. The second term in expression (7.13) represents a first-order normal derivative of u. This term does not need to be conditioned since no higher-order derivatives are prescribed. The remainder term $\omega^2 \Psi_2$ guarantees completeness of u.

Consider now a second-order boundary value problem with mixed boundary conditions

$$u_{|\partial\Omega_1} = \varphi_0, \qquad \left(\frac{\partial u}{\partial \nu} + h_0 u\right)_{|\partial\Omega_2} = \psi_0 \qquad (7.14)$$

specified on $\partial\Omega = \partial\Omega_1 \cup \partial\Omega_2$. If $\partial\Omega_i = (\omega_i = 0)$, $i = 1, 2$, and ω_i are normalized functions, the corresponding generalized Taylor series expansions are

$$u_1 = \varphi + O(\omega_1) = P_1 + O(\omega_1),$$

$$u_2 = \Psi_2 - \omega_2 D_{\nu 2}(\Psi_2) - h\omega_2\Psi_2 + \psi\omega_2 + O(\omega_2^2) = P_2 + O(\omega_2^2).$$

We could now write the corresponding solution structures for u_1 and u_2 and then interpolate them into a common solution structure using weights (7.10). Alternatively, simply substituting into the expression for the general solution structure (7.12) yields

$$u = \frac{P_1\omega_2^2 + P_2\omega_1}{\omega_1 + \omega_2^2} + \Psi\omega_1\omega_2^2 \qquad (7.15)$$

$$= \frac{1}{\omega_1 + \omega_2^2}\left(\varphi\omega_2^2 + \omega_1(\Psi_2 - \omega_2 D_{\nu 2}(\Psi_2) - h\omega_2\Psi_2 + \psi\omega_2)\right) + \Psi\omega_1\omega_2^2.$$

The above procedure for constructing solution structures can be fully automated, but the resulting expressions can often be optimized using special-case analysis. Several alternative solution structures for the above second-order boundary value problem with mixed boundary conditions were derived in Rvachev (1982). Much of the Ukrainian literature on application of R-functions is devoted to the explicit derivation and simplification of solution structures for common boundary value problems. Rvachev and Sheiko (1995) summarize the main results and give a number of illustrative examples.

The transfinite interpolation term in the solution structure (7.12) allows one to enforce all imposed boundary conditions in a boundary value problem. But the concept of the solution structure naturally generalizes to include *a priori* known behaviours of the field u by explicitly modifying the remainder term $R = \Psi \prod_{j=1}^{n} \omega_j^{k_j+1}$ in the general solution structure (7.12). Examples of such behaviour include singularities (often associated with corners and cracks), asymptotic changes (expressed as a function of distance from some boundary), and other empirical and/or postulated principles (such as St. Venant's principle in solid mechanics and Chvorinov's rule in solidification of metal castings). Computationally, it is convenient to associate the construction of the polynomial Ψ with the choice of the basis functions and numerical procedure; on the other hand, the normalized functions ω_i are the natural means for extending the boundary conditions from the boundary into the domain.

A common situation arises at corner points $\mathbf{p}_0 \in (\partial\Omega_i \cap \partial\Omega_j)$, and the field is known to behave as a function τ_j (of distances, angles, etc.) in the neighbourhood of the corner points. In this case, modifying the remainder term as $R\prod\tau_j$ captures the singular behaviour. This technique was originally applied to torsion problems in Goncharyuk, Rvachev and Shklyarov (1968), and was revised in the context of solidification in metal casting using Rvachev, Sheiko, Shapiro and Uicker (1997), using R_p-equivalence operation (5.3) in place of multiplication xy. Suppose that the domain's boundary is decomposed into n smooth segments $\partial\Omega_i$ and 'sharp features' contained in $\partial\Omega_i \cap \partial\Omega_j$. Both the smooth segments and the sharp features are represented implicitly by $(\omega_i = 0)$ with normalized functions ω_i. Then, following Theorem 4, the properties of the composite function

$$\omega = \frac{\omega_1}{m_1} \sim_p \frac{\omega_2}{m_2} \sim_p \cdots \sim_p \frac{\omega_n}{m_n}$$

are conveniently controlled by parameters m_i, because R_p-functions preserve the differential properties of the arguments up to order p. If $(\omega_i = 0)$ is a smooth boundary, then m_i defines the gradient of ω at a regular point of $\partial\Omega_i$; if $(\omega_i = 0)$ is a sharp feature, then m_i specifies the behaviour of ω as a function of the angle at the ith corner. A given normalized function ω_i may also be scaled, for example,

$$\omega_i' = \frac{\omega_i}{1 + (\beta\omega_i)^\alpha}, \quad \text{and/or} \quad \omega_i' = \omega_i \wedge^* \gamma(\mathbf{p})$$

with parameters α, β, γ controlling its magnitude within the domain, while preserving the normalization of ω_i'. Sheiko (1982) and Rvachev (1982) discuss these and several other methods for including a priori information in a solution structure for problems with corners, cracks, and interfaces. Completeness of the modified solution structures is considered in Rvachev and Mikhal' (2001).

Numerous other solution structures for many common boundary value problems have been derived: elasticity (Rvachev and Sinekop 1990), vibration and stability of plates and shells (Rvachev et al. 1973, Rvachev and Kurpa 1987), heat transfer (Rvachev and Slesarenko 1976, Rvachev, Slesarenko and Safonov 1993), fluid dynamics (Tsukanov, Shapiro and Zhang 2003, Maksimenko-Sheiko and Sheiko 2005), thermo-elasticity (Rvachev, Sinekop and Molotkov 1991), contact (Rvachev, Sinekop and Molotkov

1992), diffraction (Gulyayev, Kravchenko, Rvachev and Sizova 1995), heterogeneous media (Tsukanov 2002, Tsukanov and Shapiro 2005), time-varying domains (Shapiro and Tsukanov 1999b), and many others.

7.4. Meshfree modelling and analysis with RFM

The approach outlined above was termed the R-function method, or RFM, by Rvachev. In retrospect, it seems more appropriate to interpret RFM as 'Rvachev's function method,' in recognition that it does not directly rely on the use of R-functions, since normalized functions may be constructed by other means. RFM offers a number of computational advantages, as described in Shapiro and Tsukanov (1999b) and Rvachev et al. (2000), including the ability to satisfy all prescribed boundary conditions exactly (on the zero set of the normalized functions ω_i) without any spatial discretization. This qualifies RFM as essentially a *meshfree* method, even though background meshes may be used for integration and/or visualization purposes.

One of the very first meshfree systems based on RFM was a software system called POLYE (which means 'field' in Russian) developed in Ukraine in the 1970s and 1980s, specifically for solving two-dimensional boundary value problems (Rvachev and Shevchenko 1988, Rvachev, Manko and Shevchenko 1986, Rvachev and Manko 1983) using R-functions. In POLYE, the geometric domain, boundary conditions, and solution structure were described in a programming language RL. A typical RL program contained geometric description of the domain in terms of predefined or user-specified analytic primitives ($\omega_i = 0$) and R-functions, explicit declaration of the solution structure, as well as detailed specification of the solution procedure, including the number and type of basis functions to represent Ψ, the numerical procedure (such as Ritz or least squares), integration parameters, and so on. At run time, the solution structure was automatically differentiated and numerically integrated over the background mesh in order to assemble the corresponding linear system. The solution for Ψ was substituted back into the solution structure and visualized. Both integration and visualization algorithms utilized variants of the marching cube algorithms (Shevchenko and Tsukanov 1994, Rvachev, Shevchenko and Veretel'nik 1994).

POLYE served as an early prototype for several more advanced systems. The first fully automated system, SAGE (Shapiro and Tsukanov 1999a, Tsukanov and Shapiro 2002), algorithmically constructed all required normalized functions from boundary and/or Constructive Solid Geometry (CSG) representations using R-functions as described in Section 5, and automatically assembled the solution structure implied by the indicated boundary conditions. Greatly improved algorithms for automatic differentiation (Tsukanov and Hall 2003) and numerical integration resulted in performance that is competitive with mesh-based methods. The architecture of a general purpose meshfree system is described in detail in Tsukanov and Shapiro (2002), where three-dimensional applications of RFM are also demonstrated. A variety of basis functions may be chosen for approximating the polynomial Ψ, including multivariate polynomials, multi-resolution B-splines, trigonometric polynomials and so on. Fully automated meshfree technology has been applied to a variety of boundary value problems, ranging from thermal conduction, linear elasticity, vibration and bending, to more challenging problems such as fluid dynamics (Tsukanov et al. 2003) and thermoelasticity in domains with heterogeneous materials (Tsukanov and Shapiro 2005). The single most difficult task in three dimensions remains the automatic construction of the normalized distance functions, because the practical algorithms based *solely* on R-functions tend to be limited and inefficient. Freytag et al. (2006) recently demonstrated that RFM can be combined with approximate distance fields that are sampled directly from any three-dimensional geometric representation, using R-functions only when set operations are explicitly required.

The fundamental ideas of RFM are also used by other meshfree and meshless methods, as surveyed by Babuška, Banerjee and Osborn (2003). A major challenge for all such methods is imposition of boundary conditions, and Dirichlet boundary conditions in particular, in the absence of a mesh. The so called 'characteristic function method' for satisfying the boundary conditions relies on the Dirichlet solution structure (2.6) used by Kantorovich, and is becoming increasingly popular due to R-functions. Notably, in the WEB-splines method proposed by Höllig (2003), the undetermined polynomial Ψ is constructed using uniform multivariate B-splines that are extended based on their location with respect to the domain's boundary $\partial\Omega$ in order to ensure the stability of numerical computations. The weight function ω is constructed using R-functions on numerically constructed primitive functions of the form

$$\omega_i = 1 - \max(0, 1 - \operatorname{dist}(x, \partial D)/\delta)^\gamma.$$

With the natural neighbour Galerkin method, Laguardia, Cueto and Doblare (2005) rely on the same solution structure, but ω is constructed using R-functions on analytically defined primitives, while Ψ is constructed using Sibson's natural neighbour interpolation on a Voronoi diagram. When ($\omega = 0$) represents the geometry of small internal features (for example, cracks) and discontinuities (for example at interfaces between bonded materials), replacing ω in the Dirichlet solution structure with the Heaviside function $H(\omega)$ and Ψ with standard finite elements Dirichlet solution structure, yields a representation for an 'enhanced' solution field with built-in singularities, avoiding the usual difficulties with fine (re)meshing normally required in such problems (Belytschko, Parimi, Moes, Sukumar and Usui 2003).

More generally, Babuška *et al.* (2003) pointed out that many of the mesh-less methods can be considered to be special cases of the partition of unity or generalized finite element (GFEM) method. All such methods appear to start with a selection of basis functions (partition of unity) to represent Ψ, which is then multiplied by 'local functions' ω_i that enhance the approximating solution space, based on geometric, asymptotic, or empirical information. Considering these methods as special cases of the Dirichlet solution structure suggests a systematic and constructive approach for satisfying the imposed boundary conditions, singularities, and/or other asymptotic conditions, as discussed in Section 7.3. For example, Duarte, Kim and Quaresma (2006) recently proposed a new type of C^m non-convex finite element that is a product of the standard partition of unity Ψ and a C^m weight ω constructed using R_0^m-functions.

The concept of solution structure is also useful for adaptive refinement of the approximations and multi-resolution modelling of boundary value problems. The most obvious approach would be to build the adaptivity and multi-resolution into the undetermined polynomial function Ψ. For example, variable (non-uniform) B-splines are common, and Höllig (2003) proposed using hierarchical B-splines. Both p and h refinements may be supported, but the global problem must be solved for each refinement, and the shape of refinement regions is determined and limited by the type of basis functions used in Ψ. Another approach to refinement in Tsukanov and Shapiro (2007) advocates representing the refined solution as a series of localized structures, each requiring the solving of a local boundary value problem. Each localization is specified by a refinement window of arbitrary shape, represented implicitly by a normalized window function ($\omega_1 \geq 0$). For example, if the refinement region is contained in the interior of a domain Ω, the Dirichlet solution structure (2.6) is modified as

$$u = \varphi + \omega\Psi + \omega_1^2 H(\omega_1)\Psi_1,$$

where ω_1^2 ensures C^1-continuity of the solution field, $H(\omega_1)$ guarantees that the refined solution does not modify the solution outside the refinement window, and Ψ_1 is a refinement polynomial constructed from a set of additional basis functions. See Tsukanov and Shapiro (2007) for further details and application to more general boundary value problems and refinement windows.

8. Conclusions

Rvachev hoped that R-functions would eventually be accepted as fundamental operations on par with other elementary functions. His belief was based on the observation that R-functions seem to provide a missing link between the logic and real analysis, and that they tend to streamline and unify many computational tasks. This survey attempted to explain this link, to convey the key ideas and concepts of the theory, and to connect them to related developments in geometric modelling and engineering analysis. The survey did not attempt an in-depth analysis of any topics, and did not try to be comprehensive. The cited references (albeit many are available only in Russian) contain a wealth of additional results, techniques, open problems, and applications. The focus on semi-analytic sets is justified because their properties are well understood, and because they are assumed in many computational applications. But it should be clear that this restriction is artificial. For example, Martin (1994) proposed using R-functions with fuzzy sets.

Generalizations of R-functions were also evident to Rvachev (1982). He observed that, besides the partition of real numbers into positive and negative, there are many other choices for potentially useful partitions. For example, one can partition real numbers into rational and irrational numbers, or, say, into all real numbers in the interval $[0, 1]$, and the rest of the real numbers. It is possible to introduce several or even infinitely many gradations when subdividing the set of real numbers. In general, any such partition Δ of the set of real numbers (based on some criterion) also determines the set $R(\Delta)$ of those real functions that in some sense 'inherit' the partition criterion (sign, rationality, membership in $[0, 1]$, *etc.*) Such functions are a generalization of the concept of R-functions as described in this paper. In fact, Rvachev went a step further and defined a notion of R-mapping $f : \mathbb{X}^n \to \mathbb{X}^m$, where $\mathbb{X}$ is an arbitrary abstract space. The partition Δ of $\mathbb{X}$ into the qualitative equivalence classes is based on some multi-valued logic function $S_k : \mathbb{X} \to \mathbb{B}_k$. Then R-mapping f is identified by the existence of the companion multi-valued logic function $\Phi : \mathbb{B}_k^n \to \mathbb{B}_k^m$, satisfying the following commutative diagram:

$$
\begin{array}{ccc}
\mathbb{X}^n & \xrightarrow{\ f\ } & \mathbb{X}^m \\
\big\downarrow{\scriptstyle S_k^n} & & \big\downarrow{\scriptstyle S_k^m} \\
\mathbb{B}_k^n & \xrightarrow[\ \Phi\ ]{} & \mathbb{B}_k^m
\end{array}
$$

This and further generalizations of R-functions described by Rvachev (1982) do not appear to have found many applications so far, most likely because they remain largely unknown to the research community at large.

It is evident that R-functions are becoming more popular and are now widely used in many computational applications. It should be remembered that R-functions were invented when computational technology was in its infancy, and computational geometry and geometric modelling had not been established as disciplines. Today, semi-analytic sets may be represented or approximated by many other methods, and each method has its strength and weaknesses. The wealth of alternative methods in no way diminishes the intellectual and practical significance of Rvachev's contributions and of the theory of R-functions. Interestingly, representation of point sets by sampled distance fields appears to be growing in popularity, partly driven by advances in image processing and medical imaging (Jones and Bærentzen 2006). Such distance fields may be smoothed using interpolation or fitting techniques (Freytag *et al.* 2006), resulting in smooth approximations to normalized functions. However, R-functions must be used if such fields need to represent sharp corners and features.

It is also possible that applications of R-functions may prove to be equally or even more important than the R-functions themselves. In particular, the notion of the RFM solution structure described in Section 7 appears to be extremely useful for modelling and solving boundary value problems. With the exception of RFM itself, many modern meshfree methods appear to be struggling with modelling and approximating the Dirichlet solution structure. Meanwhile, RFM provides a systematic and accurate method for imposing any and all types of boundary conditions, without artificial topological constraints or meshing, and independently of any particular numerical scheme. It remains to be seen whether satisfaction of boundary conditions results in improved numerical properties, but there is no question that RFM provides dramatic improvement in flexibility and programmability of solvers for boundary value problems.

Acknowledgements

I am indebted to my long-time collaborator, Igor Tsukanov, for numerous technical discussions and for his help in creating the examples for this survey, to Carl de Boor for reading an earlier draft of this survey and suggesting numerous improvements, and to T. Sheiko for her help with the example demonstrating use of R-functions in modelling symmetry. I would also like to thank the editor, Arieh Iserles, for his encouragement and (seemingly) infinite patience. This work was supported in part by the National Science Foundation grants DMI-0621116, OCI-0537370, DMI-0323514, and DMI-0500380, and by the Wisconsin I&EDR Program. The responsibility for any errors and omissions lies solely with the author.

REFERENCES

K. Abdel-Malek, D. Blackmore and K. Joy (2006), 'Swept volumes: Foundations, perspectives, and applications', *Internat. J. Shape Modeling* **12**, 87–127.

E. Allgower and K. Georg (1990), *Numerical Continuation Methods: An Introduction*, Springer, New York.

D. Altiparmakov and P. Belicev (1990), 'An efficiency study of the R-function method applied as solid modeler for Monte Carlo calculations', *Progress in Nuclear Energy* **24**, 77.

C. Andradas, L. Bröcker and J. Ruiz (1996), *Constructible Sets in Real Geometry*, Springer, New York.

I. Babuška, U. Banerjee and J. Osborn (2003), Survey of meshless and generalized finite element methods: A unified approach, in *Acta Numerica*, Vol. 12, Cambridge University Press, pp. 1–125.

L. Barthe, B. Wyvill and E. De Groot (2004), 'Controllable binary CSG operators for soft objects', *Internat. J. Shape Modeling* **10**, 135–154.

S. Basu, R. Pollack and M.-F. Roy (2003), *Algorithms in Real Algebraic Geometry*, Vol. 10 of *Algorithms and Computations in Mathematics*, Springer.

S. Basu, R. Pollack and M.-F. Roy (2005), Computing the first Betti number and the connected components of semi-algebraic sets, in *STOC'05: Proc. 37th Annual ACM Symposium on Theory of Computing*, ACM Press, New York, pp. 304–312.

T. Belytschko, C. Parimi, N. Moes, N. Sukumar and S. Usui (2003), 'Structured extended finite element methods for solids defined by implicit surfaces', *Internat. J. Numer. Methods Engng.* **56**, 609–635.

A. Biswas and V. Shapiro (2004), 'Approximate distance fields with non-vanishing gradients', *Graphical Models* **66**, 133–159.

A. Biswas, V. Shapiro and I. Tsukanov (2004), 'Heterogeneous material modeling with distance fields', *Computer Aided Geometric Design* **21**, 215–242.

J. L. Blechschmidt and D. Nagasuru (1990), The use of algebraic functions as a solid modeling alternative: An investigation, in *16th ASME Design Automation Conference*, Chicago, IL, pp. 33–41.

J. Bloomenthal (1988), 'Polygonization of implicit surfaces', *Computer Aided Geometric Design* **5**, 341–355.

J. Bloomenthal (1997), *Introduction to Implicit Surfaces*, Morgan Kaufmann.

J. Bochnak, M. Coste and M.-F. Roy (1998), *Real Algebraic Geometry*, Springer.

D. Breen, S. Mauch and R. Whitaker (1998), 3D scan conversion of CSG models into distance volumes, in *IEEE Symposium on Volume Visualization 1998*, pp. 7–14.

J. Chen, M. Freytag and V. Shapiro (2007a), Shape sensitivity of constructive representations, in *ACM SPM 2007: Proc. ACM Symposium on Solid and Physical Modeling, Beijing, China, 4–6 June 2007*.

J. Chen, V. Shapiro, K. Suresh and I. Tsukanov (2007b), 'Shape optimization with topological changes and parametric control', *Internat. J. Numer. Methods Engng.*, in press.

R. Descartes (1637), *La Géométrie*.

D. Dobkin, L. Guibas, J. Hershberger and J. Snoeynik (1988), 'An efficient algorithm for finding the CSG representation of a simple polygon', *Computer Graphics* **22**, 31–40.

C. A. Duarte, D.-J. Kim and D. M. Quaresma (2006), 'Arbitrarily smooth generalized finite element approximations', *Comput. Methods Appl. Mech. Engng.* **196**, 33–56.

H. Edelsbrunner (1987), *Algorithms in Combinatorial Geometry*, Vol. 10 of *EATCS Monographs on Theoretical Computer Science*, Springer.

M. Ensz, D. Storti and M. Ganter (1998), 'Implicit methods for geometry creation', *Internat. J. Comput. Geometry Appl.* **8**, 509–536.

G. Farin, J. Hoschek and M. Kim (2002), *Handbook of Computer Aided Geometric Design*, Elsevier.

P. Fayolle, A. Pasko, B. Schmitt and N. Mirenkov (2006), 'Constructive heterogeneous object modeling using signed approximate real distance functions', *J. Comput. Inform. Sci. Engng.* **6**, 221.

Y. D. Fougerolle, A. Gribok, S. Foufou, F. Truchetet and M. A. Abidi (2005), 'Boolean operations with implicit and parametric representation of primitives using R-functions', *IEEE Trans. Visualization Computer Graphics* **11**, 529–539.

M. Freytag, V. Shapiro and I. Tsukanov (2006), 'Field modeling with sampled distances', *Computer-Aided Design* **38**, 87–100.

I. Goncharyuk, V. Rvachev and L. Shklyarov (1968), 'Torsion in bars of polygonal profile, with allowance for stress-function characteristics', *Internat. Appl. Mech.* **4**, 88–94.

Y. Gulyayev, V. Kravchenko, V. Rvachev and N. Sizova (1995), 'R-function method in investigation of elastic waves diffraction on rigid insert of arbitrary shape', *Cybernet. Systems Analysis* **344**, 457–459.

K. Höllig (2003), *Finite Element Methods with B-Splines*, Vol. 26 of *Frontiers in Applied Mathematics*, SIAM.

J. Hoschek and D. Lasser (1993), *Fundamentals of Computer Aided Geometric Design*, A. K. Peters.

H. Ilies and V. Shapiro (1999), 'Dual of sweep', *Computer Aided Design* **31**, 185–201.

M. Jones and J. Bærentzen (2006), '3D distance fields: A survey of techniques and applications', *IEEE Trans. Visualization Computer Graphics* **12**, 581–599.

L. V. Kantorovich and V. I. Krylov (1958), *Approximate Methods of Higher Analysis*, Interscience.

I. Y. Kharrik (1963), 'On approximation of functions that have zero values and derivatives on domain boundary by special functions', *Siberian Math. J.* **4**, 408–425.

Y. Kim and D. Wilde (1992), 'Local cause of non-convergence in a convex decomposition using convex hulls', *Trans. ASME J. Mech. Design* **114**, 468–476.

V. Komkov (1989), Stochastic modelling of physical processes and optimization of the domain, in *Design Theory '88: NSF Grantees Workshop on Design Theory and Methodology* (S. L. Newsome, W. R. Spillers and S. Finger, eds), Springer, New York.

L. N. Kutsenko (1990), *Computer Graphics in the Problems of Projective Nature*, Vol. 8 of *Mathematics and Cybernetics*, Znanie, Moscow.

J. Laguardia, E. Cueto and M. Doblare (2005), 'A natural neighbour Galerkin method with quadtree structure', *Internat. J. Numer. Methods Engng.* **63**, 789–812.

P. Lancaster and K. Salkauskas (1986), *Curve and Surface Fitting: An Introduction*, Academic Press.

E. Lawler (1964), 'An approach to multilevel Boolean minimization', *J. Assoc. Comput. Mach.* **11**, 283–295.

S. Lojasiewicz (1964), 'Triangulations of semi-algebraic sets', *Annali della Scuola Normale Superiore di Pisa* **18**, 449–474.

W. Lorensen and H. Cline (1987), Marching cubes: A high resolution 3D surface construction algorithm, in *Proc. 14th Annual Conference on Computer Graphics and Interactive Techniques*, pp. 163–169.

K. V. Maksimenko-Sheiko and N. Sheiko (2005), 'Simulation of incompressible viscous liquid motion in twisted tubes by R-function method', *Elektronnoe Modelirovanie* **27**, 31–43.

K. V. Maksimenko-Sheiko, A. M. Matsevityi and T. I. Sheiko (2005), 'Constructive methods of R-functions for constructing of 3D primitives', *Problemy Mashinostroyenia (Problems of Machine Building)* **8**, 59–65.

R. Martin (1994), Modelling inexact shapes with fuzzy sets, in *CSG'94: Set-Theoretic Solid Modelling Techniques and Applications*, pp. 73–100.

Y. M. Matsevity, ed. (2001), *Vladimir Logvinovich Rvachev*, Bibliography of Ukrainian Scientists, Academy of Sciences of Ukraine.

J. O'Rourke (1982), 'Polygon decomposition and switching function minimization', *Computer Graphics and Image Processing* **18**, 382–391.

J. O'Rourke (1998), *Computational Geometry in C*, Cambridge University Press, Cambridge, United Kingdom.

A. Pasko, V. Adzhiev, A. Sourin and V. Savchenko (1995), 'Function representation in geometric modeling: Concepts, implementation and applications', *The Visual Computer* **11**, 429 446.

G. Pasko, A. Pasko, M. Ikeda and T. Kunii (2002), Bounded blending operations, in *SMI'02: Proc. Shape Modeling International 2002*, IEEE Computer Society, Washington, DC, p. 95.

D. Peterson (1986), 'Boundary to constructive solid geometry mappings: A focus on 2D issues', *Computer-Aided Design* **18**, 3–14.

A. A. G. Requicha (1977), Mathematical models of rigid solid objects. Technical memo 28, Production Automation Project, University of Rochester, Rochester, NY.

A. A. G. Requicha (1980), 'Representations for rigid solids: Theory, methods, and systems', *ACM Computing Surveys* **12**, 437–464.

A. A. G. Requicha and H. B. Voelcker (1977), Constructive solid geometry. Technical memo 25, Production Automation Project, University of Rochester.

A. Ricci (1973), 'A constructive geometry for computer graphics', *Comput. J.* **16**, 157–160.

E. Rimon and D. E. Koditschek (1990), Exact robot navigation in geometrically complicated but topologically simple environment, in *IEEE International Conference on Robotics and Automation*, pp. 1937–1943.

E. Rimon and D. E. Koditschek (1992), 'Exact robot navigation using artificial potential functions', *IEEE Trans. Robotics and Automation* **8**, 501–518.

A. P. Rockwood (1989), 'The displacement method for implicit blending surfaces in solid modeling', *ACM Trans. Graphics* **8**, 279–297.

A. P. Rockwood and J. C. Owen (1987), Blending surfaces in solid modeling, in *Geometric Modeling: Algorithms and New Trends* (G. E. Farin, ed.), SIAM, Philadelphia, pp. 367–383.

J. R. Rossignac (1996), CSG formulations for identifying and for trimming faces of CSG models, in *CSG'96, Winchester, UK*, Information Geometers Ltd.

S. D. Roth (1982), 'Ray casting for modeling solids', *Computer Graphics and Image Processing* **18**, 109–144.

V. L. Rvachev (1963), 'On analytical description of some geometric objects', *Reports (Doklady) of Academy of Sciences, USSR* **153**, 765–768.

V. L. Rvachev (1967), *Geometric Applications of Logic Algebra*, Naukova Dumka. In Russian.

V. L. Rvachev (1974), *Methods of Logic Algebra in Mathematical Physics*, Naukova Dumka. In Russian.

V. L. Rvachev (1982), *Theory of R-functions and Some Applications*, Naukova Dumka. In Russian.

V. L. Rvachev (1996), Lectures at the University of Wisconsin, Madison. Video recording, Spatial Automation Laboratory; http://sal-cnc.me.wisc.edu.

V. L. Rvachev and L. V. Kurpa (1987), *R-Functions in Problems of Theory of Plates*, Naukova Dumka. In Russian.

V. L. Rvachev and G. P. Manko (1983), *Automation of Programming for Boundary Value Problems*, Naukova Dumka. In Russian.

V. L. Rvachev and E. O. Mikhal' (2001), 'Completeness of structural solutions in boundary-value problems for domains of special form', *Cybernet. Systems Analysis* **37**, 551–561.

V. L. Rvachev and V. A. Rvachev (1979), *Nonclassical Methods of Approximation Theory in Boundary Value Problems*, Naukova Dumka. In Russian.

V. L. Rvachev and T. I. Sheiko (1995), 'R-functions in boundary value problems in mechanics', *Applied Mechanics Reviews* **48**, 151–188.

V. L. Rvachev and A. N. Shevchenko (1988), *Problem-Oriented Languages and Systems for Engineering Computations*, Tekhnika, Kiev. In Russian.

V. L. Rvachev and N. S. Sinekop (1990), *R-Functions Method in Problems of the Elasticity and Plasticity Theory*, Naukova Dumka, Kiev. In Russian.

V. L. Rvachev and A. P. Slesarenko (1976), *Logic Algebra and Integral Transforms in Boundary Value Problems*, Naukova Dumka. In Russian.

V. L. Rvachev, L. V. Kurpa, N. G. Sklepus and L. A. Uchishvili (1973), *Method of R-functions in Problems on Bending and Vibrations of Plates of Complex Shape*, Naukova Dumka. In Russian.

V. L. Rvachev, G. P. Manko and A. N. Shevchenko (1986), The R-function approach and software for the analysis of physical and mechanical fields, in *Software for Discrete Manufacturing* (J. P. Crestin and J. F. McWaters, eds), North-Holland, Paris.

V. L. Rvachev, T. I. Sheiko and V. Shapiro (1999), 'The R-function method in boundary-value problems with geometric and physical symmetry', *J. Math. Sci.* **97**, 3888–3899.

V. L. Rvachev, T. I. Sheiko, V. Shapiro and I. Tsukanov (2000), 'On completeness of RFM solution structures', *Comput. Mech.* **25**, 305–316.

V. L. Rvachev, T. I. Sheiko, V. Shapiro and I. Tsukanov (2001), 'Transfinite interpolation over implicitly defined sets', *Computer Aided Geometric Design* **18**, 195–220.

V. L. Rvachev, T. I. Sheiko, V. Shapiro and J. J. Uicker (1997), 'Implicit function modeling of solidification in metal casting', *Trans. ASME J. Mech. Design* **119**, 466–473.

V. L. Rvachev, A. N. Shevchenko and V. V. Veretel'nik (1994), 'Numerical integration software for projection and projection-grid methods', *Cybernet. Systems Analysis* **30**, 154–158.

V. L. Rvachev, N. S. Sinekop and I. P. Molotkov (1991), 'R-function method for thermoelasticity problems in finite objects', *Trans. Acad. Sci. USSR (Doklady Academii nauk SSSR)* **321**, 721–725. In Russian.

V. L. Rvachev, N. S. Sinekop and I. P. Molotkov (1992), 'Solution structure for contact problems of thermoelasticity', *Trans. Ukrainian National Acad. Sci. (Dopovidi NAN Ukrainy)* **A**, 25–29. In Russian.

V. L. Rvachev, A. P. Slesarenko and N. A. Safonov (1993), 'On solution of steady non-linear heat transfer problems', *Trans. Ukrainian National Acad. Sci. (Dopovidi NAN Ukrainy)* **A**, 1015–1021. In Russian.

J. Sethian (1996), *Level Set Methods: Evolving Interfaces in Geometry, Fluid Mechanics, Computer Vision and Material Sciences*, Cambridge University Press.

V. Shapiro (1988), Theory of R-functions and applications: A primer. Technical report TR91-1219, Computer Science Department, Cornell University, Ithaca, NY. Revised 1991.

V. Shapiro (1991), Representations of semi-algebraic sets in finite algebras generated by space decompositions. PhD thesis, Cornell University, Cornell Programmable Automation, Ithaca, NY.

V. Shapiro (1994), 'Real functions for representation of rigid solids', *Computer-Aided Geometric Design* **11**, 153–175.

V. Shapiro (1997), 'Maintenance of geometric representations through space decompositions', *Internat. J. Comput. Geometry Appl.* **7**, 383–418.

V. Shapiro (1999), 'Well-formed set representations of solids', *Internat. J. Comput. Geometry Appl.* **9**, 125–150.

V. Shapiro (2001), 'A convex deficiency tree algorithm for curved polygons', *Internat. J. Comput. Geometry Appl.* **11**, 215–238.

V. Shapiro (2002), Solid modeling, in *Handbook of Computer Aided Geometric Design* (G. Farin, J. Hoschek and M.-S. Kim, eds), Elsevier Science.

V. Shapiro and I. Tsukanov (1999*a*), Implicit functions with guaranteed differential properties, in *Fifth ACM Symposium on Solid Modeling and Applications, Ann Arbor, MI*, pp. 258–269.

V. Shapiro and I. Tsukanov (1999*b*), 'Meshfree simulation of deforming domains', *Computer-Aided Design* **31**, 459–471.

V. Shapiro and D. L. Vossler (1991*a*), 'Construction and optimization of CSG representations', *Computer-Aided Design* **23**, 4–20.

V. Shapiro and D. L. Vossler (1991*b*), 'Efficient CSG representations of two-dimensional solids', *Trans. ASME J. Mech. Design* **113**, 292–305.

V. Shapiro and D. L. Vossler (1993), 'Separation for boundary to CSG conversion', *ACM Trans. Graphics* **12**, 35–55.

H. Sheffer (1913), 'A set of five independent postulates for Boolean algebras, with application to logical constants', *Trans. Amer. Math. Soc.* **14**, 481–488.

T. I. Sheiko (1982), 'Taking account of singularities at angular points and juncture points of the boundary conditions in the method of R-functions', *Internat. Appl. Mech.* **18**, 365–370.

D. Shepard (1968), A two-dimensional interpolation function for irregularly spaced data, in *Proc. 23rd ACM National Conference*, pp. 517–524.

A. N. Shevchenko and I. Tsukanov (1994), 'Software for representing results of field simulation in complex-form domains', *Control Systems and Mechanisms* 4–5, 86–89. In Russian.

A. M. Shkel (1997), Sensor-based motion planning with kinematic and dynamic constraints. PhD thesis, mechanical engineering, University of Wisconsin, Madison.

A. Sourin and A. Pasko (1995), Function representation for sweeping by a moving solid, in *Proc. Third Symposium on Solid Modeling Foundations and CAD/CAM Applications*, pp. 383–391.

A. Sourin, A. Pasko and V. Savchenko (1996), 'Using real functions with application to hair modelling', *Computers & Graphics* 20, 11–19.

Y. G. Stoian (1975), *Placement of Geometric Objects*, Naukova Dumka. In Russian.

Y. G. Stoian and A. A. Panasenko (1978), *Periodic Placement of Geometric Objects*, Naukova Dumka. In Russian.

D. Storti, M. Ganter and C. Nevrinceanu (1992), 'A tutorial on implicit solid modeling', *The Mathematica Journal* 2, 70–78.

S. Tor and A. Middleditch (1984), 'Convex decomposition of simple polygons', *ACM Trans. Graphics* 3, 244–265.

I. Tsukanov (2002), 'On the question of calculation of temperature fields in piecewise-homogeneous orthotropic media', *J. Math. Sci.* 109, 1338–1343.

I. Tsukanov and M. Hall (2003), 'Data structure and algorithms for fast automatic differentiation', *Internat. J. Numer. Methods Engng.* 56, 1949–1972.

I. Tsukanov and V. Shapiro (2002), 'The architecture of SAGE: A meshfree system based on RFM', *Engineering with Computers* 18, 295–311.

I. Tsukanov and V. Shapiro (2005), 'Meshfree modeling and analysis of physical fields in heterogeneous media', *Adv. Comput. Math.* 23, 95–124.

I. Tsukanov and V. Shapiro (2007), 'Adaptive multiresolution refinement with distance fields', *Internat. J. Numer. Methods Engng.*, to appear.

I. Tsukanov, V. Shapiro and S. Zhang (2003), 'A meshfree method for incompressible fluid dynamics problems', *Internat. J. Numer. Methods Engng.* 58, 127–158.

L. Velho, J. Gomes and L. de Figueiredo (2002), *Implicit Objects in Computer Graphics*, Springer.

D. F. Watson (1992), *Contouring: A Guide to the Analysis and Display of Spatial Data*, Pergamon Press.

H. Whitney (1934), 'Analytic extensions of differentiable functions defined in closed sets', *Trans. Amer. Math. Soc.* 36, 63–89.

H. Whitney (1957), 'Elementary structure of real algebraic varieties', *Ann. of Math.* 66, 545–556.

H. Whitney (1965), Local properties of analytic varieties, in *Differential and Combinatorial Topology*, Princeton University Press, pp. 205–44.

T. C. Woo (1982), Feature extraction by volume decomposition, in *Proc. Conference on CAD/CAM Technology in Mechanical Engineering*, Cambridge, MA, pp. 76–94.

J. R. Woodwark (1988), Blends in geometric modelling, in *Proc. Mathematics of Surfaces II*, Clarendon Press, New York, pp. 255–297.

J. R. Woodwark and A. L. Wallis (1982), Graphical input to a Boolean solid modeller, in *Proc. CAD'82*, Brighton, pp. 681–688.

B. Wyvill, A. Guy and E. Galin (1999), 'Extending the CSG tree: Warping, blending and Boolean operations in an implicit surface modeling system', *Computer Graphics Forum* **18**, 149–158.

Acta Numerica (2007), pp. 305–378
doi: 10.1017/S0962492906320016

Filters, mollifiers and the computation of the Gibbs phenomenon

Eitan Tadmor
Department of Mathematics, Institute for Physical Science & Technology
and
Center for Scientific Computation and Mathematical Modeling (CSCAMM),
University of Maryland, College Park, MD 20742, USA
E-mail: tadmor@cscamm.umd.edu

We are concerned here with processing discontinuous functions from their spectral information. We focus on two main aspects of processing such piecewise smooth data: detecting the edges of a piecewise smooth f, namely, the location and amplitudes of its discontinuities; and recovering with high accuracy the underlying function in between those edges. If f is a smooth function, say analytic, then classical Fourier projections recover f with exponential accuracy. However, if f contains one or more discontinuities, its global Fourier projections produce spurious Gibbs oscillations which spread throughout the smooth regions, enforcing local loss of resolution and global loss of accuracy. Our aim in the computation of the Gibbs phenomenon is to detect edges and to reconstruct piecewise smooth functions, while regaining the high accuracy encoded in the spectral data.

To detect edges, we utilize a general family of edge detectors based on *concentration kernels*. Each kernel forms an approximate derivative of the delta function, which detects edges by *separation of scales*. We show how such kernels can be adapted to detect edges with one- and two-dimensional discrete data, with noisy data, and with incomplete spectral information. The main feature is concentration kernels which enable us to convert global spectral moments into local information in physical space. To reconstruct f with high accuracy we discuss novel families of *mollifiers* and *filters*. The main feature here is making these mollifiers and filters *adapted* to the local region of smoothness while increasing their accuracy together with the dimension of the data. These mollifiers and filters form approximate delta functions which are properly parametrized to recover f with (root-) exponential accuracy.

CONTENTS

1. Introduction

We are interested in processing piecewise smooth functions from their spectral information. The prototype example will be one-dimensional functions that are smooth except for finitely many jump discontinuities. The locations and amplitudes of these discontinuities are not correlated. Thus, a piecewise smooth f is in fact a collection of several intervals of smoothness which do *not* communicate among themselves. The jump discontinuities can be viewed as the edges of these intervals of smoothness. Similarly, two-dimensional piecewise smooth functions consist of finitely many edges which lie along simple curves, separating two-dimensional local regions of smoothness. We are concerned here with two main aspects of processing such piecewise smooth data.

(i) **Edge detection.** Detecting the location and amplitudes of the edges. Often, these are the essential features sought in piecewise smooth data. Moreover, they define the regions of smoothness and are therefore essential for the second aspect.

(ii) **Reconstruction.** Recovering the underlying function f inside its different regions of smoothness.

There are many classical algorithms to detect edges and reconstruct the data in between those edges, based on *local* information. For example, suppose that the values of a one-dimensional f are given at equidistant grid-points, $f_\nu = f(\nu\Delta x)$. Then, the first-order differences, $\Delta f_\nu := f_{\nu+1} - f_\nu$, can detect edges where $\Delta f_\nu = \mathcal{O}(1)$, by separating them from smooth regions where $\Delta f_\nu = \mathcal{O}(\Delta x)$. Similarly, piecewise linear interpolants can recover the point values of $f(x)$ up to order $\mathcal{O}((\Delta x)^2)$. Of course, these are only asymptotic statements that may greatly vary with the dependence of the $\mathcal{O}$-terms on the *local* smoothness of f in the immediate neighbourhood of x. We may do better, therefore, by taking higher-order differences, $\Delta^r f_\nu$, where $\mathcal{O}(1)$-edges are better separated from $\mathcal{O}((\Delta x)^r)$-regions of smoothness. Similarly, reconstruction of f using r-order approximations, with $r = 2, 3, \ldots$ and so on. In practice, higher accuracy is translated into higher resolution extracted from the information on a given grid. But, as the order of accuracy increases, the stencils involved become wider and one has to be careful not to extract smoothness information *across* edges, since different regions separated by edges are completely independent of each other. An effort to extract information from one region of smoothness into another one, will result in spurious oscillations, spreading from the edges into the surrounding smooth regions, preventing uniform convergence. This is, in general terms, the *Gibbs phenomenon*, which is the starting point of the present discussion.

The prototype for spectral information we are given on f is the set of its N Fourier coefficients, $\{\widehat{f}(k)\}_{|k| \leq N}$. These are *global* moments of f. It is well known that the Fourier projection, $S_N f = \sum_{|k| \leq N} \widehat{f}(k) e^{ikx}$, forms a highly accurate approximation of f provided that f is *sufficiently smooth*. In Section 2 we revisit the classical spectral convergence statements and quantify the *actual* exponential accuracy of Fourier projections,

$$|S_N f - f| \lesssim e^{-\eta \sqrt[\alpha]{N}}.$$

Here, the root exponent α is tied to *global* smoothness of f of order $\alpha \geq 1$. But this high accuracy is lost with piecewise smooth f, due to spurious oscillations which are formed around the edges of f. It is in this context of Fourier projections that the formation of spurious oscillations became known as *the* Gibbs phenomenon, originating with Gibbs' letter of 1899. This is precisely because of the global nature of $S_N f$, which extracts smoothness information *across* the internal edges of f. The Gibbs phenomenon is also responsible for a *global* loss of accuracy: first-order oscillations spread *throughout* the regions of smoothness. The highly accurate content in the spectral data, $\{\widehat{f}(k)\}_{|k| \leq N}$, is lost in the Fourier projections, $S_N f$. The local and global effects of Gibbs oscillations are illustrated through a simple example in Section 3.

Our aim in the computation of the Gibbs phenomenon is to detect edges and reconstruct piecewise smooth functions, while regaining the high accuracy encoded in their spectral data. Here, we use two main tools.

(i) **Concentration kernels.** To detect edges, we employ a fairly general framework based on partial sums of the form

$$K_N^\sigma f(x) := \frac{\pi i}{c_\sigma} \sum_{|k| \leq N} \operatorname{sgn}(k)\sigma\left(\frac{|k|}{N}\right)\widehat{f}(k)e^{ikx}, \qquad c_\sigma := \int_0^1 \frac{\sigma(\xi)}{\xi}\, d\xi.$$

In Section 4 we show that $K_N^\sigma f(x)$ approximates the *local* jump function, $K_N^\sigma f(x) \approx f(x+) - f(x-)$. Consequently, $K_N^\sigma f$ tends to concentrate near edges, where $f(x+) - f(x-) \neq 0$, which are separated from smooth regions where $K_N^\sigma f \approx 0$. We can express $K_N^\sigma f(x)$ as a convolution with the Fourier projection of f, that is,

$$K_N^\sigma f(x) = \mathcal{K}_N^\sigma * (S_N f)(x), \qquad \mathcal{K}_N^\sigma(x) := -\frac{1}{c_\sigma} \sum_{k=1}^N \sigma\left(\frac{|k|}{N}\right) \sin kx.$$

Here, $\mathcal{K}_N^\sigma(x)$ are the corresponding *concentration kernels* which enable us to convert the global moments of $S_N f$ into local information about its edges – both their locations and their amplitudes. The choice of concentration factor σ is at our disposal. In Section 5 we discuss a few prototype examples of concentration factors and assess the different behaviour of the corresponding edge detectors, $K_N^\sigma f$. In Section 6 we turn to a series of *extensions* which show how concentration kernels apply in more general set-ups. In Section 6.1 we discuss the *discrete* framework, applying concentration kernels as edge detectors in the Fourier interpolants, $I_N f = \sum_{|k| \leq N} \hat{f}_k e^{ikx}$. In Section 6.2 we show how concentration kernels can be used to detect edges in *non-periodic* projections, $S_N f = \sum \hat{f}(k) C_k(x)$, based on general Gegenbauer expansions. In Section 6.3 we show how the concentration factors could be adjusted to deal with *noisy data*, by taking into account the noise variance, $\eta \gg 1/N$, in order to detect the underlying $\mathcal{O}(1)$-edges. Finally, Section 6.4 deals with *incomplete data*: we show how concentration kernels based on partial information, $\{\hat{f}(k)\}_{k \in K}$, can be complemented by a compressed sensing approach to form effective edge detectors.

Concentration kernels, $\mathcal{K}_N^\sigma(x)$, are approximate *derivatives* of the delta function. Convolution with such kernels, $\mathcal{K}_N^\sigma * (S_N f)$, yield edge detectors by *separation of scales*, separating between smooth and non-smooth parts of f. In Section 7 we show how to improve the edge detectors by *enhancement* of this separation of scales. In particular, in Section 7.1 we use nonlinear *limiters* which assign low- and high-order concentration kernels in regions with different characteristics of smoothness. The result is *parameter-free*, high-resolution edge detectors for one-dimensional piecewise smooth functions.

In Section 8 we turn to the two-dimensional set-up. Concentration kernels can be used to separate scales in the x_1 and x_2 directions. Enhancements and limiters are shown in Section 8.1 to greatly reduce, though not completely eliminate, the Cartesian staircasing effect. In Section 8.2 we show how concentration kernels are used to detect edges from incomplete two-dimensional data. So far, we have emphasized the role of separation of scales in edge detectors based on concentration kernels, $\mathcal{K}_N^\sigma * (S_N f)(\mathbf{x})$. But how do we actually *locate* those $\mathcal{O}(1)$ edges? In Section 8.3 we discuss the approach which seeks the zero-level set $\mathbf{x} = (x_1, x_2)$ of $\nabla_\mathbf{x} \mathcal{K}_N^\sigma * (S_N f)(\mathbf{x})$.

Depending on our choice of the concentration factors, $\sigma(\cdot)$, this leads to a large class of two-dimensional edge detectors which generalize the popular two-dimensional zero-crossing method associated with discrete Laplacian stencils.

Next, we turn our attention to highly accurate, Gibbs-free *reconstruction* of f inside its regions of smoothness from its (pseudo-) spectral content.

(ii) Mollifiers and filters. We consider two interchangeable processes to recover the values of a piecewise smooth $f(x)$ with high accuracy. These are *mollification*, carried out in the physical space, and *filtering*, carried out in the Fourier space, i.e.,

$$\Phi_{p,\delta} * (S_N f)(x) \longleftrightarrow \sum_{|k| \leq N} \varphi_{p,\delta}\left(\frac{|k|}{N}\right) \widehat{f}(k) e^{ikx}.$$

Filtering accelerates convergence when pre-multiplying the Fourier co-efficients by a *rapidly decreasing* $\varphi_{p,\delta}(|k|/N)$. The rapid decay of $\varphi_{p,\delta}(|k|/N)\widehat{f}(k)$, as $|k| \uparrow N$ in Fourier space, corresponds to mollification with *highly localized* mollifiers, $\Phi_{p,\delta}(x)$, in physical space.

Section 10 is devoted to mollifiers. There are two free parameters at our disposal. The parameter δ is chosen so that the essential support of $\Phi_{p,\delta} * (S_N f)(x)$ does *not* cross edges of f. To this end we set δ as the distance to the nearest edge, $\delta = d_x := \operatorname{dist}\{x, \operatorname{singsupp} f\}/\pi$, so that $(x - \pi\delta, x + \pi\delta)$ is the largest interval of smoothness enclosing x. It is here that we use the information on the location of the edges of f. This leads to *adaptive* mollifiers $\Phi_{p,d_x}(x)$. The parameter p is responsible for the *accuracy* of the mollifier. By properly tuning $p = p_N$ to increase with N, one obtains the *root-exponential* accurate mollifiers discussed in Section 10.2:

$$\Phi_{p_N,d_x}(x) := \frac{1}{d_x}\varphi\left(\frac{\pi x}{d_x}\right) D_{p_N}\left(\frac{x}{d_x}\right),$$

$$d_x = \frac{1}{\pi}\operatorname{dist}\{x, \operatorname{singsupp} f\}, \quad p_N \sim d_x N.$$

Here,

$$D_p(x) = \frac{\sin(p+1/2)x}{2\pi \sin(x/2)}$$

is the Dirichlet kernel of order p, which ensures accuracy by having an increasing number of (almost) vanishing moments, $\int y^n D_{p_N}(y)\, dy \approx 0$, for $p = 1, 2, \ldots, p_N$, and $\varphi = \varphi_{2q}$ is a proper $C_0^\infty(-1, 1)$ cut-off function,[1] e.g.,

$$\varphi_{2q}(y) := e^{\left(\frac{y^{2q}}{y^2-1}\right)} 1_{(-1,1)}(y),$$

---

[1] C_0^∞ is the space of compactly supported smooth functions.

which ensures that Φ_{p_N, d_x} are properly localized within the d_x-neighbourhood of the origin. The result is an adaptive mollifier with *root-exponential accuracy*

$$|\Phi_{p_N, d_x} * (S_N f)(x) - f(x)| \lesssim \mathrm{e}^{-\eta\sqrt{d_x N}}.$$

The corresponding root-exponential discrete mollifier is outlined in Section 10.3. The high accuracy of these mollifiers is adapted to the *interior* points, away from the edges where $d_x N \sim 1$. It can be modified to gain polynomial accuracy *up to* the edges. This is described in Section 10.4. In Section 10.5 we discuss mollifiers based on *Gegenbauer expansion* of $S_N f(\pi x)$, with uniform root-exponential accuracy *up to* the edges.

Section 11 is devoted to filters of the form

$$S_{p_N}^\varphi f(x) := \sum_{|k| \leq N} \varphi_{p_N}\left(\frac{|k|}{N}\right) \widehat{f}(k) \mathrm{e}^{\mathrm{i}kx}.$$

In Section 11.1 we show that, by setting $p_N \sim \sqrt{d_x N}$, the resulting filter is accurate (and hence its associated mollifier satisfies a moment condition) to order p_N. Moreover, the choice of the filter φ_{p_N} yields a highly localized mollifier which is essentially supported in the smoothness interval $(x - \pi d_x, x + \pi d_x)$. This yields the root-exponential convergence rate

$$|S_{p_N}^\varphi f(x) - f(x)| \lesssim \mathrm{e}^{-\eta\sqrt{d_x N}}.$$

We conclude, in Section 11.2, revisiting the construction of the adaptive filters and mollifiers with a better localization procedure. Instead of enforcing compactly supported φ_{2q} in either physical or Fourier space, we appeal to *optimally* space–frequency-localized filters

$$\varphi_p(\xi) = \varphi_{p,\delta}(\xi) := \mathrm{e}^{-\frac{(\delta\xi)^2}{2}} \sum_{j=0}^{p} \frac{1}{2^j j!}(\delta\xi)^{2j}.$$

We show that an adaptive parametrization, $p = p_N \sim d_x N$ and $\delta_x \sim \sqrt{d_x N}$, yields the *exponentially accurate* mollifier

$$|S_{p_N, \delta_x}^\varphi f(x) - f(x)| \lesssim \mathrm{e}^{-\eta d_x N}.$$

There is a rich literature on filters and mollifiers as effective tools for Gibbs-free reconstruction of piecewise smooth functions. Different aspects of this topic are drawn from a variety of sources, ranging from summability methods in harmonic analysis to signal processing – and, in recent years, image processing – and high-resolution spectral computations of propagation of singularities and shock discontinuities. Given the space and time limitations, we are unable to provide a complete road map but we limit ourselves to a few key references. For general references on harmonic analysis we refer to Bary (1964), Dym and McKean (1972), Folland (1992), Grafakos (2004), Katznelson (1976), Körner (1988), Stein (1993), Szegő (1958), Torchinsky (1986) and Zygmund (1959). For applications in signal processing, including wavelets, and recent exciting developments in compressed sensing, we mention Candes and Romberg (2006), Candes, Romberg and Tao (2006a), Candes, Romberg and Tao (2006b), Donoho, Elad and Temlyakov (2004), Donoho (2004), Donoho and Tanner (2005), Mallat (1989), Marr and Hildreth (1980), Tao (2005) and the references therein. General reviews on spectral methods, edge detection and the computation of Gibbs phenomenon can be found in Abarbanel, Gottlieb and Tadmor (1986), Fornberg (1996), Gelb and Tadmor (2000b), Gelb and Gottlieb (2007), Gottlieb and Hesthaven (1998), Gottlieb and Orszag (1977), Gottlieb and Shu (1997), Mhaskar and Prestin (2000), Majda, McDonough and Osher (1978), Tadmor (1989), Maday, Ould-Kaber and Tadmor (1993) and Tadmor (1994). We also had to leave out numerous other approaches for edge detection and reconstruction of piecewise smooth data. We mention, for example, Eckhoff (1995, 1998), Eckhoff and Wasberg (1995), Bruno, Han and Pohlman (2006), Srinivasa and Rajgopal (1992) and, most notably, two-dimensional reconstructions which couple Radon representation with spectral and wavelet-based ridgelets, *e.g.*, Donoho (1998) and Candes and Guo (2002).

2. Spectral accuracy

2.1. The spectral Fourier projection

Let $S_N f$ denote the Fourier projection of a 2π-periodic function,

$$S_N f(x) = \sum_{|k| \leq N} \widehat{f}(k) \mathrm{e}^{\mathrm{i}kx}, \qquad \widehat{f}(k) := \frac{1}{2\pi} \int_{-\pi}^{\pi} f(y) \mathrm{e}^{-\mathrm{i}ky} \, \mathrm{d}y. \qquad (2.1)$$

It enjoys the well-known *spectral accuracy*, that is, the decay rate of the error, $S_N f - f$, is as rapid as the global smoothness of $f(\cdot)$ permits. The error in this case amounts to the *truncation* error,

$$T_N f(x) := \sum_{|k| > N} \widehat{f}(k) \mathrm{e}^{\mathrm{i}kx},$$

which is spectrally small in the sense that for *any* $s > 1$ we have

$$|S_N f(x) - f(x)| \leq \sum_{|k| > N} |\widehat{f}(k)| \lesssim \|f\|_{C^s} \cdot \frac{1}{N^{s-1}}, \qquad \text{for all } s > 1. \qquad (2.2)$$

Here

$$\|f\|_{C^s} := \max_{k \leq s} \|f^{(k)}(\cdot)\|_{L^\infty}$$

measures the *global* smoothness of f. The interplay between the global smoothness of f and spectral convergence of its Fourier projection is reflected through Parseval's relation,

$$\|f\|_{H^s}^2 = 2\pi \sum_k \left(1 + |k|^{2s}\right) |\hat{f}(k)|^2, \quad \|f\|_{H^s}^2 := \int_{-\pi}^{\pi} \left(f(y)\right)^2 + \left(f^{(s)}(y)\right)^2 dy,$$

which, in turn, is linked to the *spectral decay* of the Fourier coefficients,

$$|\hat{f}(k)| \lesssim \|f\|_{C^s} \frac{1}{1 + |k|^s}, \quad s \geq 1. \tag{2.3}$$

Indeed, the latter follows by noting $\|f\|_{C^{s-1}} \lesssim \|f\|_{H^s} \lesssim \|f\|_{C^s}$, or by repeated integration by parts.

The spectral decay rate (2.3) and its related convergence rate (2.2) are asymptotic statements. The *actual* decay rate (as functions of k and N) depends on the growth of $\|f\|_{C^s}$. To quantify the precise spectral accuracy of C^∞-functions, it is therefore convenient to classify such functions according to the growth of their C^s-bounds: f belongs to *Gevrey class* $G_\alpha, \alpha \geq 1$ if there exists $\eta = \eta_f > 0$ such that

$$G_\alpha = \left\{ f \mid \|f\|_{C^s} \lesssim \frac{(s!)^\alpha}{\eta^s}, \quad s = 1, 2, \ldots \right\}. \tag{2.4}$$

Two examples of Gevrey classes are in order.

(i) *Analytic functions.* By Cauchy's integral formula, each analytic f belongs to G_1, with $2\eta_f$ being the width of its analyticity strip.

(ii) The C_0^∞ *cut-off functions*,

$$\rho_p(x) := e^{\left(\frac{cx^p}{x^2 - \pi^2}\right)} 1_{(-\pi,\pi)}(x), \quad c > 0, \quad p \text{ even}, \tag{2.5}$$

belong to G_2. Indeed, a straightforward computation shows that there exists a constant, $\lambda = \lambda_\rho$ (which may depend on p but is otherwise independent of s), such that

$$|\rho_2^{(s)}(x)| \lesssim \frac{s!}{(\lambda_\rho |x^2 - \pi^2|)^s} e^{\left(\frac{cx^p}{x^2 - \pi^2}\right)}, \tag{2.6}$$

and the upper bound on the right, which is maximized at $x = x_{\max}$ with $x_{\max}^2 - \pi^2 \sim -\pi^2 c/s$, implies the G_2-bound (2.4) with $\eta = c\lambda_\rho \pi^2$,

$$\sup_{x \in (-1,1)} |\rho_2^{(s)}(x)| \lesssim s! \left(\frac{s}{\eta}\right)^s e^{-s} \lesssim \frac{(s!)^2}{\eta^s}, \quad s = 1, 2, \ldots.$$

We can now combine the spectral decay (2.3) with the G_α-bound (2.4). It follows that the decay rate of the Fourier coefficients of G_α-functions is exponential to a fractional order,

$$|\widehat{f}(k)| \lesssim \min_s \left(\frac{s^\alpha}{\eta e^\alpha |k|}\right)^s \sim e^{-\alpha(\eta|k|)^{1/\alpha}}, \qquad f \in G_{\alpha \geq 1},$$

and consequently the truncation error of their Fourier projection does not exceed

$$|S_N f(x) - f(x)| \lesssim Ne^{-\alpha(\eta N)^{1/\alpha}}, \qquad f \in G_{\alpha \geq 1}.$$

In particular, an analytic $f(\cdot)$, with analyticity strip of width 2η, is *characterized* by an exponential rate corresponding to $\alpha = 1$ (see, *e.g.*, Tadmor (1986)), that is,

$$|\widehat{f}(k)| \lesssim e^{-\eta|k|}, \qquad |S_N f(x) - f(x)| \lesssim Ne^{-\eta N}, \qquad f \text{ analytic}; \qquad (2.7a)$$

while for G_2-functions, such as the cut-off $\rho_p(x)$, for example, the rate is *root-exponential*, corresponding to $\alpha = 2$:

$$|\widehat{f}(k)| \lesssim e^{-\sqrt{\eta|k|}}, \qquad |S_N f(x) - f(x)| \lesssim Ne^{-\sqrt{\eta N}}, \qquad f \in G_2. \qquad (2.7b)$$

Remark 2.1. (Notation) Throughout the paper, we use η to denote different Gevrey constants of fractional exponential orders. The same η in different equations stands for different constants. In Section 6.3, η is also used to denote the noise variance.

2.2. Optimal space–frequency decay

The previous examples tell us that if f is a C^∞ compactly supported function, then $|\widehat{f}(k)|$ decays at an exponential rate of a *fractional* order but no faster; indeed, if $|\widehat{f}(k)|$ decays exponentially fast then f is analytic, and hence it cannot decay sufficiently fast to become compactly supported. The question of optimal joint decay in both physical and Fourier spaces brings us to the classical Heisenberg *uncertainty principle*, which places a lower threshold on the joint space–frequency localization. This lower threshold manifests itself in a variety of different forms. In the context of the Fourier transform, for example, one seeks to minimize the joint variance:

$$\min_{x_0} \|(x-x_0)\Phi(x)\|_{L^2(\mathbb{R}_x)} \cdot \min_{\xi_0} \|(\xi-\xi_0)\varphi(\xi)\|_{L^2(\mathbb{R}_\xi)}, \qquad \Phi(x) := \int_{\mathbb{R}} \varphi(\xi)e^{-i\xi x}\, d\xi.$$

It admits a lower threshold which is achieved by the *quadratic* exponentials $\varphi(\xi) = e^{c(\xi-\xi_0)^2}$. For space–frequency localization in related discrete frameworks we mention recent examples of Donoho and Huo (2001), Tao (2005) and Candes and Romberg (2006). In the present context of Fourier expansions, we now construct a large family of 2π-periodic functions, $\{f_N(x)\}$, with optimal *exponential decay* in both physical and Fourier space; consult Hoffman and Kouri (2000) and the references therein.

The starting point is the family of functions with quadratic exponential decay

$$\varphi(\xi) := e^{-\frac{\xi^2}{2}} \times \left[\sum_{j=0}^{p} \frac{1}{2^j j!} \xi^{2j} \right]. \qquad (2.8\text{a})$$

Their inverse Fourier transform can be expressed in terms of Hermite polynomials, $H_{2j}(x)$, that is,

$$\Phi(x) = e^{-\frac{x^2}{2}} \times \left[\sum_{j=0}^{p} \frac{(-1)^j}{4^j j!} H_{2j}\left(\frac{x}{\sqrt{2}} \right) \right]. \qquad (2.8\text{b})$$

Observe that, with fixed p, each $\Phi(x)$ is an *entire* function and the quadratic exponential decay of its Fourier transform, $\varphi(\xi)$, corresponds to '$\eta_\Phi = \infty$'.

We need to 'tweak' $\Phi(x)$ in two ways.

(i) Dilation. We need to dilate $\Phi(x)$ in order to control its localization,

$$\Phi_\delta(x) := \frac{1}{\delta} \Phi\left(\frac{x}{\delta} \right) \longleftrightarrow \varphi_\delta(\xi) = \varphi(\delta\xi)$$

(ii) Periodization. We need a 2π-periodic version of the entire function $\Phi(x)$. To this end, fix N and set[2]

$$S_N^\varphi(x) := \frac{1}{2\pi} \sum_{k=-\infty}^{\infty} \varphi\left(\frac{|k|}{N} \right) e^{ikx}. \qquad (2.9\text{a})$$

Another way to express this 'periodization' of Φ is given by the *Poisson summation formula* (see, *e.g.*, Katznelson (1976), Torchinsky (1986))

$$S_N^\varphi(x) = \frac{N}{2\pi} \sum_{j=-\infty}^{\infty} \Phi(N(x + 2\pi j)). \qquad (2.9\text{b})$$

We can combine both dilation and periodization into one scaling involving N/δ,

$$S_N^{\varphi_\delta}(x) = \frac{1}{2\pi} \sum_{k=-\infty}^{\infty} \varphi_\delta\left(\frac{|k|}{N} \right) e^{ikx} = \frac{1}{2\pi} \sum_{k=-\infty}^{\infty} \varphi\left(\frac{\delta|k|}{N} \right) e^{ikx} \equiv S_{N/\delta}^\varphi(x).$$

We are ready to state our next result.

---

[2] Observe that the *function* $S_N^\varphi(x)$ is different from the partial sum *operation* S_N. The reason for this notation will become clear when we link these two different aspects in our discussion on mollifiers and filters in Section 11.

Lemma 2.2. (Space–frequency exponential decay) Fix p, set $\delta_N :=$ $\sqrt{\beta N}$ and consider the 2π-periodic functions,

$$f_N(x) := \mathcal{S}_N^{\varphi_{\delta_N}}(x) = \frac{1}{2\pi} \sum_{k=-\infty}^{\infty} \varphi\left(\sqrt{\frac{\beta}{N}}|k|\right) e^{ikx}, \qquad \varphi(\xi) = e^{-\frac{\xi^2}{2}} \sum_{j=0}^{p} \frac{\xi^{2j}}{2^j j!}.$$

(2.10)

Then, there exists $\eta_1, \eta_2 > 0$ such that, for all $|x| \leq \pi$,

$$|f_N(x)| = \left|\mathcal{S}_N^{\varphi_{\sqrt{\beta N}}}(x)\right| \lesssim 2^p \sqrt{\frac{N}{\beta}} \left(e^{-\eta_1 N x^2/\beta} + e^{-\eta_2 N/\beta}\right).$$

(2.11a)

Moreover, for all $|k| > N$,

$$|\widehat{f_N}(k)| \lesssim c_{p,N} \cdot e^{-\beta|k|/2}, \qquad c_{p,N} := \sum_{j=0}^{p} \frac{1}{j!}\left(\frac{\beta N}{2}\right)^j.$$

(2.11b)

Remark 2.3. We conclude, since $c_{p,N}$ has at most pth-order polynomial growth with N, that f_N should have exponential decay in both physical and frequency space. Observe that the detailed structure of the pth-order polynomial factors inside the square brackets on the right of (2.8a) and (2.8b) are not important at this stage, but they will be later on, in Section 11.2 below, when we link the increase of p with N.

Proof. To verify (2.11a) we bound $|H_{2j}(x)| \lesssim j^{j+\frac{1}{2}}(4/e)^j e^{x^2/2}$, in order to estimate the exponential decay of $\Phi_\delta(x) = \frac{1}{\delta}\Phi\left(\frac{x}{\delta}\right)$ in (2.8b),

$$|\Phi_\delta(x)| \lesssim \frac{1}{\delta} e^{-\frac{x^2}{2\delta^2}} \sum_{j=0}^{p} \frac{1}{4^j j!} \cdot \left|H_{2j}\left(\frac{x}{\sqrt{2\delta}}\right)\right| \lesssim \frac{2^p}{\delta} e^{-\frac{x^2}{4\delta^2}}.$$

(2.12)

We appeal to the Poisson representation of $\mathcal{S}_N^{\varphi_\delta}$ in (2.9b):

$$\mathcal{S}_N^{\varphi_\delta}(x) = \frac{N}{2\pi}\Phi_\delta(Nx) + \frac{N}{2\pi}\sum_{j\neq 0}\Phi_\delta(N(x + 2\pi j)).$$

(2.13)

It follows that all terms except the zeroth are exponentially negligible for $|x| \leq \pi$:

$$\sum_{j\neq 0}|\Phi_{\delta_N}(N(x + 2\pi j))| \lesssim \frac{2^p}{\delta_N}\sum_{j=1}^{\infty} e^{-\frac{((2j-1)\pi N)^2}{4\delta_N^2}}$$

$$\lesssim \frac{2^p}{\sqrt{\beta N}}e^{-\eta_2 N/\beta}, \qquad |x| \leq \pi.$$

(2.14a)

The zeroth term has double exponential decay (in x)

$$|\Phi_{\delta_N}(Nx)| \lesssim \frac{2^p}{\sqrt{\beta N}}e^{-\eta_1 N x^2/\beta},$$

(2.14b)

and the last two bounds yield the first half of (2.11).

The second half of (2.11) is straightforward:

$$|\widehat{f_N}(k)| = \left|\varphi\left(\sqrt{\frac{\beta}{N}}|k|\right)\right| \lesssim c_{p,N} \cdot e^{-\left(\frac{\beta|k|^2}{2N}\right)}, \tag{2.15}$$

$$c_{p,N} = \max_{\xi \le \delta_N} \sum_{j=0}^{p} \frac{1}{2^j j!} \xi^{2j},$$

and (2.11b) follows for $|k| > N$. $\square$

2.3. The pseudo-spectral Fourier projection

If we replace the integrals on the right of (2.1) with the quadrature sampled at the equidistant points,

$$y_\nu := -\pi + \nu h, \qquad h := \frac{2\pi}{2N+1},$$

we obtain the *discrete* Fourier coefficients $\{\widehat{f_k}\}$, which form the N-degree *trigonometric interpolant* of f at these $(2N+1)$-grid-points,[3] that is,

$$I_N f(x) := \sum_{|k| \le N} \widehat{f_k} e^{ikx}, \qquad \widehat{f_k} := \frac{h}{2\pi} \sum_{\nu=0}^{2N} f(y_\nu) e^{-iky_\nu}, \tag{2.16}$$

so that $I_N f(x_\nu) = f(x_\nu)$, $\nu = 0, \ldots, 2N$. I_N is known as the *pseudo-spectral* projection. The dual statement of interpolation in physical space is the Poisson summation formula in Fourier space, expressing the $\widehat{f_k}$ in terms of the $\widehat{f}(k)$,

$$\widehat{f_k} = \widehat{f}(k) + \sum_{j \ne 0} \widehat{f}(k + j(2N+1)). \tag{2.17}$$

Thus, the sum of all the Fourier coefficients located at $k[\text{mod } (2N+1)]$ have a discrete 'alias', $\widehat{f_k}$. This follows at once by substituting $f(y_\nu)$ in (2.16) as the sum $\sum_j \widehat{f}(j) e^{iky_\nu}$. We conclude that the interpolation error $I_N f(x) - f(x)$ consists of two contributions: the truncation error, $T_N f(x) = \sum_{|k| \ge N} \widehat{f}(k) e^{ikx}$, and the *aliasing error*,

$$A_N f(x) = \sum_{|k| \le N} \left(\sum_{|j| \ge 1} \widehat{f}(k + j(2N+1))\right) e^{ikx}. \tag{2.18}$$

---

[3] There is a slight difference between the formulae based on an even and an odd number of points; we have chosen to continue with the slightly simpler notation associated with an odd number of points.

Both $T_N f$ and $A_N f$ involve modes higher than N, and if f is sufficiently smooth they have exactly the same spectrally small size (*e.g.*, Tadmor (1994)), that is,

$$\|A_N f(x)\|_{H^s} \sim \sum_{|k| \geq N} \left(1 + |k|^{2s}\right) \left| \sum_{j \neq 0} \widehat{f}(k + j(2N+1)) \right|^2 \tag{2.19}$$

$$\leq C_s \sum_{|k| \geq N} \left(1 + |k|^{2s}\right) |\widehat{f}(k)|^2 \lesssim C_s \|T_N f(x)\|_{H^s}^2, \quad s > \frac{1}{2}.$$

We conclude with similar spectral and fractional exponential convergence rate estimates:

$$|I_N f(x) - f(x)|$$

$$\leq \sum_{|k| > N} |\widehat{f}(k)| + \sum_{|k| \leq N} \sum_{|j| \geq 1} |\widehat{f}(k + j(2N+1))| \lesssim \begin{cases} \frac{1}{N^{s-1}}, & f \in C^s, \\ N e^{-(\eta N)^{1/\alpha}}, & f \in G_\alpha. \end{cases}$$

We close this section by commenting on the discrete Fourier coefficients of the exponentially localized f_N in (2.10). The quadratic exponential decay of $\widehat{f_N}(k)$ for $|k| > N$ in (2.15) implies that the aliasing error, $A_N f_N(x)$, is exponentially negligible, and hence $(\widehat{f_N})_k \approx \widehat{f_N}(k)$, that is,

$$|(\widehat{f_N})_k| \lesssim c_{p,N} \cdot \left(e^{-\frac{\beta k^2}{2N}} + \mathcal{O}(e^{-\frac{\beta N}{2}}) \right), \quad |k| \leq N. \tag{2.20}$$

3. Piecewise smoothness and the Gibbs phenomenon

Both the spectral and the pseudo-spectral Fourier projections, $S_N f$ and $I_N f$, provide highly accurate approximations of f, whose order is limited only by the *global* smoothness of f. What happens when f lacks sufficient smoothness? This will be our main concern in the remaining sections.

We begin with a classical example. Consider an f which is *piecewise smooth* in the sense that it is sufficiently smooth except for finitely many jump discontinuities, say at $x = c_1, c_2, \ldots, c_J$, where

$$[f](c_j) := f(c_j+) - f(c_j-) \neq 0, \quad j = 1, 2, \ldots, J.$$

It is natural to measure piecewise smoothness in the space of functions of bounded variation,

$$\|f\|_{BV} := \|f'\|_{L^1[-\pi,\pi]} < \infty,$$

that is, f' is a smooth function together with finitely many Dirac masses. The finite variation of f implies (via integration by parts) the first-order decay rate of $|\widehat{f}(k)|$,

$$|\widehat{f}(k)| \lesssim \|f\|_{BV} \frac{1}{1 + |k|}. \tag{3.1}$$

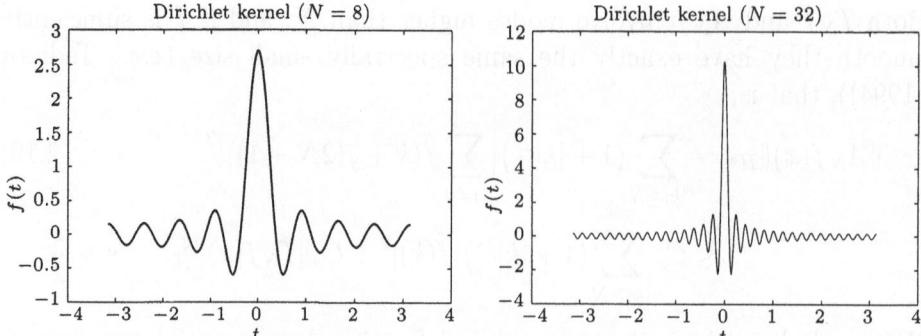

Figure 3.1. Dirichlet kernel with $N = 32$ and $N = 128$ modes.

Indeed, the decay is precisely first-order, $|\widehat{f}(k)| \sim 1/|k|$, since a faster decay would imply that f is continuous (Zygmund 1959). A similar first-order decay occurs in the discrete case: summation by parts of the discrete Fourier coefficients in (2.16) yields

$$
\widehat{f}_k = \frac{h}{2\pi} \sum_{\nu=0}^{2N} f(y_\nu) \frac{e^{-iky_\nu} - e^{-iky_{\nu+1}}}{1 - e^{-ikh}} \tag{3.2}
$$

$$
= \frac{h}{4\pi i \sin \frac{kh}{2}} \sum_{\nu=0}^{2N} \big(f(y_{\nu+1}) - f(y_\nu)\big) e^{-iky_\nu},
$$

and hence $\|\widehat{f}_k\| \lesssim \|f\|_{TV}/(1 + |k|)$, where $\|f\|_{TV}$ denotes the total variation of f.

The first-order decay of the (discrete) Fourier coefficients is too weak (as it should be!) to enforce *uniform* convergence of $S_N f(x)$ and $I_N f(x)$. Instead, we turn to examine the *local* convergence of $S_N f(x)$, which is expressed in terms of the *Dirichlet kernel*, $D_N(\cdot)$,

$$
S_N f(x) = \int_{-\pi}^{\pi} f(y) D_N(x - y), \qquad D_N(y) := \frac{1}{2\pi} \sum_{|k| \leq N} e^{iky} \equiv \frac{\sin(N + \frac{1}{2})y}{2\pi \sin(y/2)}.
$$

The Dirichlet kernel, $D_N(\cdot)$, is depicted in Figure 3.1. It has a sequence of successive local peaks at

$$
\frac{(k + 1/2)\pi}{N + 1/2}, \qquad k = 1, 2, \ldots,
$$

which accumulate a total mass of diverging order $\|D_N\|_{L^1} \sim \log N$ and which, in turn, are responsible for the failure of uniform convergence of $S_N f(x)$. As an example, consider the spectral projection of the Heaviside

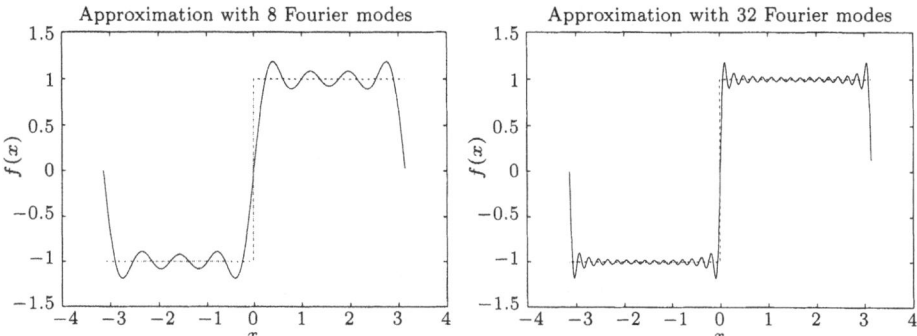

Figure 3.2. Fourier projection of the square wave
function. *Left*: $N = 32$ modes. *Right*: $N = 128$ modes.

function $H(x) := \operatorname{sgn}(x)$. Since $D_N(\cdot)$ is an even function, we have

$$S_N H(x) = \int_0^\pi \left[D_N(x - y) - D_N(x + y) \right] dy$$

$$= \frac{-i}{\pi} \sum_{|k| \le N} e^{ikx} \int_0^\pi \sin(ky)\, dy = \frac{-2i}{\pi} \sum_{\{|k| \le N : k \text{ odd}\}} \frac{e^{ikx}}{k}.$$

At $x = 0$, we find that $S_N H(x)$ assumes the average value,

$$S_N H(x)|_{x=0} = 0.$$

But the convergence is not uniform for $x \approx 0$: for example,

$$S_N H(x)\Big|_{x = \pm \frac{\pi}{N}} = \frac{\pm 2}{\pi} \sum_{\{1 \le k \le N : k \text{ odd}\}} \frac{\sin(k\pi/N)}{k\pi/N} \cdot \frac{2\pi}{N} \approx \frac{\pm 2}{\pi} \int_0^\pi \frac{\sin y}{y}\, dy.$$

Thus, the spectral projection $S_N H(x)$ magnifies the amplitude of the original jump $[H](0) = 2$, forming a 'spurious' oscillation with 18% larger amplitude:

$$S_N H(x)\Big|_{x = \frac{\pi}{N}} - S_N H(x)\Big|_{x = -\frac{\pi}{N}} \approx \frac{4}{\pi} \int_0^\pi \frac{\sin y}{y}\, dy = 1.179 \times [H](0). \quad (3.3)$$

This behaviour is called the *Gibbs phenomenon*, after Gibbs (1899) (consult Körner (1988), Carslaw (1952) or Hewitt and Hewitt (1979) for a historical perspective). The lack of uniform convergence is depicted in Figure 3.2 by spurious oscillations which concentrate near the jumps at $x = 0$ and $x = \pm\pi$. For another example, consult Figure 5.1(b) below. This is a *local* effect of the Gibbs phenomenon. But the Gibbs phenomenon also has a *global* effect: although the error $S_N H(x) - H(x)$ decays as x moves away from the jumps, the decay rate is limited to first-order, owing to a series of linearly decaying spurious peaks at $k\pi/N$, where

$$\left(S_N H(x) - H(x)\right)\big|_{x=\pm\frac{k\pi}{N}} \sim \frac{1}{N}.$$

Thus, the existence of one or more discontinuities slow down the convergence rate *throughout* the domain. Spectral accuracy is lost.

4. Detection of edges: concentration kernels

Given the Fourier coefficients, $\left\{\widehat{f}(k)\right\}_{k=-N}^{N}$, we are interested in detecting the edges of the underlying piecewise smooth f, namely, to detect their location, $c_1, \ldots, c_J$, and the amplitudes of the jumps, $[f](c_1), \ldots, [f](c_J)$. Extensions to the discrete and non-periodic set-ups will follow in the next sections.

We begin by considering the prototype case of a discontinuous f which is, say, C^2, except for a single jump at $x = c$. The fact that f experiences a jump of size $[f](c)$ dictates the decay of its Fourier coefficients: integration by parts yields

$$\widehat{f}(k) = [f](c)\frac{e^{-ikc}}{2\pi i k} + \mathcal{O}\left(\frac{1}{|k|^2}\right). \tag{4.1}$$

We want to extract information about the location of the jump from the *phase* of the leading term. To this end we use the localization of the Dirichlet kernel near the origin:

$$\frac{\pi}{N} D_N(x - c) = \frac{1}{2N} \sum_{k=-N}^{N} e^{ik(x-c)} \approx \frac{1}{1 + N|x - c|}.$$

It follows that the *derivative* of the Fourier projection $S_N f$ satisfies

$$\frac{\pi}{N} S_N(f)'(x) = \frac{\pi}{N} \sum_{|k|\leq N} ik\left\{[f](c)\frac{e^{ik(x-c)}}{2\pi i k} + \mathcal{O}\left(\frac{1}{|k|^2}\right)\right\}$$

$$= [f](c) \cdot \frac{1}{2N} \sum_{|k|\leq N} e^{ik(x-c)} + \frac{\pi}{N} \sum_{|k|\leq N} \mathcal{O}\left(\frac{1}{|k|}\right)$$

$$= \frac{[f](c)}{1 + N|x - c|} + \mathcal{O}\left(\frac{\log N}{N}\right) = \begin{cases} [f](c) + \mathcal{O}\left(\frac{\log N}{N}\right), & x \approx c, \\ \mathcal{O}\left(\frac{\log N}{N}\right), & |x - c| \gg \frac{1}{N}. \end{cases}$$

We see that $\pi S_N(f)'(x)/N$ *concentrates* in the immediate neighbourhood of $x = c$, where it approaches the desired amplitude of the jump, $[f](c) \neq 0$, while it decays to order $\mathcal{O}(\log N/N)$ as it moves away from this neighbourhood, $|x - c| \gg 1/N$. Thus, we can detect the edge at $x = c$ by separation of scales: a jump of size $|[f](c)| \gg 1/N$ is separated from the region of smoothness where $\pi S_N(f)'(x)/N \approx 1/N$. This result goes back to Fejér (Zygmund 1959, Theorem 9.3).

We turn to consider a general set-up of edge detection based on separation of scales. To this end we introduce a family of *concentration kernels*

$$\mathcal{K}_N^\sigma(y) := -\frac{1}{c_\sigma}\sum_{k=1}^N \sigma\left(\frac{k}{N}\right)\sin ky, \qquad \frac{\sigma(\xi)}{\xi} \in C^2[0,1]. \qquad (4.2a)$$

Here, c_σ is a proper normalization constant,

$$c_\sigma := \int_0^1 \frac{\sigma(\xi)}{\xi}\,d\xi, \qquad\qquad (4.2b)$$

so that, as will be shown in (4.6) below, $\int_0^\pi \mathcal{K}_N^\sigma(y)\,dy \approx -1$. We set[4]

$$K_N^\sigma f(x) := \mathcal{K}_N^\sigma * f(x) = \frac{\pi i}{c_\sigma}\sum_{|k|\le N}\operatorname{sgn}(k)\sigma\left(\frac{|k|}{N}\right)\widehat{f}(k)e^{ikx}. \qquad (4.3)$$

Our purpose is to choose the *concentration factors*, $\sigma(|k|/N)$, such that $K_N^\sigma f$ detects the $\mathcal{O}(1)$-edges, $[f](c_j)$, $j = 1,\dots,J$, by separating them from a much smaller scale of $K_N^\sigma f(x)$ in regions of smoothness. It turns out that *every* σ can serve as an admissible concentration factor.

Theorem 4.1. (Concentration kernels; Gelb and Tadmor 1999, 2000a) Assume that $f(\cdot)$ is piecewise smooth such that

$$\omega_f(y) = \omega_f(y;x) := \frac{f(x+y) - f(x-y) - [f](x)}{y} \in BV[-\pi,\pi]. \qquad (4.4)$$

Let $\mathcal{K}_N^\sigma(x)$ be an admissible concentration kernel (4.2). Then,

$$K_N^\sigma f(x) = \mathcal{K}_N^\sigma * (S_N f)(x) = \frac{\pi i}{c_\sigma}\sum_{|k|\le N}\operatorname{sgn}(k)\sigma\left(\frac{|k|}{N}\right)\widehat{f}(k)e^{ikx}$$

satisfies the concentration property

$$K_N^\sigma f(x) \sim \begin{cases} [f](c_j) + \mathcal{O}\left(\frac{\log N}{N}\right), & x \sim c_j,\ j = 1,\dots,J, \\ \mathcal{O}\left(\frac{\log N}{N}\right), & \operatorname{dist}\{x,\{c_1,\dots,c_J\}\} \gg \frac{1}{N}. \end{cases} \qquad (4.5)$$

Remark 4.2. We will show below that, up to scaling and modulo small 'manageable' residual terms, each $\mathcal{K}_N^\sigma(y)$ amounts to the same *conjugate Dirichlet kernel*,

$$\mathcal{K}_N^\sigma(y) \sim \frac{\sigma(1)}{c_\sigma}\widetilde{D}_N(y) + \text{lower-order terms}, \qquad \widetilde{D}_N(y) := \frac{\cos(N+\frac12)y}{2\sin(y/2)}.$$

---

[4] Observe that $K_N^\sigma f$ is the *operator* associated with, but otherwise different from, the concentration kernel $\mathcal{K}_N^\sigma(x)$.

Accordingly, we refer to $K_N^\sigma f(x)$ as a *conjugate sum*. The lemma shows that all these conjugate sums concentrate near the edges. Different σs yield different concentration kernels $\mathcal{K}_N^\sigma(y)$, and we will explore the role of different σs in the following sections.

Proof. The key to our proof is to observe that $\mathcal{K}_N^\sigma$ is an approximate *derivative* of the delta function. In particular, since $\mathcal{K}_N^\sigma(\cdot)$ is odd,

$$K_N^\sigma * f(x) = -\int_0^\pi \mathcal{K}_N^\sigma(y)\big(f(x+y) - f(x-y)\big)\, dy$$

$$= -\int_0^\pi \mathcal{K}_N^\sigma(y)\big(f(x+y) - f(x-y) - [f](x)\big)\, dy - [f](x)\cdot \int_0^\pi \mathcal{K}_N^\sigma(y)\, dy.$$

The rectangular quadrature rule and the normalization (4.2b) yield

$$\int_0^\pi \mathcal{K}_N^\sigma(y)\, dy = \frac{1}{c_\sigma}\sum_{k=1}^N \sigma\left(\frac{k}{N}\right)\frac{(-1)^k - 1}{k} \tag{4.6}$$

$$= -\frac{1}{c_\sigma}\sum_{k\ \mathrm{odd}\geq 1}^N \frac{\sigma(\xi_k)}{\xi_k}\frac{2}{N} = -1 + \mathcal{O}\left(\frac{1}{N^2}\right), \qquad \xi_k := \frac{k}{N},$$

and we end up with the error estimate

$$\big|K_N^\sigma f(x) - [f](x)\big| \lesssim \left|\int_0^\pi y\mathcal{K}_N^\sigma(y)\omega_f(y;x)\, dy\right| + \mathcal{O}\left(\frac{1}{N^2}\right). \tag{4.7a}$$

It remains to upper-bound the first moment of $\mathcal{K}_N^\sigma \omega_f$. To this end we use the identity $-4\sin^2(y/2)\sin(ky) \equiv \sin\big((k+1)y\big) - 2\sin(ky) + \sin\big((k-1)y\big)$ and summation by parts (twice) to find

$$4\sin^2\left(\frac{y}{2}\right)c_\sigma \mathcal{K}_N^\sigma(y) \equiv 2\sigma(1)\sin\left(\frac{y}{2}\right)\cos\left(N + \frac{1}{2}\right)y$$

$$\overbrace{+ \sum_{1\leq k\leq N-2}\big(\sigma(\xi_k) - 2\sigma(\xi_{k+1}) + \sigma(\xi_{k+2})\big)\sin\big(k+1\big)y}^{I_1(y)}$$

$$\overbrace{+ \big(\sigma(\xi_{N-1}) - \sigma(1)\big)\sin Ny}^{I_2(y)} + \overbrace{\big(\sigma(\xi_2) - 2\sigma(\xi_1)\big)\sin y}^{I_3(y)}.$$

This leads to the corresponding decomposition of $\mathcal{K}_N^\sigma(y)$ as the sum of a conjugate Dirichlet kernel, $\tilde{D}_N(y)$, plus a residual term, $R_N(y)$ (which is negligible in the precise sense to be outlined below):

$$K_N^\sigma(y) = \frac{\sigma(1)}{c_\sigma} \overbrace{\frac{\cos\left(N+\frac{1}{2}\right)y}{2\sin(y/2)}}^{\tilde{D}_N(y)} + \frac{1}{c_\sigma}R_N(y), \qquad R_N(y) := \sum_{j=1}^{3} \frac{I_j(y)}{4\sin^2(y/2)}.$$
$$(4.7b)$$

The conjugate Dirichlet kernel has a small moment due to *cancellation.* Indeed, if we let Ω_f denote

$$\Omega_f(y) \equiv \Omega(y;x) := \frac{y}{4\sin(y/2)}\omega_f(y;x),$$

then the upper bound

$$\left|\int_0^\pi y\tilde{D}_N(y)\omega_f(y)\,dy\right| = \left|\int_0^\pi \cos\left(\left(N+\frac{1}{2}\right)y\right)\Omega_f(y)\,dy\right| \lesssim \frac{\|\omega_f(\cdot)\|_{BV}}{N}$$
$$(4.8a)$$

follows from (3.1), since $\|\Omega_f\|_{BV} \lesssim \|\omega_f\|_{BV}$. The *logarithmic* upper bound of the Dirichlet kernel, $\|D_k\|_{L^1} \sim \log k$, implies

$$\left|\int_0^\pi y\frac{I_1(y)\omega_f(y)}{4\sin^2(y/2)}\,dy\right| \lesssim \frac{1}{N^2}\|\sigma\|_{C^2}\sum_{k=1}^{N-2}\log k \cdot \|\Omega_f\|_{L^\infty} \lesssim \frac{\log N}{N}, \qquad (4.8b)$$

$$\left|\int_0^\pi y\frac{I_2(y)\omega_f(y)}{4\sin^2(y/2)}\,dy\right| \lesssim \frac{1}{N}\|\sigma\|_{C^1}\cdot\log N\cdot\|\Omega_f\|_{L^\infty} \lesssim \frac{\log N}{N}. \qquad (4.8c)$$

Finally, since $|\sigma(\xi)| \lesssim \xi$, we have

$$\left|\int_0^\pi y\frac{I_3(y)\omega_f(y)}{4\sin^2(y/2)}\,dy\right| \lesssim \left(\left|\sigma\left(\frac{1}{N}\right)\right| + \left|\sigma\left(\frac{2}{N}\right)\right|\right)\|\Omega_f\|_{L^\infty} \lesssim \frac{1}{N}, \qquad (4.8d)$$

and the desired result, (4.5), follows from (4.7a), (4.7b) and (4.8). $\square$

We conclude this subsection with a couple of remarks.

Remark 4.3. (The behaviour of σ and improved concentration)
The bounds in (4.8) show that their overall error does not exceed

$$\frac{\log N}{N}\|\sigma\|_{C^2} + \left|\sigma\left(\frac{1}{N}\right)\right| + \frac{1}{N}|\sigma(1)|. \qquad (4.9)$$

Thus, the concentration error (4.5) of order $\mathcal{O}(1/N)$ becomes smaller if $\sigma(\xi)$ decays sufficiently fast at $\xi = 0$ and $\xi = 1$. This issue will be explored in the next section, in the context of the exponential concentration factors; consult (5.4) below.

Remark 4.4. (Concentration kernels: general set-up) The proof of Theorem 4.1 reveals that the concentration property holds for arbitrary kernels, $\{\mathcal{K}_N(y)\}$, as long as they satisfy the three key properties:

(i) $\mathcal{K}_N$ are odd, $\mathcal{K}_N(-y) = -\mathcal{K}_N(y)$;

(ii) $\mathcal{K}_N$ are properly normalized so that $\int_{y \geq 0} \mathcal{K}_N(y) \, dy = -1 + \varepsilon_N$; and

(iii) $\mathcal{K}_N$ has a small first moment of order

$$\left| \int y \mathcal{K}_N(y) \omega(y) \, dy \right| \lesssim \varepsilon_N \|\omega\|_{BV}. \tag{4.10}$$

Here, ε_N is a small scale associated with $\mathcal{K}_N$. If (i)–(iii) hold then we deduce, along the lines of Theorem 4.1 (consult Gelb and Tadmor (2000a, Theorem 2.1)),

$$|\mathcal{K}_N * f(x) - [f](x)| \lesssim \varepsilon_N;$$

hence, the $\mathcal{K}_N$ detect edges by separating the regions where $|\mathcal{K}_N * f(c_j) \approx [f](c_j)$ from smooth regions where $\mathcal{K}_N * f(x) \approx \varepsilon_N \ll 1$. A few examples are in order.

5. Examples

Compactly supported kernels. We consider a standard mollifier, namely $\phi_{\varepsilon_N}(y) := \frac{1}{\varepsilon_N} \phi(\frac{y}{\varepsilon_N})$, based on an even, compactly supported bump function, $\phi \in C_0^1(-1, 1)$, with $\phi(0) = 1$. We then set

$$\mathcal{K}_{\varepsilon_N}(y) = \frac{1}{\varepsilon} \phi'\left(\frac{y}{\varepsilon_N}\right). \tag{5.1}$$

Clearly, $\mathcal{K}_{\varepsilon_N}$ is an odd kernel satisfying the proper normalization

$$\int_{y \geq 0} \mathcal{K}_{\varepsilon_N}(y) \, dy = -\phi(0) = -1,$$

and its first moment is of order

$$\int_{y \geq 0} |y \mathcal{K}_{\varepsilon_N}(y)| \, dy = \varepsilon_N \int_0^1 |y| \cdot |\phi'(y)| \, dy = \mathcal{O}(\varepsilon_N).$$

The concentration property $\mathcal{K}_{\varepsilon_N} * f(x) = [f](x) + \mathcal{O}(\varepsilon_N)$ follows. As examples we mention edge detectors based on Haar and bi-orthogonal moments; e.g., Mallat (1989). Localized kernels are limited to finite order of accuracy, no matter how smooth f is.

Polynomial concentration kernels. Set $\sigma(\xi) = \xi$. Then $K_N^\sigma f$ recovers the Fejér conjugate sum

$$K_N^\sigma f(x) = \pi \sum_{|k| \leq N} \frac{ik}{N} \widehat{f}(k) e^{ikx} = \frac{\pi}{N} S_N(f)'(x), \qquad \sigma(\xi) = \xi. \tag{5.2}$$

We note in passing that in this case, $\mathcal{K}_N^\sigma(y)$ does not concentrate near the origin as do the compactly supported $\mathcal{K}_{\varepsilon N}$. Instead, (4.10) is fulfilled thanks to the more intricate property of *cancellation of oscillations*. This is the first member in the general family of *polynomial concentration factors*, $\sigma_p(\xi) = \xi^p$, introduced by Golubov (Golubov 1972, Kvernadze 1998, Gelb and Tadmor 1999). Polynomial concentration factors of odd degree, $\sigma_{2p+1}(\xi)$, correspond to differentiation in physical space, that is,

$$K_N^{\sigma_{2p+1}} f(x) = (-1)^p \frac{\pi(2p+1)}{N^{2p+1}} \frac{\mathrm{d}^{2p+1}}{\mathrm{d}x^{2p+1}} S_N f(x), \qquad \sigma_{2p+1}(\xi) = \xi^{2p+1}.$$

Polynomial factors of even degree, $\sigma_{2p}(\xi)$, yield *global* conjugate sums which convolve with

$$\widetilde{H}_N(x) := \mathrm{i} \sum_{|k| \le N} \mathrm{sgn}(k) \mathrm{e}^{\mathrm{i}kx},$$

that is,

$$K_N^{\sigma_{2p}} f(x) = (-1)^p \frac{2\pi p}{N^{2p}} \widetilde{H}_N * \frac{\mathrm{d}^{2p}}{\mathrm{d}x^{2p}} S_N f(x), \qquad \sigma_{2p}(\xi) = \xi^{2p}.$$

We shall refer to this family of kernels based on the σ_p-factors as *polynomial concentration kernels*.

Trigonometric concentration kernels. According to (3.3), the difference $S_N f(x + \pi/N) - S_N f(x - \pi/N)$ concentrates near the edges with 18% Gibbs overshoot

$$\frac{S_N f\left(x + \frac{\pi}{N}\right) - f\left(x - \frac{\pi}{N}\right)}{2\,\mathrm{Si}(\pi)/\pi} \approx [f](x), \qquad \mathrm{Si}(\pi) := \int_0^\pi \frac{\sin x}{x}\,\mathrm{d}x. \qquad (5.3)$$

The difference in the numerator amounts to concentration factors $\sigma(\xi) = \sin(\pi\xi)$

$$S_N f(x + \pi/N) - S_N f(x - \pi/N) = 2\mathrm{i} \sum_{|k| \le N} \sin\left(\frac{\pi k}{N}\right) \widehat{f}(k) \mathrm{e}^{\mathrm{i}kx},$$

and the corresponding normalization, $c_\sigma = \mathrm{Si}(\pi)$, recovers the denominator in (5.3). This edge detector was advocated by Banerjee and Geer (1997). It is the first member in the family of *trigonometric concentration factors* $\sigma_\alpha(\xi) = \sin(\alpha\xi)$. We shall have to say more on the relation between concentration factors in Fourier space and their realization as differences in the physical space when we discuss edge detection in discrete data in Section 6.1.

Exponential concentration factors. Theorem 4.1 provides us with the framework of general concentration kernels which are not necessarily limited to a realization in the physical space. In particular, we seek concentration factors $\sigma(\cdot)$, which vanish at $\xi = 0, 1$ to any prescribed order:

$$\frac{\mathrm{d}^j}{\mathrm{d}\xi^j}\sigma(\xi)\bigg|_{\xi=0} = \frac{\mathrm{d}^j}{\mathrm{d}\xi^j}\sigma(\xi)\bigg|_{\xi=1} = 0, \qquad j = 0, 1, 2, \ldots, p. \tag{5.4}$$

The higher p is, the more localized $\mathcal{K}_N^\sigma(\cdot)$ becomes, since

$$\mathcal{K}_N^\sigma(y_\ell) = -\frac{1}{c_\sigma}\sum_{k=1}^N \sigma\left(\frac{k}{N}\right)\sin\frac{2\pi k\ell}{N}, \qquad y_\ell := \frac{2\pi\ell}{N}.$$

We observe that $\mathcal{K}_N^\sigma(y_\ell)/N$ coincides with the ℓ-discrete Fourier coefficient of $\sigma(\cdot)$, and since $\sigma(\xi)$ and its first p-derivatives vanish at both ends, $\xi = 0, 1$, the C^p-regularity of σ implies the rapid decay of these discrete Fourier coefficients, $|\hat{\sigma}_\ell| \lesssim \ell^{-p}$, i.e.,

$$|K_N^\sigma(y_\ell)| \lesssim \|\sigma\|_{C^p[0,1]}\frac{1}{(Ny_\ell)^p}.$$

Thus, for y away from the origin, $K_N^\sigma(y)$ is rapidly decaying for sufficiently large N. Moreover, we can show that an increasing number of moments of $\mathcal{K}_N^\sigma(\cdot)$ vanish: consult Gelb and Tadmor (2000a, Section 2). As an example, consider the *exponential concentration factors*,

$$\sigma_{\exp}(\xi) = \xi e^{\frac{1}{\alpha\xi(\xi-1)}}, \tag{5.5}$$

for which (5.4) holds for *all* p. Indeed, since $\sigma_{\exp}$ is based on a G_2 cut-off function, then $K_N^\sigma f$ becomes root-exponentially small away from the jumps,

$$|K_N^{\sigma_{\exp}} f(x)| \lesssim e^{-\sqrt{\eta N}}, \qquad \mathrm{dist}\{x, \{c_1, \ldots, c_J\}\} \gg \frac{1}{N}. \tag{5.6}$$

This leads to an improved separation of edges from regions of smoothness, demonstrated in Figure 5.1.

Figure 5.1 compares the Fejér and exponential concentration kernels, $\mathcal{K}_N^{\sigma_1}(x)$ and $\mathcal{K}_N^{\sigma_{\exp}}(x)$ for

$$f(x) := \begin{cases} \cos\left(x - \frac{x}{2}\,\mathrm{sgn}\left(|x| - \frac{\pi}{2}\right)\right), & x < 0, \\ \cos\left(\frac{5}{2}x + x\,\mathrm{sgn}\left(|x| - \frac{\pi}{2}\right)\right), & x > 0. \end{cases} \tag{5.7}$$

In both cases, $K_N^\sigma f(x)$ concentrates near the two edges at $x = \pm\pi/2$ with amplitude $\pm\sqrt{2}$ which are separated from the remaining smooth pieces of $f(x)$. It confirms the improved localization of the exponential concentration factors.

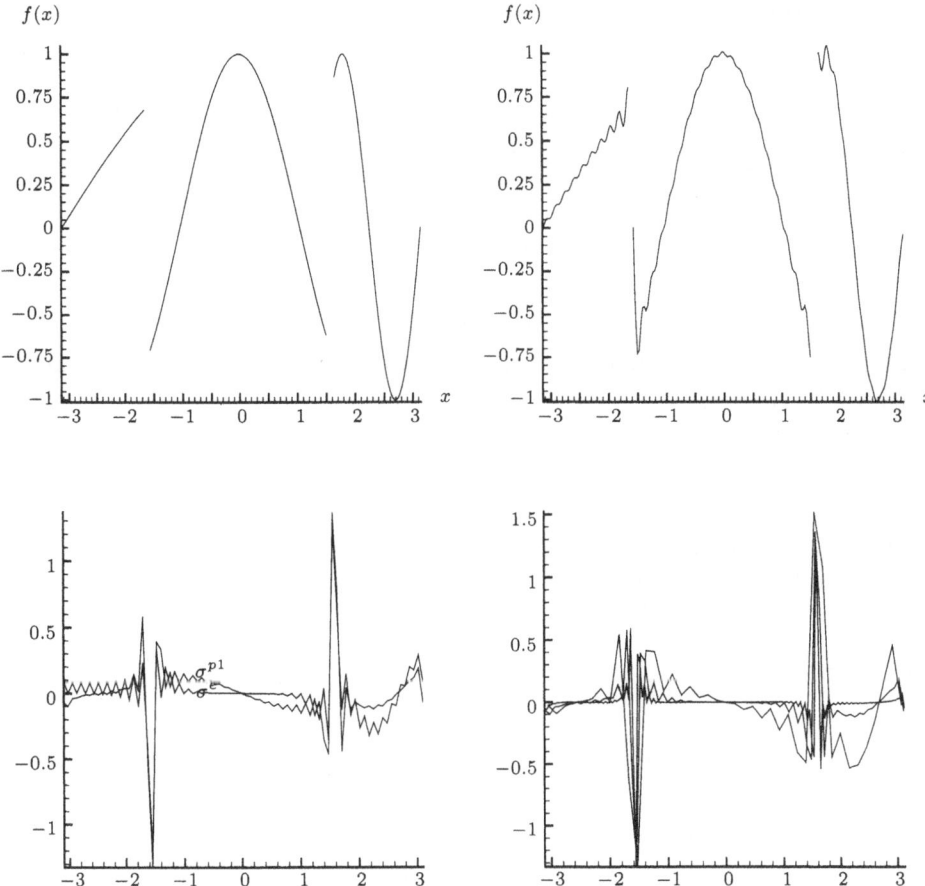

Figure 5.1. *Top left*: Piecewise smooth $f(x)$ in (5.7). *Top right*: Gibbs phenomenon for $S_{40}f(x)$. *Bottom left*: Edge detection in $S_{40}f$ using $K_{40}^\sigma f$, comparing the exponential concentration factors $\sigma_{\exp}(\xi) = \exp(\frac{1}{6\xi(\xi-1)})$ *vs* Fejér factors $\sigma(\xi) = \xi$. *Bottom right*: Exponential concentration $K_N^{\sigma_{\exp}}f(x)$ with $N = 20, 40, 80$ modes. Observe that the root-exponential decay of $K_{80}^{\sigma_{\exp}}f(x)$ becomes almost flat when x is well inside the intervals of smoothness of f, which are well separated from the neighbourhoods of the edges.

6. Extensions

6.1. Discrete data

We are interested in the recovery of the location and amplitudes of the edges, $[f](c_j)$, $j = 1, \ldots, J$, from the discrete Fourier coefficients,

$$\hat{f}_k = \frac{h}{2\pi} \sum_{\nu=0}^{2N} f(y_\nu)e^{-iky_\nu}, \qquad h = \frac{2\pi}{2N+1}. \tag{6.1}$$

As before, the regularity of f is revealed by the decay rate of $\hat{f}_k$: successive summation by parts implies the rapid decay of $\hat{f}(k)$ for smooth f, in analogy with (2.3), that is,

$$|\hat{f}_k| \lesssim \sup_\nu \frac{|\Delta^s f(y_\nu)|}{h^s} \frac{1}{1+|k|^s}, \qquad s \geq 1, \tag{6.2}$$

where $h^{-s}\Delta^s$ are the usual divided differences of order s. On the other hand, for the prototype case of an f which experiences a single jump at $x = c$, (3.2) yields

$$\hat{f}_k = \left(f(y_{\nu_c+1}) - f(y_{\nu_c}) \right) \frac{e^{-iky_{\nu_c}}}{2\pi ik} + \mathcal{O}\left(\frac{1}{|k|^2} \right),$$

where ν_c singles out the cell which encloses the location of the jump discontinuity.

In the discrete case, however, every grid value experiences a jump discontinuity: the jumps that are of order $\mathcal{O}(h)$ are acceptable as part of the smooth region, whereas the $\mathcal{O}(1)$ jumps indicate edges of the underlying function $f(x)$. Hence, in the discrete case we can identify a jump discontinuity at $x = c$ by its enclosed grid cell, $[x_{\nu_c}, x_{\nu_c+1}]$, which is characterized by the asymptotic statement

$$f(x_{\nu+1}) - f(x_\nu) = \begin{cases} [f](c) + \mathcal{O}(h), & \text{for } \nu = \nu_c : c \in [x_\nu, x_{\nu+1}], \\ \mathcal{O}(h), & \text{for other } \nu's \neq \nu_c. \end{cases} \tag{6.3}$$

Of course, this asymptotic statement (6.3) may itself serve as an edge detector based on the given *grid values*, $\{f(x_\nu)\}_{\nu=-N}^{N}$. Higher-order differences, $\Delta^p f(x_\nu)$, yield edge detectors involving increasingly larger, but finite, stencils, with improved separation between cells containing $\mathcal{O}(1)$-scale jumps and smaller, but finite, $\mathcal{O}(h^p)$-cells in regions of smoothness. We now seek alternative edge detectors based on the discrete Fourier coefficients, $\{\hat{f}_k\}_{|k|\leq N}$. Using proper concentration factors, we shall cover both local and global edge detectors. For example, global edge detectors based on the exponential concentration factors do not lend themselves to local stencils of differences: they enjoy the root-exponential accuracy encountered in (5.6).

Our starting point is the discrete conjugate sum, an analogue of (4.3):

$$I_N^\tau f(x) := \frac{\pi i}{c_\tau} \sum_{|k| \le N} \operatorname{sgn}(k) \tau\left(\frac{|k|h}{\pi}\right) \widehat{f}_k e^{ikx}, \tag{6.4a}$$

where $\tau(\xi)$ are the discrete concentration factors at our disposal and c_τ is the normalization coefficient

$$c_\tau := \frac{\pi}{2} \int_0^1 \frac{\tau(\xi)}{\sin(\pi\xi/2)} \, d\xi. \tag{6.4b}$$

It is convenient to link the discrete and continuous factors

$$\tau(\xi) = \sigma(\xi) \operatorname{sinc}\left(\frac{\pi\xi}{2}\right), \qquad \operatorname{sinc}(y) := \frac{\sin y}{y},$$

where the normalization c_τ becomes the usual $c_\tau = c_\sigma = \int_0^1 \sigma(\xi)/\xi \, d\xi$.

To gain greater insight into the behaviour of such detectors we use (6.1) to express the $\widehat{f}_k$ in terms of the $f(x_\nu)$; (6.4a) then reads

$$I_N^\tau f(x) = -\frac{h}{c_\tau} \sum_{\nu=0}^{2N} f(x_\nu) \sum_{k=1}^N \sigma\left(\frac{kh}{\pi}\right) \frac{\sin kh/2}{kh/2} \sin k(x - x_\nu)$$

$$= -\frac{1}{c_\sigma} \sum_{k=1}^N \frac{\sigma\left(\frac{kh}{\pi}\right)}{k} \sum_{\nu=0}^{2N} f(x_\nu) 2 \sin\left(\frac{kh}{2}\right) \sin k(x - x_\nu).$$

Next, we write the last product on the right as a perfect difference and sum by parts to find

$$I_N^\tau f(x) = \frac{1}{c_\sigma} \sum_{k=1}^N \frac{\sigma\left(\frac{kh}{\pi}\right)}{k} \sum_{\nu=0}^N (f(x_{\nu+1}) - f(x_\nu)) \cos k(x - x_{\nu+1/2}). \tag{6.5}$$

We claim that the second sum on the right is dominated by the discontinuous cell(s) where $f(x_{\nu_c+1}) - f(x_{\nu_c}) \sim [f](c)$, while the contributions of the 'smooth' cells are negligible, owing to cancellations of oscillations. To make this statement precise, we first identify the discontinuous cell (and, in general, finitely many like it), by its mid-point, $x_{\nu_c+\frac{1}{2}}$. We then find that

$$\sum_{\nu=0}^{2N} (f(x_{\nu+1}) - f(x_\nu)) \sin k x_{\nu+\frac{1}{2}} \tag{6.6a}$$

$$= ([f](c) + \mathcal{O}(h)) \sin k x_{\nu_c+\frac{1}{2}} + \mathcal{O}\left(\frac{h}{\sin \frac{kh}{2}}\right).$$

The first term on the right of (6.6a) is the contribution of the single jump at $\nu = \nu_c$. For the remaining terms, $\nu \ne \nu_c$, we use the identity $\sin k x_{\nu+1/2} \equiv -(\cos k x_{\nu+1} - \cos k x_\nu)/2 \sin(kh/2)$ to sum by parts once more, accumulating

$2N - 2 \sim \frac{1}{h}$ terms of order $f(x_{\nu+1}) - 2f(x_\nu) + f(x_{\nu-1}) \sim \mathcal{O}(h^2)$ and two (or finitely many) 'boundary terms' of order $f(x_{\nu+1}) - f(x_\nu) \sim \mathcal{O}(h)$. These amount to the second term on the right of (6.6a). The same argument yields

$$\sum_{\nu=0}^{2N} (f(x_{\nu+1}) - f(x_\nu)) \cos kx_{\nu+\frac{1}{2}} \qquad (6.6b)$$

$$= ([f](c) + \mathcal{O}(h)) \cos kx_{\nu+\frac{1}{2}} + \mathcal{O}\left(\frac{h}{\sin \frac{kh}{2}}\right).$$

Inserting (6.6) into (6.5) yields

$$I_N^\tau f(x) = [f](c) \times \frac{1}{c_\sigma} \sum_{k=1}^N \frac{\sigma\left(\frac{kh}{\pi}\right)}{k} \cos k(x - x_{\nu_c+1/2}) + \mathcal{O}(h |\log h|).$$

Apply Theorem 4.1 to the Heaviside function $f(x) = H(x - c)/2$: with $[f](0) = 1$ and $\widehat{f}(k) = -e^{-ikc}/(2\pi i k)$ we find

$$\frac{1}{c_\sigma} \sum_{k=1}^N \frac{\sigma\left(\frac{k}{N}\right)}{k} \cos k(x - c) = \begin{cases} 1 + \mathcal{O}(h |\log h|), & x \approx c, \\ \mathcal{O}(h |\log h|), & \mathrm{dist}\{x, c\} \gg h, \end{cases} \qquad (6.7)$$

and we obtain the following concentration property.

Theorem 6.1. (Discrete concentration kernels; Gelb and Tadmor 2000b) Assume that $f(\cdot)$ is piecewise C^2-smooth and let $I_N^\tau f(x)$ be an admissible discrete conjugate sum (6.4). Then $I_N^\tau f(x)$ satisfies the concentration property

$$I_N^\tau f(x) \sim \begin{cases} [f](c_j) + \mathcal{O}(h |\log(h)|), & x \sim c_j, \ j = 1, \ldots, J, \\ \mathcal{O}(h |\log(h)|), & \mathrm{dist}\{x, \{c_1, \ldots, c_J\}\} \gg h. \end{cases} \qquad (6.8)$$

As an example, consider the discrete concentration factors

$$\tau_{2p+1}(\xi) = \xi^{2p+1} \mathrm{sinc}\left(\frac{\pi\xi}{2}\right), \qquad \mathrm{sinc}(y) = \frac{\sin y}{y}.$$

This corresponds to $\sigma_{2p+1}(\xi) = \xi^{2p+1}$ with $c_\tau = c_\sigma = 2p+1$, and (6.5) yields

$$I_N^{\tau_{2p+1}} f(x) \qquad (6.9)$$

$$= h \sum_{\nu=0}^{2N} (f(x_{\nu+1}) - f(x_\nu)) \sum_{k=1}^N (2p+1) \left(\frac{kh}{\pi}\right)^{2p+1} \cdot \frac{\cos k(x - x_{\nu+\frac{1}{2}})}{kh}.$$

For the first-order method, (6.9) with $p = 0$ reads

$$I_N^\tau f(x) = h \sum_{\nu=0}^{2N} (f(x_{\nu+1}) - f(x_\nu)) D_N(x - x_{\nu+\frac{1}{2}}),$$

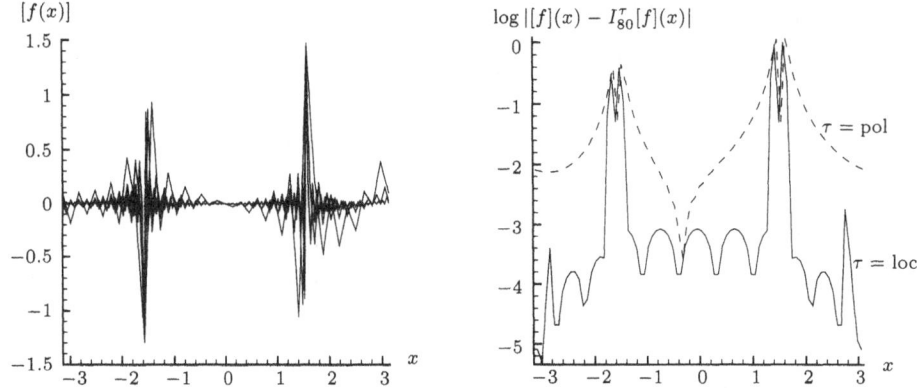

Figure 6.1. *Left*: Detection of edges in the interpolant of f in (5.7) using polynomial concentration factors, $I_N^{\tau_2} f$, with $N = 20, 40$ and 80 modes. *Right*: Logarithmic error of the fifth-order local trigonometric vs the global polynomial concentration factors $I_{80}^{\tau_5} f(x_\nu)$.

This tells us that the discrete concentration factors $\tau_1(\xi) = \xi \operatorname{sinc}(\pi\xi/2)$ amount to *interpolation* of the first-order local differences, $f(x_{\nu+1}) - f(x_\nu)$, at the intermediate grid points, $x_{\nu+\frac{1}{2}}$. In a similar fashion, concentration kernels associated with the higher-order *polynomial factors*, $\tau_{2p+1}(\xi)$, coincide with higher-order derivatives of this interpolant. As p increases, however, the global dependence of interpolation may lead to deterioration of the results, when compared with local edge detectors. An example with *even* order p is illustrated in Figure 6.1.

In contrast, if we choose the *trigonometric concentration factor*, $\tau(\xi) = \sin^3(\pi\xi/2)$, it gives the conjugate discrete sum (6.5) corresponding to $\sigma(\xi) = \xi \sin^2(\pi\xi/2)$ with $c_\tau = c_\sigma = 2$,

$$I_N^\tau f(x) = \frac{h}{2\pi} \sum_{\nu=0}^{2N} \big(f(x_{\nu+1}) - f(x_\nu)\big) \sum_{k=1}^{N} \sin^2\left(\frac{kh}{2}\right) \cos k(x - x_{\nu+\frac{1}{2}}).$$

This conjugate discrete sum coincides with the *local* cubic difference (Gelb and Tadmor 2002), $I_N^\tau f(x_{\nu+\frac{1}{2}}) = 8\Delta^3 f(x_{\nu+\frac{1}{2}})$,

$$I_N^\tau f(x_{\nu+\frac{1}{2}}) = 8\big(-f(x_{\nu+2}) + 3f(x_{\nu+1}) - 3f(x_\nu) + f(x_{\nu-1})\big).$$

The situation is analogous to the higher-order trigonometric factors $\sigma(\xi) = \xi^{2p+1}$: they have the advantage of being local but their order is finite.

6.2. Non-periodic data

We begin with the Gegenbauer expansion of a piecewise smooth $f(\cdot)$:

$$S_N f(x) = \sum_{k=0}^{N} \widehat{f}(k) C_k(x), \qquad \widehat{f}(k) := \int_{-1}^{1} f(x) C_k(x) \omega(x) \, dx. \qquad (6.10)$$

Here $\{C_k(x) = C_k^{(\alpha)}(x)\}_{k \geq 1}$ are orthogonal families of *Gegenbauer polynomials*, associated with different weight functions, $\omega(x) \equiv \omega_\alpha(x) := (1 - x^2)^{\alpha - \frac{1}{2}}$,

$$\int_{-1}^{1} C_k^{(\alpha)}(x) C_\ell^{(\alpha)}(x) \omega_\alpha(x) \, dx = 0, \qquad k \neq \ell, \quad \omega_\alpha(x) := (1 - x^2)^{\alpha - \frac{1}{2}}. \qquad (6.11)$$

They are the eigenfunctions of the singular Sturm–Liouville problem

$$((1 - x^2) \omega(x)((C_k^{(\alpha)})'(x)))' = -a_k \omega(x) C_k^{(\alpha)}(x), \qquad -1 \leq x \leq 1, \qquad (6.12)$$

with corresponding eigenvalues $a_k = a_k^{(\alpha)} = k(k + 2\alpha)$.

As in the periodic case, integration by parts against (6.12) shows that the presence of a single jump discontinuity, $[f](c)$, dictates the linear decay rate of its Gegenbauer coefficients,

$$\widehat{f}(k) = [f](c) \frac{(1 - c^2) \omega(c)}{a_k} C_k'(c) + \mathcal{O}\left(\frac{1}{a_k^2}\right). \qquad (6.13)$$

To extract information about the location of the jump, we consider the conjugate sum

$$\frac{\pi \sqrt{1 - x^2}}{N} S_N(f)'(x) = \frac{\pi \sqrt{1 - x^2}}{N} \sum_{k=1}^{N} \widehat{f}(k) C_k'(x)$$

$$= [f](c) \frac{\pi \sqrt{1 - x^2}(1 - c^2) \omega(c)}{N} \sum_{k=1}^{N} \left\{ \frac{1}{a_k} + \mathcal{O}\left(\frac{1}{a_k^2}\right) \right\} \times C_k'(c) C_k'(x).$$

This is the non-periodic analogue of the Fejèr conjugate sum (5.2) in the periodic case.

We want to quantify the localization property of the last summation. To this end, we simplify the computations by making the (non-standard) normalization $\|C_k^{(\alpha)}(x)\|_{\omega_\alpha} = 1$. Integration by parts of (6.12) against $C_k^{(\alpha)}$ then yields $(C_k^{(\alpha)})'(x) = \sqrt{a_k} C_{k-1}^{(\beta)}(x)$ with $\beta = \alpha + 1$, where the scaling factor $\sqrt{a_k}$ keeps the proper normalization $\|C_k^{(\beta)}(x)\|_{\omega_\beta} = 1$. Substituting in the leading term of the last conjugate sum, we end up with

$$\frac{\pi \sqrt{1 - x^2}}{N} S_N(f)'(x) \sim [f](c) \frac{\pi \sqrt{1 - x^2} \omega_\beta(c)}{N} \times K_N^{(\beta)}(x, c), \qquad (6.14)$$

where

$$K_N^{(\beta)}(x,y) = \sum_{k=1}^{N} C_{k-1}^{(\beta)}(x) C_{k-1}^{(\beta)}(y)$$

is the Christoffel–Darboux kernel (see, *e.g.*, Szegő (1958, Theorem 3.2.2))

$$K_N^{(\beta)}(x,y) = \frac{k_{N-1}}{k_N} \frac{C_N^{(\beta)}(x) C_{N-1}^{(\beta)}(x) - C_N^{(\beta)}(y) C_{N-1}^{(\beta)}(x)}{x-y}, \qquad \frac{k_{N-1}}{k_N} \sim \frac{1}{2}.$$

(6.15)

The concentration property now depends on the localization of $K_N(c,x)$ (see, *e.g.*, Gelb and Tadmor (2000a, Section 3)), *i.e.*,

$$\frac{\pi\sqrt{1-x^2}\omega_\beta(c)}{N} K_N^{(\beta)}(c,x) \sim \begin{cases} \dfrac{\sqrt{\omega_\alpha(c)}}{\sqrt{\omega_\alpha(x_N)}} \times \dfrac{1}{N|x-c|} \sim \dfrac{1}{N^{1-\alpha}}, & x \neq c, \\ 1, & x = c, \ |c| < 1. \end{cases}$$

We summarize.

Corollary 6.2. Let $S_N f$ denote the truncated Gegenbauer expansion (6.10) of a piecewise smooth f, associated with a weight function $\omega_\alpha = (1-x^2)^{\alpha-\frac{1}{2}}$, $|\alpha| \leq 1/2$. Then $\pi\sqrt{1-x^2}S_N(f)'(x)/N$ admits the concentration property

$$\left| \frac{\pi\sqrt{1-x^2}}{N} S_N(f)'(x) - [f](x) \right| \lesssim \frac{\log N}{N\sqrt{\omega_\alpha(x)}}, \qquad 1 - |x| \lesssim \frac{1}{N^2}.$$

We close this section with the example of a piecewise smooth f with *Chebyshev expansion*,

$$S_N f(x) \sim \sum_k \widehat{f}(k) T_k(x).$$

Using a general family of concentration factors, $\lambda(\xi) \in C^2[0,1]$ (corresponding to $\sigma(\xi)/\xi$ in the periodic case), we end up with

$$\left| \frac{\pi\sqrt{1-x^2}}{Nc_\lambda} \sum_{k=1}^{N} \lambda\left(\frac{k}{N}\right) \widehat{f}(k) T_k'(x) - [f](x) \right| \lesssim \frac{\log N}{N}, \qquad c_\lambda := \int_0^1 \lambda(\xi)\,d\xi.$$

(6.16)

6.3. Noisy data

We consider the problem of detecting edges in a piecewise smooth f from its spectral content, which is assumed to be corrupted by noise. We begin with the simple case of an f which experiences a single jump discontinuity, $[f](c)$. As in (4.1), this implies first-order decay of the Fourier coefficients:

$$\widehat{f}(k) = [f](c)\frac{e^{-ikc}}{2\pi ik} + \widehat{g}(k) + \widehat{n}(k).$$

(6.17)

Here, the $\widehat{g}(k)$ are associated with the regular part of f after extracting the jump $[f](c)$; their decay is of order $\sim |k|^{-2}$ or faster, depending on the smoothness of the regular part $g(\cdot)$. The new aspect of the problem enters through the $\widehat{n}(k)$, which are the Fourier coefficients of the noisy part corrupting the smooth part of the data; we assume $n(\cdot)$ to be white noise with variance $E(|\widehat{n}(k)|^2) = \eta$. With (6.17), the conjugate sum (4.3) becomes

$$K_N^\sigma f(x) = [f](c)\frac{2\pi i}{c_\sigma}\sum_{k=1}^N \frac{\sigma\left(\frac{k}{N}\right)}{k}\cos k(x-c)$$

$$-\frac{2\pi}{c_\sigma}\sum_{k=1}^N \sigma\left(\frac{k}{N}\right)\widehat{g}(k)\sin kx - \frac{2\pi}{c_\sigma}\sum_{k=1}^N \sigma\left(\frac{k}{N}\right)\widehat{n}(k)\sin kx.$$

We quantify the 'energy' of each of the three sums on the right. E_J and E_R are associated with the discontinuous and regular parts of f,

$$E_J := \sum_{k=1}^N \left(\frac{\sigma\left(\frac{k}{N}\right)}{k}\right)^2 \approx \frac{1}{N}\int_0^1 \left(\frac{\sigma(\xi)}{\xi}\right)^2 d\xi, \qquad (6.18a)$$

$$E_R := \sum_{k=1}^N \sigma^2\left(\frac{k}{N}\right)|\widehat{g}(k)|^2 \ll \frac{1}{N^3}\int_0^1 \frac{\sigma^2(\xi)}{\xi^4} d\xi, \qquad (6.18b)$$

and E_η is associated with the noisy part of f which was assumed to have variance η:

$$E_\eta := \sum_{k=1}^N \sigma^2\left(\frac{k}{N}\right)E(|\widehat{n}(k)|^2) \approx \eta N\int_0^1 \sigma^2(\xi)\,d\xi. \qquad (6.18c)$$

Following Engelberg and Tadmor (2007), the key to detection of edges in such noisy data is to treat the problem as a constrained minimization. We seek a linear combination $a_J E_J + a_R E_R + a_\eta E_\eta$ which minimizes the total energy, thus making the conjugate sum $K_N^\sigma f$ as localized as possible, subject to a prescribed normalization constraint (4.2b)

$$\min\left\{a_J E_J + a_R E_R + a_\eta E_\eta \,\bigg|\, \int_0^1 \frac{\sigma(\xi)}{\xi}\,d\xi = c_\sigma\right\}. \qquad (6.19)$$

This yields

$$\sigma(\xi) = \frac{C\xi^{-1}}{a_J N^{-1}\xi^{-2} + a_R N^{-3}\xi^{-4} + \eta a_\eta N} = \frac{CN^3\xi^3}{a_J N^2\xi^2 + a_R + \eta a_\eta N^4\xi^4}.$$

We ignore the relatively negligible contribution of the regular part which becomes even smaller as $g(\cdot)$ becomes smoother. Setting $a_R = 0$ we end up with concentration factors of the form

$$\sigma(\xi) = \frac{C}{a_J} \cdot \frac{N\xi}{1 + \eta\beta^2 N^2\xi^2}, \qquad \beta := \sqrt{\frac{a_\eta}{a_J}}. \qquad (6.20)$$

It is worthwhile noting that the resulting concentration factor depends on three parameters.

(i) The *relative* size of the amplitudes $\beta = a_\eta/a_J$. Indeed, the normalization factor is given by

$$c_\sigma = \int_0^\pi \frac{\sigma(\xi)}{\xi}\, d\xi = \frac{C}{a_J\sqrt{\eta}\beta} \tan^{-1}(\sqrt{\eta}\beta N).$$

The corresponding concentration factor

$$\sigma \equiv \sigma_\eta = \frac{\sqrt{\eta}\beta N}{\tan^{-1}(\sqrt{\eta}\beta N)} \cdot \frac{\xi}{1 + \eta\beta^2 N^2\xi^2} \qquad (6.21a)$$

yields

$$K_N^{\sigma_\eta} f(x) = \frac{\pi i\sqrt{\eta}\beta}{\tan^{-1}(\sqrt{\eta}\beta N)} \sum_{|k|\leq N} \frac{|k|}{1 + \eta\beta^2 k^2} \hat{f}(k) e^{ikx}, \qquad \beta = \sqrt{\frac{a_\eta}{a_J}}. \qquad (6.21b)$$

(ii) The number of modes, N. General concentration factors may depend on the wave number k and the number of modes N, $\sigma = \sigma_{k,N}$. It is useful to rearrange this dependence, emphasizing the dependence on the relative wave number, $\sigma = \sigma_N(\frac{|k|}{N})$. Clearly, Theorem 4.1 (and likewise, Theorem 6.1) applies to such $\sigma_N(\xi)$. Here, one has to verify the precise dependence of the error bound (4.5) on σ_N. In particular, in the present context of noisy data this leads us to consider the following parametrization of the noise.

(iii) The variance of the noise, η. We now have *three* scales involved: the small 'smoothness' of order $h \sim 1/N$, the noise scale $\sim \eta$ and the $\mathcal{O}(1)$ scale of jump discontinuities. We distinguish between two cases. If η is sufficiently small, $\eta \ll 1/N$ so that $\sqrt{\eta}\beta N \ll 1$, then the noise can be sought as part of the smooth variation of f; indeed, (6.21a) recovers Fejér concentration factor for noise-free data, $\sigma_\eta(\xi) \approx \xi$ (and in particular, $\sigma_\eta(\xi) = \xi$ at the limit of $\eta \downarrow 0$). If, on the other hand, $\eta \gtrsim 1/N$, then the $\mathcal{O}(1/N)$-smoothness scale is dominated by the $\mathcal{O}(\eta)$-noise scale, which we assume to be still well below the $\mathcal{O}(1)$-scale of the jumps

$$\frac{1}{N} \lesssim \sqrt{\eta}\beta \ll \mathcal{O}(1).$$

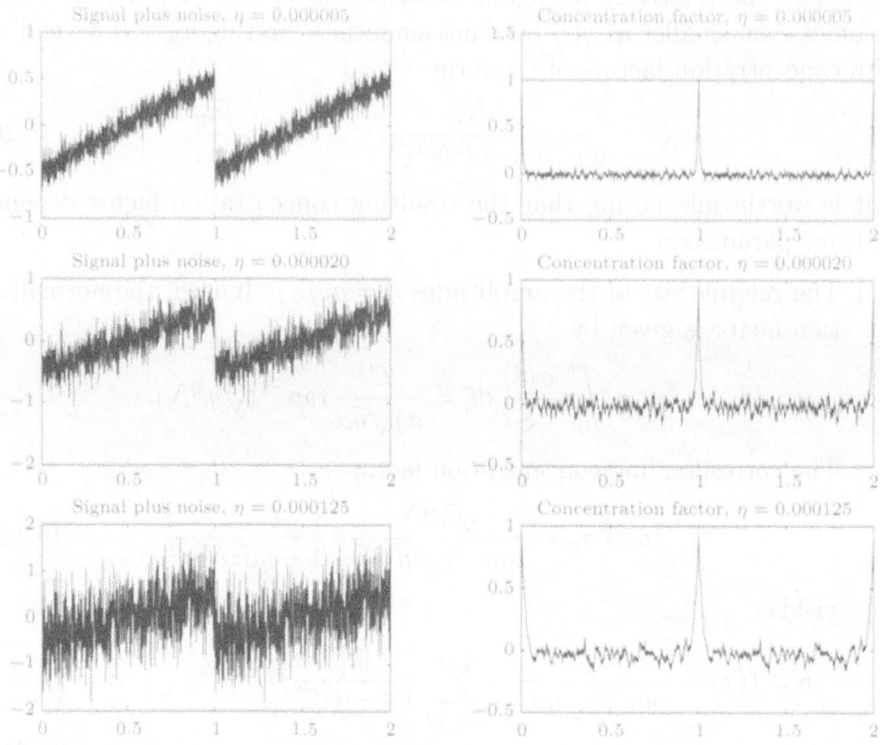

Figure 6.2. Detection of edges in noisy sawtooth function
corrupted with various values of η, using the concentration
kernel (6.21a) with $\beta = \pi\eta^{-1/6}$.

In this case, we can ignore the bounded factor $1/\tan^{-1}(\sqrt{\eta}\beta N)$ and,
using (4.9), we then find an error bound of order

$$\frac{\log N}{N}\|\sigma_\eta\|_{C^2} + \left|\sigma_\eta\left(\frac{1}{N}\right)\right| + \frac{1}{N}|\sigma_\eta(1)| \lesssim \sqrt{\eta}\beta|\log(\sqrt{\eta}\beta)|.$$

A careful examination of the various error bounds involved in Theorem 4.1 (*e.g.*, Engelberg and Tadmor (2007)), shows that all other
σ_η-dependent contributions to the error do not exceed the small scale
of order $\sqrt{\eta}\beta|\log(\sqrt{\eta}\beta)|$,

$$|K_N^{\sigma_\eta} f(x) - [f](x)| \lesssim \sqrt{\eta}\beta|\log(\sqrt{\eta}\beta)|.$$

The resulting concentration kernel, $K_N^{\sigma_\eta} f$, tends to de-emphasize both
the low frequencies, which are 'corrupted' by the jump discontinuity
(-ies), and the high frequencies, which are corrupted by the noise. Different procedures yield different policies for the choice of β. Figure 6.2
demonstrates the edge detected in noisy data using the concentration
kernel (6.21) with the advocated $\beta \sim \eta^{-1/6}$.

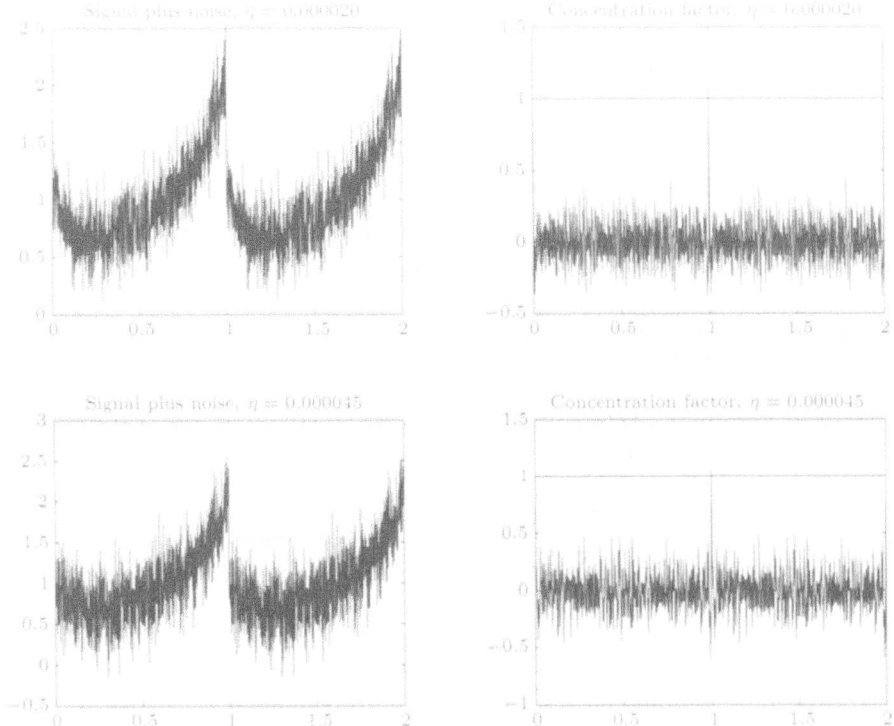

Figure 6.3. Detection of edges in noisy sawtooth function corrupted with various values of η, using the concentration factors (6.22) with $\beta = \pi\eta^{-1/6}$.

As an alternative approach, we may replace the L^2-'averaged' effect of the regular part taken in (6.18b) by the BV-like quantity

$$E_R := \sum_{k=1}^{N} \left| \sigma\left(\frac{k}{N}\right) \right| \cdot |\widehat{g}(k)|.$$

In this case, the constrained minimization (6.19) with $|\widehat{g}(k)| \sim 1/|k|^2$ yields

$$\sigma_\eta(\xi) = \frac{1}{c_\sigma} \cdot \frac{(N\xi - a_R)_+}{1 + \eta\beta^2 N^2 \xi^2}. \tag{6.22}$$

Figure 6.3 quotes the results of Engelberg and Tadmor (2007), with the detection of edges in noisy data using these concentration factors which were tuned with $a_R = 6\pi$, $c_\sigma \sim 3$.

6.4. Incomplete data: compressed sensing

We are interested in the detection of edges in a piecewise smooth f from an *incomplete* set of its spectral content, that is, we have access only to the $\hat{f}(k)$ (or $\hat{f}_k$) for $k \in K$, where K is a strict subset of $\{-N, \ldots, N\}$. Our methodology for edge detection in such cases is motivated by the *compressive sensing* approach (Donoho and Tanner 2005, Candes, Romberg and Tao 2006a, 2006b). Equipped with the partial information of $\hat{f}(k), k \in K$, one can form the incomplete concentration kernel

$$K_N^\sigma f(x) = \sum_{k \in K} \widetilde{f}(k) e^{ikx}, \qquad \widetilde{f}(k) := \frac{\pi i}{c_\sigma} \operatorname{sgn}(k) \sigma\left(\frac{|k|}{N}\right) \hat{f}(k). \qquad (6.23)$$

We follow Tadmor and Zou (2007), seeking to recover a 'complete' concentration kernel, $g(x) \sim K_N^\sigma f(x)$, of the form

$$g(x) = \overbrace{\sum_{k \in K} \widetilde{f}(k) e^{ikx}}^{\text{prescribed data}} + \overbrace{\sum_{k \notin K} \hat{g}(k) e^{ikx}}^{\text{missing data}}.$$

Here, $\widetilde{f}(k)$ are the conjugate coefficients corresponding to the prescribed data for $k \in K$, while the missing conjugate coefficients $\{\hat{g}(k) | k \notin K\}$ at our disposal are sought as minimizers of the total variation $\|g(x)\|_{TV}$,

$$g(x) = \operatorname{argmin}\left\{ \|g\|_{TV} \,\middle|\, g(x) = K_N^\sigma f(x) + \sum_{k \notin K} \hat{g}(k) e^{ikx} \right\}. \qquad (6.24)$$

Similarly, in the discrete case we seek a 'complete' concentration kernel

$$g(x) = \frac{\pi i}{c_\tau} \sum_{k \in K} \operatorname{sgn}(k) \tau\left(\frac{k}{N}\right) \hat{f}_k e^{ikx} + \sum_{k \notin K} \hat{g}_k e^{ikx},$$

which is selected by the TV minimization principle,

$$g(x) = \operatorname{argmin}\left\{ \|g\|_{TV} = \sum_\nu |g(x_{\nu+1}) - g(x_\nu)| \,\middle|\, \hat{g}_k = \widetilde{f}_k, \ k \in K \right\}. \qquad (6.25)$$

The complete concentration kernel $g(x)$ can be viewed as an approximation to the 'ultimate' jump function

$$\Gamma_f(x) := \sum_{j=1}^{J} [f](c_j) 1_{[x_{\nu_{c_j}}, x_{\nu_{c_j}+1}]}(x),$$

where the missing $\{\hat{g}_k\}_{|k \notin K}$ complement the prescribed $\{\widetilde{f}_k\}_{|k \in K}$ as the approximate Fourier coefficients $(\widehat{\Gamma_f})_k$. The rationale behind the TV minimization in (6.24), (6.25) is to enforce the ℓ_1-minimization of the differences, which imposes sparsity in the sense of maximizing the number of zero differences (Candes, Romberg and Tao 2006a, 2006b). Hence, it yields $g(x)$ as an approximate jump function with a minimal number of piecewise components. The optimization model (6.25) can be solved by the second-order cone programs, which takes time $\mathcal{O}(N^3 \log N)$.

The compressed sensing approach can be extended to noisy data (Donoho, Elad and Temlyakov 2004, Candes, Romberg and Tao 2006a). Following Tadmor and Zou (2007), we assume that the observed (pseudo-) spectral data may be contaminated by white noise with variance $\leq \eta$. To recover edges from such noisy and incomplete data, the following compressed sensing model is sought:

$$g(x) = \operatorname{argmin}\left\{ \|g\|_{TV} \mid g = \sum_{k=1}^{N} \widehat{g}_k e^{ikx} \quad \text{s.t. } \|\widehat{g}_k - \widetilde{f}_k\|_{\ell^2(k \in K)} \leq \eta \right\}. \quad (6.26)$$

7. Enhancements

The detection of edges in Theorems 4.1 and 6.1 is based on the asymptotic behaviour of the concentration kernels K_N^σ which separate between the large and small scales as $\varepsilon_N \sim 1/N \downarrow 0$. To improve the edge detection, we want to enhance the separation of scales in (4.5). To this end we consider

$$N^q \big(K_N^\sigma f(x)\big)^{2q} = \begin{cases} \sim N^q \big([f](c_j)\big)^{2q}, & x \approx \{c_1, c_2, \ldots, c_J\}, \\ \mathcal{O}(N^{-q}), & \operatorname{dist}\{x, \{c_1, c_2, \ldots, c_J\}\} \gg \frac{1}{N}. \end{cases}$$

The exponent $q \geq 1$ is at our disposal: by increasing q, we enhance the separation between the vanishing scale at the points of smoothness (of order $\mathcal{O}(N^{-q})$) and the amplified scale at the jumps (of order $\mathcal{O}(N^q)$).

Next, one must introduce a *critical threshold* to eliminate the unacceptable jumps: only those edges with amplitudes larger than the critical threshold $[f](x) > J_c^{1/2q}/\sqrt{N}$ will be detected. Here, J_c is a measure which defines the small scale in our computation of edge detection. We note that J_c is data-dependent and is typically related to the variation of the smooth part of f.

Given this critical threshold, we form our enhanced concentration kernel

$$K_{N,J_c}^\sigma f(x) = \begin{cases} K_N^\sigma f(x), & \text{if } N^q \big|K_N^\sigma f(x)\big|^{2q} > J_c, \\ 0, & \text{otherwise.} \end{cases} \quad (7.1)$$

Clearly, with sufficiently large q, one ends up with a sharp edge detector where $K_{N,J_c}^\sigma f(x) = 0$ at all but $\mathcal{O}(1/N)$-neighbourhoods of the jumps $x = c_1, c_2 \ldots$. In practical applications, $q \leq 3$ will suffice. For example, enhancing the local concentration kernel (5.1) $K_{\varepsilon_N}(y) = \phi'_{\varepsilon_N}(y)$ with

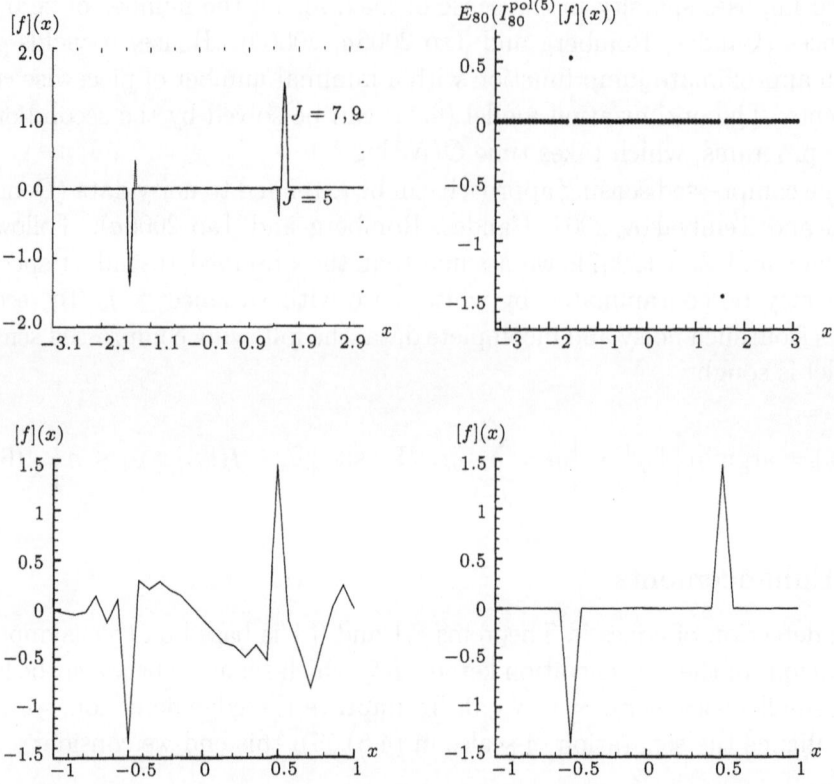

Figure 7.1. *Top left*: Jump value obtained by $K_{40,J_c}^\sigma f$ for $f(x)$ in (7.1) with $q = 1$ and $\sigma(\xi) = \sin(\xi)$. *Top right*: $I_{80,J_c}^\tau f$ with $\tau(\xi) = \xi^5$ in (7.2). *Bottom*: Detection of edges in Chebyshev expansion of $f(x/\pi)$ before and after enhancement with $q = 1$ and $J_c = 5$.

$q = 1$ leads to the *quadratic filter* (*e.g.*, Firoozye and Sverak (1996)), where $(K_{\varepsilon_N} f(x))^2 = (\phi_{\varepsilon_N}' * f(x))^2 \to [f]^2(x)$.

We can apply this nonlinear enhancement in conjunction with discrete concentration kernels $I_N^\tau(y)$. The corresponding *enhanced spectral concentration kernel* amounts to

$$I_{N,J_c}^\tau = \begin{cases} I_N^\tau f(x), & \text{if } N^q |I_N^\tau f(x)|^{2q} > J_c, \\ 0, & \text{otherwise.} \end{cases} \qquad (7.2)$$

Observe that the use of concentration kernels, (4.5) and (6.8), actually detects the $\mathcal{O}(\varepsilon_N)$-*neighbourhoods* of jump discontinuities rather than the discontinuities themselves. Figure 7.1 demonstrates how the nonlinear enhancement of concentration kernels helps to pinpoint the location of edges in the discrete and non-periodic set-ups.

7.1. Nonlinear limiter: minmod edge detection

Implementation of the enhanced edge detectors $K_{N,J_c}^\sigma f(x)$ and $I_{N,J_c}^\tau f(x)$ requires an outside threshold parameter, J_c, which should be properly chosen to separate the specific scales associated with f. This becomes an impediment for detecting edges in both small-scale problems and problems with steep gradients and high variation. A second, related difficulty arises when oscillations are formed in the neighbourhood of the jump discontinuities. The particular behaviour of these oscillations depends on the specific concentration factors used, and it can be difficult to distinguish between a true jump discontinuity and an oscillating artifact, particularly when several jump discontinuities are located in the same neighbourhood, *i.e.*, when there is limited resolution for the problem. Wrong parametrization may lead to misidentification of jump discontinuities that are located 'too' close together. We discuss an improved enhancement procedure based on the non-linear limiting of low- and high-order concentration factors. The rationale, outlined in Gelb and Tadmor (2006), is as follows.

Edge detectors based on a low-order concentration kernel K_N^σ with polynomial factors $\sigma_p(\xi), p \sim 1$, or trigonometric factors $\sigma_\alpha(\xi)$, have a relatively slow, $\mathcal{O}(\log N/N)$ decay away from the discontinuities, yet they yield only a few spurious oscillations (if any) in the immediate neighbourhoods of the discontinuities. In contrast, highly accurate kernels such as $K_N^{\sigma_{\exp}}$ rapidly converge to zero away from the neighbourhoods of discontinuities, but suffer from severe oscillations within these immediate neighbourhoods. The loss of monotonicity with increasing order is, of course, the canonical situation in many numerical algorithms; the passage from the first-order, monotone Fejér kernel (see Section 9 below) to the spurious oscillations in the spectrally accurate Dirichlet kernel is a prototypical case.

We therefore take advantage of the different behaviour of low- and high-order edge detectors. Away from the jump discontinuities, we let the high-order, possibly exponentially small, kernel dominate, by taking the (signed) minimum,

$$K_N f(x) = s \times \min\{|K_N^{\sigma_{\text{high}}} f(x)|, |K_N^{\sigma_{\text{low}}} f(x)|\}, \quad s := \text{sgn}\{K_N^{\sigma_{\text{high}}} f(x)\}.$$

As we approach the jump discontinuity, however, high-order methods produce spurious oscillations which should be rejected: this could be achieved through comparison with essentially monotone profiles produced by low-order detectors. Thus, when the two profiles disagree in sign – indicating spurious oscillations – then our detector is set to zero:

$$K_N f(x) = 0, \quad \text{if } \text{sgn}\{K_N^{\sigma_{\text{high}}} f(x)\} \neq \text{sgn}\{K_N^{\sigma_{\text{low}}}\}.$$

We end up with the so-called minmod limiter

$$K_N^{\sigma_{mm}} f(x) := \text{minmod}\{K_N^{\sigma_{\exp}} f(x), K_N^{\sigma_1} f(x)\}, \tag{7.3}$$

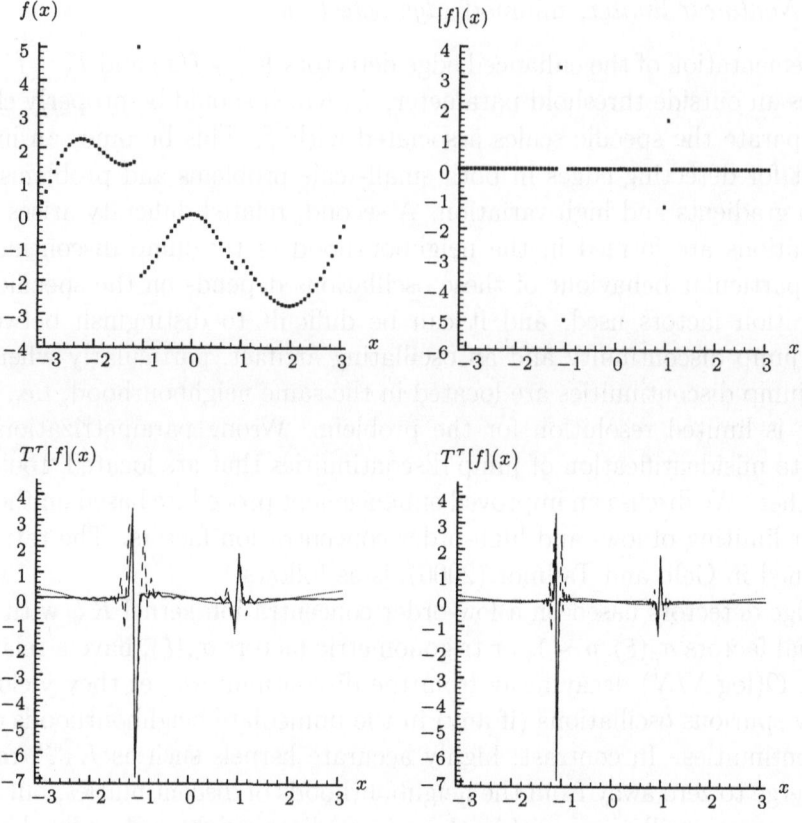

Figure 7.2. *Top left*: Piecewise f with adjacent edge.
Top right: $K_{80}^{\sigma mm} f(x)$. *Bottom left*: Edge detection using
first-order $I_N^{T\text{pol}} f(x)$ (dotted), $I_N^{T\text{exp}} f(x)$ (dashed) and the
$I_N^{Tmm} f(x)$ algorithm (solid) with $N = 80$ grid-points.
Bottom right: The same with $N = 160$ points.

which plays a central role in non-oscillatory reconstruction of high-resolution
methods for nonlinear conservation laws (see, *e.g.*, Harten (1983), Tadmor
(1998) and the references therein). This adaptive algorithm can be extended
to include several concentration factors, *e.g.*,

$$I_N^{Tmm} f(x) := \text{minmod}\{I_N^{T\text{exp}} f(x), I_N^{T\text{pol}} f(x), I_N^{T\text{trig}} f(x)\}, \qquad (7.4)$$

where the k-tuple minmod limiter takes the form

$$\text{minmod}\{a_1, \ldots, a_k\}$$
$$:= \begin{cases} s \times \min_{1 \leq j \leq k} |a_j|, & \text{if } \text{sgn}(a_1) = \cdots = \text{sgn}(a_k) := s, \\ 0, & \text{otherwise.} \end{cases}$$

It retains the high order in smooth regions while 'limiting' the high-order

spurious oscillations in the neighbourhoods of the jumps by the less oscillatory low-order detectors. By incorporating such a mixture of low-order and
high-order methods in different regimes of the computation, the resulting
minmod-based adaptive detection provides a parameter-free edge detector,
which in turn enables more robust nonlinear enhancements.

Figure 7.2 illustrates the improvement in using the minmod edge detector (7.4) when applied to a piecewise smooth f exhibited in the top part of
Figure 7.2. There is a 'clean' detection of the jump discontinuities which
are located close together. It also compares the results of application of
the concentration kernels and the minmod algorithm. It is evident that the
polynomial factor τ_{2p+1} does not converge to zero sufficiently fast away from
the discontinuities; hence the steep gradients of the function might be mis-
interpreted as jump discontinuities. On the other hand, the concentration
method using $\tau_{\exp}$ causes interfering oscillations in the neighbourhoods of
the discontinuities, making it difficult to determine where the true jumps
are. The minmod algorithm ensures the convergence to the jump function
without interference of the oscillations. An early application of the minmod
enhancement to non-negative band pass filters can be found in Bauer (1995,
Section 4).

8. Edge detection in two-dimensional spectral data

8.1. Two-dimensional concentration kernels

Given the two-dimensional spectral data

$$\widehat{f}(\mathbf{k}) := \frac{1}{(2\pi)^2} \int_{-\pi}^{\pi} \int_{-\pi}^{\pi} f(\mathbf{y}) e^{-i\mathbf{k} \cdot \mathbf{y}} \, dy_1 \, dy_2,$$

we are interested in detecting the edges of the underlying piecewise smooth
$f(\cdot)$. We assume the generic case where these edges lie along simple curves,
and we proceed with a straightforward application of one-dimensional concentration kernels which apply dimension-by-dimension. Accordingly, we
have a 2-vector concentration kernel $\mathbf{K}_N^\sigma$, of the form

$$\mathbf{K}_N^\sigma f(\mathbf{x}) = \begin{bmatrix} K_{N,x_1}^\sigma \\ K_{N,x_2}^\sigma \end{bmatrix} f(\mathbf{x}) := \frac{\pi i}{c_\sigma} \sum_{|k_1| \le N} \sum_{|k_2| \le N} \begin{bmatrix} \mathrm{sgn}(k_1) \\ \mathrm{sgn}(k_2) \end{bmatrix} \sigma\left(\frac{|\mathbf{k}|}{N}\right) \widehat{f}(\mathbf{k}) e^{i\mathbf{k} \cdot \mathbf{x}}.$$

(8.1)

Edges along the x_1-axis with each fixed $x_2 \in [-\pi, \pi]$ are sought as extremal
values of the first component, $K_{N,x_1}^\sigma f(x_1, \cdot)$, while $K_{N,x_2}^\sigma f(\cdot, x_2)$ process
edges along the x_2-axis. Similarly, the discrete set-up is based on the two-
dimensional conjugate sums

$$\mathbf{I}_N^\tau f(\mathbf{x}) = \frac{\pi i}{c_\tau} \sum_{|k_1| \le N} \sum_{|k_2| \le N} \begin{bmatrix} \mathrm{sgn}(k_1) \\ \mathrm{sgn}(k_2) \end{bmatrix} \tau\left(\frac{|\mathbf{k}| h}{\pi}\right) \widehat{f}(\mathbf{k}) e^{i\mathbf{k} \cdot \mathbf{x}}.$$

(8.2)

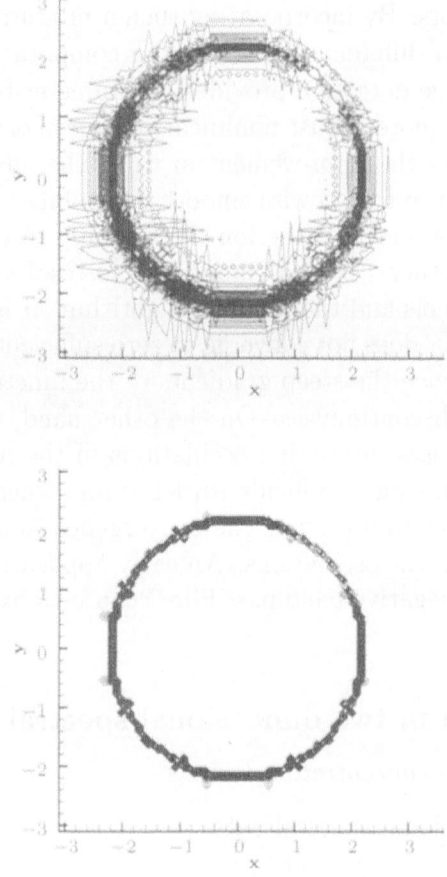

Figure 8.1. Two-dimensional detection $\mathbf{I}_{80}^{T_{\exp}} f(\mathbf{x})$
for a circular edge, before and after enhancement.

The approach is simple to implement, although it may suffer from the Cartesian preference when the edges lie along curves which do not align with the axis, as illustrated in Figure 8.1 for $f(\mathbf{x})$ whose edges lie along the circle $|\mathbf{x}| = 0.7\pi$. We observe the familiar Cartesian-based phenomenon of staircasing; much of it is removed by nonlinear enhancement.

A parameter-free enhancement based on the minmod limiter (7.4) yields improved results for edge detection in the Shepp–Logan brain image, shown in Figure 8.2.

8.2. Incomplete data

The extension of edge detection for *incomplete* data in two dimensions, $\{\widehat{f}(\mathbf{k})\}_{\mathbf{k} \in K}$ where $K \subsetneq [-N, N]^2$, is straightforward. We shall focus on the discrete case, where we set a rectangular grid $\mathbf{x}_{\nu,\mu} := (\nu \Delta x_1, \mu \Delta x_2)$.

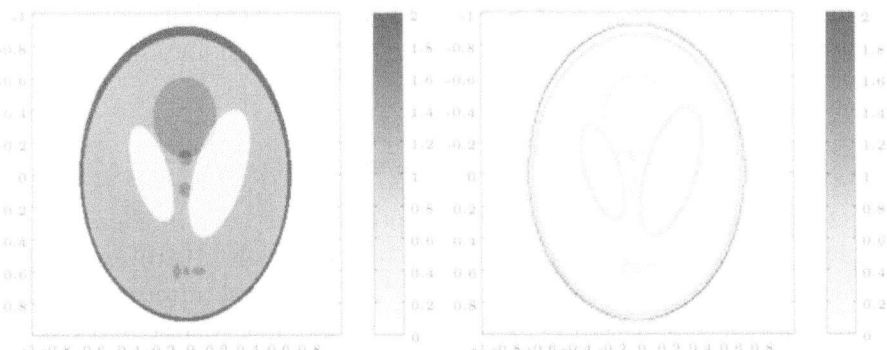

Figure 8.2. *Left*: Contour plot of the Shepp–Logan brain image. *Right*: Nonlinear enhancement procedure (7.4) applied to the Shepp–Logan brain phantom image.

Figure 8.3. Recovered phantom image from incomplete spectral data. *Left*: Result by the back projection. *Right*: Recovered edges of Shepp phantom graph by compressive sensing edge detection.

We seek $\widehat{\mathbf{g}}_{\mathbf{k}}|\mathbf{k} \notin K$ which produce an approximate concentration kernel

$$\mathbf{g}(\mathbf{x}) = \frac{\pi \mathrm{i}}{c_\tau} \sum_{\mathbf{k} \in K} \begin{bmatrix} \mathrm{sgn}(k_1) \\ \mathrm{sgn}(k_2) \end{bmatrix} \tau\left(\frac{|\mathbf{k}|}{N}\right) \widehat{f}_{\mathbf{k}} \mathrm{e}^{\mathrm{i}\mathbf{k}\cdot\mathbf{x}} + \sum_{\mathbf{k} \notin K} \widehat{\mathbf{g}}_{\mathbf{k}} \mathrm{e}^{\mathrm{i}\mathbf{k}\cdot\mathbf{x}}, \qquad (8.3\mathrm{a})$$

with minimal total variation

$$\|\mathbf{g}\|_{TV} = \sum_{\nu,\mu} |\mathbf{g}(\mathbf{x}_{\nu+1,\mu}) - \mathbf{g}(\mathbf{x}_{\nu,\mu})|\Delta x_1 + |\mathbf{g}(\mathbf{x}_{\nu,\mu+1}) - \mathbf{g}(\mathbf{x}_{\nu,\mu})|\Delta x_2. \quad (8.3\mathrm{b})$$

Figure 8.3 illustrates edge detection for use of the compressive sensing model of $\mathbf{K}_N^\sigma f(\mathbf{x})$ for incomplete data of the two-dimensional Shepp–Logan phantom image. Here, $N = 256$, we use Fejér concentration factors, $\sigma(\xi) = \xi$ and partial data are gathered along each of 100 radial lines in the spectral domain.

8.3. Concentration kernels and zero-crossing

The zero-crossing method is one of the popular methods in edge detection in two-dimensional data: consult Marr and Hildreth (1980) and the references therein. It searches for zero-crossings in the discrete Laplacian of the function f, in order to find the underlying edges. This is intimately connected with the conjugate kernels. To clarify this point, we begin with the one-dimensional example of the Fejér conjugate sum (5.2), $K_N^{\sigma_1} f = \pi S_N(f)'(x)/N$. Recall that edges are sought as extremal values of $K_N^{\sigma_1} f(x)$, and we therefore seek the zeros of $K_N^{\sigma_1}(f)''(x)$, i.e.,

$$\left\{ c_j \mid \frac{\mathrm{d}}{\mathrm{d}x} K_N^{\sigma_1} f(x)|_{x=c_j} \propto \frac{\mathrm{d}^2}{\mathrm{d}x^2} S_N f(x)|_{x=c_j} = 0 \right\}. \tag{8.4}$$

This is the one-dimensional zero-crossing. We note that, unlike the edges sought as the extrema of $K_N^{\sigma_1} f(x)$, the zero-crossing in (8.4) may introduce additional spurious inflection points. A similar situation occurs with general concentration factors. Setting $\sigma(\xi) := \xi\lambda(\xi)$, then (4.3) amounts to

$$K_N^{\sigma} f(x) = \frac{\pi}{Nc_\lambda} \sum_{|k| \le N} ik\lambda\left(\frac{|k|}{N}\right) \widehat{f}(k) e^{ikx}, \qquad c_\lambda = \int_0^1 \lambda(\xi)\,\mathrm{d}\xi.$$

Thus $K_N^{\sigma} f(x)$ with $\sigma(\xi) = \xi\lambda(\xi)$ is merely the derivative of a mollified version of $S_N(f)(x)$,

$$K_N^{\sigma} f(x) = \frac{\mathrm{d}}{\mathrm{d}x} \Lambda_N * S_N(f)(x), \qquad \Lambda_N(x) := \frac{1}{2Nc_\lambda} \sum_{|k| \le N} \lambda\left(\frac{|k|}{N}\right) e^{ikx},$$

and the zero-crossing procedure amounts to identifying edges as zeros of this mollified spectral projection,

$$\left\{ c_j \mid \frac{\mathrm{d}^2}{\mathrm{d}x^2} \Lambda_N * S_N f(x)\Big|_{x=c_j} = 0, \quad \Lambda_N(x) := \frac{1}{2Nc_\lambda} \sum_{|k| \le N} \lambda\left(\frac{|k|}{N}\right) e^{ikx} \right\}. \tag{8.5}$$

By suitable choice of λ, we obtain a large class of 'regularized' zero-crossings. But once again, we need to augment (8.5) with a procedure to rule out inflection points which otherwise could be detected as spurious edges.

A similar set-up holds in the two-dimensional case. Here, we consider the generic case of a piecewise smooth $f(\mathbf{x})$ whose edges lie along simple curves to be detected by the 2-vector of concentration kernels, $\mathbf{K}_N^{\sigma} f(\mathbf{x})$, in (8.1). To simplify matters, we set $\sigma(\xi) = \xi\lambda(\xi)$. The 2D-detection-based concentration approach now seeks the edges as extremal values of

$$\nabla_{\mathbf{x}}\Lambda_N * S_N f(\mathbf{x}), \qquad \Lambda_N(\mathbf{x}) = \frac{1}{2Nc_\lambda}\sum_{|\mathbf{k}|\leq N}\lambda\left(\frac{|\mathbf{k}|}{N}\right)e^{i\mathbf{k}\cdot\mathbf{x}}. \qquad (8.6)$$

The zero-crossing method realizes these extremal values as the zeros of

$$\left\{\mathbf{c}\;\middle|\;\Delta_{\mathbf{x}}\Lambda_N * S_N f(\mathbf{x})\big|_{\mathbf{x}=\mathbf{c}} = 0\right\}, \qquad (8.7a)$$

We observe that the two-dimensional zero-crossing could add a considerable number of spurious edges: *e.g.*, Ulupinar and Medioni (1988) and Clark (1989). We can improve this deficiency by augmenting (8.7a) with a more careful zero-crossing criterion, *e.g.*,

$$\left\{\mathbf{c}\;\middle|\;\frac{\partial^2}{\partial x_1^2}\Lambda_N * S_N f(\mathbf{x})\;\text{ or }\;\frac{\partial^2}{\partial x_2^2}\Lambda_N * S_N f(\mathbf{x})\;\text{ changes sign at }\;\mathbf{x}=\mathbf{c}\right\}. \qquad (8.7b)$$

Following Tadmor and Zou (2007), we can now combine the improved zero-crossing (8.7b) with compressed sensing, in order to deal with incomplete (and possibly noisy ...) data. Given the partial spectral information, $\left\{\widehat{f}(\mathbf{k})\right\}_{\mathbf{k}\in K}$, we seek to complement the missing data by the usual TV-minimization

$$g(\mathbf{x}) = \mathrm{argmin}\left\{\|g\|_{TV}\;\middle|\;g(\mathbf{x}) = \sum_{\mathbf{k}\in K}|\mathbf{k}|^2\widehat{f}(\mathbf{k})e^{i\mathbf{k}\cdot\mathbf{x}} + \sum_{\mathbf{k}\notin K}|\mathbf{k}|^2\widehat{g}(\mathbf{k})e^{i\mathbf{k}\cdot\mathbf{x}}\right\}.$$

$$(8.8)$$

Observe the *sparsity* of the TV-based compressive sensing of the zero-crossings in the minimizer (8.8): it is tied to minimizing the number of

Figure 8.4. Edge detection in incomplete spectral data by zero-crossing. *Left*: Original image. *Centre*: Image recovered from incomplete data zero-crossing. *Right*: Zero-crossing combined with compressive sensing (8.8), with Gaussian concentration factor Λ^1 and threshold $\zeta = 3$.

zero components of $g \approx \Delta_{\mathbf{x}} K_N^\sigma f(\mathbf{x})$, that is, minimizing the number of piecewise linear components of $K_N^\sigma f(\mathbf{x})$.

Figure 8.4, from Tadmor and Zou (2007), illustrates edge detection from incomplete data using (improved) zero-crossing with compressed sensing. Here we use the normalized Gaussian,

$$\Lambda^\beta(\mathbf{x}) := \frac{1}{2\pi\beta^2} e^{-\frac{|\mathbf{x}|^2}{2\beta^2}},$$

which is a typical choice of zero-crossing mollifier, Λ_N.

9. Reconstruction of piecewise smooth data

We want to reconstruct a piecewise smooth f from its spectral coefficients $\{\widehat{f}(k)\}_{|k|\leq N}$. To avoid the spurious Gibbs oscillations formed by the spectral projection $S_N f$, one may consider the classical Fejér partial sums

$$S_N^F f(x) := \sum_{|k|\leq N} \left(1 - \frac{|k|}{N}\right) \widehat{f}(k) e^{ikx},$$

which amount to a convolution against the *Fejér kernel*,

$$S_N^F f(x) = (F_N * f)(x),$$

$$F_N(y) := \frac{1}{2\pi} \sum_{|k|\leq N} \left(1 - \frac{|k|}{N}\right) e^{iky} \equiv \frac{1}{2\pi N} \left(\frac{\sin\left(\frac{Ny}{2}\right)}{\sin\left(\frac{y}{2}\right)}\right)^2.$$

Since $F_N \geq 0$, it follows that $-1 \leq S_N^F H(x) \leq 1$; moreover, since $H'(x) \geq 0$ implies $S_N^F(H)'(x) \geq 0$, it follows that $S_N^F H(x)$ increases *monotonically* between -1 and 1. Thus, Fejér partial sums avoid spurious oscillations; in fact they are monotone, and converge uniformly whenever f is continuous. But the monotonicity of Fejér partial sums comes at a price: according to a classical theorem of Korovkin (*e.g.*, DeVore and Lorentz (1993)), every family of linear *positive operators* such as the S_N^F is at most second-order accurate, that is,

$$|S_N^F f(x) - f(x)| \lesssim \frac{1}{N^2}, \qquad f \in C^1,$$

and this second-order convergence rate estimate does *not* improve for more regular f (since it is essentially dictated by $f(x) = 1, x$ and x^2).

It is possible to utilize the Fejér sums to regain spectral accuracy while still avoiding Gibbs oscillations. To this end, we consider the partial sum

$$S_N^\varphi f(x) := \sum_{|k|\leq N} \varphi\left(\frac{|k|}{N}\right) \widehat{f}(k) e^{ikx}, \qquad \varphi\left(\frac{|k|}{N}\right) = \begin{cases} 1, & |k| \leq \frac{N}{2}, \\ 2 - \frac{2|k|}{N}, & \frac{N}{2} \leq |k| \leq N. \end{cases}$$

Although S_N^φ is no longer positive, it is the difference of two positive Fejér sums,

$$S_N^\varphi f(x) \equiv 2S_N^F f(x) - S_{N/2}^F f(x),$$

and as such, it converges uniformly whenever f is merely continuous. At the same time, the convergence rate of S_N^φ increases together with the global smoothness of f, and we have the spectral error estimate

$$|S_N^\varphi f(x) - f(x)|$$

$$\leq \sum_{\frac{N}{2} \leq |k| \leq N} \left| 1 - \frac{2|k|}{N} \right| \cdot |\widehat{f}(k)| + \sum_{|k|>N} |\widehat{f}(k)| \lesssim \|f\|_{C^s} \frac{1}{N^{s-1}}, \quad \text{for } s > 1.$$

Observe that the spectral accuracy of S_N^φ lies in the fact that the first $N/2$ coefficients in S_N^φ are left unchanged,

$$\varphi(\xi) = \begin{cases} 1, & 0 \leq \xi \leq \frac{1}{2}, \\ 2 - 2\xi, & \frac{1}{2} \leq \xi \leq 1. \end{cases} \tag{9.1}$$

But what happens when we apply $S_N^\varphi f$ to piecewise smooth f? As we shall explore in the next few sections, the answer lies with the smoothness of $\varphi(\cdot)$ or, equivalently, the decay behaviour of the mollifier associated with (9.1), $\mathcal{S}_N^\varphi(x) = 2F_N(x) - F_{\frac{N}{2}}(x)$.

The preceding examples demonstrate two interchangeable processes which are available for recovering the rapid convergence in the piecewise smooth case. These are *mollification*, carried out in the physical space, and *filtering*, carried out in the Fourier space, *i.e.*,

$$\Phi * (S_N f)(x) \longleftrightarrow \sum_{|k| \leq N} \varphi\left(\frac{|k|}{N}\right) \widehat{f}(k) e^{ikx}.$$

Filtering accelerates convergence when pre-multiplying the Fourier coefficients by a *rapidly decreasing* $\varphi(|k|/N)$, as $|k| \uparrow N$. This rapid decay in Fourier space corresponds to mollification with *highly localized* mollifiers, $\Phi(x) = \mathcal{S}_N^\varphi(x)$, in physical space:[5]

$$\mathcal{S}_N^\varphi(x) = \frac{1}{2\pi} \sum_{|k| \leq N} \varphi\left(\frac{|k|}{N}\right) e^{ikx}.$$

There is a rich literature on filters and mollifiers as effective tools for Gibbs-free reconstruction of piecewise smooth functions. Different aspects of this topic are drawn from a variety of sources, ranging from summability

---

[5] Observe that $\mathcal{S}_N^\varphi(x)$ is the mollifier function associated with, but otherwise different from, the filtered sum $S_N^\varphi f$.

methods in harmonic analysis to signal processing – and, in recent years, image processing – and high-resolution spectral computations of propagation of singularities and shock discontinuities.

Classical mollifiers of finite polynomial order, $\mathcal{O}(N^{-p})$, are dictated by a moment condition of order p, (10.1), discussed in Section 10 below. By properly tuning $p = p_N$ to increase with N, one obtains *spectrally accurate* mollifiers (Gottlieb and Tadmor 1985) and spectrally accurate filters (Majda, McDonough and Osher 1978, Vandeven 1991). Improved results are obtained by a further adaptation of p_N to the distance from the edges (Boyd 1995, 1996). By carefully tuning p_N together with proper G_2 cut-off functions, we obtain improved *root-exponential* accurate mollifiers (Tadmor and Tanner 2002), which are discussed in Section 10.2 below, together with the corresponding discrete mollifiers in Section 10.3. The root-exponential accuracy of these mollifiers is adapted to the *interior* points, away from the vicinity of the edges. It can be modified to gain polynomial accuracy *up to* the edges; the details are outlined in Section 10.4. Finally, in Section 10.5 we discuss mollifiers based on Gegenbauer expansion (Gottlieb, Shu, Solomonoff and Vandeven 1992, Gottlieb and Shu 1998, Gelb and Tanner 2006), with uniform root-exponential accuracy *up to* the edges. In Section 11 we revisit the construction of accurate mollifiers based on the corresponding filters. We conclude in Section 11.2 with *exponentially* accurate mollifiers (Tanner 2006) based on optimally space–frequency-localized filters.

We now turn to discuss those mollifiers and filters which enable highly accurate, Gibbs-free reconstruction of f from its (pseudo-) spectral content.

10. Spectral mollifiers

10.1. Compactly supported mollifiers

We begin with classical compactly supported mollifiers. Fix $p < q$ and let $\Phi = \Phi_p \in C_0^q(-\pi, \pi)$ be a unit mass kernel which possesses $p - 1$ *vanishing moments*,

$$\int_{-\pi}^{\pi} x^n \Phi(x)\, dx = \begin{cases} 1, & n = 0, \\ 0, & n = 1, \dots, p. \end{cases} \tag{10.1}$$

Example 10.1. (Mollifiers satisfying the moment conditions) It is easy to construct such Φ satisfying the moment constraints for small p. As an example for arbitrary p, we can set Φ_p to be the w_α-weighted Gegenbauer polynomial of degree p (see (6.12)),

$$\Phi_p(x) = c_{\alpha,p} \left(1 - \left(\frac{x}{\pi}\right)^2\right)^{\alpha - \frac{1}{2}} C_p^{(\alpha)}\left(\frac{x}{\pi}\right) 1_{(-\pi,\pi)}(x), \quad \alpha > q.$$

(Clearly, such a $\Phi_p(x)$ is a C_0^p-function, which can be normalized to have a unit mass by a proper choice of $c_{\alpha,p}$.) The ω_α-orthogonality of the $C_k^{(\alpha)}$ (see (6.11)) implies that $\Phi_p^{(\alpha)}$ satisfies the moment conditions (10.1).

Next, given a mollifier $\Phi(x) = \Phi_p(x)$ satisfying (10.1), we form the family of *dilated mollifiers*

$$\Phi_{p,\delta}(x) := \frac{1}{\delta} \Phi_p\left(\frac{x}{\delta}\right),$$

with δ being a free dilation parameter at our disposal; by tuning δ we can adjust the support of $\Phi_{p,\delta}$ over the symmetric interval $(-\pi\delta, \pi\delta)$. Observe that $\Phi_{p,\delta}$ retains the same p vanishing moments (10.1) over its restricted support $(-\pi\delta, \pi\delta)$. To reconstruct f from its spectral projection, we consider the mollified Fourier projection

$$\Phi_{p,\delta} * (S_N f)(x) \approx f(x).$$

We now examine the error $\Phi_{p,\delta} * (S_N f) - f$. By orthogonality, $\Phi_{p,\delta} * (S_N f) = S_N(\Phi_{p,\delta}) * f$, hence we can express the error as the sum of two terms:

$$\Phi_{p,\delta} * (S_N f)(x) - f(x) \equiv \overbrace{\left(S_N(\Phi_{p,\delta}) - \Phi_{p,\delta}\right) * f(x)}^{\text{truncation} = T_N(\Phi_{p,\delta}) * f} + \overbrace{\left(\Phi_{p,\delta} * f(x) - f(x)\right)}^{\text{regularization}}.$$
$$(10.2)$$

The first term on the right is the usual *truncation error*, which, by (2.2), does not exceed

$$|T_N(\Phi_{p,\delta}) * f(x)| \lesssim \|f\|_{L^1} \cdot \|\Phi_{p,\delta}\|_{C^q} \frac{1}{N^{q-1}} \lesssim \alpha_{p,q} \frac{1}{\delta^{q+1} N^{q-1}}, \qquad (10.3a)$$

$$\text{where } \alpha_{p,q} = \|\Phi_p\|_{C^q}.$$

The second term on the right of (10.2) represents the *regularization error*

$$\Phi_{p,\delta} * f(x) - f(x) = \int_{-\pi}^{\pi} \left[f(x - \delta y) - f(x)\right] \Phi_p(y) \, dy.$$

It does not involve any spectral content of f but depends solely on the regularity of f in the interval $(x - \pi\delta, x + \pi\delta)$. The moment conditions imply that Φ_p is orthogonal to the first p terms in the Taylor expansion of

$$f(x - \delta y) - f(x) = \sum_{n=1}^{p} \frac{(-1)^n}{n!} \delta^n f^{(n)}(x) y^n + \frac{1}{(p+1)!} \delta^{(p+1)} f^{(p+1)}(\cdots) y^{p+1},$$

and we are left with the following bound on the *regularization error*:

$$|\Phi_{p,\delta} * f(x) - f(x)| \lesssim \beta_p \|f\|_{C^{p+1}(x-\delta,x+\delta)} \delta^{p+1}, \qquad (10.3b)$$

$$\text{where } \beta_p = \frac{1}{p!} \int_{-\pi}^{\pi} |y|^p |\Phi_p(y)| \, dy.$$

There are two ways to make both error bounds (10.3a) and (10.3b) small.

(i) Fix $p \sim q$ and set $\delta = \delta_N \sim 1/\sqrt{N}$. In this case, we recover $f(x)$ from the δ_N-neighbourhood of its spectral projection $S_N f(x)$:

$$|\Phi_{p,\delta_N} * (S_N f)(x) - f(x)| \lesssim \gamma_p \big(1 + \|f\|_{C^p(x-\delta_N, x+\delta_N)}\big) \frac{1}{N^{p/2}}, \qquad \delta_N = \frac{1}{\sqrt{N}}.$$

Here, $\gamma_p = \alpha_{p,p} + \beta_p$. This yields the locally supported mollifiers, Φ_{p,δ_N}. Their finite accuracy is determined by the finitely many vanishing moments in (10.1). To gain spectral accuracy requires an increasingly smooth mollifier Φ_p with an increasing number of (almost) vanishing moments. The result is a family of *local* mollifiers, Φ_{p_N, δ_N}, whose degree $p = p_N$ is adjusted as an increasing function of N. Local mollifiers do not make use of all the information available in the interval of smoothness enclosing x, and are therefore replaced by global mollifiers.

(ii) Fix δ by setting

$$\delta = d_x, \quad d_x := \frac{1}{\pi} \operatorname{dist}\{x, \{c_1, \ldots, c_J\}\}[\operatorname{mod} \pi], \qquad (10.4)$$

so that $(x - \pi\delta, x + \pi\delta)$ is the largest interval of smoothness enclosing x. We pause here to make the following remark.

Remark 10.2. Note that d_x can be calculated from the given (pseudo-) spectral data. It is here that we use the information about the edges $\{c_1, \ldots, c_J\}$ detected from $S_N f$ and $I_N f$.

Once we set $\delta = d_x$, we let p vary as an increasing function of N: by choosing $p = p_N$ (which necessarily has an increasing order of smoothness, $q_N > p_N$), we can try to enforce a spectrally small β_{p_N} in (10.3b) while balancing a spectrally small ratio $\alpha_{p_N, q_N} N^{1-q_N}$ in (10.3a). This balancing act depends of course on a careful study of the asymptotic behaviour of $\|f\|_{C^p}$ and Φ_p, as p increases. The result is a family of *adaptive* mollifiers, Φ_{p_N, d_x}, whose degree $p = p_N$ is adapted as an increasing function of N, while their support, d_x, is adapted to the largest interval of smoothness enclosing x. The Φ_{p_N, d_x} are *global* mollifiers; they achieve (root-) exponential convergence rate by *cancellation*.

10.2. Adaptive mollifiers: root-exponential accuracy

Following Gottlieb and Tadmor (1985), we consider the compactly supported mollifiers

$$\Phi_p(x) := \rho_2(x) D_p(x), \qquad \rho_2(x) = e^{\left(\frac{cx^2}{x^2 - \pi^2}\right)} 1_{(-\pi, \pi)}(x), \qquad (10.5)$$

where ρ_2 is the G_2 cut-off function imported from (2.5) to *localize* the Dirichlet kernel, D_p. Recall that, after dilation, this family of mollifiers takes the form

$$\Phi_{p,d_x}(x) = \frac{1}{d_x}\rho_2\left(\frac{x}{d_x}\right)D_p\left(\frac{x}{d_x}\right),$$

where d_x is set by (10.4) and the degree p is at our disposal.

Let us estimate the error, $\Phi_{p,d_x} * (S_N f)(x) - f(x)$. Following (10.2), we proceed in two steps, starting with the *truncation error*, $T_N(\Phi_{p,d_x})(x)$. According to (10.3a), its decay is controlled by the C^q-regularity of Φ_{p,d_x}. As the product of a G_2-function and the analytic Dirichlet kernel, we deduce $\Phi_{p,d_x} \in G_2$. But we still need to quantify the dependence of its G_2-bound on *both* p and d_x, as we are going to let p increase with N. To this end, we use the Leibniz rule and (2.6):

$$|\Phi_p^{(s)}(x)| \le \sum_{k=0}^{s}\binom{s}{k}|\rho_2^{(k)}(x)| \cdot |D_p^{(s-k)}(x)|$$

$$\lesssim s!\left(\sum_{k=0}^{s}\frac{p^{s-k}}{(s-k)!}\frac{1}{(\eta|x^2-\pi^2|)^k}\right) \cdot e^{\left(\frac{cx^2}{x^2-\pi^2}\right)}$$

$$\lesssim \frac{s!}{(\lambda_\rho|x^2-\pi^2|)^s}c^{\left(p\lambda_\rho|x^2-\pi^2|+\frac{cx^2}{r^2-\pi^2}\right)},$$

which after dilation reads

$$|\Phi_{p,d_x}^{(s)}(x)| \lesssim s!\left(\frac{d_x}{|c(x)|}\right)^s \cdot e^{\left(\frac{p|c(x)|}{d_x^2}+\frac{c\lambda_\rho x^2}{c(x)}\right)}, \quad c(x) := \lambda_\rho\left(x^2-\pi^2 d_x^2\right).$$

The upper bound on the right-hand side is maximized at $x = x_{max}$ with $x_{max}^2 - \pi^2 d_x^2 \sim -c\pi^2 d_x^2/s$, which leads to the G_2-regularity bound for Φ_{p,d_x} (here, $\eta := c\lambda_\rho\pi^2$):

$$\sup_{x\in(-1,1)}|\Phi_{p,d_x}^{(s)}(x)| \lesssim s!\left(\frac{s}{\eta d_x e}\right)^s e^{\left(\frac{p\eta}{s}\right)} \lesssim \frac{(s!)^2}{(\eta d_x)^s}e^{\left(\frac{p\eta}{s}\right)} \quad s = 1, 2, \dots. \quad (10.6)$$

Equipped with (10.6), we find that the truncation error (10.3a) does not exceed

$$|T_N(\Phi)_{p,d_x}(x)| \lesssim \frac{(s!)^2}{(\eta d_x N)^s}e^{\left(\frac{p\eta}{s}\right)} \sim \left(\frac{s^2}{\eta d_x e^2 N}\right)^s e^{\left(\frac{p\eta}{s}\right)} =: M(s,p), \quad (10.7)$$

for all $s > 1$. We seek the minimizer, $s = s_{min}$, such that

$$\partial_s(\log M(s,p))|_{s=s_{min}} = \log\left(\frac{s_{min}^2}{\eta d_x N}\right) - \frac{p\eta}{s_{min}^2} = 0.$$

This yields a rather precise bound on s_{min} which turns out to be essentially independent of p. Indeed, for the first expression on the right to be positive we set $s_{min} = \sqrt{\beta\eta d_x N}$ with a free $\beta > 1$ at our disposal. The corresponding optimized $p = p_N(x) = \frac{s^2}{\eta} \log \frac{s^2}{\eta \delta N}_{|s=s_{\min}}$ amounts to

$$p_N(x) = \beta \log \beta \cdot d_x N. \qquad (10.8)$$

We conclude with an optimized choice of $p = p_N(x)$ of order $\mathcal{O}(d_x N)$, which is *adapted* to the distance between x and the singular support of f. The resulting exponentially small truncation error bound, (10.7), now reads

$$|T_N(\Phi_{p_N, d_x})(x)| \lesssim \frac{(s!)^2}{(\eta d_x N)^s} e^{\left(\frac{p_N \eta}{s}\right)}_{|s=\sqrt{\beta \eta d_x N}} \sim \sqrt{d_x N} \left(\frac{\beta}{e}\right)^{2\sqrt{\beta \eta d_x N}}.$$

$$(10.9a)$$

In the second step, we turn our attention to the *regularization error*

$$\Phi_{p, d_x} * f(x) - f(x)$$

$$= \int_{-\pi}^{\pi} \overbrace{f(x - d_x y) \rho_2(y)}^{F(y;x)} D_p(y) \, dy - f(x) \equiv S_p F(y; x)_{|y=0} - F(0; x)$$

in (10.2). Assume that $f(\cdot)$ is piecewise smooth: to simplify matters, we match its piecewise smoothness with that of ρ_2, assuming that f is a piecewise-G_2 function. Our choice of $\delta = d_x$ in (10.4) guarantees that $f(x - \delta y)$ is G_2 in the range $|y| \leq \pi$ and, hence, so is its product with $\rho_2(y)$, implying the G_2-regularity of $F(y; \cdot) = f(\cdot - d_x y)\rho_2(y)$. When dealing with local mollifiers, we use the moment conditions (10.1) to bound their regularization error in (10.3b). Instead, dealing with global mollifiers, we now use the global root-exponential decay (2.7b), whence $\eta = \eta_{\rho, f}$ such that

$$|\Phi_{p_N, d_x} * f(x) - f(x)| = |(S_p F(y; x) - F(y; x))_{|y=0}| \lesssim p \, e^{-\alpha\sqrt{\eta p}}.$$

The same choice of an adaptive $p = p_N$ made in (10.8) yields essentially the same exponentially small bound on the regularization error,

$$|\Phi_{p_N, d_x} * f(x) - f(x)| \lesssim d_x N \cdot e^{-2\sqrt{\beta \log \beta \cdot \eta d_x N}}. \qquad (10.9b)$$

The free β can now be further optimized by equilibrating the truncation and regularization error bounds (10.9a) and (10.9b), whence $\beta_{\text{opt}} \log \beta_{\text{opt}} \sim 1/\sqrt{e}$. We summarize with the following theorem.

Theorem 10.3. (Root-exponential accurate mollifiers; Tadmor and Tanner 2002) Given the Fourier projection, $S_N f(\cdot)$, of a piecewise smooth function, $f(\cdot) \in$ piecewise-G_2, we consider the 2-parameter family of spectral mollifiers

$$\Phi_{p, \delta}(x) := \frac{1}{\delta} \rho_2\left(\frac{x}{\delta}\right) D_p\left(\frac{x}{\delta}\right), \quad \rho_2 := e^{\left(\frac{cx^2}{x^2 - \pi^2}\right)} 1_{(-\pi, \pi)}(x), \quad c > 0. \quad (10.10a)$$

Fix x inside one of the smoothness intervals of f and set the adaptive

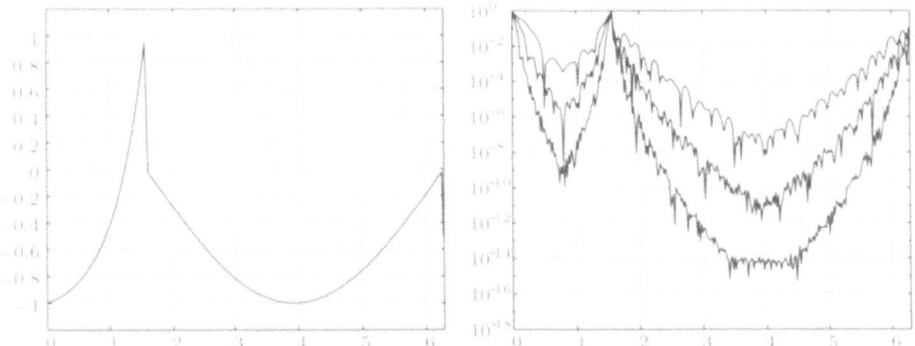

Figure 10.1. *Left*: Function $f(x)$ in (10.12). *Right*: Log-error of its reconstruction from $S_N f$, $N = 32, 64, 128$ using the adaptive mollifier $\Phi_{p_N,\delta}$ in (10.10) with $p_N = d_x N/\sqrt{e}$.

parametrization

$$\delta = d_x := \frac{1}{\pi}\,\mathrm{dist}\{x,\{c_1,\dots,c_J\}\}[\mathrm{mod}\ \pi], \tag{10.10b}$$

$$p = p_N(x) \sim d_x N/\sqrt{e}. \tag{10.10c}$$

Then, there exists a constant $\eta = \eta_{\rho,f}$ such that $\Phi_{p_N,d_x} * S_N f$ recovers $f(x)$ with the following root-exponential accuracy:

$$|\Phi_{p_N,d_x} * (S_N f)(x) - f(x)| \lesssim d_x N e^{-0.84\sqrt{\eta d_x N}}. \tag{10.11}$$

Figure 10.1 illustrates the reconstruction of

$$f(x) = \begin{cases} (2e^{2x} - 1 - e^{\pi})/(e^{\pi} - 1), & x \in [0, \pi/2), \\ -\sin(2x/3 - \pi/3), & x \in [\pi/2, 2\pi), \end{cases} \tag{10.12}$$

using the mollifier (10.10). Although the error estimates which lead to Theorem 10.3 serve only as upper bounds for the errors, it is still remarkable that the (close to) optimal parametrization of the adaptive mollifier is found to be essentially independent of the properties of $f(\cdot)$.

We conclude this section with several remarks on the root-exponential accuracy behind the mollifiers Φ_{p_N,d_x} in (10.10).

Remark 10.4. (Spectral *vs* root-exponential decay) The two-parameter family of mollifiers, $\Phi_{p,\delta}$, was introduced by Gottlieb and Tadmor (1985). They used *spectral decay* bounds of the regularization and truncation errors,

$$|\Phi_{p,\delta} * f(x) - f(x)| \lesssim \|\rho_2\|_{C^s}\|f\|_{C^s(x-\delta,x+\delta)}\left(\frac{2}{p}\right)^s,$$

$$|T_N \Phi_{p,\delta}(x)| \lesssim \|\rho_2\|_{C^s}\left(\frac{1+p}{\delta N}\right)^s,$$

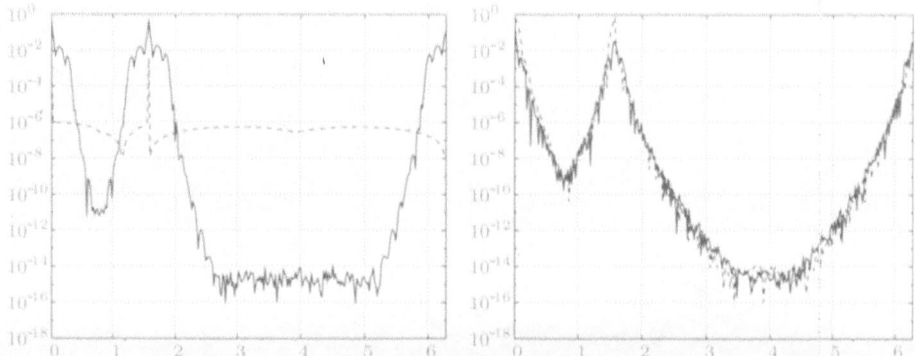

Figure 10.2. *Left*: Regularization errors (dashed) and
truncation errors (solid) using the spectral mollifier $\Phi_{p,\delta}$ of
degree $p \sim \sqrt{N}$ with δ. *Right*: The same using the adaptive
mollifier Φ_{p_N,d_x} with $p_N = d_x N/\sqrt{e}$.

which led to the spectral decay

$$\left| \Phi_{p,\delta} * S_N f(x) - f(x) \right|_{p \sim \sqrt{N}} \leq \mathrm{Const}_{s,d_x} \frac{1}{N^{s/2}}. \tag{10.13}$$

Although this estimate yields the desired spectral convergence rate sought
for by Gottlieb and Tadmor (1985), it suffers from coupling the regularization and truncation through the *same* dependence of p on s and on δ, which
in turn leads to their balance at the pessimistic estimate of $p_N \sim \sqrt{N}$. Indeed, the results in Figure 10.2 clearly indicate that the contributions of
the truncation and regularization terms are equilibrated only when $p \sim N$.
Moreover, Figure 10.2 illustrates that a non-adaptive choice of $p = p_N$
which is independent of d_x, e.g., $p \sim \sqrt{N}$, leads to a loss of convergence
in a larger $\mathcal{O}(N^{-1/2})$-neighbourhood of the discontinuity, compared with
the adaptive parametrization $p_N \sim d_x N$, which achieves, in Figure 10.1(b),
root-exponential accuracy up to the immediate, $\mathcal{O}(1/N)$, vicinity of these
discontinuities.

Remark 10.5. (Piecewise smooth f)   The root-exponential error estimate (10.11) originates with the (piecewise-) smoothness of f and ρ_2 measures in G_2. Similar results apply in general cases when $f \in$ piecewise-G_α
or $f \in$ piecewise-C^s. In this case, the convergence rate is, respectively,
root-α exponential and s-order polynomial.

Remark 10.6. (Gevrey regularity)   Optimal parametrization of Φ_{p_N,d_x}
depends in an essential way on the *Gevrey regularity* of the cut-off function
$\rho_2(\cdot) \in G_2$. This G_2-regularity is reflected, for example, in the log-error in
Figure 10.1.

Remark 10.7. (Approximate moment conditions) To achieve root-exponential accuracy, we give up the *exact* moment conditions. Instead, it is satisfied up to an exponentially small error (see (10.17)). Relaxing the side constraints imposed on an 'approximate identity' will turn out to be a key feature which enables us to maintain (root-) exponential accuracy.

10.3. Root-exponential accurate reconstruction of pseudo-spectral data

We are interested in reconstruction of a piecewise smooth $f(x)$ from its Fourier interpolant, $I_N f(x)$. To this end we consider the discrete convolution

$$\Phi_{p_N, d_x} * I_N f(x) = h \sum_{\nu=0}^{2N} \Phi_{p_N, d_x}(x - y_\nu) f(y_\nu), \qquad (10.14)$$

corresponding to (10.11). The overall error, $\Phi_{p_N, d_x} * I_N f(x) - f(x)$, consists of aliasing and regularization errors. According to (2.18), the former is upper-bounded by the truncation of f', which retains the same piecewise analyticity properties as f does. We conclude with the following result.

Theorem 10.8. (Reconstruction of piecewise smooth discrete data; Tadmor and Tanner 2002) Given equidistant grid values, $\{f(x_\nu)\}_{\nu=0}^{2N}$ of $f(\cdot) \in$ piecewise-G_2, we consider the spectral mollifiers (10.10),

$$\Phi_{p,\delta}(x) := \frac{1}{\delta} \rho_2\left(\frac{x}{\delta}\right) D_p\left(\frac{x}{\delta}\right), \qquad \rho_2 := e^{\left(\frac{cx^2}{x^2 - \pi^2}\right)} 1_{(-\pi,\pi)}(x), \ c > 0,$$

with adaptive parametrization, $\delta = d_x := \frac{1}{\pi} \text{dist}\{x, \{c_1, \ldots, c_J\}\}[\text{mod } \pi]$ and $p_N(x) \sim d_x N / \sqrt{e}$. Then, there exists a constant $\eta = \eta_{\rho,f}$, such that the discrete convolution (10.14) recovers $f(x)$ with the following root-exponential accuracy:

$$\left| h \sum_{\nu=0}^{2N} \Phi_{p_N, d_x}(x - y_\nu) f(y_\nu) - f(x) \right| \lesssim (d_x N)^2 e^{-0.84\sqrt{\eta d_x N}}. \qquad (10.15)$$

Observe that, by forming the discrete convolution (10.14), we completely bypass the need to compute the discrete Fourier coefficients $\widehat{f}_k$. Instead, we recover $f(x)$ with root-exponential accuracy directly from the given grid values in the d_x-smooth neighbourhood enclosing x,

$$\{f(y_\nu) \mid |x - y_\nu| \le d_x\}.$$

Thus, by relaxing the property of exact interpolation, $I_N f(x)|_{x=x_\nu} = f(x_\nu)$, the Fourier interpolant, $I_N f(x)$, is replaced here by what we might call the Fourier 'expolant', $\Phi_{p_N, d_x} * I_N f$, a root-exponentially accurate approximant which recovers smooth, as well as piecewise smooth, functions.

10.4. Polynomial accuracy up to the edges

What happens in the neighbourhood of jump discontinuities where $d_x N \sim 1$? We observe that the error bound (10.11) is of order $\mathcal{O}(1)$ as reflected in Figure 10.1. To understand the source of this loss of accuracy, we note that the two ingredients involved in $\Phi_{p,\delta}$, namely, $\rho_2(x)$ and $D_p(x)$, have essentially different roles, associated with the two independent parameters δ and p: the role of $\rho_2(\frac{x}{\delta})$ is to *localize* the support of $\Phi_{p,\delta}(x)$ to the δ-neighbourhood of x; the Dirichlet kernel $D_p(x)$ is charged, by varying p, with controlling the increasing number of *near-vanishing moments* of $\Phi_{p,\delta}$, and hence the overall superior accuracy of our mollifier. Indeed, since

$$\rho(0) = 1, \tag{10.16}$$

we find that the moments of Φ_{p_N,d_x} are of order

$$\int_{-\pi d_x}^{\pi d_x} y^n \Phi_{p_N,d_x}(y) \, dy \tag{10.17}$$

$$= \int_{-\pi}^{\pi} (y d_x)^n \rho_2(y) D_{p_N}(y) \, dy = (d_x)^n D_{p_N} * (y^n \rho_2)(y)|_{y=0}$$

$$\approx \delta_{n0} + (d_x)^n \inf_n \left\{ \|y^n \rho_2(y)\|_{C^n} \frac{1}{p_N^{n-1}} \right\} \approx \delta_{n0} + (d_x)^n e^{-\sqrt{\eta d_x N}}.$$

Consequently, Φ_{p_N,d_x} possesses exponentially small moments at all x, except for the immediate vicinity of the jumps where $d_x N \sim 1$, the same $\mathcal{O}(1/N)$ neighbourhoods where the error bound (10.11) is of order $\mathcal{O}(1)$. To enforce a faster convergence in these neighbourhoods of the jumps, we ask that finitely many moments of $S_N \Phi_{p_N,d_x}$ vanish *exactly*,

$$\int_{-\pi}^{\pi} y^n (S_N \Phi_{p_N,d_x})(y) \, dy = \begin{cases} 1, & n = 0, \\ 0, & n = 1, 2, \ldots, r. \end{cases}$$

This amounts to the vanishing moment constraint

$$\overbrace{\int_{-\pi}^{\pi} \Phi_{p_N}(y) \, dy = 1}^{\text{unit mass}}, \quad \overbrace{\int_{-\pi}^{\pi} S_N(y^n) \Phi_{p_N}(y) \, dy = 0}^{\text{vanishing moments}}, \quad n = 1, 2, \ldots, r. \tag{10.18}$$

It follows that adaptive mollifiers satisfying (10.18) recover $f(x)$ with the desired polynomial order $\mathcal{O}(d_x)^r$, *i.e.*,

$$\Phi_{p_N,d_x} * S_N f(x) = f(x) + \log(p_N) \mathcal{O}(d_x)^{r+1}.$$

The point to note here is that this error estimate holds *up to the edges*. Indeed, noting that, for each x, the function $f(x - d_x y)$ remains smooth in the neighbourhood $|y| \leq \pi$, the vanishing moments (10.18) imply

$$\Phi_{p_N, d_x} * S_N f(x) - f(x)$$

$$= \int_{-d_x}^{d_x} \Phi_{p_N, d_x} S_N f(x - y) \, dy - f(x)$$

$$= \int_{-\pi}^{\pi} \Phi_{p_N}(y) S_N f(x - d_x y) \, dy - f(x)$$

$$= \int_{-\pi}^{\pi} [f(x - d_x y) - f(x)](S_N \Phi_{p_N})(y) \, dy$$

$$\sim (d_x)^{r+1} \int_{-\pi}^{\pi} S_N(y^{r+1}) \Phi_{p_N}(y) \, dy \lesssim \log(p_N)(d_x)^{r+1}.$$

To enforce (10.18) we modify the cut-off ρ_2, setting

$$\widetilde{\rho}_2(x) = \mathcal{M}_0 \rho_2(x), \qquad \mathcal{M}_0 := \frac{1}{\int_{-\pi}^{\pi} \Phi_{p_N}(y) \, dy}. \qquad (10.19a)$$

Observe that this normalizes $\widetilde{\rho}_2$ so that

$$\widetilde{\Phi}_{p_N, d_x} := \frac{1}{d_x} \left(\widetilde{\rho}_2 \left(\frac{x}{d_x} \right) D_{p_N} \left(\frac{x}{d_x} \right), \qquad \widetilde{\rho}_2(x) - \mathcal{M}_0 \rho_2(x), \qquad (10.19b)$$

has a unit mass and hence (10.18) holds, at the expense of an exponentially negligible rescaling of ρ_2 in (10.16):

$$\widetilde{\rho}_2(0) = \mathcal{M}_0 = \frac{1}{(D_{p_N} * \rho_2)(0)} = 1 + d_x N e^{-2\sqrt{\eta p_N}}, \qquad p_N \sim d_x N / \sqrt{e}.$$

Moreover, since Φ_{p_N, d_x} is even, its odd moments vanish, *i.e.*, (10.18) holds for $r = 1$. We end up with the following corollary.

Corollary 10.9. (Uniformly quadratic, root-exponential mollifiers)
The adaptive mollifier $\widetilde{\Phi}_{p_N, d_x}$ in (10.19) recovers piecewise smooth $f(x)$ with root-exponential accuracy at interior points of smoothness, and with quadratic accuracy in the vicinity of jump discontinuities, *i.e.*,

$$\left| \widetilde{\Phi}_{p_N, d_x}(x) * (S_N f)(x) - f(x) \right| \lesssim \log(d_x N)(d_x)^2 \cdot e^{-0.84\sqrt{\eta d_x N}}. \qquad (10.20)$$

In a similar manner, we can enforce higher vanishing moments by proper *normalization* of the cut-off $\rho_2(\cdot)$. To enforce (10.18) with $r = 2$, for example, we use a rescaled cut-off, $\widetilde{\rho}_2(x)$, given by

$$\widetilde{\rho}_2(x) = \mathcal{M}_2 \rho_2(x), \qquad \mathcal{M}_2(x) := \frac{m_2(x)}{\int_{-\pi}^{\pi} m_2(y) \Phi_{p_N}(y) \, dy}, \qquad (10.21a)$$

where

$$m_2(x) = 1 + a_2 x^2, \qquad a_2 := \frac{-\int_{-\pi}^{\pi} S_N(y^2) \Phi_{p_N}(y) \, dy}{\int_{-\pi}^{\pi} S_N(y^2) y^2 \Phi_{p_N}(y) \, dy}. \qquad (10.21b)$$

As before, the resulting mollifier

$$\widetilde{\Phi}_{p_N, d_x}(x) := \frac{1}{d_x} \tilde{\rho}_2 \left(\frac{x}{d_x} \right) D_{p_N} \left(\frac{x}{d_x} \right), \qquad \tilde{\rho}_2(x) = \mathcal{M}_2(x) \rho_2(x), \quad (10.21c)$$

is admissible in the sense of satisfying the normalization (10.16) modulo an exponentially small error term, since the pre-factor $\mathcal{M}_2(0) - 1$ is equally negligible. Now, $\widetilde{\Phi}_{p_N, d_x}$ has a unit mass; moreover, the S_N-projection of $\widetilde{\Phi}_{p_N, d_x}$ satisfies the *exact* second vanishing moment (10.18) with $r = 2$, for

$$\int_{-\pi}^{\pi} S_N \left(y^2 \right) \left(1 + a_2 y^2 \right) \Phi_{p_N}(y) \, dy$$

$$= \int_{-\pi}^{\pi} S_N \left(y^2 \right) \Phi_{p_N}(y) \, dy + a_2 \int_{-\pi}^{\pi} S_N \left(y^2 \right) y^2 \Phi_{p_N}(y) \, dy = 0.$$

Finally, since $S_N \widetilde{\Phi}_{p_N, d_x}$ is even, its third moment vanishes as well, which implies the normalized mollifier (10.21).

Corollary 10.10. (Uniformly quartic, root-exponential mollifiers) The adaptive mollifier $\widetilde{\Phi}_{p_N, d_x}$ in (10.21) recovers piecewise smooth $f(x)$ with root-exponential accuracy at interior points of smoothness while maintaining a fourth-order convergence rate in the immediate vicinity of the jump discontinuities,

$$\left| \widetilde{\Phi}_{p_N, d_x}(x) * (S_N f)(x) - f(x) \right| \lesssim \log(d_x N)(d_x)^4 \cdot e^{-0.84 \sqrt{\eta d_x N}}. \quad (10.22)$$

We can implement a similar upgrade up to the edges in the discrete case. To this end, we consider the normalized mollifier

$$\widetilde{\Phi}_{p_N, d_x}(x) = \frac{1}{d_x} \tilde{\rho}_2 \left(\frac{x}{d_x} \right) D_{p_N} \left(\frac{x}{d_x} \right), \qquad \tilde{\rho}_2(x) := \mathcal{M}_r(x) \rho_2(x). \quad (10.23a)$$

Here, $\mathcal{M}_r(x)$ is a pre-factor of the form

$$\mathcal{M}_r(x) = \frac{m_r(x)}{\sum_\nu m_r \left(\frac{x - y_\nu}{d_x} \right) \Phi_{p_N, d_x}(x - y_\nu) h}, \qquad m_r(x) = 1 + a_1 x + \cdots + a_r x^r,$$

$$(10.23b)$$

whose r free parameters are sought so that the first r *discrete* moments of $\widetilde{\Phi}_{p_N, d_x}(y)$ vanish,

$$\sum_{\{y_\nu \, : \, |x - y_\nu| \leq d_x\}} (x - y_\nu)^n \Phi_{p_N, d_x}(x - y_\nu) h = \begin{cases} 0, & n = 0, \\ 0, & n = 1, \ldots, r. \end{cases} \quad (10.23c)$$

Observe that, unlike the moment constraint (10.18) associated with the continuous case, the discrete constraint (10.23c) is not translation-invariant

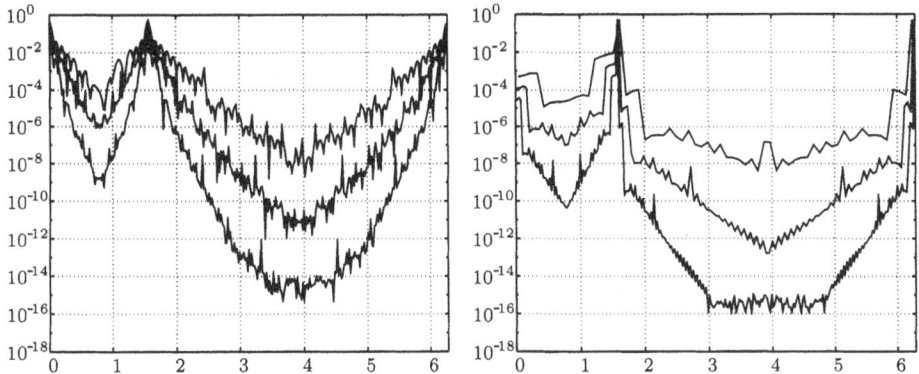

Figure 10.3. *Left*: Log-error of reconstructing f from $S_N f$, $N = 32, 64, 128$ using the 4th-order normalized adaptive mollifier (10.21c) with $p_N = d_x N/\sqrt{e}$. *Right*: Reconstruction of f from its discrete data, $I_N f$, $N = 32, 64, 128$, using the 4th-order normalized mollifier (10.23).

and hence requires x-dependent normalizations. The additional computational effort is minimal, however, due to the discrete summations which are localized in the immediate vicinity of x.

Then, (10.15) is replaced by the improved error estimate

$$\left| h \sum_{\nu=0}^{2N} \widetilde{\Phi}_{p_N, d_x}(x - y_\nu) f(y_\nu) - f(x) \right| \lesssim (d_x)^{r+1} e^{-0.84\sqrt{\eta d_x N}}, \quad r \sim N d_x. \quad (10.24)$$

Figure 10.3 illustrates the improvement using the normalized adaptive mollifier (10.21) and its discrete version (10.23), in reconstructing the same $f(x)$ used in (10.12):

$$f(x) = \begin{cases} (2e^{2x} - 1 - e^\pi)/(e^\pi - 1), & x \in [0, \pi/2), \\ -\sin(2x/3 - \pi/3), & x \in [\pi/2, 2\pi). \end{cases}$$

Compared with the adaptive mollifier (10.10a) used in Figure 10.1, the improvement of the error *up to* the edges is evident.

10.5. *Gegenbauer-based mollifiers: exponential accuracy up to the edges*

We want to recover the values of a piecewise analytic $f(x)$ inside each interval of smoothness, with exponential accuracy, *uniformly* in $x \in (c_{j-1}, c_j)$, $j = 1, \dots, J$. After a proper translation and dilation of each interval, we may assume that f experiences a single jump discontinuity at $|x| = \pi$ and we seek exponential recovery of $f(x)$, $|x| \leq \pi$ *up to* the boundary. We have now come full circle back to our starting point, the Gegenbauer polynomials $C_k^{(\alpha)}(x)$, which formed the moment-satisfying mollifiers in Example 10.1.

Let

$$G_N^{(\alpha)} f(x) := \sum_{k=0}^{N} \langle f, C_k^{(\alpha)} \rangle_{\omega_\alpha} C_k^{(\alpha)}(x)$$

denote the truncated Gegenbauer expansion of $f(x)$, $x \in (-1,1)$, where $\langle f, C_k^{(\alpha)} \rangle_{\omega_\alpha}$ are normalized moments of f with respect to the weight function $\omega_\alpha(x) = (1-x^2)^{\alpha-\frac{1}{2}}$. The *Gegenbauer reconstruction* of f (see Gottlieb and Shu (1998) and the references therein) is the reprojection of $S_N f(x)$,

$$G_p^{(\alpha)}(S_N f)_\pi(x), \qquad g_\pi := g(\pi x),$$

with a proper parametrization of $p = p_N$ and $\alpha = \alpha_N$.

Remark 10.11. Observe that the Gegenbauer reconstruction can be evaluated in terms of a (non-translatory) convolution with the corresponding Christoffel–Darboux mollifier (6.15),

$$G_p^{(\alpha)}(S_N f)_\pi(x) \sim \int_{-1}^{1} K_p^{(\alpha)}(x,y)(S_N f)_\pi(y)\, dy.$$

To determine these parameters, we upper-bound the error in the standard fashion (10.2), by the sum of regularization and truncation errors,

$$G_p^{(\alpha)}(S_N f)_\pi(x) - f_\pi(x) = \overbrace{G_p^{(\alpha)} f_\pi(x) - f_\pi(x)}^{\text{regularization}} + \overbrace{G_p^{(\alpha)}(S_N f)_\pi(x) - G_p^{(\alpha)} f_\pi(x)}^{\text{truncation}}.$$

The regularization error does not involve any spectral information of f, but depends solely on the regularity of $f(x)$ over the interval $(-\pi, \pi)$. Since f_π is assumed to be analytic inside $(-1,1)$, its Gegenbauer projection is exponentially accurate *up to* the boundary,

$$|G_p^{(\alpha)} f_\pi(x) - f_\pi(x)| \le c_\alpha e^{-\eta p}, \qquad \text{for all } x \in (-1,1). \qquad (10.25)$$

We now come to the truncation error which was shown to be upper-bounded by Gottlieb, Shu, Solomonoff and Vandeven (1992):

$$\left\| G_p^{(\alpha)}(S_N f)_\pi(x) - G_p^{(\alpha)} f_\pi(x) \right\|_{L^\infty(-1,1)} \lesssim \left(\frac{\eta p}{N} \right)^\alpha.$$

Thus, to upgrade this polynomial decay in N, one has to increase $\alpha = \alpha_N$ while carefully balancing the growth of c_{α_N} in (10.25) by adjusting $p = p_N$. To this end, one sets $\alpha = \theta p \sim N$, to obtain exponentially small regularization and truncation errors (Gottlieb and Shu 1997).

The superior accuracy of the resulting Gegenbauer reconstruction is illustrated in Figure 10.4, from Gelb and Gottlieb (2007). It comes with a price, however: a sufficiently small θ needs to be *carefully tuned* (e.g., Gottlieb and Shu (1997)) so that $\theta^{-\theta}\eta \lesssim 1$, where $\eta = \eta_f$ measures the width of the *ellipse* of analyticity of f in the complex plane (corresponding to the width of the

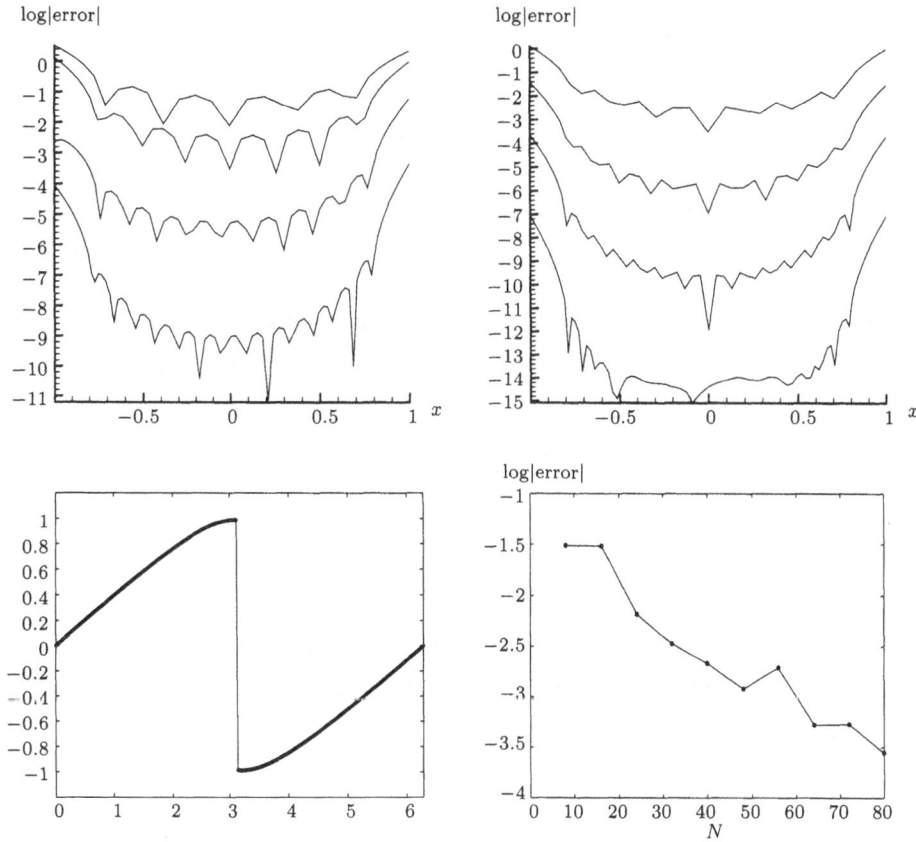

Figure 10.4. *Top*: Error in log-log scale of the Gegenbauer reconstruction $G_p^{(\alpha)}(S_N f)_\pi(x)$ with $N = 16, 24, 36$ and 52 modes for $f(x) = \cos(1.4\pi(x - 1))$, $x \in (-1, 1)$. *Top left*: $\alpha_N = p_N = N/4$. *Top right*: $\alpha_N = p_N = N/5$. *Bottom left*: Gegenbauer reconstruction of $I_{40} f(x)$, where $I_N f = u_N(x, t = 1.5)$ is a steady discontinuous solution of the inviscid Burgers equation, computed using a smoothed pseudo-spectral Fourier projection, $\partial_t u_N(x_\nu, t) + \partial_x \left(\frac{I_N(u_N)^2}{2} \right)(x_\nu, t) = 0$ and subject to $u_N(x_\nu, t = 0) = \sin(x_\nu)$. *Bottom right*: Log-log plot of the error.

analyticity *strip* in the periodic case: *e.g.*, Tadmor (1986)). This translates into tuning of p_N, α_N, depending on the different analyticity regions for each smoothness interval of f. The overall Gegenbauer reconstruction method becomes rather sensitive to its parametrization (Boyd 2005). The superior accuracy is achieved at the expense of losing the robustness we had with the reconstruction methods based on adaptive mollifiers. A more robust reconstruction was offered recently in Gelb and Tanner (2006), where the strongly peaked Gegenbauer weight, $\omega_\alpha(x) = (1 - x^2)^{\alpha - \frac{1}{2}}$, is replaced by the Froud weight $\omega_m(x) = e^{-cx^{2m}}$.

11. Spectral filters

11.1. Adaptive filters: root-exponential accuracy

In Section 10 we showed how to parametrize an optimal mollifier, $\Phi_{p_N, d_x}(x)$, in order to gain the root-exponential convergence for piecewise analytic f. The key ingredient in our approach was *adaptivity*, where the $\delta = d_x$ and $p_N \sim d_x N$ were adapted to the maximal region of local smoothness. Here we continue the same line of thought by introducing *adaptive filters*, which allow the same root-exponential recovery of piecewise analytic functions.

We consider a family of general filters $\varphi(\cdot) \in C^q(\mathbb{R})$, operating in Fourier space:

$$S_N^\varphi(x) := \sum_{|k| \leq N} \varphi\left(\frac{|k|}{N}\right) \widehat{f}(k) e^{ikx}. \tag{11.1}$$

They are characterized by two main properties.

(i) First, we seek the *rapid smooth decay* of $\varphi(\xi)$ as ξ moves away from the origin. Translated from Fourier to physical space, the operation of $S_N^\varphi f$ corresponds to mollification against the smoothing kernel $\mathcal{S}_N^\varphi(x)$:[6]

$$S_N^\varphi f(x) = \mathcal{S}_N^\varphi * (S_N f)(x), \qquad \mathcal{S}_N^\varphi(x) := \frac{1}{2\pi} \sum_{k=-\infty}^{\infty} \varphi\left(\frac{|k|}{N}\right) e^{ikx}.$$

Then the rapid smooth decay of $\varphi(\cdot)$ is responsible for $\mathcal{S}_N^\varphi(x)$, which is *strongly localized* around $x = 0$.

(ii) Second, the mollifier $\mathcal{S}_N^\varphi(x)$ associated with the filter $\varphi(\xi)$ is required to satisfy the moment conditions (10.1). This property drives the *accuracy* by annihilating an increasing number of the moments of $\mathcal{S}_N^\varphi$. As observed in Remark 10.7, however, a key ingredient in the construction of exponentially accurate mollifiers in Section 10.2 was giving up the exactness of (10.1). In a similar manner, in our quest for exponentially accurate filters, the moment conditions (10.1) are replaced by the *accuracy condition*

$$\varphi^{(n)}(0) = \delta_{n0}, \qquad n = 0, 1, \ldots, p. \tag{11.2}$$

It follows that if the filter φ is p-order accurate, then its associated mollifier, $\mathcal{S}_N^\varphi$, satisfies the moment conditions to order p.

We can quantify the above statements in a more precise manner, with a couple of examples which are summarized in the following two claims.

---

[6] As before, the *function* $\mathcal{S}_N^\varphi(x)$ represents a *smoothing kernel* associated with but otherwise different from the corresponding filtering *operator* $S_N^\varphi f$.

Claim 11.1. (Rapid decay of $\mathcal{S}_N^\varphi(x)$) Let φ be a $C_0^q[-1,1]$-filter. Then its associated mollifier,

$$\mathcal{S}_N^\varphi(x) := \frac{1}{2\pi} \sum_{|k|\leq N} \varphi\left(\frac{|k|}{N}\right) e^{ikx},$$

is strongly localized near the origin in the sense that

$$|\mathcal{S}_N^\varphi(x)| \lesssim N\|\varphi\|_{C^q}\frac{1}{(|x|N)^q}, \qquad 0 < |x| \leq \pi. \tag{11.3a}$$

Thus, the smoother φ is, the better $\mathcal{S}_N^\varphi$ is localized. As an example, we state an immediate consequence of (11.3a).

Example 11.2. If $\varphi \in G_\alpha$ then $\mathcal{S}_N^\varphi(x)$ experiences the root-exponential decay, namely, there exists $\eta = \eta_1$ (depending on φ) such that, for all $|x| \leq \pi$,

$$|\mathcal{S}_N^\varphi(x)| \lesssim N \min_q \frac{(q!)^\alpha}{(\eta_\varphi |x|N)^q} \lesssim (1+|x|N)e^{-\eta_1 \sqrt[\alpha]{|x|N}}, \qquad \varphi \in G_\alpha. \tag{11.3b}$$

At the end of the 'smoothness scale', we find the entire function φ with quadratic exponential decay (2.8a); the mollifier $\mathcal{S}_N^{\varphi_{\delta_N}}(x)$, with $\delta_N = \sqrt{\beta N}$, admits exponential decay (2.11a).

We turn to verify Claim 11.1 in two different ways. First, we rewrite $\mathcal{S}_N^\varphi(x)$ in the form

$$\mathcal{S}_N^\varphi(x) = \sum_{|k|\leq N} \varphi(\xi_k)\frac{e^{i(k+1)x} - e^{ikx}}{e^{ix} - 1}, \qquad \xi_k := kh, \ h = \frac{1}{N}.$$

Summation by parts yields

$$\mathcal{S}_N^\varphi(x) = \frac{1}{e^{ix} - 1} \sum_{|k|\leq N} \Delta_h\varphi(|\xi_k|)e^{ikx} + \text{ a couple of boundary terms,}$$

and by repeating this argument,

$$\mathcal{S}_N^\varphi(x) = \frac{h^q}{(e^{ix} - 1)^q} \sum_{|k|\leq N} h^{-q}\Delta_h^q\varphi(|\xi_k|)e^{ikx} + 2q \text{ boundary terms.}$$

We can safely neglect the small boundary terms (precisely because of (11.2)) and the C_0^q-regularity of $\varphi(\cdot)$ implies (11.3a).

A second approach is to use the Poisson summation formula (2.9b), expressing $\mathcal{S}_N^\varphi(x)$ in terms of Φ, the inverse Fourier transform of φ,

$$\mathcal{S}_N^\varphi(x) \equiv \frac{N}{2\pi} \sum_{j=-\infty}^{\infty} \Phi(N(x + 2\pi j)), \qquad \Phi(x) = \int_{\mathbb{R}} \varphi(\xi)e^{i\xi x}\, d\xi.$$

This, together with the spectral decay estimate (2.3), yields (11.3a)

$$|S_N^\varphi(x)| \lesssim N^{1-q}\|\varphi\|_{C^q} \sum_{j=-\infty}^{\infty} \frac{1}{|x+2\pi j|^q} \lesssim N\|\varphi\|_{C^q} \frac{1}{(N|x|)^q}, \qquad 0 < |x| \le \pi.$$

$\square$

Claim 11.3. (Accuracy and moment conditions) If $\varphi \in C_0^q[-1,1]$ satisfies the *accuracy condition* of order $p < q$,

$$\varphi^{(n)}(0) = \delta_{n0}, \qquad n = 0, 1, \ldots, p,$$

then $S_N^\varphi(x)$ satisfies the moment conditions to order p:

$$\int_{-\pi}^{\pi} y^n S_N^\varphi(y)\, dy = \delta_{n0}, \qquad n = 0, 1, \ldots, p. \tag{11.4a}$$

Moreover, $S_N^\varphi(x)$ concentrates in a neighbourhood of the origin in the sense that the contribution to its moments outside such a neighbourhood is negligible,

$$\left| \int_{|y|\ge r} y^n S_N^\varphi(y)\, dy \right| \lesssim \|\varphi\|_{C^p} \frac{1}{(rN)^{p-1}}, \qquad n = 0, 1, \ldots, p. \tag{11.4b}$$

Thus, the more accurate φ is, the better S_N^φ satisfies the moment conditions. As an example we state the following immediate consequence of (11.4).

Example 11.4. If $\varphi = \varphi_p$ is a G_α-filter which is accurate of order $p = p_N \sim (rN)^{1/\alpha}$, then the unit mass S_N^φ has vanishing moments to order p_N. Moreover, there exists $\eta_2 > 0$ (depending on φ) such that

$$\int_{|y|\ge r} y^n S_N^\varphi(y)\, dy = \delta_{n0} + \mathcal{O}\left(\min_{p \le \sqrt[\alpha]{rN}} \frac{(p!)^\alpha}{(\eta_2 rN)^p} \right) = \delta_{n0} + \mathcal{O}\left(e^{-\eta_2 p_N}\right),$$

$$\text{for } n \le p_N \sim (rN)^{1/\alpha}, \quad \varphi \in G_\alpha. \tag{11.5}$$

To verify the first part of Claim 11.3, we appeal again to the Poisson formula (2.9b), expressing $S_N^\varphi(x)$ in terms of translates of $\Phi(x)$:

$$\int_{-\pi}^{\pi} y^n S_N^\varphi(y)\, dy = \underbrace{\frac{N}{2\pi} \int_{-\pi}^{\pi} y^n \Phi(Ny)\, dy}_{\mathcal{I}_1} + \underbrace{\frac{N}{2\pi} \sum_{j\ne 0} \int_{-\pi}^{\pi} y^n \Phi(N(y+2\pi j))\, dy}_{\mathcal{I}_2}$$

$$= \underbrace{\frac{N}{2\pi} \int_{-\infty}^{\infty} y^n \Phi(Ny)\, dy}_{\mathcal{I}_3} - \frac{N}{2\pi} \int_{|y|\ge\pi} y^n \Phi(Ny)\, dy$$

$$+ \frac{N}{2\pi} \sum_{j\ne 0} \int_{-\pi}^{\pi} y^n \Phi(N(y+2\pi j))\, dy.$$

The first term on the right equals $\mathcal{I}_1 = (\mathrm{i}N)^{-n}\varphi^{(n)}(0)$, since $\Phi(x)$ is the inverse Fourier transform of $\varphi(\xi)$, and therefore, since φ is p-order accurate, $\mathcal{I}_1 = \delta_{n0}$, $n \le p$. The second and third terms cancel, and (11.4a) follows.

To verify the second part of the claim, we use the Poisson summation formula again to write

$$
\int_{r \le |y| \le \pi} y^n S_N^\varphi(y)\,\mathrm{d}y = \overbrace{\frac{N}{2\pi} \int_{r \le |y| \le \pi} y^n \Phi(Ny)\,\mathrm{d}y}^{\mathcal{I}_1}
$$

$$
+ \overbrace{\frac{N}{2\pi} \int_{r \le |y| \le \pi} y^n \left(\sum_{j \ne 0} \Phi(N(y + 2\pi j)) \right) \mathrm{d}y}^{\mathcal{I}_2} .
$$

The usual spectral decay rate $|\Phi(y)| \lesssim \|\varphi\|_{C^p} \cdot |y|^{-p}$ implies

$$
|\mathcal{I}_1| \lesssim N^{1-p} \|\varphi\|_{C^p} \int_{|y| \ge r}^{\pi} y^{n-p}\,\mathrm{d}y \lesssim \|\varphi\|_{C^p} \frac{1}{(rN)^{p-1}}, \qquad n = 0, 1, \ldots, p.
$$

Similarly,

$$
|\mathcal{I}_2| \lesssim N \|\varphi\|_{C^p} \sum_{j \ne 0} \frac{\pi^n}{((2j-1)N\pi)^p} \lesssim \|\varphi\|_{C^p} \frac{1}{N^{p-1}}, \qquad n = 0, 1, \ldots, p,
$$

and (11.4b) follows. $\qquad\qquad\qquad\qquad\qquad\qquad\qquad\qquad\qquad\qquad\qquad\square$

We note that it is rather simple to construct admissible filters satisfying the last requirement for an *arbitrary* p; a prototype example is given by the G_2-filters

$$
\varphi_p(\xi) = \mathrm{e}^{\left(\frac{\xi^p}{\xi^2 - 1}\right)} 1_{(-1,1)}(\xi). \tag{11.6}
$$

This should be contrasted with the more intricate construction of mollifiers satisfying the exact moment conditions in Example 10.1. We are now ready to state a key result.

Theorem 11.5. (Root-exponential filters; Tadmor and Tanner 2005)
Assume that $f(\cdot)$ is piecewise analytic and let $S_N^{\varphi_p}$ denote the filtered sum

$$
S_N^{\varphi_p} f(x) := \sum_{|k| \le N} \varphi_p\left(\frac{|k|}{N}\right) \hat{f}(k) \mathrm{e}^{\mathrm{i}kx}, \qquad \varphi_p(\xi) = \mathrm{e}^{\left(\frac{\xi^p}{\xi^2 - 1}\right)} 1_{(-1,1)}(\xi). \tag{11.7}
$$

We set the order $p = p_N(x) \sim \sqrt{d_x N}$ (p_N even) where, as usual,

$$
d_x := \frac{1}{\pi} \operatorname{dist}\{x, \{c_1, \ldots, c_J\}\}[\operatorname{mod} \pi],
$$

so that $(x - \pi d_x, x + \pi d_x)$ is the largest interval of analyticity enclosing x.

Then, the adaptive filter $S_N^{\varphi_{pN}} f$ recovers the point values $f(x)$ within the following root-exponential accuracy:

$$|S_N^{\varphi_{pN}} f(x) - f(x)| \lesssim d_x N \cdot e^{-\eta \sqrt{d_x N}}. \tag{11.8}$$

Here, the constant $\eta = \eta_{\varphi,f}$ is dictated by the specific Gevrey and piecewise analyticity properties of φ and f.

Proof. We begin by decomposing the filtering error into the usual truncation and regularization term (compare (10.2)),

$$S_N^{\varphi_{pN}} f(\cdot) - f(\cdot) = \overbrace{S_{p_N}^\varphi * S_N f - S_{p_N}^\varphi * f}^{\text{truncation}} + \overbrace{S_{p_N}^\varphi * f - f}^{\text{regularization}}.$$

Here, $S_{p_N}^\varphi(x) \equiv S_N^{\varphi_{pN}}(x)$ is an abbreviated notation for the mollifier associated with φ_{pN},

$$S_{p_N}^\varphi(x) := \frac{1}{2\pi} \sum_{|k| \leq N} \varphi_{pN}\left(\frac{|k|}{N}\right) e^{ikx}.$$

Since $S_{p_N}^\varphi$ is a trigonometric polynomial of degree $\leq N$, the truncation error vanishes:

$$S_{p_N}^\varphi * S_N f - S_{p_N}^\varphi * f = (S_N S_{p_N}^\varphi - S_{p_N}^\varphi) * f \equiv 0.$$

We remain with the regularization error, which we split into two terms:

$$\overbrace{S_{p_N}^\varphi * f(x) - f(x)}^{\text{regularization}} = \overbrace{\int_{\theta d_x \leq |y| \leq \pi} S_{p_N}^\varphi(y) [f(x) - f(x-y)] \, dy}^{\mathcal{I}_1} \tag{11.9}$$

$$+ \underbrace{\int_{|y| \leq \theta d_x} S_{p_N}^\varphi(y) [f(x) - f(x-y)] \, dy}_{\mathcal{I}_2}.$$

Here $\theta < 1$ is a free parameter at our disposal. The first term on the right of (11.9) is straightforward: the G_2-regularity of φ_p implies the root-exponential decay of $S_{p_N}^\varphi$, namely, (11.3b) with $p_N = \sqrt{d_x N}$ implies

$$|\mathcal{I}_1| \lesssim (1 + \theta d_x N) \cdot e^{-\eta_1 \sqrt{\theta d_x N}}, \qquad \eta_1 = \eta_\varphi > 0. \tag{11.10}$$

We turn to the second error term, $\mathcal{I}_2$. As before, it will be shown to be small due to *cancellation* of oscillations with the increasing order p of $S_{p_N}^\varphi$.

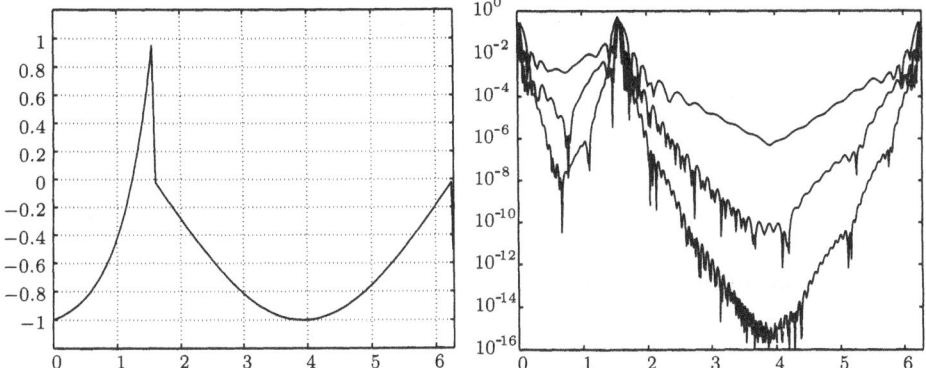

Figure 11.1. Function $f(x)$ in (10.12) and the log-error in its reconstruction from $S_N f$, $N = 32, 64, 128$ using the adaptive filter $S_N^{\varphi_{p_N}}$ in (11.7) with $p_N = \max(2, \sqrt{d_x N})$.

To this end we use Taylor's expansion of $f(\cdot) - f(\cdot - y)$ to express $\mathcal{I}_2$ as

$$
\mathcal{I}_2 - \overbrace{\sum_{1 \leq n < \theta p_N} \frac{(-1)^n}{n!} f^{(n)}(x) \int_{|y| \leq \theta d_x} y^n \mathcal{S}_{p_N}^\varphi(y)\,\mathrm{d}y}^{\mathcal{I}_{21}}
$$

$$
+ \overbrace{\frac{(-1)^{\theta p_N}}{(\theta p_N)!} f^{(\theta p_N)}(\cdot) \int_{|y| \leq \theta d_x} y^{\theta p_N} \mathcal{S}_{p_N}^\varphi(y)\,\mathrm{d}y}^{\mathcal{I}_{22}}. \tag{11.11}
$$

But since φ_{p_N} is accurate to order p_N, (11.4a) and (11.5) with $r = \theta d_x$ and $p_N = \sqrt{rN}$ tell us that

$$
\int_{|y| \leq \theta d_x} y^n \mathcal{S}_{p_N}^\varphi(y)\,\mathrm{d}y = -\int_{|y| \geq r} y^n \mathcal{S}_{p_N}^\varphi(y)\,\mathrm{d}y = \delta_{n0} + \mathcal{O}(e^{-\eta_2 \sqrt{\theta d_x N}}),
$$

and hence

$$
|\mathcal{I}_{21}| \lesssim \sum_1^{\theta p_N} \frac{\pi^n}{\eta_f^n} e^{-\eta_2 \sqrt{d_x N}} \lesssim e^{\sqrt{d_x N}(\kappa_1 \theta - \eta_2 \sqrt{\theta})}, \qquad \kappa_1 := \log(\pi/\eta_f). \tag{11.12a}
$$

Finally, the term $\mathcal{I}_{22}$ is exponentially small since near the origin, $|\mathcal{S}_N^\varphi(y)| \lesssim 2^{p_N}$, $e.g.$, by (2.14), and by choosing sufficiently small θ,

$$
|\mathcal{I}_{22}| \lesssim \frac{1}{(\eta_f)^{\theta p_N}} (\theta d_x)^{\theta p_N} 2^{p_N} \lesssim \left(2 \left(\frac{\theta \pi}{\eta_f}\right)^\theta\right)^{p_N} \lesssim e^{-\eta \sqrt{d_x N}}. \tag{11.12b}
$$

Result (11.8) follows from (11.10) and (11.12) by choosing appropriately small $\theta = \theta(\eta_f, \eta_1, \eta_2)$. $\qquad\square$

Figure 11.1 illustrates the use of the adaptive filter (11.7) to reconstruct the same function dealt with earlier using the adaptive mollifier (10.10) illustrated in Figure 10.1.

11.2. Optimal filters: exponentially accurate reconstruction

To reconstruct piecewise analytic f with *exponential* accuracy, we turn to the filters based on the exponential optimality of space–frequency localization discussed in Section 2.2, *i.e.*,

$$\varphi_{p,\delta}(\xi) := e^{-\frac{(\delta\xi)^2}{2}} \sum_{j=0}^{p} \frac{1}{2^j j!}(\delta\xi)^{2j}.$$

The $\varphi_{p,\delta}$ are the truncated Hermite expansion of the weighted Gaussian, so that they form $(2p+1)$-order accurate filters in the sense that (11.2) holds. Since we are going to use adaptive parametrization where both $\delta = \delta_x$ and p increase with N, we now explicitly specify the dependence of φ on both. The corresponding $\varphi_{p,\delta}$-filter reads

$$S_N^{\varphi_{p,\delta}} f(x) = \sum_{|k|\leq N} \varphi_{p,\delta}\left(\frac{|k|}{N}\right)\hat{f}(k)e^{ikx}.$$

It can be expressed in terms of the associated mollifier, $S_N^{\varphi_{p,\delta}}(x)$,

$$S_N^{\varphi_{p,\delta}} f(x) = S_N^{\varphi_{p,\delta}} * (S_N f)(x), \qquad S_N^{\varphi_{p,\delta}} := \frac{1}{2\pi} \sum_{k=-\infty}^{\infty} \varphi_{p,\delta}\left(\frac{|k|}{N}\right)e^{ikx}$$

We observe that, in this case, neither the filter nor its associated mollifier are compactly supported. Relaxing the constraint of having compact support in either physical space – as for $\Phi_p = \rho_2 D_p$ in (10.5) – or the Fourier space – as for $\Phi_p \leftrightarrow \varphi_p$ in (11.7) – will enable us to obtain exponential accuracy after appropriate *adaptive* choice of the free parameters,

$$\delta_x := \sqrt{\theta d_x N}, \quad p_N := \theta^2 d_x N, \tag{11.13}$$

where d_x in (10.4) defines the usual analytic neighbourhood enclosing x, and with $\theta < 1$ at our disposal. We use $S_{p_N,\delta_x}^{\varphi}(x)$ to abbreviate the notation of the corresponding mollifier

$$S_{p_N,\delta_x}^{\varphi}(x) \equiv S_N^{\varphi_{p_N,\delta_x}}(x) = \frac{1}{2\pi} \sum_{k=-\infty}^{\infty} \varphi_{p_N,\delta_x}\left(\frac{|k|}{N}\right)e^{ikx}.$$

To estimate the error,

$$S_N^{\varphi_{p_N,\delta_x}} f(x) - f(x) = S_{p_N,\delta_x}^{\varphi} * (S_N f)(x) - f(x),$$

we first need to quantify the exponentially rapid decay of $S_{p_N,\delta_x}^{\varphi}(x)$ in both physical *and* Fourier space.

We appeal to (2.11). Our choice of $\delta_x = \sqrt{\theta d_x N}$ corresponds to $\beta = \theta d_x$ and the spatial exponential decay in (2.11a) yields

$$|S^{\varphi}_{p_N, \delta_x}(x)| \lesssim 2^{p_N} \sqrt{\frac{N}{d_x}} \left(e^{-\eta_1 \frac{N x^2}{\theta d_x}} + e^{-\eta_2 \frac{N}{\theta d_x}} \right), \qquad |x| \le \pi.$$

Since $p_N = \theta^2 d_x N$, we have $2^{p_N} \le e^{\kappa_2 \theta^2 d_x N}$ with $\kappa_2 := \log(2)$, and the last inequality confirms the exponential decay of $\Phi_{p_N, \delta_x}(x)$ outside the d_x-neighbourhood of the origin. Indeed, by choosing sufficiently small $\theta < 1$,

$$|S^{\varphi}_{p_N, \delta_x}(x)| \lesssim \sqrt{\frac{N}{d_x}} \left(e^{(\kappa_2 \theta^2 - \eta_1 \theta) d_x N} + e^{\left(\kappa_2 \theta^2 - \frac{\eta_2}{\theta d_x^2} \right) d_x N} \right)$$

$$\lesssim \sqrt{\frac{N}{d_x}} e^{-\eta d_x N}, \qquad \theta d_x \le |x| < \pi. \qquad (11.14)$$

Next, we consider the Fourier space decay of $\varphi_{p_N, \delta_x}\left(\frac{|k|}{N} \right)$. Appealing to (2.11b) with $\beta = \theta d_x$, we find

$$\left| \varphi_{p_N, \delta_x}\left(\frac{|k|}{N} \right) \right| \lesssim c_{p_N, N} e^{-\frac{\theta d_x}{2} |k|}, \qquad c_{p_N, N} = \sum_{j=0}^{p_N} \frac{1}{j!} \left(\frac{\delta_x^2}{2} \right)^j.$$

The pre-factor $c_{p,N}$ has exponential growth of order

$$c_{p_N, N} = \sum_{j=0}^{\theta^2 d_x N} \frac{1}{j!} \left(\frac{\theta d_x N}{2} \right)^j \lesssim e^{\eta_\theta \frac{\theta d_x}{2} N},$$

but with a coefficient η_θ, which can be made sufficiently small by decreasing θ. Consequently, the last two inequalities imply that $|\varphi_{p_N, \delta_x}\left(\frac{|k|}{N} \right)|$ decay exponentially fast for $|k| > N$, i.e.,

$$\left| \varphi_{p_N, \delta_x}\left(\frac{|k|}{N} \right) \right| \lesssim e^{-\left(1 - \eta_\theta \frac{N}{|k|} \right) \frac{\theta d_x}{2} |k|} \lesssim e^{-\eta d_x |k|}, \qquad |k| > N. \qquad (11.15)$$

Equipped with (11.14) and (11.15), we are ready to prove the following theorem.

Theorem 11.6. (Exponentially accurate filter; Tanner 2006)
Assume that $f(\cdot)$ is piecewise analytic, and let

$$S^{\varphi_{p, \delta}}_N f(x) := \sum_{|k| \le N} \varphi_{p, \delta}\left(\frac{|k|}{N} \right) \widehat{f}(k) e^{ikx}$$

denote the filtered Fourier projection, based on the quadratic exponential filter

$$\varphi_{p, \delta}(\xi) = \varphi_p(\delta \xi) := e^{-\frac{(\delta \xi)^2}{2}} \sum_{j=0}^{p} \frac{1}{j!} \left(\frac{(\delta \xi)^2}{2} \right)^j, \qquad (11.16a)$$

of degree $p = p_N := \theta^2 d_x N$, with adaptive scaling $\delta = \delta_x := \sqrt{\theta d_x N}$. Here,

$$d_x = \frac{1}{\pi} \operatorname{dist}\{x, \{c_1, \ldots, c_J\}\}[\operatorname{mod} \pi]$$

defines a πd_x-neighbourhood of analyticity around x. Then, for sufficiently small $\theta < 1$, there exists $\eta = \eta_{\theta,f} > 0$ such that the adaptive filter $S_N^{\varphi_{p_N}, \delta_N} f(x)$ recovers $f(x)$ with the following exponential accuracy:

$$|S_N^{\varphi_{p_N}, \delta_x} f(x) - f(x)| \lesssim \sqrt{\frac{N}{d_x}} e^{-\eta d_x N}. \tag{11.16b}$$

The constant $\eta = \eta_{\theta,f} > 0$ is dictated by the specific piecewise analyticity properties of f. The exponential adaptive filter takes the final form

$$S_N^{\varphi_{p_N}, \delta_x} f(x) = \sum_{|k| \leq N} \left[\sum_{j=0}^{[\theta^2 d_x N]} \frac{1}{j!} \left(\frac{\theta d_x k^2}{2N} \right)^j \right] e^{-\frac{\theta d_x k^2}{2N}} \widehat{f}(k) e^{ikx}.$$

Proof. We proceed with an error decomposition similar to (11.9):

$$S_N^{\varphi_{p_N}, \delta_x} f(x) - f(x) = S_{p_N, \delta_x}^\varphi * f(x) - f(x) + S_{p_N, \delta_x}^\varphi * S_N f(x) - S_{p_N, \delta_x}^\varphi * f(x)$$

$$= \overbrace{\int_{\theta d_x \leq |y| \leq \pi} S_{p_N, \delta_x}^\varphi(y) [f(x) - f(x-y)] \, dy}^{\mathcal{I}_1} \tag{11.17}$$

$$+ \overbrace{\int_{|y| \leq \theta d_x} S_{p_N, \delta_x}^\varphi(y) [f(x) - f(x-y)] \, dy}^{\mathcal{I}_2}$$

$$+ \overbrace{\left(S_N S_{p_N, \delta_x}^\varphi - S_{p_N, \delta_x}^\varphi \right) * f(x)}^{\mathcal{I}_3 = \text{truncation}}.$$

To recall, $S_N^{\varphi_{p,\delta}}(x)$ abbreviates the mollifier associated with the filter $\varphi_{p,\delta}$:

$$S_N^{\varphi_{p,\delta}}(x) = \frac{1}{2\pi} \sum_{k=-\infty}^{\infty} \varphi_{p,\delta}\left(\frac{|k|}{N} \right) e^{ikx}.$$

We first observe the addition of a truncation error term, $\mathcal{I}_3$, which is due to the fact that $\varphi_{p,\delta}(\xi)$ is no longer compactly supported on $(-1, 1)$, i.e., $S_N^{\varphi_{p,\delta}}(x)$ is no longer a trigonometric polynomial of degree $\leq N$. But the truncation error term is exponentially small because $|\varphi_{p,\delta}(|k|/N)|$ are; indeed, by (11.15) we have

$$|\mathcal{I}_3| \lesssim \| S_N S_{p_N, \delta_x}^\varphi - S_{p_N, \delta_x}^\varphi \|_{L^\infty} \lesssim \sum_{|k| > N} e^{-\eta d_x |k|} \lesssim e^{-\eta d_x N}. \tag{11.18}$$

The first term in the error decomposition can be made exponentially small because of the rapid decay of $S^{\varphi}_{p_N,\delta_x}(x)$. Indeed, since the support of the first integrand is bounded θd_x from the origin, we find, thanks to (11.14),

$$|\mathcal{I}_1| \lesssim \int_{|y| \geq \theta d_x}^{\pi} |S^{\varphi}_{p_N,\delta_x}(x)|\, dy \lesssim \sqrt{\frac{N}{d_x}} e^{-\eta d_x N}. \tag{11.19}$$

We now come to the second term $\mathcal{I}_2$. It can be made small because of the accuracy of $\varphi_{p_N,\delta_x}(\xi)$, which in turn implies that $S^{\varphi}_{p_N,\delta_x}(x)$ has vanishing moments to order p_N, so that the *local* moments of $\Phi_{p_N,\delta_x}(x)$ equals

$$\overbrace{\int_{|y| \leq \theta d_x} y^n S^{\varphi}_{p_N,\delta_x}(y)\, dy}^{\text{moments associated with } \mathcal{I}_2} = -\int_{|y| \geq \theta d_x}^{\pi} y^n S^{\varphi}_{p_N,\delta_x}(y)\, dy;$$

but the rapid decay of $S^{\varphi}_{p_N,\delta_x}(y)$ in (11.14) implies

$$\left| \int_{|y| \geq \theta d_x}^{\pi} y^n S^{\varphi}_{p_N,\delta_x}(y)\, dy \right| \lesssim \sqrt{\frac{N}{d_x}} \pi^{p_N} e^{-\eta d_x N} \lesssim \sqrt{\frac{N}{d_x}} e^{d_x N (\kappa_3 \theta^2 - \eta)},$$

$$\kappa_3 = \log(\pi).$$

The estimate of $\mathcal{I}_2$ now follows the lines of Theorem 11.5 using a similar decomposition into two terms, $\mathcal{I}_{21} + \mathcal{I}_{22}$, each of which is exponentially small due to the rapid decay of $S^{\varphi}_{p_N,\delta_x}(y)$ outside the origin (11.14).   □

Remark 11.7. (Exponentially accurate mollifier)   Observe that the mollifier $S^{\varphi_p,\delta}_N(x)$ associated with the filter $\varphi_{p,\delta_N}(\xi)$ in (11.16a) is exponentially close to $\Phi_{\delta,p}(Ny)$; consult (2.13) and (2.14a). Accordingly, we find the exponentially accurate *mollifier* $\Phi_{\delta,p}(Ny)$, with $\delta = \delta_x := \theta^2 d_x N$ and $p = p_N := \sqrt{\theta d_x N}$:

$$\Phi_{\delta_x,p_N}(Ny) = \frac{1}{\sqrt{\theta d_x N}} e^{-\frac{Ny^2}{2\theta d_x}} \times \sum_{j=0}^{[\theta^2 d_x N]} \frac{(-1)^j}{4^j j!} H_{2j}\left(\frac{\sqrt{N}y}{\sqrt{2\theta d_x}} \right). \tag{11.20a}$$

It is particularly useful to implement in the discrete case, where we end up with the exponentially accurate discrete mollifier (Tanner 2006, Theorem 4.2)

$$\left| h \sum_{\nu=0}^{2N} \Phi_{\delta_x,p_N}\left(N(x - y_\nu) \right) f(y_\nu) - f(x) \right| \lesssim \sqrt{\frac{N}{d_x}} e^{-\eta d_x N}. \tag{11.20b}$$

Figure 11.2, from Tanner (2006), illustrates the superior convergence rate of the optimal filter (11.16b) (with $\theta \sim 1/4$) and its associated mollifier (11.20b).

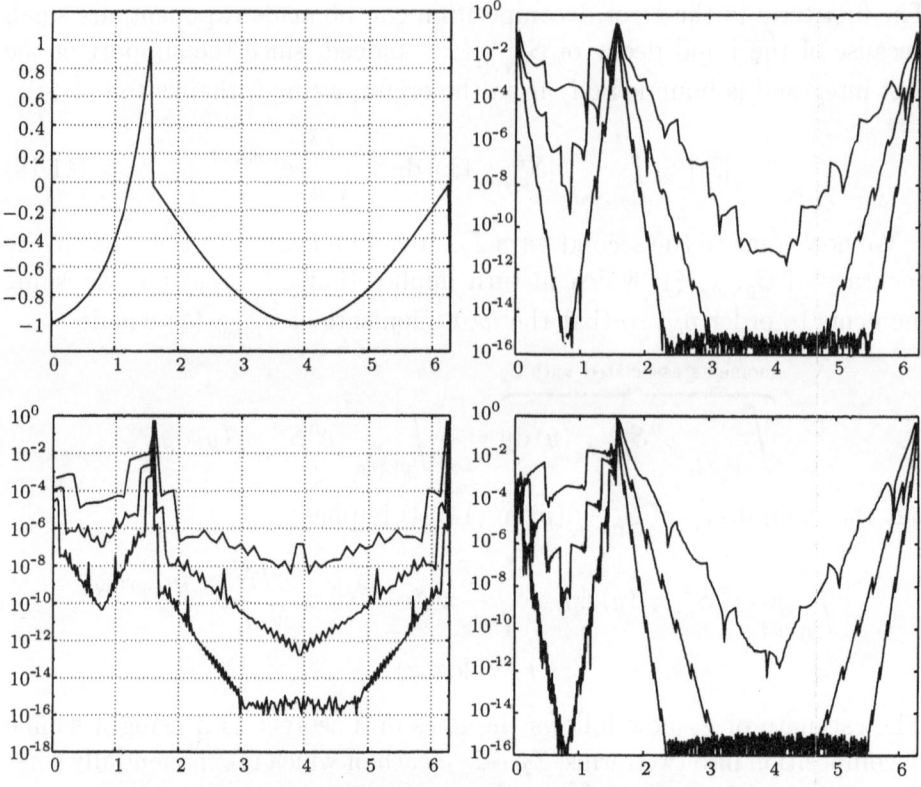

Figure 11.2. *Top left*: Function $f(x)$ in (10.12). *Top right*:
Log-error in its reconstruction from $S_N f$, $N = 32, 64, 128$ using
the optimal filter (11.16b). *Bottom left*: Log-error in
reconstruction of f from $I_N f$, $N = 32, 64, 128$ using the
4th-order normalized adaptive mollifier (10.21c). *Bottom right*:
The same using the optimal pseudo-spectral mollifier (11.20b).

Acknowledgements

A large portion of the material covered in this review is based on the work
of the author with various collaborators, and it is my pleasure to acknowl-
edge here Shlomo Engelberg, Anne Gelb, David Gottlieb, Jared Tanner
and Jing Zou. I am indebted in particular to David Gottlieb, for years of
friendship and collaboration which began with our first paper, Gottlieb and
Tadmor (1985), the forerunner of a large body of subsequent work. Research
was supported by NSF grant DMS04-07704 and by ONR grant N00014-91-
J-1076.

REFERENCES

S. Abarbanel, D. Gottlieb and E. Tadmor (1986) 'Spectral methods for discontinuous problems', in *Numerical Methods for Fluid Dynamics II: Proc. 1985 Conference on Numerical Methods for Fluid Dynamics* (K. W. Morton and M. J. Baines, eds), Clarendon Press, Oxford, pp. 129–153.

N. S. Banerjee and J. Geer (1997) Exponential approximations using Fourier series partial sums. ICASE Report No. 97-56, NASA Langley Research Center.

N. Bary (1964) *Treatise of Trigonometric Series*, Macmillan, New York.

R. Bauer (1995) Band filters for determining shock locations. PhD thesis, Applied Mathematics, Brown University.

J. Bourgain (1988) 'A remark on the uncertainty principle for Hilbertian basis', *J. Funct. Anal.* **79**, 136–143.

J. P. Boyd (1995) 'A lag-averaged generalization of Euler's method for accelerating series', *Appl. Math. Comput.* 143–166.

J. P. Boyd (1996) 'The Erfc-log filter and the asymptotic of the Euler and Vandeven sequence accelerations', in *Proc. Third International Conference on Spectral and High Order Methods*, pp. 267–276.

J. P. Boyd (2005) 'Trouble with Gegenbauer reconstruction for defeating Gibbs' phenomenon: Runge phenomenon in the diagonal limit of Gegenbauer polynomial approximations', *J. Comput. Phys.* **204**, 253–264.

O. Bruno, Y. Han and M. Pohlman (2006), 'Accurate, high-order representation of complex three-dimensional surfaces via Fourier-continuation analysis'. Preprint.

E. Candes and F. Guo (2002) 'New multiscale transform, minimum total variation synthesis: Applications to edge-preserving image reconstruction', *Signal Process.* **82**, 1519–1543.

E. Candes and J. Romberg (2006) 'Quantitative robust uncertainty principles and optimally sparse decompositions', *Found. Comput. Math.* **6**, 227–254

E. Candes, J. Romberg and T. Tao (2006*a*) 'Robust uncertainty principles: Exact signal reconstruction from highly incomplete frequency information', *IEEE Trans. Inform Theory* **52**, 489–509.

E. Candes, J. Romberg and T. Tao (2006*b*) 'Stable signal recovery from incomplete and inaccurate measurements', *Comm. Pure Appl. Math.* **59**, 1208–1223.

J. Canny (1986) 'A computational approach to edge detection', *IEEE Trans. Pattern Analysis and Machine Intelligence* **8**, 679–714.

H. S. Carslaw (1952) *Introduction to the Theory of Fourier's Series and Integrals*, Dover.

J. Clark (1989) 'Authenticating edges produced by zero-crossing algorithms', *IEEE Trans. Pattern Analysis and Machine Intelligence* **11**, 43–57.

R. DeVore and B. Lucier (1992) 'Wavelets', in *Acta Numerica*, Vol. 1, Cambridge University Press, pp. 1–56.

R. DeVore and Lorentz (1993) *Constructive Approximation*, Springer.

D. Donoho (1998) 'Orthonormal ridgelets and linear singularities'. www-stat.stanford.edu/~donoho/Reports/1998/ridge-lin-sing.pdf

D. Donoho (2004) 'Compressed sensing'. www-stat.stanford.edu/~donoho/Reports/2004/CompressedSensing091604.pdf

D. Donoho and X. Huo (2001) 'Uncertainty principles and ideal atomic decomposition', *IEEE Trans. Inform. Theory* **47**, 2845–2862.

D. Donoho and J. Tanner (2005) 'Sparse nonnegative solutions of underdetermined linear equations by linear programming'. Preprint.

D. Donoho, M. Elad and V. Temlyakov (2004) 'Stable recovery of sparse overcomplete representations in the presence of noise'. Preprint.

H. Dym and H. McKean (1972) *Fourier Series and Integrals*, Academic Press.

K. S. Eckhoff (1995) 'Accurate reconstructions of functions of finite regularity from truncated series expansions', *Math. Comp.* **64**, 671–690.

K. S. Eckhoff (1998) On a high order numerical method for functions with singularities, *Math. Comp.* **67**, 1063–1087.

K. S. Eckhoff and C. E. Wasberg (1995) On the numerical approximation of derivatives by a modified Fourier collocation method. Report No. 99, Department of Mathematics, University of Bergen, Norway.

S. Engelberg and E. Tadmor (2007) 'Recovery of edges from noisy spectral data: A new perspective'. Preprint.

N. Firoozye and V. Sverak (1996) 'Measure filters: An extension of Weiner's theorem', *Indiana Univ. Math. J.* **45**, 695–707.

G. Folland (1992), *Fourier Analysis and its Applications*, Brooks/Cole.

B. Fornberg (1996) *A Practical Guide to Pseudospectral Methods*, Cambridge University Press.

A. Gelb (2004) 'Parameter optimization and reduction of round-off error for the Gegenbauer reconstruction method', *J. Sci. Comput.* **20**, 433–459.

A. Gelb and S. Gottlieb (2007) 'The resolution of the Gibbs phenomenon for Fourier spectral methods'. Preprint.

A. Gelb and E. Tadmor (1999) 'Detection of edges in spectral data', *Appl. Comput. Harmon. Anal.* **7**, 101–135.

A. Gelb and E. Tadmor (2000a) 'Detection of edges in spectral data II: Nonlinear enhancement', *SIAM J. Numer. Anal.* **38**, 1389–1408.

A. Gelb and E. Tadmor (2000b) 'Enhanced spectral viscosity approximations for conservation laws', *Appl. Numer. Math.* **33**, 3–21.

A. Gelb and E. Tadmor (2002) 'Spectral reconstruction of one- and two-dimensional piecewise smooth functions from their discrete data', *Math. Model. Numer. Anal.* **36**, 155–175.

A. Gelb and E. Tadmor (2006) 'Adaptive edge detectors for piecewise smooth data based on the minmod limiter', *J. Sci. Comput.* **28**, 279–306.

A. Gelb and J. Tanner (2006) 'Robust reprojection methods for the resolution of the Gibbs phenomenon', *Appl. Comput. Harmon. Anal.* **20**, 3–25.

J. W. Gibbs (1899) 'Fourier Series', *Nature* **59**, 606.

B. I. Golubov (1972) 'Determination of the jump of a function of bounded p-variation by its Fourier series', *Math. Notes* **12**, 444–449.

D. Gottlieb and J. Hesthaven (2001) 'Spectral methods for hyperbolic problems', *J. Comput. Appl. Math.* **128**, 83–131.

D. Gottlieb and S. Orszag (1977) *Numerical Analysis of Spectral Methods: Theory and Applications*, Vol. 26 of *CBMS-NSF*, SIAM.

D. Gottlieb, C.-W. Shu, A. Solomonoff and H. Vandeven (1992) 'On the Gibbs phenomenon I: Recovering exponential accuracy from the Fourier partial sum of a non-periodic analytic function', *J. Comput. Appl. Math.* **43**, 81–98.

D. Gottlieb and C.-W. Shu (1995a) 'On the Gibbs phenomenon IV: Recovering exponential accuracy in a sub-interval from a Gegenbauer partial sum of a piecewise analytic function', *Math. Comp.* **64**, 1081–1095.

D. Gottlieb and C.-W. Shu (1995b) 'On the Gibbs phenomenon V: Recovering exponential accuracy from collocation point values of a piecewise analytic function', *Numer. Math.* **71**, 511–526.

D. Gottlieb and C.-W. Shu (1997) 'The Gibbs phenomenon and its resolution', *SIAM Review* **39**, 644–668.

D. Gottlieb and C.-W. Shu (1998) 'General theory for the resolution of the Gibbs phenomenon', Accademia Nazionale Dei Lincey, *ATTI Dei Convegni Lincey* **147**, 39–48.

D. Gottlieb and E. Tadmor (1985) 'Recovering pointwise values of discontinuous data within spectral accuracy', in *Progress and Supercomputing in Computational Fluid Dynamics: Proc. 1984 US–Israel Workshop*, Vol. 6 of *Progress in Scientific Computing* (E. M. Murman and S. S. Abarbanel, eds), Birkhäuser, Boston, pp. 357–375.

L. Grafakos (2004) *Classical and Modern Fourier Analysis*, Pearson Education.

A. Harten (1983) 'High resolution schemes for hyperbolic conservation laws', *J. Comput. Phys.* **49**, 357–393.

E. Hewitt and R. Hewitt (1979) 'The Gibbs–Wilbraham phenomenon: An episode in Fourier analysis', *Arch. Hist. Exact Sci.* **21**, 129–160.

D. K. Hoffman and D. J. Kouri (2000) 'Hierarchy of local minimum solutions of Heisenberg's uncertainty principle', *Phys. Rev. Lett.* **85**, 5263–5267.

Y. Katznelson (1976) *An Introduction to Harmonic Analysis*, Dover.

T. W. Körner (1988) *Fourier Analysis*, Cambridge University Press.

G. Kvernadze (1998) 'Determination of the jump of a bounded function by its Fourier series', *J. Approx. Theory* **92**, 167–190.

Y. Maday. S. Ould-Kaber and E. Tadmor (1993) 'Legendre pseudospectral viscosity method for nonlinear conservation laws', *SIAM J. Numer. Anal.* **30**, 321–342.

A. Majda, J. McDonough and S. Osher (1978) 'The Fourier method for nonsmooth initial data', *Math. Comp.* **30**, 1041–1081.

S. Mallat (1989) 'Multiresolution approximations and wavelets orthonormal bases of $L^2(R)$', *Trans. Amer. Math. Soc.* **315**, 69–87.

D. Marr and E. Hildreth (1980) 'Theory of edge detection', *Proc. Roy. Soc. Lond.* **B207**, 187–217.

H. N. Mhaskar and J. Prestin (2000) 'On the detection of singularities of a periodic function', *Adv. Comput. Math.* **12**, 95–131.

S. Pilipovic and N. Teofanov (2002) 'Wilson bases and ultramodulation spaces', *Math. Nachr.* **242**, 179–196.

N. Srinivasa and K. Rajgopal (1992) 'Detection of edges from projections', *IEEE Trans. Medical Imaging* **11**, 76–80.

E. Stein (1993) *Harmonic Analysis: Real-Variable Methods, Orthogonality and Oscillatory Integrals*, Vol. 32 of *Princeton Mathematical Series*, Princeton University Press.

G. Szegő (1958) *Orthogonal Polynomials*, AMS, Providence, RI.

E. Tadmor (1986) 'The exponential accuracy of Fourier and Chebyshev differencing methods', *SIAM J. Numer. Anal.* **23**, 1–10.

E. Tadmor (1989) 'Convergence of spectral methods for nonlinear conservation laws', *SIAM J. Numer. Anal.* **26**, 30–44.

E. Tadmor (1994) *Spectral Methods for Hyperbolic Problems*, Lecture Notes delivered at Ecole des Ondes, INRIA, Rocquencourt, January 24–28: www.cscamm.umd.edu/people/faculty/tadmor/pub/spectral-approximations/Tadmor.INRIA-94.pdf.

E. Tadmor (1998) 'Approximate solutions of nonlinear conservation laws', in *Advanced Numerical Approximation of Nonlinear Hyperbolic Equations*, Vol. 1697 of *Lecture Notes in Mathematics*, 1997 CIME course in Cetraro, Italy (A. Quarteroni, ed.), Springer, pp. 1–149.

E. Tadmor and J. Tanner (2002) 'Adaptive mollifiers: High resolution recovery of piecewise smooth data from its spectral information', *J. Found. Comput. Math.* **2**, 155–189.

E. Tadmor and J. Tanner (2003) 'An adaptive order Godunov type central scheme', In *Hyperbolic Problems: Theory, Numerics, Applications: Proc. 9th International Conference in Pasadena, March 2002* (T. Hou and E. Tadmor, eds), Springer, pp. 871–880.

E. Tadmor and J. Tanner (2005) 'Adaptive filters for piecewise smooth spectral data', *IMA J. Numer. Anal.* **25**, 635–647.

E. Tadmor and J. Zou (2007) 'Novel edge detection methods for incomplete and noisy spectral data'. Preprint.

J. Tanner (2006) 'Optimal filter and mollifier for piecewise smooth spectral data', *Math. Comp.*

T. Tao (2005) 'An uncertainty principle for cyclic groups of prime order', *Math. Res. Lett.* **12**, 121–127.

A. Torchinsky (1986) *Real-Variable Methods in Harmonic Analysis*, Academic Press.

F. Ulupinar and G. Medioni (1988) 'Refining edges detected by the LoG operator', in *CVPT'88: Proc. Computer Society Conference on Computer Vision and Pattern Recognition*, pp. 202–207.

H. Vandeven (1991) 'Family of spectral filters for discontinuous problems', *J. Sci. Comput.* **6**, 159–192.

A. Zygmund (1959) *Trigonometric Series*, Cambridge University Press.

Acta Numerica (2007), pp. 379–478
doi: 10.1017/S0962492906330012

Numerical aspects of special functions

Nico M. Temme

Centrum voor Wiskunde en Informatica (CWI),
Kruislaan 413, NL-1098 SJ Amsterdam,
The Netherlands
E-mail: Nico.Temme@cwi.nl

This paper describes methods that are important for the numerical evaluation of certain functions that frequently occur in applied mathematics, physics and mathematical statistics. This includes what we consider to be the basic methods, such as recurrence relations, series expansions (both convergent and asymptotic), and numerical quadrature. Several other methods are available and some of these will be discussed in less detail. Examples will be given on the use of special functions in certain problems from mathematical physics and mathematical statistics (integrals and series with special functions).

CONTENTS

1. Introduction

Special functions arise in various branches of applied mathematics, mathematical statistics, physics, and engineering in the form of integrals, as solutions of differential or difference equations, as integrands of integrals, as terms of infinite series, and so on. In which form they arise is not very important for their numerical evaluation, because for all the common special functions many analytic representations exist.

In this paper we discuss elements of a selection of methods that we consider to be the most important for designing algorithms for special functions. Series expansions (convergent or asymptotic) are important, but usually the parameter domain cannot be completely covered by convergent or asymptotic series. For the intermediate area we consider numerical quadrature and recurrence relations to be the most useful tools.

We start the paper with hypergeometric series, mainly for notational purposes, in particular the Gauss hypergeometric function $_2F_1(a, b; c; z)$. We give information on the use of the several power series representations of this function in the complex z-plane, and give an alternative power series for domains that cannot be reached when using power series for the Gauss function.

The second basic method is numerical quadrature and we give an overview of the simplest, but also most efficient, quadrature rule for special functions, namely the trapezoidal rule. We consider this rule for both finite and infinite intervals, and present the remainders of this rule in several forms. Because the functions representing the integrands are always analytic in this area, we can apply the powerful results of complex analysis. The effectiveness of the rule and the elegant form of the remainders are mainly based on Cauchy's integral for analytic functions,

$$f(z) = \frac{1}{2\pi i} \int_{\mathcal{C}} \frac{f(\zeta)}{\zeta - z}\, d\zeta,$$

where $\mathcal{C}$ is a circle around the point z inside the domain where f is analytic. Viewing this from the perspective of numerical quadrature, with some imagination, we can say that Cauchy's integral gives the relation between an integral and just one function value, and this rule is exact.

Recurrence relations form the third important tool. We give the theory that is of practical use for computing special functions, for example, the Miller algorithm. We give details for Bessel functions, Legendre functions, and Gauss hypergeometric functions.

In the remaining sections we give information on how to deal with so-called uniform asymptotic expansions, in which the coefficients are usually difficult to handle numerically. We use uniform asymptotic expansions for the incomplete gamma function for the asymptotic and numerical inversion of these functions for large parameters. In this way we describe a method for the inversion of cumulative distribution functions, which is an important topic in mathematical statistics.

In a final section we discuss two problems in which special functions play a role in series expansions: a distribution function and the solution of a singular perturbation problem.

Our approach in almost all discussions is to keep an eye out for the large-parameter case. Many published algorithms work fine for small or medium-sized parameters. There are big challenges in extending existing algorithms, or developing completely new algorithms, for large real or complex parameters for rather common special functions. A second aspect of our approach is to consider the stability of the numerical method as the main issue. In particular, when several real or complex parameters occur, a strict error analysis is far beyond the daily practice of efficient numerical algorithms.

Almost all functions considered in this paper are defined in the *Handbook of Mathematical functions* (Abramowitz and Stegun 1964), which has now been updated to form the *Digital Library of Mathematical Functions*, of which the book and web version will be published in the very near future; see also http://dlmf.nist.gov/.

For further properties of special functions we refer to Olver (1997), where, just as in Wong (2001), information on asymptotic analysis can be found. For an introduction to special functions, and for details of some examples considered in this paper, we refer to Temme (1996). For a web link with many definitions and descriptions of special functions, as well as for generalized functions such as the Fox H-function, the Meijer G-function, the Kampé de Fériet function, the MacRobert E-Function, and the Appell functions, we refer to http://mathworld.wolfram.com/.

The topics mentioned in this paper, and several other topics, will be discussed extensively, with examples of software, in a new book entitled *Numerical Methods for Special Functions*, written by Amparo Gil, Javier Segura, and the present author. This book will be published by SIAM in 2007.

2. Series expansions

Many special functions can be defined by power series that are of hypergeometric type. That is, they can be defined by power series of the form

$$f(z) = \sum_{n=0}^{\infty} c_n z^n, \tag{2.1}$$

where c_{n+1}/c_n is a rational function of n. Examples are

$$e^z = \sum_{n=0}^{\infty} \frac{z^n}{n!}, \qquad (1+z)^a = \sum_{n=0}^{\infty} \binom{a}{n} z^n. \tag{2.2}$$

A useful framework for working with these functions is the class of generalized hypergeometric functions. We define

$$_pF_q\left(\begin{matrix} a_1, \ldots, a_p \\ b_1, \ldots, b_q \end{matrix}; z\right) = \sum_{n=0}^{\infty} \frac{(a_1)_n \cdots (a_p)_n}{(b_1)_n \cdots (b_q)_n} \frac{z^n}{n!}, \tag{2.3}$$

where $(a)_n$ is the Pochhammer symbol, also called the shifted factorial, defined by

$$(a)_0 = 1, \quad (a)_n = a(a+1)\cdots(a+n-1) \quad (n \geq 1). \tag{2.4}$$

In terms of the gamma function we have

$$(a)_n = \frac{\Gamma(a+n)}{\Gamma(a)}, \quad n = 0, 1, 2, \ldots. \tag{2.5}$$

The series in (2.3) defines an entire function in z if $p \leq q$.

In the case $p = q+1$ the infinite series converges if $|z| < 1$, and defines an analytic function in this disk. This function can be continued analytically outside the disk, with a branch cut from 1 to $+\infty$. Let

$$\gamma_q = (b_1 + \cdots + b_q) - (a_1 + \cdots + a_{q+1}). \tag{2.6}$$

Then on the circle $|z| = 1$ the series (2.3) is absolutely convergent if $\operatorname{Re}\gamma_q > 0$, convergent except at $z = 1$ if $-1 < \operatorname{Re}\gamma_q \leq 0$, and divergent if $\operatorname{Re}\gamma_q \leq -1$.

The binomial coefficient in (2.2) can be written in several forms,

$$\binom{a}{n} = \frac{\Gamma(a+1)}{n!\,\Gamma(a+1-n)} = (-1)^n \frac{\Gamma(n-a)}{n!\,\Gamma(-a)} = (-1)^n \frac{(-a)_n}{n!}, \tag{2.7}$$

and we find that the series in (2.2) can be written as

$$e^z = {}_0F_0\left(\begin{matrix} - \\ - \end{matrix}; z\right), \quad (1+z)^a = {}_2F_1\left(\begin{matrix} -a, & - \\ - \end{matrix}; -z\right) = {}_1F_0\left(\begin{matrix} -a \\ - \end{matrix}; -z\right). \tag{2.8}$$

The second relation holds for $|z| < 1$. When $a = m$, a non-negative integer, the binomial function in (2.2) becomes Newton's binomial formula, with only $m + 1$ terms. Also from (2.7) we see that $(-m)_n$ equals 0 when $n \geq m + 1$. In general, the power series in (2.3) terminates when one of the a_j equals a non-positive integer. In that case p and q can be any non-negative integer.

On the other hand, the $_pF_q$-function of (2.3) is not defined if one of the b_j equals a non-positive integer, except in the following typical case. Let $a_j = -m$ and $b_j = -m - \ell$, with ℓ, m nonnegative integers. Then we have (cf. (2.7))

$$\frac{(a_j)_n}{(b_j)_n} = \frac{(-m)_n}{(-m-\ell)_n} = \begin{cases} \dfrac{m!}{(m-n)!}\,\dfrac{(m+\ell-n)!}{(m+\ell)!}, & m \geq n, \\[2mm] 0, & m < n, \end{cases} \tag{2.9}$$

Other examples of hypergeometric functions are the Bessel functions, with the special case

$$\Gamma(\nu+1)(\tfrac{1}{2}z)^{-\nu}J_\nu(z) = {}_0F_1\left(\begin{matrix}-\\\nu+1\end{matrix}; -\tfrac{1}{4}z^2\right) = e^{-iz}{}_1F_1\left(\begin{matrix}\nu+\tfrac{1}{2}\\2\nu+1\end{matrix}; 2iz\right),$$

(2.10)

where J_ν denotes the ordinary Bessel function of the first kind.

The ${}_1F_1$-function is also denoted by

$$M(a,c,z) = {}_1F_1\left(\begin{matrix}a\\c\end{matrix}; z\right),$$

(2.11)

and M is also called the confluent hypergeometric function. This entire function is a solution of the Kummer differential equation

$$zw'' + (c-z)w' - aw = 0.$$

(2.12)

A second solution of this equation is denoted by $U(a;c;z)$ and can be defined by

$$U(a;c;z) = \frac{1}{\Gamma(a)}\int_0^\infty e^{-zt}t^{a-1}(1+t)^{c-a-1}\,dt.$$

(2.13)

This function cannot be written as a convergent ${}_pF_q$-function. However, see (2.18).

The Airy function $\mathrm{Ai}(z)$, a solution of the differential equation $w''-zw=0$, is an entire function with power series representation

$$\mathrm{Ai}(z) = c_1 f(z) - c_2 g(z),$$
$$f(z) = 1 + \tfrac{1}{3!}z^3 + \tfrac{1\cdot 4}{6!}z^6 + \tfrac{1\cdot 4\cdot 7}{9!}z^9 + \cdots,$$
$$g(z) = z + \tfrac{2}{4!}z^4 + \tfrac{2\cdot 5}{7!}z^7 + \tfrac{2\cdot 5\cdot 8}{10!}z^{10} + \cdots,$$
$$c_1 = 3^{-\frac{2}{3}}\Gamma(\tfrac{2}{3}), \qquad c_2 = 3^{-\frac{1}{3}}\Gamma(\tfrac{1}{3}).$$

(2.14)

The functions f and g can be written as

$$f(z) = {}_0F_1\left(\begin{matrix}-\\\tfrac{2}{3}\end{matrix}; \tfrac{1}{9}z^3\right), \qquad g(z) = z\,{}_0F_1\left(\begin{matrix}-\\\tfrac{4}{3}\end{matrix}; \tfrac{1}{9}z^3\right).$$

(2.15)

Remark 2.1. For large complex values with $|\mathrm{ph}\, z| < \pi$, the Airy function behaves as follows:

$$\mathrm{Ai}(z) \sim \tfrac{1}{2}\pi^{-\frac{1}{2}}z^{-\frac{1}{4}}e^{-\frac{2}{3}z^{\frac{3}{2}}}.$$

(2.16)

We have $\mathrm{Ai}(5) = 0.000108344\ldots$. This value cannot be obtained by using the representation of $\mathrm{Ai}(z)$ in the first line of (2.14) and the power series for f and g, working to 10 digits. This is a nice example in which the power series, although convergent for all $z \in \mathbb{C}$, have a limited range of validity in numerical algorithms.

Notation for asymptotic expansions
When γ_q of (2.6) satisfies $\gamma_q \leq -1$ the definition in (2.3) has no meaning because the series diverges. However, the notation may be used in an asymptotic sense. For example, the confluent hypergeometric function $U(a; c; z)$ defined in (2.13) has the large z asymptotic expansion

$$U(a; c; z) \sim z^{-a} \sum_{n=0}^{\infty} \frac{(a)_n (1 + a - c)_n}{n!} (-z)^{-n}, \quad |\mathrm{ph}\, z| < \tfrac{3}{2}\pi, \qquad (2.17)$$

which we can write as

$$U(a; c; z) \sim z^{-a} {}_2F_0\left(\begin{matrix} a, \ 1 + a - c \\ - \end{matrix}; -\frac{1}{z}\right). \qquad (2.18)$$

This cannot be interpreted as an identity.

Remark 2.2. It is confusing that Maple 9.5 identifies the hypergeometric function in (2.18) with the Kummer U-function. For example, the Maple function *KummerU* gives the same numerical output as an evaluation based on the hypergeometric function shown on the right-hand side of (2.18), even when z is small.

2.1. The Gauss hypergeometric function

The Gauss hypergeometric function is the case $p = 2$, $q = 1$, that is,

$$ {}_2F_1\left(\begin{matrix} a, \ b \\ c \end{matrix}; z\right) = \sum_{n=0}^{\infty} \frac{(a)_n (b)_n}{(c)_n\, n!} z^n = 1 + \frac{ab}{c\, 1!} z + \frac{a(a+1)b(b+1)}{c(c+1)\, 2!} z^2 + \cdots, $$

$$(2.19)$$

where $c \neq 0, -1, -2, \ldots$ and $|z| < 1$. It is a solution of the hypergeometric differential equation

$$z(1-z)w'' + [c - (a + b + 1)z]w' - abw = 0. \qquad (2.20)$$

It is not difficult to verify that

$$z^{1-c} {}_2F_1\left(\begin{matrix} a - c + 1, \ b - c + 1 \\ 2 - c \end{matrix}; z\right) \qquad (2.21)$$

is a second solution of (2.20).

If $\mathrm{Re}\,(c - a - b) > 0$, the value of the Gauss hypergeometric function at $z = 1$ is given by

$$ {}_2F_1\left(\begin{matrix} a, \ b \\ c \end{matrix}; 1\right) = \frac{\Gamma(c)\Gamma(c - a - b)}{\Gamma(c - a)\Gamma(c - b)}. \qquad (2.22)$$

The power series for the Gauss hypergeometric function provides a simple and efficient means for the computation of this function, when z is properly inside the unit disk. The terms can easily be computed by the recursion in the representation

$$
{}_2F_1\left(\begin{matrix} a,\ b \\ c \end{matrix}; z\right) = \sum_{n=0}^{\infty} T_n, \quad T_{n+1} = \frac{(a+n)(b+n)}{(c+n)(n+1)} T_n, \quad n \geq 0, \quad T_0 = 1.
$$

$$(2.23)$$

Other power series for the Gauss hypergeometric function
The power series in (2.19) converges inside the unit disk and for numerical computations we can use only the disk $|z| \leq \rho < 1$. Other power series are available, however, to extend this domain.

The ${}_2F_1$-function with argument z can be written in terms of one or two other ${}_2F_1$-functions with argument

$$
\frac{1}{z}, \quad 1-z, \quad \frac{1}{1-z}, \quad \frac{z}{z-1}, \quad \frac{z-1}{z}. \tag{2.24}
$$

For a useful set of relations we refer to Abramowitz and Stegun (1964, p. 559) and Temme (1996, pp. 110 and 113).

When we restrict the absolute values of the quantities in (2.24) to the bound ρ, with again $0 < \rho < 1$, we find, writing $z = x + iy$,

$$
|z| \leq \rho, \quad \Longrightarrow \quad x^2 + y^2 \leq \rho^2,
$$

$$
\left|\frac{1}{z}\right| \leq \rho, \quad \Longrightarrow \quad x^2 + y^2 \geq \frac{1}{\rho^2},
$$

$$
|1 - z| \leq \rho, \quad \Longrightarrow \quad (x-1)^2 + y^2 \leq \rho^2,
$$

$$
\frac{1}{|1-z|} \leq \rho, \quad \Longrightarrow \quad (x-1)^2 + y^2 \geq \frac{1}{\rho^2}, \tag{2.25}
$$

$$
\left|\frac{z}{1-z}\right| \leq \rho, \quad \Longrightarrow \quad \left(x - \frac{\rho^2}{1-\rho^2}\right)^2 + y^2 \leq \frac{\rho^2}{(1-\rho^2)^2},
$$

$$
\left|\frac{z}{1-z}\right| \geq \frac{1}{\rho}, \quad \Longrightarrow \quad \left(x - \frac{1}{1-\rho^2}\right)^2 + y^2 \geq \frac{\rho^2}{(1-\rho^2)^2}.
$$

The domains defined by these inequalities do not cover the entire z-plane. The points $z = e^{\pm \pi i/3}$ do not satisfy these six conditions, for any $\rho \in (0, 1)$. When $\rho \to 1$, the domain of points not satisfying the six conditions shrinks to the exceptional points $z = e^{\pm \pi i/3}$. See Figure 2.1, where these points are indicated with black dots, for the cases $\rho = \frac{1}{2}$ and $\rho = \frac{3}{4}$. In the light area none of the inequalities of (2.25) hold.

To compute the ${}_2F_1$-functions in a neighbourhood of the points $z = e^{\pm \pi i/3}$ many other methods are available. One very useful method is discussed now.

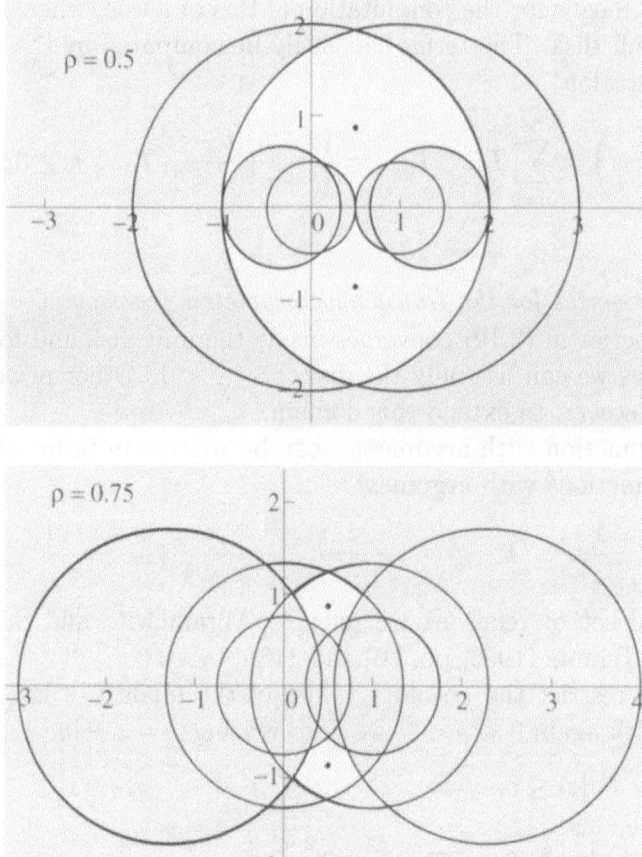

Figure 2.1. In the light domains none of the inequalities of (2.25) are satisfied. For $\rho \to 1$ these domains shrink to the points $e^{\pm \pi i/3}$, which are indicated by black dots.

Bühring's analytic continuation formula

Bühring (1987) derived power series expansions of the Gauss function, which enable computations near these special points. Bühring's expansion reads as follows. If $b - a$ is not an integer, we have for $|\mathrm{ph}(z_0 - z)| < \pi$ the continuation formula

$$
{}_2F_1\left(\begin{matrix} a,\, b \\ c \end{matrix}; z\right) = \frac{\Gamma(c)\Gamma(b-a)}{\Gamma(b)\Gamma(c-a)}(z_0 - z)^{-a} \sum_{n=0}^{\infty} d_n(a, z_0)(z - z_0)^{-n} \quad (2.26)
$$

$$
+ \frac{\Gamma(c)\Gamma(a-b)}{\Gamma(a)\Gamma(c-b)}(z_0 - z)^{-b} \sum_{n=0}^{\infty} d_n(b, z_0)(z - z_0)^{-n},
$$

where both series converge outside the circle $|z - z_0| = \max(|z_0|, |z_0 - 1|)$ and the coefficients are given by the three-term recurrence relation

$$d_n(s, z_0) = \frac{n + s - 1}{n(n + 2s - a - b)} \left[P_n d_{n-1}(s, z_0) + Q_n d_{n-2}(s, z_0) \right], \quad (2.27)$$

where $n = 1, 2, 3, \ldots$ and

$$P_n = (n+s)(1-2z_0)+(a+b+1)z_0-c, \qquad Q_n = z_0(1-z_0)(n+s-2), \quad (2.28)$$

with starting values

$$d_{-1}(s, z_0) = 0, \qquad d_0(s, z_0) = 1. \quad (2.29)$$

For the case that $b - a$ is an integer, a limiting process is needed (as in the next section). Details of the case $a = b$ are given in Bühring (1987), where different representations of the coefficients in the series of (2.26) are also given.

When we take $z_0 = \frac{1}{2}$ then the series in (2.26) converge outside the circle $|z - \frac{1}{2}| = \frac{1}{2}$, and both points $z = e^{\pm \pi i/3}$ discussed earlier are inside the domain of convergence.

Removable singularities

As mentioned earlier, there are several connection formulas that enable computation of the $_2F_1-$ outside the unit disk. One example is

$$_2F_1 \left(\begin{matrix} a, \ b \\ c \end{matrix}; z \right) = \frac{\Gamma(c)\Gamma(c - a - b)}{\Gamma(c - a)\Gamma(c - b)} \, _2F_1 \left(\begin{matrix} a, \ b \\ a + b - c + 1 \end{matrix}; 1 - z \right) \quad (2.30)$$

$$+ \frac{\Gamma(c)\Gamma(a + b - c)}{\Gamma(a)\Gamma(b)} (1 - z)^{c-a-b} \, _2F_1 \left(\begin{matrix} c - a, \ c - b \\ c - a - b + 1 \end{matrix}; 1 - z \right).$$

For some combinations of the parameters a, b and c the relation in (2.30) cannot be used straightforwardly. For example, when $c = a + b$, the gamma functions $\Gamma(c - a - b)$ and $\Gamma(a + b - c)$ are not defined, and the two Gauss functions become the same. From an analytical point of view the limiting form of (2.30) exists when $c \to a + b$, but from a numerical point of view instabilities arise when the relation in (2.30) is used in that case.

To see what happens in the limit $c \to a + b$, we write $c = a + b + \varepsilon$. We expand the Gauss functions in (2.30) in powers of $1 - z$ and obtain for the nth term

$$\frac{\Gamma(a + b + \varepsilon)\Gamma(1 + \varepsilon)\Gamma(1 - \varepsilon)(1 - z)^n}{\Gamma(a)\Gamma(b)\Gamma(a + \varepsilon)\Gamma(b + \varepsilon)\Gamma(1 - \varepsilon + n)\Gamma(1 + \varepsilon + n)n!} f(\varepsilon), \quad (2.31)$$

where

$$f(\varepsilon) = \frac{1}{\varepsilon} (\Gamma(a + n)\Gamma(b + n)\Gamma(1 + \varepsilon + n) \quad (2.32)$$

$$- (1 - z)^{\varepsilon} \Gamma(a + \varepsilon + n)\Gamma(b + \varepsilon + n)\Gamma(1 - \varepsilon + n)).$$

Taking the limit $\varepsilon \to 0$ in $f(\varepsilon)$, we find

$$
{}_2F_1\left(\begin{matrix} a,\ b \\ a+b \end{matrix}; z\right) = \frac{\Gamma(a+b)}{\Gamma(a)\Gamma(b)} \tag{2.33}
$$

$$
\times \sum_{n=0}^{\infty} \frac{(a)_n(b)_n}{n!\,n!} \left(2\psi(n+1) - \psi(a+n) - \psi(b+n) - \ln(1-z)\right)(1-z)^n,
$$

where $\psi(z)$ is the logarithmic derivative of the gamma function:

$$
\psi(z) = \frac{\Gamma'(z)}{\Gamma(z)}. \tag{2.34}
$$

This expansion holds for $|z - 1| < 1$ with $|\mathrm{ph}(1 - z)| < \pi$. Thus, there is, as usual, a branch cut from $z = 1$ to $z = +\infty$, and z is not on this cut. The logarithm $\ln(1 - z)$ assumes its principal branch, which is real for $z < 1$.

Gauss hypergeometric functions: Concluding remarks
We conclude with the following observations.

- The power series (2.19) of the Gauss hypergeometric function is a very important representation for numerical evaluations of this function.

- Non-trivial problems may arise, even for real values of the parameters and argument.

- With the transformation formulas that write the function in terms of functions with arguments shown in (2.24), we cannot reach all points in the complex z-plane.

- These formulas may cause numerical difficulties for certain combinations of the parameters a, b and c because of removable singularities in the formulas.

- In Forrey (1997) many details are discussed for the numerical use of the transformation formulas, and details of a Fortran program are given.

- Other instabilities will arise in the power series evaluation when a and/or b and/or c assume large (complex) values.

- Some of the problems with these large parameters can be resolved by using recurrence relations for these functions. See Section 4 for more details, in particular Section 4.3.

2.2. Chebyshev expansions of special functions

A function f that is continuous and of finite variation in $[-1, 1]$ can be expanded in terms of Chebyshev polynomials, and the expansion given by

$$f(x) = \tfrac{1}{2}c_0 + \sum_{k=1}^{\infty} c_k T_k(x), \quad -1 \leq x \leq 1. \tag{2.35}$$

By using the orthogonality of $T_n(x)$, the coefficients of the expansion can be given as integrals in the form

$$c_k = \frac{2}{\pi} \int_{-1}^{1} f(x) T_k(x)(1 - x^2)^{-1/2} \, dx \tag{2.36}$$

$$= \frac{2}{\pi} \int_{0}^{\pi} f(\cos \theta) \cos k\theta \, d\theta \approx \frac{2}{n} \sum_{j=0}^{n}{}'' f\left(\cos \frac{\pi j}{n}\right) \cos \frac{\pi k j}{n},$$

where the approximation is for sufficiently large n. The double prime over the summation indicates that the first and last terms are to be multiplied by $\tfrac{1}{2}$. Chebyshev coefficients are thus Fourier cosine transform coefficients of the function evaluated at non-uniformly spaced points.

As Clenshaw (1957) explained, the coefficients in a Chebyshev expansion can also be obtained from recurrence relations when the function satisfies a linear differential equation with polynomial coefficients; see also Gil, Segura and Temme (2007b). All special functions of hypergeometric type satisfy such a differential equation. However, for many special functions we can obtain expansions in which the coefficients can be expressed in terms of known special functions. As an example, we have (see Luke (1969a, p. 37))

$$J_0(ax) = \sum_{n=0}^{\infty} \varepsilon_n (-1)^n J_n^2(a/2) T_{2n}(x),$$

$$J_1(ax) = 2 \sum_{n=0}^{\infty} (-1)^n J_n(a/2) J_{n+1}(a/2) T_{2n+1}(x), \tag{2.37}$$

where $-1 \leq x \leq 1$ and $\varepsilon_0 = 1$, $\varepsilon_n = 2$ if $n > 0$. The parameter a can be any complex number. Similar expansions are available for J-Bessel functions of any complex order, in which the coefficients are $_1F_2$-hypergeometric functions, and explicit recurrence relations are available for computing the coefficients. For integer orders, the coefficients are a product of two J-Bessel functions. Again, see Luke (1969a).

Remark 2.3. The complexity of computing the coefficients of the expansions in (2.37) seems to be greater than the computation of the function that has been expanded. In some sense this is true, but as we will see in the next section, the coefficients in (2.37), and those of many other examples for special functions, satisfy linear recurrence relations, and the coefficients satisfying such relations can usually be computed very efficiently by the so-called backward recursion algorithm. Details for the second-order recurrence relation will be given in the next section.

Another example is the expansion for the error function:

$$e^{a^2x^2}\mathrm{erf}(ax) = \sqrt{\pi}e^{\frac{1}{2}a^2}\sum_{n=0}^{\infty}I_{n+\frac{1}{2}}\left(\tfrac{1}{2}a^2\right)T_{2n+1}(x), \quad -1 \le x \le 1, \quad (2.38)$$

in which the modified Bessel function is used. Again, a can be any complex number.

The expansions in (2.37) and (2.38) can be viewed as expansions near the origin. Other expansions are available that can be viewed as expansions at infinity, and these may be considered as alternatives for asymptotic expansions of special functions. For example, for the confluent hypergeometric U-function defined in (2.13) we have the convergent expansion in terms of shifted Chebyshev polynomials $T_n^*(x) = T_n(2x - 1)$:

$$(\omega z)^a U(a; c; \omega z) = \sum_{n=0}^{\infty} C_n(z)T_n^*(1/\omega), \quad z \ne 0, \quad |\mathrm{ph}\,z| < \tfrac{3}{2}\pi, \quad 1 \le \omega \le \infty.$$

$$(2.39)$$

Furthermore, $a, 1 + a - c \ne 0, -1, -2, \dots$. When equalities hold for these values of a and c the Kummer function reduces to a Laguerre polynomial. This follows from

$$U(a; c; z) = z^{1-c}U(1 + a - c; 2 - c; z) \tag{2.40}$$

and

$$U(-n; \alpha + 1; z) = (-1)^n n! L_n^\alpha(z), \quad n = 0, 1, 2, \dots. \tag{2.41}$$

The expansion (2.39) is given in Luke (1969a, p. 25). The coefficients can be represented in terms of generalized hypergeometric functions, in fact, Meijer G-functions, and they can be computed from the recurrence relation

$$\frac{2C_n(z)}{\varepsilon_n} = 2(n + 1)A_1 C_{n+1}(z) + A_2 C_{n+2}(z) + A_3 C_{n+3}(z), \tag{2.42}$$

where $b = a + 1 - c$, $\varepsilon_0 = \frac{1}{2}$, $\varepsilon_n = 1(n \ge 1)$, and

$$A_1 = 1 - \frac{(2n + 3)(n + a + 1)(n + b + 1)}{2(n + 2)(n + a)(n + b)} - \frac{2z}{(n + a)(n + b)},$$

$$A_2 = 1 - \frac{2(n + 1)(2n + 3 - z)}{(n + a)(n + b)}, \tag{2.43}$$

$$A_3 = -\frac{(n + 1)(n + 3 - a)(n + 3 - b)}{(n + 2)(n + a)(n + b)}.$$

Again, the coefficients satisfying this third-order recurrence relation can be computed by a backward recursion algorithm. For applying the backward recursion it is important to know that we have the normalization relation

$$\sum_{n=0}^{\infty}(-1)^n C_n(z) = 1, \quad |\text{ph } z| < \tfrac{3}{2}\pi. \tag{2.44}$$

This follows from

$$\lim_{\omega\to\infty}(\omega z)^a U(a;c;\omega z) = 1, \quad \text{and} \quad T_n^*(0) = (-1)^n, \tag{2.45}$$

and from the asymptotic expansion given in (2.18). The standard backward recursion scheme for computing the coefficients $C_n(z)$ only works for $|\text{ph } z| < \pi$. For $\text{ph } z = \pm\pi$ a modification seems to be possible; see Luke (1969a, p. 26).

Although the expansion in (2.39) converges for all $z \neq 0$ in the indicated sector, it is better to avoid small values of the argument of the U-function. Luke gives an estimate of the coefficients $C_n(z)$ of which the dominant factor that determines the speed of convergence is given by

$$C_n(z) = \mathcal{O}\big(n^{2(2a-c-1)/3}\,e^{-3n^{\frac{2}{3}}z^{\frac{1}{3}}}\big), \quad n\to\infty, \tag{2.46}$$

and we see that large values of $\text{Re } z^{\frac{1}{3}}$ improve the convergence.

The expansion in (2.39) can be used for all special cases of the Kummer U-function, that is, for Bessel functions (Hankel functions and K-modified Bessel functions), for the incomplete gamma function $\Gamma(a, z)$, with as special cases the complementary error function and exponential integrals.

Example 2.4. (Airy function) For the Airy function $\text{Ai}(x)$ we have the relations

$$\xi^{\frac{1}{6}}\,U(\tfrac{1}{6};\tfrac{1}{3};\xi) = 2\sqrt{\pi}x^{\frac{1}{4}}e^{\frac{1}{2}\xi}\text{Ai}(x),$$
$$\xi^{-\frac{1}{6}}\,U(-\tfrac{1}{6};-\tfrac{1}{3};\xi) = -2\sqrt{\pi}x^{-\frac{1}{4}}e^{\frac{1}{2}\xi}\text{Ai}'(x), \tag{2.47}$$

where $\xi = \tfrac{4}{3}x^{\frac{3}{2}}$. We take $\omega = (x/7)^{3/2}$ and $z = \tfrac{4}{3}7^{3/2} = 24.69\ldots$.

To generate the coefficients with this value of z we determine the smallest value of n for which the exponential factor in (2.46) is smaller than 10^{-15}. This gives $n = 8$. Next we generate, for both U-functions in (2.47), the coefficients $C_n(z)$ by using (2.42) in the backward direction. Details of this method for second-order recurrence relations are given in Section 4.

We take starting values

$$\tilde{C}_{19}(z) = 1, \quad \tilde{C}_{20}(z) = 0, \quad \tilde{C}_{21}(z) = 0. \tag{2.48}$$

Table 2.1. Coefficients of the Chebyshev expansion (2.51).

n	$C_n(z)$	$D_n(z)$
0	0.99727 33955 01425	1.00383 55796 57251
1	−0.00269 89587 07030	0.00380 27374 06686
2	0.00002 71274 84648	−0.00003 22598 78104
3	−0.00000 05043 54523	0.00000 05671 25559
4	0.00000 00134 68935	−0.00000 00147 27362
5	−0.00000 00004 63150	0.00000 00004 97977
6	0.00000 00000 19298	−0.00000 00000 20517
7	−0.00000 00000 00938	0.00000 00000 00989
8	0.00000 00000 00052	−0.00000 00000 00054
9	−0.00000 00000 00003	0.00000 00000 00003

We also compute for normalization

$$S = \sum_{n=0}^{18} (-1)^n \widetilde{C}_n(z) = -0.902363242772764\,{}_{10}{}^{25}, \qquad (2.49)$$

where the numerical value is for the Ai-case. Finally we compute

$$C_n(z) = \widetilde{C}_n(z)/S, \quad n = 0, 1, 2, \ldots, 10. \qquad (2.50)$$

Using this scheme for the expansions (2.39) of both U-functions in (2.47), we obtain the coefficients $C_n(z)$ and $D_n(z)$ of the expansions

$$2\sqrt{\pi}x^{\frac{1}{4}}e^{\frac{2}{3}x^{3/2}}\,\mathrm{Ai}(x) \approx \sum_{n=0}^{9} C_n T_n^*\big((7/x)^{3/2}\big),$$

$$\qquad\qquad (2.51)$$

$$-2\sqrt{\pi}x^{\frac{1}{4}}e^{\frac{2}{3}x^{3/2}}\,\mathrm{Ai}'(x) \approx \sum_{n=0}^{9} D_n T_n^*\big((7/x)^{3/2}\big),$$

for $x \geq 7$, for which numerical values are given in Table 2.1.

3. Numerical quadrature

The many special functions that we meet in applications can usually be defined by integral representations. For straightforward use these integrals may not be the optimal choice for numerical quadrature. In many cases we like to transform the integral into a different one with a better condition for numerical computations. The condition can be improved by considering two important points.

- The original integral may behave badly because of strong oscillations, and the new representation may be free of strong oscillations.

- The integral may be very large or very small (because of large parameters) and with the transformation a dominant factor has been placed in front of the new integral.

For an example of the first point, consider the ordinary Bessel function of the first kind with integer order, which can be defined as (Abramowitz and Stegun 1964, equation 9.1.21)

$$J_n(z) = \frac{1}{2\pi} \int_{-\pi}^{\pi} \cos(z \sin \theta - n\theta) \, d\theta, \quad n \in \mathbb{Z}, \quad z \in \mathbb{C}. \tag{3.1}$$

This function has, for fixed z, the asymptotic behaviour (Abramowitz and Stegun 1964, 9.3.1)

$$J_n(z) \sim \frac{1}{\sqrt{2\pi n}} \left(\frac{ez}{2n} \right)^n, \quad n \to \infty, \tag{3.2}$$

and, for obtaining a high relative precision, the integral in (3.1) becomes useless when n is large. There is no simple transformation possible for this integral to extract the dominant asymptotic factor, and a different method should be used, based on contour integrals in the complex plane, to obtain an integral with a better condition.

In the case of the gamma function

$$\Gamma(z) = \int_0^{\infty} t^{z-1} e^{-t} \, dt, \quad \operatorname{Re} z > 0, \tag{3.3}$$

a simple transformation $t = zs$ (first assuming that $z > 0$) gives

$$\Gamma(z) = z^z e^{-z} \int_0^{\infty} e^{-z\phi(s)} \frac{ds}{s}, \quad \phi(s) = s - \log s - 1, \tag{3.4}$$

and the integral is now $\mathcal{O}(1/\sqrt{z})$ as $z \to \infty$.

Another important decision when using numerical quadrature for the computation of special functions is the choice of the quadrature rule or method. Because the integrals for special functions are always given in terms of integrands that are analytic functions (except possibly at the endpoints, as in the case of (3.3) and (3.4) at the origin) the conditions for the remainders are usually quite favourable.

In addition, we like to use an adaptive rule that can easily be applied with any number of nodes and weights, without calculating these quantities *a priori*. Of all known quadrature rules the trapezoidal rule is one of the simplest to apply and it has the property that no nodes and weights have to be computed. Moreover, when halving the stepsize for the nodes, earlier function values can be used in the finer rule.

3.1. The finite interval

We recall two forms of the n-point extended trapezoidal rule (Davis and Rabinowitz 1984). Let $n = 1, 2, 3, \ldots$ and

$$h = \frac{b - a}{n}. \tag{3.5}$$

Then the standard rule reads

$$\int_a^b f(t)\, dt = \tfrac{1}{2}h[f(a) + f(b)] + h \sum_{j=1}^{n-1} f(jh) + R_n, \tag{3.6}$$

and the modified or shifted rule is given by

$$\int_a^b f(t)\, dt = h \sum_{j=0}^{n-1} f(a + d + jh) + R_n^d, \quad 0 < d < h. \tag{3.7}$$

In both cases the standard theory gives that the error terms have the form

$$R_n, R_n^d = -\frac{n\, h^3}{12} f''(\xi), \tag{3.8}$$

for some point $\xi \in (a, b)$, and for functions with continuous second derivative on $[a, b]$.

Because of this form of the remainder R_n the standard trapezoidal rule is not a high precision rule, but under certain conditions the rule is extremely accurate. We will describe these conditions and will give several examples to show the efficiency of the rule.

We apply the trapezoidal rule to functions belonging to two classes, one for finite and one for infinite intervals. For finite integrals the class is defined as follows.

Definition 3.1.   Let $C_p^\infty([a, b])$ denote the class of C^∞-functions

$$f \colon [a, b] \mapsto \mathbb{C}, \quad -\infty < a < b < \infty, \tag{3.9}$$

with equal derivatives of all orders at a and b:

$$f^{(n)}(a) = f^{(n)}(b), \quad n = 0, 1, 2, \ldots. \tag{3.10}$$

Remark 3.2.   The functions of $C_p^\infty([a, b])$ met in this paper can often be continued analytically so as to be single-valued and regular in a region $D \subset \mathbb{C}$ containing (a, b). They can be continued outside the interval $[a, b]$ to become C^∞-periodic functions on $\mathbb{R}$ with period $b - a$.

As we will show later, the trapezoidal rule on $[a, b]$ is extremely efficient for functions belonging to $C_p^\infty([a, b])$. The integrand $\cos(z \sin \theta - n\theta)$ of (3.1) belongs to $C_p^\infty([-\pi, \pi])$. For moderate values of n and z the trapezoidal is indeed very efficient, as follows from the following example.

Table 3.1. Values of the remainders in the trapezoidal rule for the Bessel function $J_0(x)$ by using (3.11) and (3.12). The rightmost column gives the upper bound on $|R_n|$ based on (3.13).

n	R_n	$R_n^{h/2}$	bound of $|R_n|$
4	$-0.12_{10}(-000)$	$0.12_{10}(-000)$	$0.92_{10}(-000)$
8	$-0.48_{10}(-006)$	$0.48_{10}(-006)$	$0.27_{10}(-005)$
16	$-0.11_{10}(-021)$	$0.11_{10}(-021)$	$0.50_{10}(-021)$
32	$-0.13_{10}(-062)$	$0.13_{10}(-062)$	$0.56_{10}(-062)$
64	$-0.13_{10}(-163)$	$0.13_{10}(-163)$	$0.54_{10}(-163)$
128	$-0.53_{10}(-404)$	$0.53_{10}(-404)$	$0.21_{10}(-403)$

Example 3.3. (Bessel function $J_0(x)$) We take (3.1) over the interval $[0, \pi]$. Using (3.6) and (3.7) with $d = \frac{1}{2}h$, we obtain

$$\pi J_0(x) = \int_0^\pi \cos(x \sin t)\, dt = h + h \sum_{j=1}^{n-1} \cos\left[x \sin(jh)\right] + R_n, \qquad (3.11)$$

$$\pi J_0(x) = \int_0^\pi \cos(x \sin t)\, dt = h \sum_{j=0}^{n-1} \cos\left[x \sin\left(\tfrac{1}{2}h + jh\right)\right] + R_n^{h/2}, \qquad (3.12)$$

where $h = \pi/n$. We take $x = 5$ and have the results as shown in Table 3.1.

We observe that the error terms R_n and $R_n^{h/2}$ are much smaller than the upper bound that can be obtained from (3.8). Also, comparison of the two remainders (and considering more digits) shows that the differences in accuracy is negligible. In addition, R_n and $R_n^{h/2}$ have opposite signs. This phenomenon often occurs and will be explained later.

For more details we refer to Davis and Rabinowitz (1984). Luke (1969b, p. 218) considered this Bessel function integral in detail, and from this reference and Krumhaar (1965) we derive an upper bound for R_n,

$$|R_n| \le 2e^{x/2} \frac{(x/2)^{2n}}{(2n)!}, \qquad (3.13)$$

which is quite realistic for the value of x we have chosen. For a different representation of the remainder see Example 3.8.

Remainder estimate with higher derivatives
Other representations of the remainders which are completely different from those in (3.8) are related to Euler's summation formula; see Temme (1996, Section 1.1.3).

First we introduce a function $\widetilde{B}_m(x)$ that is related to the generalized Bernoulli polynomial $B_m(x)$, which is defined by the generating function relation

$$\frac{ze^{xz}}{e^z - 1} = \sum_{n=0}^{\infty} \frac{B_m(x)}{m!} z^m, \quad |z| < 2\pi. \tag{3.14}$$

Explicit representations of $B_m(x)$ are (see Abramowitz and Stegun (1964, equation 23.1.17–23.1.18))

$$B_{2m}(x) = \frac{2(-1)^{m+1}(2m)!}{(2\pi)^{2m}} \sum_{k=1}^{\infty} \frac{\cos(2\pi kx)}{k^{2m}}, \quad 0 \le x \le 1, \quad m \ge 1 \tag{3.15}$$

and

$$B_{2m+1}(x) = \frac{2(-1)^{m+1}(2m+1)!}{(2\pi)^{2m+1}} \sum_{k=1}^{\infty} \frac{\sin(2\pi kx)}{k^{2m+1}}, \quad 0 \le x \le 1, \quad m \ge 0. \tag{3.16}$$

The expansion for $B_1(x)$ holds for $0 < x < 1$.

The functions $\widetilde{B}_m(x)$ are defined as the periodic continuations of $B_m(x)$ outside the basic interval $[0,1]$. That is,

$$\widetilde{B}_m(x) = \begin{cases} B_m(x) & \text{if } 0 \le x \le 1; \\ \widetilde{B}_m(x-1) & \text{if } x \in \mathbb{R}. \end{cases} \tag{3.17}$$

In other words, $\widetilde{B}_m(x)$, $m \ge 2$, can also be defined by (3.15) and (3.16), with $x \in \mathbb{R}$.

Theorem 3.4. Suppose that f is differentiable in $[a, b]$ up to and including order $2m + 1$. Then we have

$$R_n = -\sum_{k=1}^{m} \frac{h^{2k} B_{2k}}{(2k)!} \left[f^{(2k-1)}(b) - f^{(2k-1)}(a) \right] + R_{n,m}, \tag{3.18}$$

where

$$R_{n,m} = -\frac{h^{2m+1}}{(2m+1)!} \int_a^b f^{(2m+1)}(t) \widetilde{B}_{2m+1}\left(\frac{s-a}{h}\right) ds, \tag{3.19}$$

Proof. See Temme (1996, Exercise 1.7, p. 23). □

This result corresponds to Euler's summation formula on finite intervals, in which a finite sum is expressed in terms of an integral plus terms and remainder of the form just given.

Corollary 3.5. (Euler's summation formula) Suppose that f is differentiable in $[0, n]$ up to and including order $2m + 1$, $n \geq 1$. Then

$$\sum_{j=0}^{n} f(j) = \int_{0}^{n} f(x)\,dx + \tfrac{1}{2}[f(0) + f(n)] \tag{3.20}$$

$$+ \sum_{j=1}^{m} \frac{B_{2j}}{(2j)!}\left[f^{(2j-1)}(n) - f^{(2j-1)}(0)\right] + R_{n,m},$$

where

$$R_{n,m} = \frac{1}{(2m+1)!} \int_{0}^{n} f^{(2m+1)}(x)\tilde{B}_{2m+1}(x)\,dx. \tag{3.21}$$

For the modified rule (3.7) we have the following result.

Theorem 3.6. Suppose that f is differentiable in $[a, b]$ up to and including order $2m + 1$. Let $d = \tfrac{1}{2}h$. Then

$$R_n^{h/2} = -\sum_{k=1}^{m} \frac{h^{2k} B_{2k}(\tfrac{1}{2})}{(2k)!}\left[f^{(2k-1)}(b) - f^{(2k-1)}(a)\right] + R_{n,m}^{h/2}, \tag{3.22}$$

where

$$R_{n,m}^{h/2} = -\frac{h^{2m+1}}{(2m+1)!} \int_{a}^{b} f^{(2m+1)}(t)\tilde{B}_{2m+1}\left(\frac{s-a}{h} + \frac{1}{2}\right) ds. \tag{3.23}$$

Proof. The proof is as for Theorem 3.4 and also follows from Luke (1969b, pp. 218–219) (with modifications and a few corrections). $\qquad\square$

Now, for $f \in C_p^\infty[a, b]$ (see Definition 3.1), we see that all terms in the sums of (3.18) and (3.22) vanish, and from (3.16) we infer

$$|R_{n,m}|, |R_{n,m}^{h/2}| \leq 2\left(\frac{h}{2\pi}\right)^{2m+1} \int_{a}^{b} \left|f^{(2m+1)}(t)\right| dt, \quad m = 1, 2, 3, \ldots . \tag{3.24}$$

This shows that for $f \in C_p^\infty[a, b]$ the trapezoidal rule has order h^{2m+1} for any $m \geq 1$. Because (3.16) implies that

$$\tilde{B}_{2m+1}(x + \tfrac{1}{2}) = \frac{2(-1)^{m+1}(2m+1)!}{(2\pi)^{2m+1}} \sum_{k=1}^{\infty} (-1)^k \frac{\sin(2\pi k x)}{k^{2m+1}}, \quad x \in \mathbb{R}, \tag{3.25}$$

we have also explained why the remainders $R_{n,m}$ and $R_{n,m}^{h/2}$ are almost equal, and may have opposite signs.

Adding the two remainders we have

$$R_{n,m} + R_{n,m}^{h/2} = \tag{3.26}$$

$$-\frac{h^{2m+1}}{(2m+1)!} \int_{a}^{b} f^{(2m+1)}(t)\left[\tilde{B}_{2m+1}\left(\frac{s-a}{h}\right) + \tilde{B}_{2m+1}\left(\frac{s-a}{h} + \frac{1}{2}\right)\right] ds,$$

and using (3.16) and (3.25) we see that

$$\tilde{B}_{2m+1}(x)+\tilde{B}_{2m+1}(x+\tfrac{1}{2}) = \frac{1}{2^{2m}}\frac{2(-1)^{m+1}(2m+1)!}{(2\pi)^{2m+1}}\sum_{k=1}^{\infty}\frac{\sin(4\pi kx)}{k^{2m+1}}. \quad (3.27)$$

It is obvious that when we add the standard trapezoidal rule (3.6) and the modified rule (3.7) with $d = \tfrac{1}{2}h$, we obtain the standard rule with n replaced by $2n$ (and with h replaced by $h/2$). From (3.26) and (3.27) we see that the sum of the remainders of both rules follows the same pattern. From Table 3.1 we see that doubling the number of terms initially triples the number of correct decimal digits.

Representation of the remainder based on Fourier series
In the next theorem we express the remainder of the trapezoidal rule in terms of the Fourier series of f.

Theorem 3.7. Assume that f is continuous on $[a, b]$ with an absolutely and uniformly convergent Fourier series

$$f(x) = \sum_{k=-\infty}^{\infty} a_k e^{i\omega kx}, \quad a_k = \frac{\omega}{2\pi}\int_a^b f(x)e^{-i\omega kx}\,dx, \quad (3.28)$$

where $\omega = 2\pi/(b-a)$. Then R_n of (3.6) is given by

$$R_n = -(b-a)\sum_{\substack{m=-\infty\\m\neq 0}}^{\infty} e^{i\omega amn} a_{mn}. \quad (3.29)$$

For R_n^d of (3.7) we have

$$R_n^d = -(b-a)\sum_{\substack{m=-\infty\\m\neq 0}}^{\infty} e^{i\omega mn(a+d-h)} a_{mn}. \quad (3.30)$$

Proof. First consider R_n for a single Fourier term. That is, we consider (3.6) with $f_k(x) = e^{i\omega kx}$. The integral vanishes, and consequently

$$R_h = \frac{1}{2}h[f_k(b) - f_k(a)] - he^{i\omega ak}\sum_{j=1}^{n} e^{i\omega hjk}, \quad (3.31)$$

where $f_k(a) = f_k(b)$. From the well-known result

$$\sum_{j=1}^{n} e^{ij\theta} = e^{\frac{1}{2}i(n+1)\theta}\frac{\sin(n\theta/2)}{\sin(\theta/2)}, \quad (3.32)$$

we easily find, with $\theta = \omega h k = 2\pi k/n$ for $m = \pm 1, \pm 2, \ldots$,

$$R_n = \begin{cases} 0 & \text{if } k \neq mn, \\ -(b-a)e^{i\omega amn} & \text{if } k = mn. \end{cases} \tag{3.33}$$

Using this result for $f_k(x) = e^{i\omega k x}$ for all terms in the Fourier series of (3.28), we find the claim in (3.29). The result (3.30) for R_n^d follows easily. $\qquad\square$

When $d = \frac{1}{2}h$ in (3.30), we have

$$R_n^d = -(b-a) \sum_{\substack{m=-\infty \\ m\neq 0}}^{\infty} (-1)^m e^{i\omega mna} a_{mn}. \tag{3.34}$$

and when the Fourier expansion converges fast and we consider the first term approximations $m = \pm 1$ of (3.29) and (3.30), we have

$$R_n \sim -(b-a)\left[e^{i\omega an}a_n + e^{-i\omega an}a_{-n}\right], \quad R_n^{h/2} \sim -R_n. \tag{3.35}$$

Example 3.8. (Bessel function $J_0(x)$) Consider the Fourier series

$$\cos(z\sin x) = \sum_{k=-\infty}^{\infty} J_{2k}(z)\cos(2kx) \tag{3.36}$$

(Abramowitz and Stegun 1964, equation 9.1.42). We have $a = 0$, $b = \pi$, $\omega = 2$, $a_k = J_{2k}(z)$, and $J_{-k}(z) = (-1)^k J_k(z)$. The integral of the left-hand side equals $\pi J_0(z)$ (see (3.12)). For the remainder we have

$$R_n = -2\pi \sum_{m=1}^{\infty} J_{2mn}(z). \tag{3.37}$$

This gives an exact form of the remainder R_n of (3.11).

Contour integrals for the remainders
Representations of the remainders in (3.6) and (3.7) can be given in the form of contour integrals in the complex plane in which f but no derivatives of f are used. For details we refer to Takahasi and Mori (1970) and Mori (1974).

3.2. The infinite interval

For the trapezoidal rule on $\mathbb{R}$, we consider, for a certain class of f,

$$\int_{-\infty}^{\infty} f(t)\,dt = \sum_{j=-\infty}^{\infty} f(d+jh) + R_d(h), \quad 0 \le d < h. \tag{3.38}$$

This class is described as follows.

Definition 3.9. Let

$$G_a = \{z = x + iy \mid x \in \mathbb{R},\ |y| < a\} \tag{3.39}$$

be the strip in the complex domain of width $2a > 0$. Let H_a denote the class of bounded holomorphic functions $f\colon G_a \mapsto \mathbb{C}$ such that $\int_{-\infty}^{\infty} f(x+iy)\,dx$ converges, uniformly with respect to $y \in [-a, a]$, with $\lim_{x\to\pm\infty} f(x+iy) = 0$, uniformly with respect to $y \in [-a, a]$, and

$$M_{\pm a}(f) = \int_{-\infty}^{\infty} |f(x \pm ia)|\,dx = \lim_{b\uparrow a} \int_{-\infty}^{\infty} |f(x \pm ib)|\,dx < \infty. \tag{3.40}$$

For the remainder $R_d(h)$ in (3.38) we have the following result.

Theorem 3.10. Let $f \in H_a$ for some $a > 0$. Let $h > 0$ and $0 \le d < h$. Then

$$R_d(h) = \int_{-\infty}^{\infty} \frac{f(x+iy)}{1 - e^{-2\pi i(x+iy-d)/h}}\,dx + \int_{-\infty}^{\infty} \frac{f(x-iy)}{1 - e^{2\pi i(x-iy-d)/h}}\,dx, \tag{3.41}$$

for any $y \in (0, a)$.

Proof. The known proof is based on residue calculus for evaluating integrals. Observe that

$$\frac{1}{2\pi i} \int_{\partial G_y} f(z)\cot(\pi(z-d)/h)\,dz = h \sum_{j=-\infty}^{\infty} f(d+jh), \tag{3.42}$$

where ∂G_y is the boundary of G_y (see (3.39)), and the integration in (3.42) is in the positive direction. Furthermore,

$$\int_{-\infty}^{\infty} f(x)\,dx + \int_{\infty}^{-\infty} f(x+iy)\,dx = 0 \tag{3.43}$$

and

$$\int_{-\infty}^{\infty} f(x)\,dx + \int_{\infty}^{-\infty} f(x-iy)\,dx = 0. \tag{3.44}$$

Combining these results we arrive at (3.41). $\qquad\square$

Corollary 3.11. Let $f \in H_a$ for some $a > 0$ and let f be even (which is not a restriction for integrals over $\mathbb{R}$). Then $R_d(h)$ in (3.38) can be bounded in the following way:

$$|R_d(h)| \le \frac{e^{-\pi a/h}}{\sinh(\pi a/h)} M_a(f), \tag{3.45}$$

where $M_a(f)$ is given in (3.40).

From (3.45) it follows that the error in the trapezoidal rule for small h is $\mathcal{O}\big(\exp(-2\pi a/h)\big)$. Also, large values of a result in small errors, but the quantity $M_a(f)$ may influence the error if a is large. Often, it is advisable to choose h and a such that the right-hand side of (3.45) is minimized.

For example, when f has the form

$$f(x) = e^{-\omega x^2} g(x), \quad \omega > 0, \quad g \in H_a, \tag{3.46}$$

for some $a > 0$ and g an even function, (3.45) can be written as

$$|R_d(h)| \leq \frac{e^{\omega a^2 - \pi a/h}}{\sinh(\pi a/h)} M_a(g). \tag{3.47}$$

If g allows, we may choose a value of a that makes the exponential function very large. However, the function $-2\pi a/h + \omega a^2$, considered as a function of a, is minimal for

$$a = \frac{\pi}{\omega h}. \tag{3.48}$$

When we can take this value (again, if g allows), and we suppose that $M_a(g) = \mathcal{O}(1)$, we conclude that the error satisfies

$$|R_d(h)| \leq C e^{-\pi^2/(\omega h^2)}, \quad \text{for some } C > 0. \tag{3.49}$$

Remark 3.12. In applications we may be interested in large values of ω. If we apply the bound in (3.49) to make $|R_d(h)| < \varepsilon$, we see that h should satisfy $h \sim \pi/\sqrt{\omega \log(1/\varepsilon)}$, which may be quite small, when ω is large. On the other hand, we need to consider the rate of convergence of the series in (3.38) with this value of h and ω for the function in (3.46). We see that $f(jh) = \exp(-\omega j^2 h^2) g(jh)$ and that $\omega j^2 h^2 = \pi^2 j^2 / \log(1/\varepsilon)$. So, the value of j that should be taken to neglect terms in the infinite series is not dependent on ω, when we take h as mentioned earlier. It is of some value to know the stepsize h, but for efficiency the number of terms is the most important issue. Of course, we can also take a new variable of integration $x = t/\sqrt{\omega}$ when f has the form of (3.46). In that case the number of terms in the series may again become independent of ω, and the singularities of g are moving off the real axis as ω increases.

Another representation of the remainder

Another representation for $R_d(h)$ can be obtained by using Poisson's summation formula, which we write in the general form

$$h \sum_{j=-\infty}^{\infty} e^{ij\alpha} f(d+jh) = \sum_{k=-\infty}^{\infty} F\left(\frac{2\pi k + \alpha}{h}\right) e^{-id(2\pi k + \alpha)/h}, \tag{3.50}$$

where F is the Fourier transform of f:

$$F(y) = \int_{-\infty}^{\infty} f(x) e^{ixy} \, dx. \tag{3.51}$$

Details on the Poisson summation formula and condition of validity can be found in Andrews, Askey and Roy (1999, Section D.4).

For $\alpha = 0$ we obtain a representation for $R_d(h)$ of (3.38):

$$R_d(h) = \sum_{k \neq 0} F\left(\frac{2\pi k}{h}\right) e^{-2\pi i k d/h}. \tag{3.52}$$

For even functions this becomes

$$R_d(h) = 2 \sum_{k=1}^{\infty} F\left(\frac{2\pi k}{h}\right) \cos(2\pi k d/h). \tag{3.53}$$

For functions $f \in H_a$ this result can also be obtained by expanding in (3.41) the integrands in geometric series with exponential functions.

Remark 3.13. Observe that, when the series converges fast, and the first term gives a good estimate of the error, the remainders $R_0(h)$ and $R_{h/2}(h)$ are almost equal in modulus and have opposite sign.

Including series with derivatives

An interesting generalization of (3.38) is based on the idea of Hermite interpolation. The result is summarized in the following theorem.

Theorem 3.14. Let $f \in H_a$, for some $a > 0$, be an even function. For any even positive integer p, let numbers $a_{q,p}$ be determined by the identity

$$c_{0,p} + c_{2,p}z^2 + \cdots + c_{p,p}z^p = \prod_{q=1}^{\frac{1}{2}p} \left((1 + (z/q))^2\right). \tag{3.54}$$

Then, for $h > 0$,

$$\int_{-\infty}^{\infty} f(x)\,dx = T_{p,h}(f) + R_{p,h}(f), \tag{3.55}$$

where

$$T_{p,h}(f) = h \sum_{\substack{q=0 \\ q\ \text{even}}}^{p} c_{q,p} \left(\frac{h}{2\pi}\right)^q \sum_{j=-\infty}^{\infty} f^{(q)}(jh), \tag{3.56}$$

and $R_{p,h}(f)$ is bounded as follows:

$$|R_{p,h}(f)| \leq \frac{e^{-\pi a/h}}{\sinh^{p+1}(\pi a/h)} M_a(f). \tag{3.57}$$

Proof. For a proof we refer to Kreß (1972). □

We give the rules with derivatives for $p = 0, 2, 4$:

$$T_{0,h}(f) = h \sum_{j=-\infty}^{\infty} f(jh),$$

$$T_{2,h}(f) = T_{0,h}(f) + \frac{h^3}{4\pi^2} \sum_{j=-\infty}^{\infty} f''(jh), \tag{3.58}$$

$$T_{4,h}(f) = T_{0,h}(f) + \frac{5h^3}{16\pi^2} \sum_{j=-\infty}^{\infty} f''(jh) + \frac{h^5}{64\pi^4} \sum_{j=-\infty}^{\infty} f''''(jh).$$

The coefficients $c_{q,p}$, of which the first few appear in (3.58), can easily be obtained as follows. By differentiating (3.38) with respect to d and by using (3.52) for $R_d(h)$, we obtain (by taking $d = 0$ afterwards) for even functions f:

$$\int_{-\infty}^{\infty} f(x)\, \mathrm{d}x = h \sum_{j=-\infty}^{\infty} f(jh) \quad - \quad 2 \sum_{k=1}^{\infty} F\left(\frac{2\pi k}{h}\right),$$

$$0 = h \sum_{j=-\infty}^{\infty} f''(jh) + \frac{8\pi^2}{h^2} \sum_{k=1}^{\infty} k^2 F\left(\frac{2\pi k}{h}\right), \tag{3.59}$$

$$0 = h \sum_{j=-\infty}^{\infty} f''''(jh) - \frac{32\pi^4}{h^4} \sum_{k=1}^{\infty} k^4 F\left(\frac{2\pi k}{h}\right),$$

and so on. By taking linear combinations of these equations we can eliminate $F(2\pi/h), F(4\pi/h), \ldots$, and the coefficients for these linear combinations are the numbers $c_{q,p}$. This procedure reminds us of the Romberg method, where terms in the error representation are eliminated by taking linear combinations of the trapezoidal rule with $h, h/2, h/4, \ldots$.

Example 3.15. ($f(x) = \mathrm{e}^{-x^2}$) The remainder $R_d(h)$ follows from (3.52) and is given by

$$R_d(h) = -2\sqrt{\pi} \sum_{k=1}^{\infty} \mathrm{e}^{-\pi^2 k^2/h^2} \cos(2\pi k d/h). \tag{3.60}$$

The first term can be compared with the estimate in (3.49) when $\omega = 1$. For an accuracy of 10^{-10}, $R_d(h)$ is negligible for h smaller than (approximately) $\pi/\sqrt{10\log(10)} = 0.65\cdots$. With this value we can estimate the number of terms in the infinite sum in (3.38). The terms are negligible for j larger than $\sqrt{10\log(10}/h = 7.3\cdots$.

To see the effect of including the series with the second derivative, we derive

$$R_{p,h}(f) = 2\sqrt{\pi} \sum_{k=2}^{\infty} (k^2 - 1)e^{-\pi^2 k^2/h^2}. \tag{3.61}$$

To obtain the same accuracy as for (3.60) we can take $h = 1.31$ and neglect terms in the infinite series for j larger than 3.66. Hence, by using the second derivative the number of terms in the infinite series can be halved in this example.

Example 3.16. $(f(x) = 1/(1 + x^2))$ This case is only of theoretical interest, because the series in (3.38) does not converge fast enough. In any case, we have $f \in H_a$, $a \in (0, 1)$, and we can evaluate

$$R_d(h) = -2\pi \sum_{k=1}^{\infty} e^{-2\pi k/h} \cos(2\pi k d/h) = -\frac{2\pi}{e^{2\pi/h} - 1}, \tag{3.62}$$

where the last result holds when $d = 0$. So, although the rule is useless, the error bound is perfect.

Slowly convergent integrals can become more rapidly convergent by using simple transformations. In the present case we use $x = \sinh t$ and obtain

$$\int_{-\infty}^{\infty} \frac{dx}{1 + x^2} = \int_{-\infty}^{\infty} \frac{dt}{\cosh t}. \tag{3.63}$$

The new integrand belongs to H_a, $a \in (0, \frac{1}{2}\pi)$, and

$$R_d(h) = -2\pi \sum_{k=1}^{\infty} \frac{\cos(2\pi k d/h)}{\cos(\pi^2 k/h)}. \tag{3.64}$$

Taking into account a pole singularity
Kress and Martensen (1970) consider the trapezoidal rule for Cauchy principal value integrals of the form

$$\int_{-\infty}^{\infty} \frac{f(x)}{x - \xi} \, dx, \quad \xi \in \mathbb{R}, \tag{3.65}$$

with $f \in H_a$ for some $a > 0$. Even more interesting is a generalization of this taking ξ to be a complex number inside the strip G_a. The integral is no longer interpreted as a Cauchy principal value. The pole at $x = \xi$ gives an extra residue term in the right-hand side of (3.38).

A standard integral for this case is the Hilbert transform of the Gaussian (in physics, the plasma dispersion function)

$$w(z) = \frac{1}{\pi i} \int_{-\infty}^{\infty} \frac{e^{-t^2}}{t - z} \, dt, \quad \text{Im } z > 0, \tag{3.66}$$

which is an entire function after analytic continuation and is related to the complementary error function. We have (see also (5.12))

$$w(z) = e^{-z^2} \operatorname{erfc}(-iz), \quad z \in \mathbb{C}. \tag{3.67}$$

Matta and Reichel (1971) give several interesting examples for the complementary error function and related functions, such as the Fresnel integral and the Voigt functions. For the plasma dispersion function in (3.66) the trapezoidal rule becomes

$$w(z) = \frac{h}{\pi i} \sum_{j=-\infty}^{\infty} \frac{e^{-(jh)^2}}{jh - z} + \theta \frac{2e^{-z^2}}{1 - e^{-2\pi i z/h}} + R(h), \tag{3.68}$$

where $R(h) = \mathcal{O}(\exp(-\pi^2/h^2))$ $(cf.$ (3.49)) and

$$\theta = \begin{cases} 1 & \text{if } \operatorname{Im} z < \pi/h, \\ \frac{1}{2} & \text{if } \operatorname{Im} z = \pi/h, \\ 0 & \text{if } \operatorname{Im} z > \pi/h. \end{cases} \tag{3.69}$$

To avoid numerical instability for the case that z is close to jh for some j, we can change h or use a shifted rule. However, when we combine the particular j-term with the extra term in (3.68), the limit exists if $z \to jh$. For example, when $z \to 0$, we have

$$\lim_{z \to 0} \left[-\frac{h}{\pi i z} + \frac{2e^{-z^2}}{1 - e^{-2\pi i z/h}} \right] = 1, \tag{3.70}$$

which corresponds to $w(0) = 1$ (this value follows from (3.67)).

For $\operatorname{Im} z < 0$ we can use

$$w(-z) = 2e^{-z^2} - w(z), \tag{3.71}$$

although (3.68) also holds when $\operatorname{Im} z \leq 0$.

By using (3.68) and by taking into account the combination of singular terms or a proper choice of h, a uniform numerical algorithm for the function $w(z)$ and related functions can be constructed.

The function $w(z)$ is important in uniform asymptotic expansions of certain integrals when a pole and saddle point are close together; see Section 5.2, where the complementary error function is used in asymptotic approximations of the incomplete gamma functions.

3.3. Contour integrals

We discuss examples in which the starting point is a contour integral in the complex plane. We describe how simple transformations, inspired by the saddle point method for contour integrals, produce new integrals that are suitable for applying the trapezoidal rule on finite or infinite intervals.

First we compare two simple integrals,

$$F(\lambda) = \int_{-\infty}^{\infty} e^{-t^2 + 2i\lambda\sqrt{t^2+1}} \, dt \qquad (3.72)$$

and

$$G(\lambda) = \int_{-\infty}^{\infty} e^{-t^2 + 2i\lambda t} \, dt. \qquad (3.73)$$

When we compute $F(\lambda)$ by using Maple 9.5 with Digits $= 10$, we obtain

$$F(10) = -0.1837516481 + 0.53053428931. \qquad (3.74)$$

With Digits $= 40$, the answer is

$$F(10) = -0.18375164805320696644188906630534087900017 \qquad (3.75)$$
$$+ 0.53053428925506068760950289282504487400201.$$

So, the first answer seems to be correct in all shown digits.

The integral in (3.73) is slightly different and we repeat the computations for $G(\lambda)$. Maple 9.5, Digits $= 10$, for $\lambda = 10$, gives

$$G(10) = -0.1257674520_{10}{}^{-15}. \qquad (3.76)$$

With Digits $= 40$, the answer is

$$G(10) = 0.16_{10}{}^{-43}. \qquad (3.77)$$

The correct answer is

$$G(\lambda) = \sqrt{\pi} e^{-\lambda^2}, \qquad (3.78)$$

and for $\lambda = 10$ we have

$$G(10) = 0.6593662989_{10}{}^{-43}. \qquad (3.79)$$

Maple can evaluate the exact answer, but we forced Maple to use numerical quadrature to produce (3.76) and (3.77).

We also used Mathematica to compute $G(10)$ and we received the message

> *NIntegrate* failed to converge to prescribed accuracy after 7
> recursive bisections in t near $t = 2.9384615384615387$.

Obtaining no result is better than obtaining a completely wrong result.

The lesson is: One should have some feeling about the computed result when dealing with oscillatory integrals, otherwise a completely incorrect answer can be accepted. And $\lambda = 10$ is not really large.

From some points of view the integrals $F(\lambda)$ and $G(\lambda)$ are quite different, but an occasional user may not see this difference. The fact that $G\lambda)$ is so small compared with $F(\lambda)$ can be explained by observing that the 'phase function' of $F(\lambda)$, that is, $-t^2 + 2i\lambda\sqrt{t^2+1}$ has a stationary point (saddle

point) inside the domain of integration (at $t = 0$), whereas the saddle point
of $G(\lambda)$ (at $t = i\lambda$) is outside the domain.

By shifting the path upwards in the complex plane, through the saddle
point at $t = i\lambda$, or by just substituting $t = s + i\lambda$, we obtain

$$G(\lambda) = e^{-\lambda^2} \int_{-\infty}^{\infty} e^{-s^2} \, ds. \qquad (3.80)$$

Now we have the dominant factor in front of the integral, and the integral
(in this case we know its answer) can be calculated easily by numerical
quadrature, say, by using the trapezoidal rule.

This procedure will be carried out in several less trivial cases in which
a given integral is transformed by selecting a new contour in the complex
plane such that, for large values of parameters, a dominant factor is taken
in front of the new integral. We use the standard methods of asymptotic
analysis (saddle point methods) of which details can be found in Wong
(2001).

The evaluation of contour integrals is also important in the numerical
inversion of Laplace transforms, when the initial vertical line of integration
can be deformed into the left half-plane (Talbot 1979, Murli and Rizzardi
1990, Rizzardi 1995, Trefethen, Weideman and Schmelzer 2005, Weideman
2005).

Example 3.17. (The reciprocal gamma functions) The Maclaurin
series of the exponential function, $e^w = \sum_{n=0}^{\infty} w^n/n!$, implies the Cauchy
integral

$$\frac{1}{n!} = \frac{1}{2\pi i} \oint \frac{e^s}{s^{n+1}} \, ds, \qquad (3.81)$$

where the integral can be taken over a circle in the complex plane. A more
general form of this integral is Hankel's loop integral for the reciprocal
gamma function:

$$\frac{1}{\Gamma(z)} = \frac{1}{2\pi i} \int_{\mathcal{L}} \frac{e^s}{s^z} \, ds, \qquad (3.82)$$

where $\mathcal{L}$ is a contour as shown in Figure 3.1. It runs from $-\infty$ (with
ph $s = -\pi$), encircles the origin in the positive direction, and terminates at
$s - \infty$, now with ph $s = +\pi$. The many-valued function s^{-z} is assumed to
be equal to 1 at $s = 1$.

We consider $z > 0$ and substitute $s = zt$ and obtain

$$\frac{1}{\Gamma(z)} = \frac{e^z z^{1-z}}{2\pi i} \int_{\mathcal{L}} e^{z\phi(t)} \, dt, \quad \phi(t) = t - 1 - \log t, \qquad (3.83)$$

where the contour is the same as in Figure 3.1. Compare this with (3.3).
We choose a special contour on which the imaginary part of $\phi(t) = C$, where
C is a constant. In addition, we take this contour through the saddle point

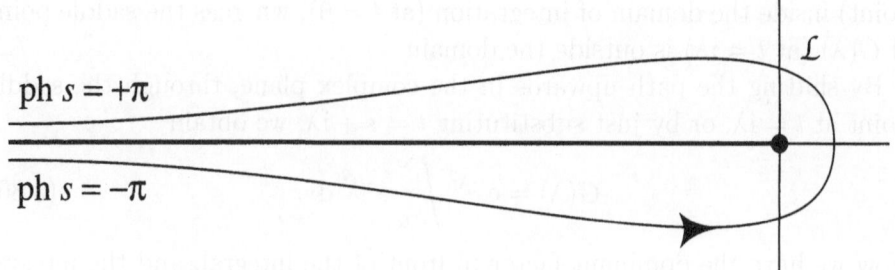

$ph\ s = +\pi$

$ph\ s = -\pi$

$\mathcal{L}$

Figure 3.1. Hankel contour for the integral in (3.82).

of the integrand at $t = 1$, where $\phi'(t) = 0$. Since $\phi(1) = 0$ we have $C = 0$. Writing $t = re^{i\theta}$, we find that the equation $\mathrm{Im}\,\phi(t) = r\sin\theta - \theta = 0$ is satisfied when the polar coordinates of t satisfy

$$r = \frac{\theta}{\sin\theta}, \quad -\pi < \theta < \pi. \tag{3.84}$$

On this path, the steepest descent path, the the function $\phi(t)$ of (3.83) is real and negative. The standard way to deal with integrals of this form in asymptotics is the transformation

$$\phi(t) = -\tfrac{1}{2}w^2, \tag{3.85}$$

in which the lower branch of the path corresponds to $w < 0$ and the upper part to $w > 0$. Locally at $t = 1$ we have

$$w = \mathrm{i}(t - 1)\big[1 - \tfrac{1}{3}(t - 1) + \cdots\big]. \tag{3.86}$$

The transformation (3.85) gives

$$\frac{1}{\Gamma(z)} = \frac{e^z z^{1-z}}{2\pi} \int_{-\infty}^{\infty} e^{-\frac{1}{2}zw^2} g(w)\,\mathrm{d}w, \tag{3.87}$$

where

$$g(w) = \frac{1}{\mathrm{i}}\frac{\mathrm{d}t}{\mathrm{d}w} = \frac{tw}{\mathrm{i}(1 - t)} = 1 + \tfrac{2}{3}\mathrm{i}w - \tfrac{1}{12}w^2 + \cdots. \tag{3.88}$$

The integrand in (3.87) is of the type (3.46), and we like to know the singularities of g in order to determine the strip G_a of Definition 3.9 around the real axis. The conformal mapping defined in (3.85) is regular at $t = 1$, but $t = \exp(2\pi i n)$, $n = \pm 1, \pm 2, \ldots$, which are important because of the many-valued logarithm in $\phi(t)$, give singular points. These points correspond to w-values of which those with $n = \pm 1$ are closest to the real w-axis. The singularities that define the strip G_a satisfy $\tfrac{1}{2}w^2 = \pm 2\pi i$, which have imaginary parts equal to $\sqrt{2\pi} \doteq 2.51$. We also see that the series in (3.88) has radius of convergence $2\sqrt{\pi} \doteq 3.54$

Table 3.2. Values of h and the remainders in the trapezoidal rule for the reciprocal gamma function $1/\Gamma(z)$ by using (3.87) and (3.38). The last column gives the sum of the relative errors to show that the remainders are almost equal with opposite sign.

z	h	rel. error, $d = 0$	rel. error, $d = \frac{1}{2}h$	sum of rel. errors
1	0.926	$-0.41\,_{10}(-7)$	$0.41\,_{10}(-7)$	$0.23\,_{10}(-13)$
2	0.656	$0.46\,_{10}(-9)$	$-0.46\,_{10}(-9)$	$-0.30\,_{10}(-14)$
3	0.536	$0.44\,_{10}(-9)$	$-0.44\,_{10}(-9)$	$0.60\,_{10}(-14)$
4	0.466	$0.39\,_{10}(-9)$	$-0.39\,_{10}(-9)$	$0.20\,_{10}(-14)$
5	0.415	$0.36\,_{10}(-9)$	$-0.36\,_{10}(-9)$	$0.18\,_{10}(-13)$
6	0.376	$0.33\,_{10}(-9)$	$-0.33\,_{10}(-9)$	$0.11\,_{10}(-13)$
7	0.350	$0.31\,_{10}(-9)$	$-0.31\,_{10}(-9)$	$-0.25\,_{10}(-13)$
8	0.327	$0.30\,_{10}(-9)$	$-0.30\,_{10}(-9)$	$0.10\,_{10}(-13)$
9	0.309	$0.29\,_{10}(-9)$	$-0.29\,_{10}(-9)$	$-0.18\,_{10}(-13)$
10	0.293	$0.28\,_{10}(-9)$	$-0.28\,_{10}(-9)$	$-0.55\,_{10}(-13)$

In Table 3.2 we give details on the computation of $1/\Gamma(z)$ by using (3.87) and (3.38) with $d = 0$ and $d = \frac{1}{2}h$. We have chosen (see Remark 3.12) $h = \pi/\sqrt{\omega \log(1/\varepsilon)}$, $\omega = \frac{1}{2}z$ with $\varepsilon = 10^{-10}$, and sum the series for $j = -j_0, -j_0+1, \ldots, j_0 - 1, j_0$, where j_0 follows from the smallest j that satisfies $\exp(-\omega j^2 h^2) < \varepsilon$. For all z-values, $j_0 = 8$ (as explained in Remark 3.12, j_0 will not depend on z). We observe that the values of the relative errors $R_0(h)\Gamma(z) - 1$ and $R_{h/2}(h)\Gamma(z) - 1$ are rather uniform for the values of z shown, and that they have opposite signs. The sum of these errors shows that they are nearly equal in modulus. Computations are done in Maple with Digits $= 15$.

Remark 3.18. The representation in (3.87) can also be used when z is complex. When $|\mathrm{ph}\, z| < \frac{1}{2}\pi$ the integral remains convergent, and the exponential function $\exp(-\frac{1}{2}zw^2)$ is no longer real on the path. By turning the path of integration over an angle less than $\frac{1}{4}\pi$ we can repair this. We can even control convergence for $|\mathrm{ph}\, z| < \pi$. However, when we turn the path of integration, the singularities of g mentioned earlier approach the real axis, and this makes the trapezoidal rule less efficient.

Because t, as a function of w, is not explicitly known in terms of standard functions, the function g of (3.88) needs an inversion procedure to compute the complex number t, when $w \in \mathbb{R}$ is given. In terms of the Lambert W-function, which is the inverse function of We^W, we can write $t(w) = -W(-\exp(\frac{1}{2}w^2 - 1))$.

It may be of interest to use a different representation, in which such inversion is not needed. This can be found as follows.

When we use this parametrization of the contour and integrate with respect to θ, we have

$$dt = \frac{d(re^{i\theta})}{d\theta}\, d\theta = [i + h(\theta)]\, d\theta, \qquad (3.89)$$

where $h(\theta) = (\cos\theta\sin\theta - \theta)/\sin^2\theta$ is an odd function of θ. It follows that

$$\frac{1}{\Gamma(z)} = \frac{e^z z^{1-z}}{2\pi} \int_{-\pi}^{\pi} e^{-z\Phi(\theta)}\, d\theta, \qquad (3.90)$$

where

$$\Phi(\theta) = -\operatorname{Re}\phi(t) = 1 - \theta\cot\theta + \log\frac{\theta}{\sin\theta}. \qquad (3.91)$$

To evaluate $\Phi(\theta)$ for small values of θ we have

$$\Phi(\theta) = \tfrac{1}{2}\theta^2 + \tfrac{1}{36}\theta^4 + \tfrac{1}{405}\theta^6 + \tfrac{1}{4200}\theta^8 + \tfrac{1}{42525}\theta^{10} + \cdots. \qquad (3.92)$$

All terms in the series are positive, as follows from the general form

$$\Phi(\theta) = \sum_{n=1}^{\infty} (-1)^{n+1} \frac{2n+1}{2n} \frac{2^{2n} B_{2n}}{(2n)!} \theta^{2n}, \quad |\theta| < \pi, \qquad (3.93)$$

where B_{2n} are the Bernoulli numbers, and $(-1)^{n+1} B_{2n} > 0$.

The integrand belongs to $C_p^{\infty}([-\pi, \pi])$ (see Definition 3.1), with vanishing derivatives of all orders at the end points $\pm\pi$.

Example 3.19. (Bessel functions $J_{\nu}(x)$)   In Example 3.3 we have seen that for the Bessel function $J_0(x)$ a very simple integral can be used to apply the trapezoidal rule. For general integer values we can use the more general integral in (3.1), but when n is large the many oscillations make the representations very unstable. In the present example we consider an integral representation that can be used to compute $J_{\nu}(x)$ for $\nu \geq x$, and large values of the parameters cause no problems.

We use the contour integral

$$J_{\nu}(x) = \frac{1}{2\pi i} \int_{\mathcal{C}} e^{x\sinh t - \nu t}\, dt, \qquad (3.94)$$

where $\mathcal{C}$ starts at $\infty - i\pi$ and terminates at $\infty + i\pi$; see Temme (1996, p. 222). On this contour oscillations will occur, but we will select a special contour that is free of oscillations for the case $x \leq \nu$.

We write

$$\nu = x\cosh\mu, \quad \mu \geq 0. \qquad (3.95)$$

The saddle point of

$$x \sinh t - \nu t = x(\sinh t - t \cosh \mu) \tag{3.96}$$

occurs at $t = \mu$, and at this saddle point the imaginary part of $x \sinh t - \nu t$ equals zero.

A path free of oscillations (a steepest descent path through the saddle point) can be described by the equation

$$\mathrm{Im}\,(x \sinh t - \nu t) = 0. \tag{3.97}$$

Writing $t = \sigma + i\tau$ we obtain for the path the equation

$$\cosh \sigma = \cosh \mu \frac{\tau}{\sin \tau}, \quad -\pi < \tau < \pi, \tag{3.98}$$

and on this path we have

$$\mathrm{Re}\,(x \sinh t - \nu t) = x(\sinh \sigma \cos \tau - \sigma \cosh \mu). \tag{3.99}$$

Integrating with respect to τ, using $\mathrm{d}t/\mathrm{d}\tau = (\mathrm{d}\sigma/\mathrm{d}\tau + i)$ (where $\mathrm{d}\sigma/\mathrm{d}\tau$ is an odd function of τ), we obtain

$$J_\nu(x) = \frac{1}{2\pi} \int_{-\pi}^{\pi} e^{x(\sinh \sigma \cos \tau - \sigma \cosh \mu)} \, \mathrm{d}\tau, \quad 0 < x \le \nu. \tag{3.100}$$

The integrand is analytic and vanishes with all its derivatives at the end points of the interval $[-\pi, \pi]$.

When $\nu \gg x$ the Bessel function becomes very small and we can put the value of the integrand at $\tau = 0$ as the dominant part in front of the integral. This gives the representation

$$J_\nu(x) = \frac{e^{x(\sinh \mu - \mu \cosh \mu)}}{2\pi} \int_{-\pi}^{\pi} e^{-x\psi(\tau)} \, \mathrm{d}\tau, \tag{3.101}$$

where

$$\psi(\tau) = \sinh \mu - \sinh \sigma \cos \tau + (\sigma - \mu) \cosh \mu. \tag{3.102}$$

For small values of τ we have the expansion

$$\psi(\tau) = \tfrac{1}{2} \sinh \mu \tau^2 + \frac{2 \cosh^2 \mu + 3}{72 \sinh \mu} \tau^4 + \mathcal{O}(\tau^6). \tag{3.103}$$

We see that this breaks down when $\mu \to 0$, that is, $x \uparrow \nu$. This happens because the the phase function $x \sinh t - \nu t = x(\sinh t - t)$ (when $x = \nu$) has a cubic character at $t = 0$. When $\mu = 0$ the expansion of $\psi(\tau)$ reads

$$\psi(\tau) = \frac{4\sqrt{3}}{27} \tau^3 + \frac{8\sqrt{3}}{14175} \tau^7 + \mathcal{O}(\tau^9). \tag{3.104}$$

In fact, the representation in (3.101) should not be used when x and ν are nearly equal. For all $\mu > 0$ the contour defined by (3.98) cuts the real t-axis under a right angle; when $\mu = 0$ this angle is $\tfrac{1}{3}\pi$.

Remark 3.20. The change of behaviour of the integrand when $\nu \sim x$ is related to the turning point character of $J_\nu(x)$ when argument and order are equal. The function $\sqrt{z}J_\nu(\nu z)$ satisfies the differential equation

$$w'' + \left[\nu^2\left(\frac{z^2-1}{z^2}\right) - \frac{1}{4z^2}\right]w = 0. \tag{3.105}$$

In particular, when ν is large the solutions of this equation change behaviour from $z < 1$ (monotonic character) to $z > 1$ (oscillatory character). The Airy function is the main approximant in the large ν asymptotic approximation that holds in a neighbourhood of $z = 1$. See Olver (1997, Chapter 11).

To avoid this type of discontinuous behaviour of the contour we may choose a different path, which should run through the saddle point and end in the valleys when $t \to \infty \pm i\pi$, but for which the parametrization does not become discontinuous when $\mu \downarrow 0$. For example, we can take

$$\sigma = \mu + \frac{\tau^2}{\pi^2 - \tau^2}, \quad -\pi < \tau < \pi. \tag{3.106}$$

When using this path the imaginary part of the phase function $x \sinh t - \nu t$ will not vanish identically, but the oscillations of the integrand will have only a minor influence on the stability of the representation. That is, the function $\psi(\tau)$ in (3.101) will become different, but the dominant factor in front of the integral can still be extracted from the integrand.

When $x > \nu$ (the oscillatory case), the Bessel function can be represented in a similar way, now by using two integrals (coming from the Hankel functions). First we consider

$$H_\nu^{(1)}(x) = \frac{1}{\pi i}\int_{-\infty}^{\infty+\pi i} e^{x\sinh t - \nu t}\, dt, \tag{3.107}$$

now with $\nu = x\cos\mu$, $0 < \mu < \frac{1}{2}\pi$. The saddle points are now complex, and the relevant saddle point is $i\mu$. The path of steepest descent follows from the equation

$$\mathrm{Im}\left[\sinh(\sigma + i\tau) - \cos\mu(\sigma + i\tau)\right] = \sin\mu - \mu\cos\mu, \tag{3.108}$$

which gives the equation

$$\cosh\sigma = \frac{\tau\cos\mu + \sin\mu - \mu\cos\mu}{\sin\tau}, \quad 0 < \tau < \pi, \tag{3.109}$$

where $\sigma \leq 0$ when $\tau \leq \mu$ and $\sigma \geq 0$ when $\tau \geq \mu$. Using this path we obtain

$$H_\nu^{(1)}(x) = \frac{e^{ix(\sin\mu - \mu\cos\mu)}}{\pi i}\int_0^\pi e^{-x\psi(\tau)}g(\tau)\, d\tau, \tag{3.110}$$

where

$$\psi(\tau) = \sigma \cos \mu - \sinh \sigma \cos \tau, \qquad g(\tau) = \frac{\mathrm{d}t}{\mathrm{d}\tau} = \frac{\mathrm{d}\sigma}{\mathrm{d}\tau} + \mathrm{i}. \qquad (3.111)$$

After computing the Hankel function $H_\nu^{(1)}(x)$ the result for $J_\nu(x)$ follows from $J_\nu(x) = \mathrm{Re}\,[H_\nu^{(1)}(x)]$ (which holds when $x > 0$ and $\nu \in \mathbb{R}$).

Example 3.21. (Modified Bessel functions $K_\nu(x)$) For the modified Bessel functions the representations are quite simple when we consider real parameters. In that case there is no turning point.

First consider the function (we take $x > 0$ and $\nu \in \mathbb{R}$)

$$K_\nu(x) = \int_0^\infty \mathrm{e}^{-x \cosh t} \cosh \nu t \, \mathrm{d}t = \tfrac{1}{2} \int_{-\infty}^\infty \mathrm{e}^{-x \cosh t + \nu t} \, \mathrm{d}t. \qquad (3.112)$$

This integral has no oscillations and we can apply the trapezoidal rule if we wish without further steps. When the parameters are large it may be convenient to scale by extracting the dominant factor; we also like to have the peak at the origin, as in previous examples. We write

$$\nu = x \sinh \mu, \qquad (3.113)$$

and see that the integrand of the second integral in (3.112) has a saddle point at $t = \mu$. We put $t = \mu + s$ and obtain

$$K_\nu(x) = \tfrac{1}{2} \mathrm{e}^{-x(\cosh \mu - \mu \sinh \mu)} \int_{-\infty}^\infty \mathrm{e}^{-x\psi(s)} \, \mathrm{d}s, \qquad (3.114)$$

where

$$\psi(s) = \cosh \mu(\cosh s - 1) + \sinh \mu(\sinh s - s). \qquad (3.115)$$

By evaluating the Fourier integrals in (3.53), we obtain, for the remainder in

$$K_\nu(x) = \tfrac{1}{2} h \mathrm{e}^{-x(\cosh \mu - \mu \sinh \mu)} \sum_{j=-\infty}^\infty \mathrm{e}^{-x\psi(jh)} + R_0(h), \qquad (3.116)$$

the series

$$R_0(h) = - \sum_{\substack{m=-\infty \\ m \neq 0}}^\infty K_{\nu+2\pi \mathrm{i}m/h}(x). \qquad (3.117)$$

Example 3.22. (Modified Bessel functions $I_\nu(x)$) For the modified Bessel function $I_\nu(z)$ the starting point is

$$I_\nu(x) = \frac{1}{2\pi \mathrm{i}} \int_C \mathrm{e}^{x \cosh t - \nu t} \, \mathrm{d}t, \qquad (3.118)$$

where the contour C starts at $\infty - \mathrm{i}\pi$ and ends at $\infty + \mathrm{i}\pi$. The relevant saddle point is $t = \mu$, where μ is as in (3.113). We write $t = \sigma + \mathrm{i}\tau$. The path of steepest descent through $t = \mu$ is given by $\mathrm{Im}\,[\cosh(\sigma + i\tau) - \mu(\sigma + i\tau)] = 0$, which gives the parametrization

$$\sinh \sigma = \frac{\mu\tau}{\sin \tau}, \quad -\pi < \tau < \pi. \tag{3.119}$$

We put the dominant exponential factor in front of the integral and obtain

$$I_\nu(x) = \frac{e^{x(\cosh \mu - \mu \sinh \mu)}}{2\pi} \int_{-\pi}^{\pi} e^{-x\psi(\tau)} \, d\tau, \tag{3.120}$$

where

$$\psi(\tau) = \cosh \mu - \cosh \sigma \cos \tau + (\sigma - \mu) \sinh \mu. \tag{3.121}$$

For small values of τ we have

$$\psi(\tau) = \frac{2\mu^2 + 3 + \mu \sinh \mu}{6 \cosh \mu} \tau^2 + \mathcal{O}(\tau^4). \tag{3.122}$$

For $\nu = 0$, that is, $\mu = 0$, this representation reduces to

$$I_0(x) = \frac{1}{2\pi} \int_{-\pi}^{\pi} e^{x \cos \tau} \, d\tau, \tag{3.123}$$

which is a standard integral for this function (Abramowitz and Stegun 1964, equation 9.6.16).

3.4. Some numerical aspects

We discuss the main aspects in error handling of the trapezoidal rule (3.7) and (3.38) for the finite and infinite interval.

The size of the remainders R_n^d and $R_d(h)$
In many cases where special functions are involved we can estimate these remainders. In the simple Examples 3.15 and 3.16, the series clearly show what happens. Also, in other cases we can compute or estimate the remainders, for example by evaluating the Fourier transforms needed in the series (3.52) or (3.53); see Example 3.21 where the remainder is given in (3.117) as an infinite sum of K-Bessel functions with complex order. For these functions asymptotic estimates are available.

The truncation error when taking the relevant terms of the series
This point is very relevant for infinite series. The smallest integer number j_0, such that in

$$I(f) = \int_{-\infty}^{\infty} f(x) \, dx = h \left[f(0) + 2 \sum_{j=1}^{j_0} f(d + jh) \right] + S_{j_0}, \tag{3.124}$$

assuming that f is even, $|S_{j_0}|$ or $|S_{j_0}/I(f)|$ is smaller than the desired accuracy ε, is called the number of relevant function evaluations corresponding to h and ε.

In the finite case, as in Example 3.3, all terms in the sum (3.6) or (3.7) may be relevant, but sometimes the integrand in the finite integral becomes exponentially small at the end points (see (3.90)), and in that case the number of relevant function evaluations is important, and there is no need to compute all the terms in the finite sum.

How to choose h?

Another point is a stopping criterion when testing whether the number h is small enough. When it is difficult to obtain useful and realistic analytic bounds for the remainders it may be necessary during computations to verify if the number h is small enough. Let us denote the series in (3.38) by

$$T_d(h) = h \sum_{j=-\infty}^{\infty} f(d + jh). \tag{3.125}$$

We can compare $T_0(h)$ and $T_{h/2}(h)$ for a given h. When they agree within our desired accuracy, we are finished, and can take one of the two as our final result. However, as follows from Example 3.17 (see also Table 3.2), it is better to add both results, because

$$T_0(\tfrac{1}{2}h) = \tfrac{1}{2}\big[T_0(h) + T_{h/2}(h)\big], \tag{3.126}$$

and the error in this result may be much smaller than the desired accuracy.

So, because of the fast convergence of the trapezoidal rule we may compute too many terms, and to control efficiency a different stopping criterion may be useful.

In the case of a finite interval the same fast convergence may occur, and to control both accuracy and efficiency a simple stopping criterion is a matter of trial and error for a certain problem.

In many problems that we have tested, in which, say, N correct decimal digits had to be obtained, we have verified whether h is sufficiently small to ensure that

$$\big|T_0(h) - T_{h/2}(h)\big| < \varepsilon, \quad \varepsilon = 10^{-\frac{3}{4}N}, \tag{3.127}$$

and used (3.126) for the final result. In many cases this stopping criterion gave the results we wanted.

3.5. Other aspects of numerical quadrature

We mention other quadrature methods and conclude with further observations.

Gauss quadrature

Gauss quadrature is an efficient method for evaluating integrals. It is based on choosing the zeros of a class of orthogonal polynomials as the interpolation nodes. Usually, these zeros have to be precomputed, as well as the weights, which are also associated with the polynomials. This makes the method less attractive in adaptive algorithms, where we like to increase the number of nodes as often as we please. On the other hand, Gauss quadrature has proven to have its merits for certain types of integrals, and the underlying theory is very elegant. In a recent book, Gautschi (2004), the aspects of computation and approximation of orthogonal polynomials, and in particular Gauss quadrature rules, are discussed in great detail.

Filon's method for oscillatory integrals

In previous examples we have discussed how to deal with integrals on contours in the complex plane and how, particularly when the parameters are large, strong oscillations can be handled by choosing appropriate contours through saddle points.

In this section we discuss another method of how to deal with oscillatory integrals. For further discussions we refer to Blakemore, Evans and Hyslop (1976), where a number of related methods for the evaluation of oscillatory integrals over infinite ranges are compared.

Oscillatory integrals of the form

$$I(f;p) = \int_a^b f(x)e^{ipx}\,dx \tag{3.128}$$

can be evaluated using Filon's method (Filon 1928). The method, especially useful when p is large, is based on the piecewise approximation of $f(x)$ on the interval of integration by low-degree polynomials. We give the following details.

The interval of integration $[a, b]$ is divided into $2N$ equally spaced subintervals

$$a = x_0 < x_1 < \cdots < x_{2N} = b, \quad x_k = a + hk, \quad k = \frac{b-a}{2N}. \tag{3.129}$$

On each subinterval $[x_{2k-2}, x_{2k}]$ the function $f(x)$ is locally approximated by a polynomial $P_k(x)$ of degree at most 2, and the corresponding integral on that interval

$$\int_{2k-2}^{2k} P_k(x)e^{ipx}\,dx \tag{3.130}$$

is evaluated exactly. This gives

$$I(f;p) \approx h\Big\{i\alpha\big(e^{ipa}f(a) - e^{ipb}f(b)\big) + \beta E_{2N} + \gamma E_{2N-1}\Big\}, \tag{3.131}$$

with

$$\theta^3 \alpha = \theta^2 + \theta \sin\theta \cos\theta - 2\sin^2\theta,$$
$$\theta^3 \beta = 2\big[\theta(1 + \cos^2\theta) - 2\sin\theta\cos\theta\big], \qquad (3.132)$$
$$\theta^3 \gamma = 4\big[\sin\theta - \theta\cos\theta\big],$$

and

$$E_{2N} = \sum_{k=0}^{N}{}'' f(x_{2k})e^{iwx_{2k}}, \qquad E_{2N-1} = \sum_{k=1}^{N} f(x_{2k-1})e^{iwx_{2k-1}}. \qquad (3.133)$$

The double prime over the first summation indicates that the first and last terms are to be multiplied by $\frac{1}{2}$.

The quantities α, β, and γ defined in (3.132) need to be recomputed when we change h, or for different p. When θ is small the right-hand sides in (3.132) should be expanded in powers of θ to preserve accuracy. We have

$$\alpha = \tfrac{2}{45}\theta^3 - \tfrac{2}{315}\theta^5 + \tfrac{2}{4725}\theta^7 + \cdots,$$
$$\beta = \tfrac{2}{3} + \tfrac{2}{15}\theta^2 - \tfrac{4}{105}\theta^4 + \tfrac{2}{567}\theta^6 + \cdots, \qquad (3.134)$$
$$\gamma = \tfrac{4}{3} - \tfrac{2}{15}\theta^2 + \tfrac{1}{210}\theta^4 - \tfrac{1}{11340}\theta^6 + \cdots.$$

It can be easily verified that for $p = 0$, that is, $\theta = 0$, Filon's method becomes Simpson's extended rule. For this quadrature rule we refer to Abramowitz and Stegun (1964, equation 25.4.6).

For recent investigations of Filon-type quadrature in connection with highly oscillatory integrals with extensive numerical and asymptotic analysis, see Iserles (2004, 2005) and Iserles and Nørsett (2005).

Asymptotic expansion
We consider integrals of the form

$$I(f;p) = \int_0^\infty f(x)e^{ipx}\,dx, \qquad (3.135)$$

in which p may be a complex parameter, and f is a function sufficiently smooth on $(0,\infty)$ and with sufficient decay at ∞ to ensure convergence of the integral.

First we make some observations concerning integrals of the type (3.135). Earlier we gave the details of Filon's method, which can be used on a finite interval.

When derivatives of f are available and when p is large we can first integrate by parts to obtain the main contribution to the integral. In this way,

$$I(f;p) = \frac{i}{p}f(0) - \frac{1}{p^2}f'(0) - \frac{1}{p^2}\int_0^\infty f''(x)e^{ipx}\,dx, \qquad (3.136)$$

which may be continued in order to obtain higher approximations, as far as the smoothness and growth conditions of the derivatives of f allow.

Odd or even functions

When f is an even function, all its odd derivatives at the origin vanish; hence, all terms with even powers of p^{-1} vanish when we continue the expansion in (3.136). For example, when we take $f(x) = 1/(1+x^2)$, we have, when $p > 0$,

$$I(f;p) = \int_0^\infty \frac{e^{ipx}}{1+x^2}\,dx = \tfrac{1}{2}\pi e^{-p} + i\int_0^\infty \frac{\sin(px)}{1+x^2}\,dx, \qquad (3.137)$$

in which the real part of $I(f;p)$ is exponentially small and the imaginary part is $\mathcal{O}(p^{-1})$ when p is large. So, computing $I(f;p)$ of (3.135) by using a quadrature rule when f is a real even function may give a large relative error (but a small absolute error) in the real part of $I(f;p)$, and similarly when f is an odd function.

Analytic functions

When f is slowly decreasing at ∞, convergence of a quadrature rule may be rather poor. When f is analytic in the right half-plane, and p is positive, we may investigate if turning the path of integration up into the complex plane is possible. In that case convergence of the integral and of the quadrature rule may be improved.

Orthogonal polynomials

For integrals of the form

$$I(f;p) = \int_{-1}^1 e^{ipx} f(x)\,dx, \qquad (3.138)$$

we can try to expand $f(x)$ in terms of Legendre polynomials, that is, $f(x) = \sum_{n=0}^\infty c_n P_n(x)$, and obtain

$$I(f;p) = \sqrt{\frac{2\pi}{p}} \sum_{n=0}^\infty (-i)^n c_n J_{n+\frac{1}{2}}(p), \qquad (3.139)$$

in terms of spherical Bessel functions; see Temme (1996, equation (6.64)).

In the same manner we can expand the function $f(x)$ in

$$I(f;p) = \int_{-1}^1 e^{ipx} \frac{f(x)}{\sqrt{1-x^2}}\,dx \qquad (3.140)$$

in terms of Chebyshev polynomials, $f(x) = \sum_{n=0}^\infty c_n T_n(x)$, and obtain

$$I(f;p) = \pi \sum_{n=0}^\infty i^n c_n J_n(p). \qquad (3.141)$$

in terms of ordinary Bessel functions, which follows from Temme (1996, equation (9.20)).

For further examples on the use of orthogonal polynomials (also on infinite intervals), see Patterson (1976/77).

More general forms
Oscillatory integrals also occur in the form

$$I(f;p) = \int_0^\infty f(x)\Phi(xp)\,\mathrm{d}x, \qquad (3.142)$$

where Φ is an oscillatory function. For example, in the case of Bessel functions we have the class of Hankel transforms

$$I_{\mu,\nu}(f;p) = \int_0^\infty x^{\mu-1}f(x)J_\nu(px)\,\mathrm{d}x, \qquad (3.143)$$

which play an important role in applied mathematics. In Wong (1982) it is explained how quadrature rules of Gauss type can be constructed for these integrals and also for integrals of the type (3.135). In the latter case Wong gives a modification of the Gauss–Laguerre rule, and the method works for functions f that are analytic in the right half-plane.

Reducing the interval
Because the exponential function in (3.135) is periodic, with interval of periodicity $[0, 2\pi/p]$, we can write $I(f;p)$ in the form

$$I(f;p) = \frac{1}{p}\int_0^{2\pi} e^{\mathrm{i}t}S_p(t)\,\mathrm{d}t, \qquad S_p(t) = \sum_{k=0}^\infty f\left(\frac{2\pi k + t}{p}\right). \qquad (3.144)$$

Another summation method
When in (3.135) the exponential function is written in terms of the sine and cosine function, the resulting integrals can written as alternating series of positive and negative subintegrals that are computed individually (for example, when f is positive). A similar method can also be used for (3.142) and (3.143) by using subintervals with endpoints the zeros of $\Phi(x)$ or $J_\nu(px)$. See Longman (1956).

Convergence acceleration schemes, for instance Levin's or Weniger's transformations (see Weniger (1989)), can be used when evaluating the series. For further information see Clendenin (1966), Lyness (1985) and Lucas and Stone (1995).

3.6. Other curious exercises with integrals

Needless to say, it will be evident that Maple and Mathematica are great tools when working with special functions, and in other areas of pure and

applied mathematics. Numerical quadrature with these computer algebra
packages can be very efficient, although, as mentioned at the beginning of
Section 3.3, the user may not always get what he or she wants. We consider
another example, a simple integral, where Maple and Mathematica give
answers that are confusing and/or wrong (perhaps different results might
have been obtained using later versions of Maple and Mathematica).

Consider

$$F(u) = \int_0^\infty e^{uit} \frac{dt}{t - 1 - i}, \quad u > 0. \tag{3.145}$$

Numerical quadrature gives $F(2) = -0.934349 - 0.70922i$.

Mathematica 4.1 gives for $u = 2$, in terms of the Meijer G-function,

$$F(2) = \pi G_{2,3}^{2,1} \left(\begin{matrix} 0, \frac{1}{2} \\ 0, 0, \frac{1}{2} \end{matrix} ; 2 - 2i \right). \tag{3.146}$$

For workers in special functions this may be a useful answer, but for those
not familiar with this rather generalized hypergeometric function some confusion may arise. However, Mathematica can evaluate this answer numerically, and gives the result

$$F(2) = -0.547745 - 0.532287i,$$

which is not the same as that obtained earlier by using numerical quadrature.

Next, ask Mathematica to evaluate $F(u)$, that is, with a general argument.
Surprisingly, the answer is much simpler:

$$F(u) = e^{iu-u} \Gamma(0, iu - u), \tag{3.147}$$

in terms of the incomplete gamma function. Again, asking Mathematica to
evaluate numerically this result for $u = 2$, we obtain

$$F(2) = -0.16114 - 0.355355i.$$

So, we have three numerical results:

$$\begin{aligned} F_1 &= -0.934349 - 0.70922i, \\ F_2 &= -0.547745 - 0.532287i, \\ F_3 &= -0.16114 - 0.355355i. \end{aligned} \tag{3.148}$$

Observe that $F_2 = (F_1 + F_3)/2$. So, in some sense these answer have something in common. It turns out that F_1 is correct.

Turning to Maple 9.5, we obtain

$$F(u) = e^{iu-u} \text{Ei}(1, iu - u), \tag{3.149}$$

in terms of the exponential integral. We can write this in terms of the incomplete gamma functions[1] and obtain the same result (3.147) as Mathematica. This is the wrong answer.

Next, Maple 9.5, with $u = 2$:

$$F(2) = e^{2i-2}\text{Ei}(1, 2i - 2) + 2\pi i e^{2i-2}. \tag{3.150}$$

We see that the answer has an extra term now, and, in Maple, numerical evaluation gives

$$F(2) = -0.9343493872 - 0.70921951102i, \tag{3.151}$$

which is the correct answer.

The extra term in (3.150) equals $2\pi i$ times the residue of the integrand of (3.145) at the pole $t = 1 + i$. This residue arises when we turn the path of integration in (3.145) to the positive imaginary axis. The residue is missing in the answer F_3 of (3.148) and in F_2 it is taken into account incorrectly (only πi times the residue is added, which is why $F_2 = (F_1 + F_3)/2$).

The confusing part is that both Maple and Mathematica also give different symbolic answers for $F(u)$ and $F(2)$.

3.7. Numerical quadrature: Concluding remarks

The first paper that mentioned the superiority of the trapezoidal rule for the integration of analytic functions on an infinite interval seems to be Goodwin (1949). Many papers worked out this idea, of which we mention important contributions from the Japanese school: Takahasi and Mori (1970, 1971, 1973/74) and Mori (1974). In these papers several transformations are discussed and error terms for a number of quadrature rules are represented as derivative-free contour integrals in the complex plane.

The trapezoidal rule may give the exact value of the integral when h is less than some fixed value. For example (Rice 1973), if m and n are positive integers such that $m - n = 0$, or $2, 4, \ldots$, the integral

$$\int_{-\infty}^{\infty} \frac{\sin^m x}{x^n} \, dx \tag{3.152}$$

is exactly integrated to the trapezoidal rule sum when $h < 2\pi/m$ (h can equal $2\pi/m$ when $n \geq 2$). The proof follows from the fact that all of the terms in the series (3.53) for the remainder vanish. Rice gives several other interesting examples.

---

[1] Maple's *convert* function gives $convert(F(u), GAMMA) = e^{iu-u}\Gamma(0, iu - u)$; the notation Ei$(a, z)$ is not widely used in the standard works for special functions. A better notation is $E_a(z)$, the generalized exponential integral.

For further papers on the use of the trapezoidal rule on finite and infinite intervals, also with special transformations and interesting examples, we mention Elliott (1998/99), Eggert and Lund (1989), Haber (1977), Squire (1976a, 1976b) and Weideman (2002).

4. Recurrence relations

Consider the recurrence relation

$$y_{n+1} + a_n y_n + b_n y_{n-1} = 0, \quad n = 1, 2, 3, \ldots, \tag{4.1}$$

where a_n and b_n are given, with $b_n \neq 0$. Many special functions of mathematical physics satisfy such a relation. Equation (4.1) is also called a linear homogeneous difference equation of the second order.

In analogy with the theory of differential equations, two linearly independent solutions f_n, g_n exist in general, with the property that any solution y_n of (4.1) can be written in the form

$$y_n = A f_n + B g_n, \tag{4.2}$$

where A and B do not depend on n. We are interested in the special case that the pair $\{f_n, g_n\}$ satisfies

$$\lim_{n \to \infty} \frac{f_n}{g_n} = 0. \tag{4.3}$$

Then, for any solution (4.2) with $B \neq 0$, we have $f_n/y_n \to 0$ as $n \to \infty$. When $B = 0$ in (4.2), we call y_n a minimal solution; when $B \neq 0$, we call y_n a dominant solution. When we have two initial values y_0, y_1, assuming that f_0, f_1, g_0, g_1 are known as well, then we can compute A and B. That is,

$$A = \frac{g_1 y_0 - g_0 y_1}{f_0 g_1 - f_1 g_0}, \qquad B = \frac{y_0 f_1 - y_1 f_0}{g_0 f_1 - g_1 f_0}. \tag{4.4}$$

The denominators are different from 0 when the solutions f_n, g_n are linearly independent.

When we assume that the initial values y_0, y_1 are to be used for generating a dominant solution, then A may, or may not, vanish; B should not vanish: $y_0 f_1 \neq y_1 f_0$. When, however, the initial values are to be used for the computation of a minimal solution, then the much stronger condition $y_0 f_1 = y_1 f_0$ should hold. It follows that, in this case, one and only one initial value can be prescribed; the other one follows from the relation $y_0 f_1 = y_1 f_0$. In the numerical approach this leads to well-known instability phenomena for the computation of minimal solutions. The fact is that, when our initial values y_0, y_1 are not specified to an infinite precision, and consequently B does not vanish exactly, the computed solution (4.2) always contains a fraction of g_n, the dominant solution. Hence, in the long run, our solution y_n does not behave as a minimal solution, although we assumed that we were computing a minimal solution. This happens even if all further computations are done exactly.

In applications it is important to know whether a given equation (4.1) has dominant and minimal solutions. Often this can be easily concluded from the asymptotic behaviour of the coefficients a_n and b_n. The following useful theorem is due to Perron and taken from Gautschi (1967), which paper contains a wealth of information.

Theorem 4.1. (Perron)   Assume that for large values of n the coefficients a_n, b_n behave as follows:

$$a_n \sim an^\alpha, \qquad b_n \sim bn^\beta, \qquad ab \neq 0, \tag{4.5}$$

with α and β real; assume that t_1, t_2 are the zeros of the characteristic polynomial $\Phi(t) = t^2 + at + b$ with $|t_1| \geq |t_2|$.

[1] If $\alpha > \frac{1}{2}\beta$ then the difference equation (4.1) has two linearly independent solutions $y_{n,1}$ and $y_{n,2}$, with the property

$$\frac{y_{n+1,1}}{y_{n,1}} \sim -an^\alpha, \qquad \frac{y_{n+1,2}}{y_{n,2}} \sim -\frac{b}{an}^{\beta-\alpha}, \qquad n \to \infty. \tag{4.6}$$

[2] If $\alpha = \frac{1}{2}\beta$ then the difference equation (4.1) has two linearly independent solutions $y_{n,1}$ and $y_{n,2}$, with the property

$$\frac{y_{n+1,1}}{y_{n,1}} \sim t_1 n^\alpha, \qquad \frac{y_{n+1,2}}{y_{n,2}} \sim t_2 n^\alpha, \qquad n \to \infty, \tag{4.7}$$

assuming that $|t_1| > |t_2|$. If $|t_1| = |t_2|$ then we have

$$\limsup_{n \to \infty} \left[|y_n|(n!)^{-\alpha} \right]^{\frac{1}{n}} = |t_1| \tag{4.8}$$

for each non-trivial solution of (4.1).

[3] If $\alpha < \frac{1}{2}\beta$ then

$$\limsup_{n \to \infty} \left[|y_n|(n!)^{-\beta/2} \right]^{\frac{1}{n}} = \sqrt{|b|} \tag{4.9}$$

for each non-trivial solution of (4.1).

Proof.   For a proof we refer to the cited literature in Gautschi (1967) or to Elaydi (2005). □

In case [1] and the first part of case [2] $f_n = y_{n,2}$ is a minimal solution of (4.1). In addition, in the first part of [2],

$$\lim_{n \to \infty} \frac{y_{n+1}}{n^\alpha y_n} = t_r, \qquad r = 1 \quad \text{or} \quad r = 2, \tag{4.10}$$

where $r = 2$ holds for the minimal solution and $r = 1$ for any other solution. To verify this, we derive from [1]:

$$\frac{y_{n+1,2}}{y_{n+1,1}} \Big/ \frac{y_{n,2}}{y_{n,1}} \sim \frac{b}{a^2} n^{\beta - 2\alpha}, \quad n \to \infty. \tag{4.11}$$

The right-hand side converges to 0, since $\beta - 2\alpha < 0$. It follows that $y_{n,2}/y_{n,1}$ converges to 0. In the first part of [2] we have

$$\frac{y_{n+1,2}}{y_{n+1,1}} \Big/ \frac{y_{n,2}}{y_{n,1}} \sim \frac{t_2}{t_1}, \quad n \to \infty. \tag{4.12}$$

Since $|t_1| > |t_2|$ we again conclude that $y_{n,2}/y_{n,1}$ converges to 0.

The second part of case [2] and case [3] of the theorem do not give information on the minimal and dominant solutions. As can be seen from the examples below we then need extra asymptotic information about the solutions of the difference equation (4.1). For the general asymptotic theory we refer to Wong and Li (1992b) and (1992a).

Example 4.2. (Bessel and Legendre functions) We give the details of important recurrence relations for special functions and the stability aspects including the maximal and minimal solutions of the particular relation. The quantities f_n, g_n denote the minimal and maximal solutions, respectively. For recent computer programs for Legendre functions and toroidal harmonics (a subclass of the Legendre functions) we refer to Gil and Segura (1997, 2000).

(1) Bessel functions.
  Recurrence relation:

$$y_{\nu+1} - \frac{2\nu}{z} y_\nu + y_{\nu-1} = 0, \quad z \neq 0. \tag{4.13}$$

Solutions:

$$f_\nu = J_\nu(z), \qquad g_\nu = Y_\nu(z). \tag{4.14}$$

This is covered by case [1] of the theorem, with

$$a = -\frac{2}{z}, \qquad \alpha = 1, \qquad b = 1, \qquad \beta = 0. \tag{4.15}$$

Claim of the theorem:

$$\frac{f_{\nu+1}}{f_\nu} \sim \frac{z}{2\nu}, \qquad \frac{g_{\nu+1}}{g_\nu} \sim \frac{2\nu}{z}. \tag{4.16}$$

Known asymptotic behaviour:

$$f_\nu \sim \frac{1}{\sqrt{2\pi\nu}} \left(\frac{ez}{2\nu}\right)^\nu, \qquad g_\nu \sim -\sqrt{\frac{2}{\pi\nu}} \left(\frac{ez}{2\nu}\right)^{-\nu}, \qquad \nu \to \infty. \tag{4.17}$$

Similar results hold for the recurrence relation for the modified Bessel functions

$$y_{\nu+1} + \frac{2\nu}{z}y_\nu - y_{\nu-1} = 0, \quad z \neq 0, \tag{4.18}$$

with solutions $I_\nu(z)$ (minimal) and $e^{\pi i \nu} K_\nu(z)$ (dominant).

(2) Legendre functions, recursion with respect to the order.
Recurrence relation:

$$y_{m+1} + \frac{2mz}{\sqrt{z^2-1}}y_m + (m+\nu)(m-\nu-1)y_{m-1} = 0. \tag{4.19}$$

Solutions:

$$f_m = P_\nu^m(z), \qquad g_m = Q_\nu^m(z), \tag{4.20}$$

where

$$\operatorname{Re} z > 0, \quad \nu \in \mathbb{C} \quad \nu \neq -1, -2, \ldots, \quad z \notin (0, 1]. \tag{4.21}$$

This is covered by case [2] of the theorem, with

$$a = \frac{2z}{\sqrt{z^2-1}}, \qquad \alpha = 1, \qquad b = 1, \qquad \beta = 2, \tag{4.22}$$

$$t_1 = -\sqrt{\frac{z+1}{z-1}}, \qquad t_2 = \frac{1}{t_1}, \qquad |t_1| > 1 > |t_2|. \tag{4.23}$$

Claim of the theorem:

$$\lim_{m \to \infty} \frac{f_{m+1}}{m f_m} = t_2, \qquad \lim_{m \to \infty} \frac{g_{m+1}}{m g_m} = t_1. \tag{4.24}$$

(3) Legendre functions, recursion with respect to the degree.
Recurrence relation:

$$y_{n+1} - z\frac{2n+2\nu+1}{n+\nu-\mu+1}y_n + \frac{n+\nu+\mu}{n+\nu-\mu+1}y_{n-1} = 0. \tag{4.25}$$

Solutions:

$$f_n = Q_{\nu+n}^\mu(z), \qquad g_n = P_{\nu+n}^\mu(z), \qquad \operatorname{Re} z > 0. \tag{4.26}$$

This is covered by case [2] of the theorem, with

$$a = -2z, \qquad \alpha = 0, \qquad b = 1, \qquad \beta = 0, \tag{4.27}$$

$$t_1 = z + \sqrt{z^2-1}, \qquad t_2 = \frac{1}{t_1}, \qquad |t_1| > 1 > |t_2|. \tag{4.28}$$

Claim of the theorem:

$$\lim_{n \to \infty} \frac{f_{n+1}}{f_n} = t_2, \qquad \lim_{n \to \infty} \frac{g_{n+1}}{g_n} = t_1. \tag{4.29}$$

4.1. Miller's algorithm

From the previous discussion, it appears that the numerical computation of the minimal solution of a recurrence relation (4.1) with initial values f_0 and f_1 is quite problematic. One has to accept that the results are completely wrong after a few recursion steps. Of course, it depends on the required absolute or relative accuracy as to how much risk can be incurred, but in general one should be very careful.

From the asymptotic behaviour of the minimal and a dominant solution, one can usually conclude whether recursion for the minimal solution is dangerous.

We give details of an algorithm for computing a sequence of values

$$f_0, f_1, \ldots, f_N \tag{4.30}$$

of a minimal solution; N is a non-negative integer. Obviously, we can apply (4.1) in the backward direction; in that case f_n becomes a dominant solution and g_n the minimal solution. Then we need two initial values f_N and f_{N-1}. Miller's algorithm does not need these values, and uses a smart idea for the computation of the required sequence (4.30). The algorithm works for many interesting cases and gives an efficient method for computing the sequence (4.30).

Assume we have a relation of the form

$$\sum_{n=0}^{\infty} \lambda_n f_n = s, \quad s \neq 0. \tag{4.31}$$

The series should be convergent and λ_n and s should be known. The series in (4.31) plays a role in normalizing the required minimal solution. The series may be finite; we only require that at least one coefficient λ_n is different from zero. When just one coefficient, say λ_j, is different from zero, we assume that the value f_j is available.

In Miller's algorithm a starting value ν is chosen, $\nu > N$, and a solution $\{y_n^{(\nu)}\}$ of (4.1) is computed with the false initial values

$$y_{\nu+1}^{(\nu)} = 0, \qquad y_\nu^{(\nu)} = 1. \tag{4.32}$$

The right-hand sides may be replaced by other values; at least one value should be different from zero. In some cases a judicious choice of these values may improve the convergence of the algorithm. The computed solution, with (4.32) as initial values, is a linear combination of the solutions f_n and g_n introduced earlier. A simple computation gives

$$y_n^{(\nu)} = \frac{g_{\nu+1}f_n - f_{\nu+1}g_n}{g_{\nu+1}f_\nu - f_{\nu+1}g_\nu}, \quad n = 0, 1, \ldots, \nu + 1. \tag{4.33}$$

This can be verified by checking the relations in (4.32). We write this in the form

$$y_n^{(\nu)} = p_\nu f_n + q_\nu g_n. \tag{4.34}$$

We observe that $y_n^{(\nu)}/p_\nu = f_n - [f_{\nu+1}/g_{\nu+1}]g_n$, and from (4.3) it follows that

$$\lim_{\nu \to \infty} \frac{y_n^{(\nu)}}{p_\nu} = f_n, \quad 0 \le n \le N. \tag{4.35}$$

Apparently, when ν is large enough, an approximation of f_n can be obtained from the quantities $y_n^{(\nu)}$ and p_ν. However, in general, p_ν is not known. At this moment the normalizing relation (4.31) becomes of interest. We compute

$$s^{(\nu)} = \sum_{n=0}^{\nu} \lambda_n y_n^{(\nu)}, \quad f_n^{(\nu)} = \frac{s}{s^{(\nu)}} y_n^{(\nu)}. \tag{4.36}$$

Replacing $y_n^{(\nu)}$ in the series with $p_\nu f_n$, on account of (4.35), we then obtain $p_\nu \sim s^{(\nu)}/s$. It follows that we can consider $f_n^{(\nu)}$ as an approximation to f_n, if ν is large enough. That is, we assume that the circumstances are favourable, and that we can conclude that

$$f_n = \lim_{\nu \to \infty} f_n^{(\nu)}, \quad n = 0, 1, \ldots, N. \tag{4.37}$$

This claim will be founded by introducing extra conditions.

From (4.36) we obtain for the relative error (when $f_n \ne 0$)

$$\varepsilon_n^{(\nu)} = \frac{f_n^{(\nu)} - f_n}{f_n} = \frac{s/s^{(\nu)} \, y_n^{(\nu)} - f_n}{f_n} = \frac{s(p_\nu + q_\nu g_n/f_n) - s^{(\nu)}}{s^{(\nu)}}. \tag{4.38}$$

We rewrite this in the form

$$\varepsilon_n^{(\nu)} = \frac{\sigma_\nu - \rho_{\nu+1}/\rho_n + \tau_\nu}{1 - \sigma_\nu - \tau_\nu} \tag{4.39}$$

with

$$\sigma_\nu = \frac{1}{s} \sum_{m=\nu+1}^{\infty} \lambda_m f_m, \quad \rho_n = \frac{f_n}{g_n}, \quad \tau_\nu = \frac{\rho_{\nu+1}}{s} \sum_{m=0}^{\nu} \lambda_m g_m. \tag{4.40}$$

When introducing (4.31) we assumed that the series converges. Hence, $\sigma_\nu \to 0$ as $\nu \to \infty$. Also (see (4.3)), we assumed that $\rho_\nu \to 0$. From this we infer that the relative error $\varepsilon_n^{(\nu)}$ of (4.39) converges to zero (as $\nu \to \infty$), if and only if τ_ν converges to zero. Under this final condition, the limit in (4.37) holds.

For the numerical part of the method it is important to obtain an estimate of $\varepsilon_n^{(\nu)}$ for large values of ν. In many cases it is not easy to obtain a strict estimate; usually some terms in (4.40) can be approximated by replacing the series with their dominant terms. Taking in the first series only the first term, and in the second series the final term, we obtain

$$\sigma_\nu \simeq \frac{1}{s}\lambda_{\nu+1}f_{\nu+1}, \qquad \tau_\nu \simeq \frac{1}{s}\rho_{\nu+1}\lambda_\nu g_\nu. \tag{4.41}$$

With these approximations (4.39) reads

$$\varepsilon_n^{(\nu)} \simeq \frac{1}{s}\lambda_{\nu+1}f_{\nu+1} + \frac{f_{\nu+1}}{g_{\nu+1}}\frac{\lambda_\nu g_\nu}{s} - \frac{f_{\nu+1}}{g_{\nu+1}}\frac{g_n}{f_n} \tag{4.42}$$

$$\simeq \frac{1}{s}\lambda_{\nu+1}f_{\nu+1} - \frac{f_{\nu+1}}{g_{\nu+1}}\frac{g_n}{f_n},$$

since the second term on the first right-hand side is usually less important than the first term. A further step is to replace in this estimate n by N, because, when the Nth element in the sequence in (4.30) is accurate, the situation will only improve for the remaining values. Reasoning in this way, we finally arrive at

$$\varepsilon_n^{(\nu)} \simeq \frac{1}{s}\lambda_{\nu+1}f_{\nu+1} - \frac{f_{\nu+1}}{g_{\nu+1}}\frac{g_N}{f_N}. \tag{4.43}$$

By using asymptotic estimates of the dominant and minimal solutions, the estimation of ν can be executed, perhaps numerically. The estimate of the error in (4.43) reflects two aspects of the algorithm for favourable convergence. The first term on the right-hand side of (4.43) indicates that the series in (4.31) should converge quickly. The second term indicates that the extent of dominance of g_n with respect to f_n is very significant.

In Gautschi (1967) this algorithm is discussed in great detail (in a slightly different form). Gautschi estimates the starting point of the backward recursion by using asymptotic estimates of the special functions involved. In Olver (1967) a direct numerical approach is used for obtaining a good starting point ν. Olver also considers inhomogeneous recurrence equations. An excellent monograph for the numerical aspects of recurrence relations, including Miller's algorithm, is Wimp (1984).

Example 4.3. (Computing $I_n(x)$) In Bickley, Comrie, Miller, Sadler and Thompson (1952) the above method was introduced for computing the modified Bessel functions $I_n(x)$. The recurrence relation for these functions reads

$$I_{n+1}(x) + \frac{2n}{x}I_n(x) - I_{n-1}(x) = 0. \tag{4.44}$$

A normalizing condition (4.31) is

$$e^x = I_0(x) + 2I_1(x) + 2I_2(x) + 2I_3(x)\dots. \tag{4.45}$$

Table 4.1. Computing the modified Bessel functions $I_n(x)$ for $x = 1$ by using (4.44) in the backward direction. The underlined digits in the third column are correct.

n	y_n before normalization	$y_n \doteq I_n(1)$ after normalization
0	$2.2879\ 49300_{10}(+8)$	$\underline{1.26606\ 587801}_{10}(-0)$
1	$1.0213\ 17610_{10}(+8)$	$\underline{5.65159\ 104106}_{10}(-1)$
2	$2.4531\ 40800_{10}(+7)$	$\underline{1.35747\ 669794}_{10}(-1)$
3	$4.0061\ 29000_{10}(+6)$	$\underline{2.21684\ 249288}_{10}(-2)$
4	$4.9434\ 00000_{10}(+5)$	$\underline{2.73712\ 022160}_{10}(-3)$
5	$4.9057\ 00000_{10}(+4)$	$\underline{2.71463\ 156012}_{10}(-4)$
6	$4.0640\ 00000_{10}(+3)$	$\underline{2.24886\ 614761}_{10}(-5)$
7	$2.8900\ 00000_{10}(+2)$	$\underline{1.59921\ 829887}_{10}(-6)$
8	$1.8000\ 00000_{10}(+1)$	$\underline{9.96052\ 919710}_{10}(-8)$
9	$1.0000\ 00000_{10}(+0)$	$\underline{5.53362\ 733172}_{10}(-9)$
10	$0.0000\ 00000_{10}(+0)$	$0.00000\ 000000_{10}(-0)$

That is, $s = e^x$, $\lambda_0 = 1$, $\lambda_n = 2\,(n \geq 1)$. We take $x = 1$ and initial values (4.32) with $\nu = 9$ and obtain the results given in Table 4.1.

The rightmost column in Table 4.1 is obtained by dividing the results of the middle column by (see (4.35) and (4.36))

$$p_9 \simeq \sum_{n=0}^{9} \lambda_n y_n^{(9)}/e^1 = 1.8071328986_{10}(+8). \tag{4.46}$$

When we take $N = 5$, which means we want to compute the sequence $I_0(1), I_1(1), \ldots, I_5(1)$, we see that these quantities are computed with at least 10 correct decimal digits. For the present values of x, ν, N the estimate of the relative error ε_n^ν of (4.43) is $0.20_{10}{}^{-9}$, which is quite realistic.

4.2. Recurrence relations and continued fractions

For computing minimal solutions of three-term recurrence relations, the continued fraction for the ratios of consecutive solutions is also a useful tool. The basic result to be considered is Pincherle's theorem.

Theorem 4.4. (Pincherle) Given a three-term recurrence relation

$$y_{n+1} + b_n y_n + a_n y_{n-1} = 0, \tag{4.47}$$

then the continued fraction

$$\frac{-a_k}{b_k+} \frac{-a_{k+1}}{b_{k+1}+} \cdots \tag{4.48}$$

converges if and only if the recurrence relation possesses a minimal solution. Furthermore, if f_n is a minimal solution, then the continued fraction converges to f_k/f_{k-1}.

Proof. See, for instance, Gautschi (1967) and Gil *et al.* (2007b). □

4.3. The Gauss hypergeometric family

Legendre functions are are already considered in Example 4.2 and are a special case of the Gauss hypergeometric functions, which are defined by the series given in (2.19) for $|z| < 1$. By using integral representations and connection formulas this function can be continued analytically on and outside the unit circle, with the general exception of the point $z = 1$, which is usually a branch point. In this way the standard domain for this function is $|\mathrm{ph}(1 - z)| < \pi$ with a branch cut from 1 to $+\infty$.

There are many formulas (contiguous relations) that connect a Gauss function with parameters a, b, c to two other functions with parameters $a \pm 1, b \pm 1, c \pm 1$. Of special interest are the recurrence relations for the functions

$$y_n(z) = {}_2F_1\left(\begin{matrix} a + \varepsilon_1 n, \ b + \varepsilon_2 n \\ c + \varepsilon_3 n \end{matrix}; z\right) \tag{4.49}$$

where $\varepsilon_j = 0, \pm 1$ are fixed (not all ε_j equal to zero). Each family y_n satisfies a second-order linear recurrence relation (difference equation) of the form

$$A_n y_{n-1} + B_n y_n + C_n y_{n+1} = 0. \tag{4.50}$$

All combinations of ε_j in (4.49) constitute 26 cases of recursions but because of symmetry relations and functional relations for the Gauss functions, many of these 26 cases can be transformed into each other.

For example, we have the trivial symmetry relation

$${}_2F_1\left(\begin{matrix} a, \ b \\ c \end{matrix}; z\right) = {}_2F_1\left(\begin{matrix} b, \ a \\ c \end{matrix}; z\right). \tag{4.51}$$

In addition, the following relations can be used (Abramowitz and Stegun 1964, p. 559)

$${}_2F_1\left(\begin{matrix} a, \ b \\ c \end{matrix}; z\right) = (1 - z)^{-a}\, {}_2F_1\left(\begin{matrix} a, \ c - b \\ c \end{matrix}; \frac{z}{z-1}\right), \tag{4.52}$$

$${}_2F_1\left(\begin{matrix} a, \ b \\ c \end{matrix}; z\right) = (1 - z)^{-b}\, {}_2F_1\left(\begin{matrix} c - a, \ b \\ c \end{matrix}; \frac{z}{z-1}\right), \tag{4.53}$$

$${}_2F_1\left(\begin{matrix} a, \ b \\ c \end{matrix}; z\right) = (1 - z)^{c-a-b}\, {}_2F_1\left(\begin{matrix} c - a, \ c - b \\ c \end{matrix}; z\right). \tag{4.54}$$

By using these four relations it is possible to reduce the set of 26 recursions to just five cases from which the properties of the remaining 21 recursions can be obtained (by using other considerations we can eliminate one of these cases, but here we consider the five cases).

For each of the five basic recursions, the following points have to be considered.

(1) To find the domains in the z-plane where we can identify minimal and dominant solutions of the recurrence relation (luckily, these domains do not depend on a, b, or c). These domains follow from Theorem 4.1, and are defined by by the equation $|t_1| = |t_2|$, where t_1 and t_2 are the zeros of the characteristic polynomial.

(2) To identify a pair of linearly independent solutions of each recursion in each domain. This means that we want to know which functions, next to $_2F_1(a + \varepsilon_1 n, b + \varepsilon_2 n; c + \varepsilon_3 n; z)$ for a given set ε_j, are solutions of the same recurrence relation, and which of these are minimal or dominant.

These points are considered in great detail in the recent papers by Gil, Segura and Temme (2006, 2007a), of which the second paper contains the full proofs. It is quite remarkable that a systematic study of the set of recursions for the Gauss function was neglected in the literature so far. Of course, an important subclass, the Legendre functions, was considered earlier (Gil and Segura 1997, 2000); Gautschi (1967) has paid attention to recursions for the incomplete beta functions, which are also a special case. Furthermore, Wimp (1984) has discussed the Miller algorithm with some examples of Gauss hypergeometric functions.

For each of the five basic recursions the coefficients A_n, B_n and C_n of the recurrence relation (4.50) are available now (Gil, Segura and Temme 2006, 2007a). A nice property of these recurrences is that the ratios of the coefficients A_n/C_n and B_n/C_n have finite limits as $n \to +\infty$ except, perhaps, at some singular points in the z-plane. We use the following notation:

$$\alpha = \lim_{n \to \infty} \frac{A_n}{C_n}, \qquad \beta = \lim_{n \to \infty} \frac{B_n}{C_n}. \tag{4.55}$$

According to Perron's theorem the existence of minimal solutions is guaranteed when the roots of the characteristic equation $t^2 + \beta t + \alpha = 0$ have different moduli. The equation $|t_1| = |t_2|$ gives curves in the complex z-plane, and these curves are the boundaries of domains where we can distinguish minimal and dominant solutions of the recurrence relations.

To denote the cases we use the notation $(\text{sign}(\varepsilon_1) \, \text{sign}(\varepsilon_2) \, \text{sign}(\varepsilon_3))$. For example, the recursion related to

$$y_n = {}_2F_1\left(\begin{matrix} a + n, \ b \\ c + n \end{matrix} ; z \right). \tag{4.56}$$

will be denoted by $(+0+)$. By using (4.53) we can write this as

$$y_n = (1-z)^{-b}{}_2F_1\left(\begin{matrix} c-a,\ b \\ c+n \end{matrix}; \zeta\right), \qquad \zeta = \frac{z}{z-1}. \tag{4.57}$$

If we have a pair $\{F_n, G_n\}$ of minimal and dominant solutions of recursions related to the basic form $(00+)$, we can use the relation (4.53) to obtain a pair $\{f_n, g_n\}$ for the recursion related to $(+0+)$.

We give the details of two basic recursions, namely the cases $(+00)$ and $(++0)$. For the other basic forms $(++-)$, $(+0-)$, and $(00+)$ we refer to Gil, Segura and Temme (2006, 2007a).

We give six solutions of the recurrence relation for the two cases considered here. These solutions are related to the three pairs of Gauss functions that are linearly independent solutions in the neighbourhood of the singular points $0, 1, \infty$ of the Gauss hypergeometric differential equation (2.20).

The $(+\ 0\ 0)$ recursion

The recurrence relation reads

$$A(a+n)y_{n-1} + B(a+n)y_n + C(a+n)y_{n+1} = 0, \tag{4.58}$$

where

$$A(a) = c - a,$$
$$B(a) = 2a - c - (a - b)z, \tag{4.59}$$
$$C(a) = a(z - 1).$$

The zeros of the characteristic polynomial of the recurrence relation (4.58) are

$$t_1 = 1, \qquad t_2 = \frac{1}{1-z}. \tag{4.60}$$

The equation $|t_1| = |t_2|$ holds when $|1 - z| = 1$, which defines a circle with centre $z = 1$ and radius 1. Inside the circle we have $|t_2| > |t_1|$.

Solutions of the recurrence relation (4.58) are

$$y_{1,n} = {}_2F_1\left(\begin{matrix} a+n,\ b \\ c \end{matrix}; z\right),$$

$$y_{2,n} = \frac{\Gamma(a+n+1-c)}{\Gamma(a+n)}{}_2F_1\left(\begin{matrix} 1+a+n-c,\ 1+b-c \\ 2-c \end{matrix}; z\right),$$

$$y_{3,n} = \frac{\Gamma(a+n+1-c)}{\Gamma(a+b+n+1-c)}{}_2F_1\left(\begin{matrix} a+n,\ b \\ a+b+n+1-c \end{matrix}; 1-z\right), \tag{4.61}$$

$$y_{4,n} = (1-z)^{-n}\frac{\Gamma(n+a+b-c)}{\Gamma(n+a)}{}_2F_1\left(\begin{matrix} c-a-n,\ c-b \\ c+1-a-b-n \end{matrix}; 1-z\right),$$

$$y_{5,n} = (-z)^{-n} \frac{\Gamma(n+1-c+a)}{\Gamma(1+a+n-b)} {}_2F_1 \left(\begin{matrix} a+n, \ a+n+1-c \\ a+n+1-b \end{matrix} ; \frac{1}{z} \right),$$

$$y_{6,n} = \frac{\Gamma(a+n-b)}{\Gamma(a+n)} {}_2F_1 \left(\begin{matrix} b, \ b+1-c \\ b+1-a-n \end{matrix} ; \frac{1}{z} \right).$$

All these functions are solutions of the recurrence relation (4.58). Only one of these is the minimal solution. Any solution different from the minimal solution cannot be minimal, because the minimal solution is unique (up to multiplicative factors not depending on n).

In the following scheme we give the properties of the six solutions:

	$\|z-1\|<1$	$\|z-1\|>1$	
y_1	dominant	dominant	
y_2	dominant	dominant	
y_3	*minimal*/dominant	dominant	(4.62)
y_4	dominant/*minimal*	dominant	
y_5	dominant	*minimal*	
y_6	dominant	dominant/*minimal*	

where, when two possibilities appear, the first one corresponds to $n \to +\infty$ and the second one to $n \to -\infty$.

The $(+ + 0)$ recursion
The recurrence relation reads

$$A(a+n, b+n)y_{n-1} + B(a+n, b+n)y_n + C(a+n, b+n)y_{n+1} = 0, \quad (4.63)$$

where

$$A(a,b) = (c-a)(c-b)(c-a-b-1),$$

$$B(a,b) = (c-a-b)\{c(a+b-c)+c-2ab$$
$$+ z[(a+b)(c-a-b)+2ab+1-c]\},$$

$$C(a,b) = ab(c-a-b+1)(1-z)^2.$$

The coefficients of characteristic equation $\lambda^2 + \beta\lambda + \alpha = 0$ are

$$\alpha = 1/(1-z)^2, \qquad \beta = -2(1+z)/(1-z)^2, \quad (4.65)$$

with roots

$$t_1 = \frac{1}{(1-\sqrt{z})^2}, \qquad t_2 = \frac{1}{(1+\sqrt{z})^2}. \quad (4.66)$$

The equation $|t_1| = |t_2|$ holds when $z \le 0$, otherwise $|t_1| > |t_2|$. In this case, the region $|t_1| \ne |t_2|$ is one connected region.

We give six solutions of the recurrence relation (4.63):

$$y_{1,n} = {}_2F_1\left(\begin{matrix} a+n, b+n \\ c \end{matrix}; z\right), \tag{4.67}$$

$$y_{2,n} = \frac{\Gamma(1+a-c+n)\Gamma(1+b-c+n)}{\Gamma(a+n)\Gamma(b+n)} {}_2F_1\left(\begin{matrix} 1+a-c+n, 1+b-c+n \\ 2-c \end{matrix}; z\right),$$

$$y_{3,n} = \frac{\Gamma(1+a-c+n)\Gamma(1+b-c+n)}{\Gamma(1+a+b-c+2n)} {}_2F_1\left(\begin{matrix} a+n, b+n \\ 1+a+b-c+2n \end{matrix}; 1-z\right),$$

$$y_{4,n} = (1-z)^{-2n}\frac{\Gamma(a+b-c+2n)}{\Gamma(a+n)\Gamma(b+n)} {}_2F_1\left(\begin{matrix} -a+c-n, -b+c-n \\ 1-a-b+c-2n \end{matrix}; 1-z\right),$$

$$y_{5,n} = (-z)^{-n}\frac{\Gamma(1+a-c+n)}{\Gamma(b+n)} {}_2F_1\left(\begin{matrix} a+n, 1+a-c+n \\ 1+a-b \end{matrix}; \frac{1}{z}\right),$$

$$y_{6,n} = (-z)^{-n}\frac{\Gamma(1+b-c+n)}{\Gamma(a+n)} {}_2F_1\left(\begin{matrix} b+n, 1+b-c+n \\ 1-a+b \end{matrix}; \frac{1}{z}\right).$$

In the following scheme we give the properties of the six solutions; the properties hold in $\mathbb{C} \setminus \{z \leq 0\}$:

$$
\begin{array}{ll}
y_{1,n} & \text{dominant} \\
y_{2,n} & \text{dominant} \\
y_{3,n} & \textit{minimal}/\text{dominant} \\
y_{4,n} & \text{dominant}/\textit{minimal} \\
y_{5,n} & \text{dominant} \\
y_{6,n} & \text{dominant}
\end{array}
\tag{4.68}
$$

where, when two possibilities appear, the first one corresponds to $n \to +\infty$ and the second one to $n \to -\infty$.

This case has applications for Jacobi polynomials. We have

$$P_n^{(\alpha,\beta)}(x) = \binom{n+\alpha}{n}\left(\frac{1+x}{2}\right)^n {}_2F_1\left(\begin{matrix} -n, -\beta-n \\ \alpha+1 \end{matrix}; z\right), \quad z = \frac{x-1}{x+1}. \tag{4.69}$$

A representation with $+n$ at the a and b places follows from applying (4.54). The interval of orthogonality is $[-1,1]$, and if $x \in [-1,1]$ we have $z \leq 0$. We see that in and outside the orthogonality interval the Jacobi polynomial $P_n^{(\alpha,\beta)}(x)$ is a dominant solution of its recurrence relation and this relation can be used for computing these polynomials in the forward direction. Only the usual rounding errors should be taken into account.

Remark 4.5. Observe that $y_{3,n}$ and $y_{4,n}$ are related to recursion schemes of the form $(+ + 2+)$ and $(- - 2-)$, which fall outside the set of 26 forms considered here.

4.4. Anomalous behaviour of some second-order homogeneous and first-order inhomogeneous recurrences

It is usually assumed that asymptotic information is enough for predicting stable directions for recursion, at least for second-order homogeneous equations. This, however, is not always true and there exist examples for which, for finite orders n, a minimal solution interchanges its role with certain dominant solutions. This, as a consequence, implies the anomalous convergence of the associated continued fraction to a value different from the ratio of consecutive minimal solutions, a phenomenon first observed (Gautschi 1977) in connection with a recurrence relation for confluent hypergeometric functions. See also Deaño and Segura (2007) and Gil *et al.* (2007*b*).

5. Uniform asymptotic expansions

Writing efficient algorithms for special functions may become problematic when several large parameters are involved. In particular, problems arise when functions suddenly change their behaviour, say from monotonic to oscillatory behaviour. For example, the Bessel function $J_\nu(x)$ has a turning point at $x = \nu$ (see Remark 3.20), and for large ν the function is oscillatory for $x > \nu$ and monotonic for $x < \nu$. See Figure 5.1. To describe the asymptotic behaviour of $J_\nu(x)$ for large ν and $x \sim \nu$ we need the Airy function $\mathrm{Ai}(z)$, which is a solution of the simple turning point equation $w'' - zw = 0$.

For many other special functions of mathematical physics, powerful uniform asymptotic expansions are available, which describe precisely how

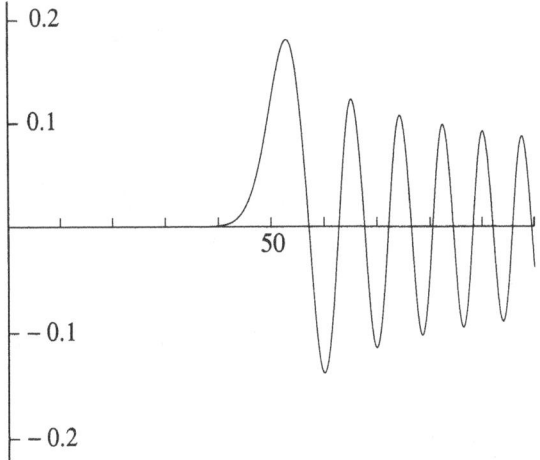

Figure 5.1. The Bessel function $J_{50}(x)$, $0 \leq x \leq 100$.

the functions behave, which are valid for large domains of the parameters, and which provide tools for designing high-performance computational algorithms.

In this section we discuss examples of uniform asymptotic expansions and their numerical aspects. We explain why these expansions are useful, and why they are usually difficult to handle in numerical algorithms. First we consider uniform expansions of the incomplete gamma functions. These functions are important in probability theory, but also in physical problems. Another important class concerns the functions having a turning point in their defining differential equation, in which case Airy-type expansions arise, as mentioned earlier. We give details of Airy-type expansions for Bessel functions.

5.1. Asymptotic expansions for the incomplete gamma functions

We recall the definitions of the incomplete gamma functions

$$\gamma(a, z) = \int_0^z t^{a-1} e^{-t} \, dt, \qquad \Gamma(a, z) = \int_z^\infty t^{a-1} e^{-t} \, dt, \qquad (5.1)$$

where for the first integral we need $\operatorname{Re} a > 0$ and for both integrals we assume that $|\operatorname{ph} z| < \pi$.

Integrating by parts in the second integral gives

$$\Gamma(a, z) = -\int_z^\infty t^{a-1} \, de^{-t} = z^{a-1} e^{-z} + (a - 1) \int_z^\infty t^{a-2} e^{-t} \, dt. \qquad (5.2)$$

Repeating this we find for $n = 1, 2, 3, \ldots$

$$\Gamma(a, z) = z^{a-1} e^{-z} \left[1 + \frac{a - 1}{z} + \frac{(a - 1)(a - 2)}{z^2} + \cdots \right. \qquad (5.3)$$

$$\left. + \frac{(a - 1)(a - 2) \cdots (a - n + 1)}{z^{n-1}} \right] + R_n(a, z),$$

where

$$R_n(a, z) = (a - 1)(a - 2) \cdots (a - n) \int_z^\infty t^{a-n-1} e^{-t} \, dt. \qquad (5.4)$$

For positive a and z we can easily find a bound for the remainder. If $a > n+1$ the integrand has a maximum at $t_0 = a - n - 1$. If $a \leq n+1$ the integrand is decreasing on $t > 0$. In any case, if $z > a - n$, we can integrate in the integral in (5.4) with respect to the variable $p = t + (n - a) \ln t$, which gives

$$R_n(a, z) = (a - 1)(a - 2) \cdots (a - n) \int_{p_0}^\infty e^{-p} \frac{dp}{t + n - a}, \qquad (5.5)$$

where $p_0 = z + (n-a) \ln z$. Because $t \geq z$ in (5.4) we have $t+n-a \geq z+n-a$, and we obtain

$$R_n(a, z) \leq \frac{(a-1)(a-2)\cdots(a-n)}{z^n} \frac{z}{(z+n-a)} z^{a-1}e^{-z}, \quad z > a - n.$$
(5.6)

This shows the asymptotic character of the expansion (5.3) when $n > a$, as $z \to \infty$. However, the condition $z > a - n$ is not enough to make it a useful expansion. When a is also large, say $a \sim z$, then the ratios of successive terms in the expansion (5.3) are of order $\mathcal{O}(1)$, and, hence, the terms are not even becoming small. We say that the expansion in (5.3) does not hold uniformly with respect to $a > 0$. However, it holds uniformly for a in compact intervals.

For the function $\gamma(a, z)$ we can also obtain an asymptotic representation. Integration by parts now starts with

$$\gamma(a, z) = \frac{1}{a} \int_0^z e^{-t} \, dt^a = \frac{1}{a} z^a e^{-z} + \frac{1}{a} \int_0^z t^a e^{-t} \, dt.$$
(5.7)

This is the beginning of the convergent expansion

$$\gamma(a, z) = \frac{1}{a} z^u e^{-z} \sum_{n=0}^{\infty} \frac{z^n}{(a+1)(a+2)\cdots(a+n)}.$$
(5.8)

This expansion has an asymptotic character when a is large, and again we see that the asymptotic property does not hold uniformly with respect to $z > 0$ (although the expansion is convergent for all finite z). Both expansions in (5.3) and (5.8) have their limitations with respect to in which domains we can use them for numerical computations. But they share one nice property: the coefficients can be computed very easily.

5.2. Uniform asymptotic expansions

The asymptotic expansions of the incomplete gamma functions $\Gamma(a, z)$ and $\gamma(a, z)$ given in the previous section become useless when both parameters a and z are of the same size. The representation for $\Gamma(a, z)$ in (5.3) is valid for any a and z (with the usual condition $|\mathrm{ph}\, z| < \pi$), but we can use it as an asymptotic representation only when $|z| \gg |a|$. As mentioned after (5.6), the ratio of successive terms in the representation (5.3) are of order $\mathcal{O}(1)$ when $z \sim a$.

We give details of the uniform asymptotic expansions for the incomplete gamma functions. One essential feature of such expansions is the role of certain special functions in the expansions. In the standard, non-uniform, expansions usually only elementary functions occur, such as exponential and trigonometric functions. In uniform expansions we usually need higher transcendental functions, such as Airy functions, error functions,

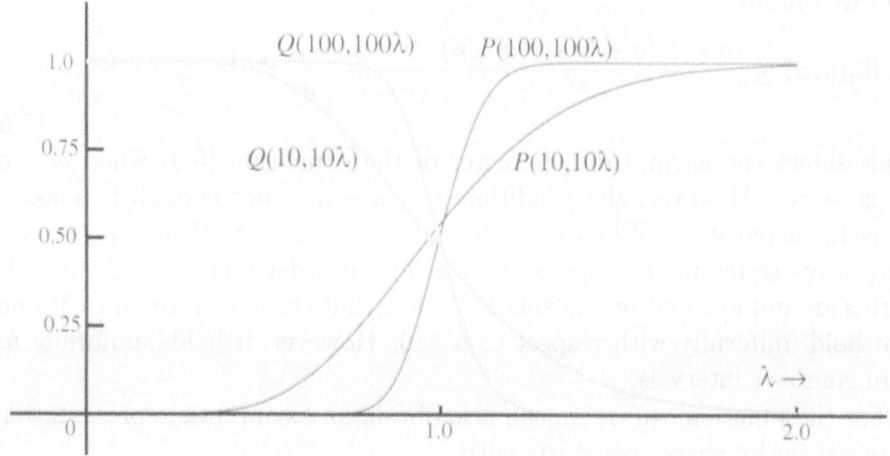

Figure 5.2. The function $P(a, \lambda a)$ and $Q(a, \lambda a)$
for $\lambda \in [0, 2]$ and $a = 10$ and $a = 100$.

Fresnel integrals, and so on. The proper choice of these special functions is not always clear without a further study of asymptotic analysis.

We consider a uniform expansion that can be used for both $\Gamma(a, z)$ and $\gamma(a, z)$, and for all large values of a and z, as well as for complex values, but we continue the discussion for positive real parameters.

The incomplete gamma functions are related to several cumulative distribution functions of probability theory, with the underlying distribution being the gamma distribution. In particular, the incomplete gamma functions appear in the form of the chi-square probability functions. For several reasons it is convenient to work with the normalized functions

$$P(a, z) = \frac{\gamma(a, z)}{\Gamma(a)}, \qquad Q(a, z) = \frac{\Gamma(a, z)}{\Gamma(a)}, \tag{5.9}$$

for example, because of their role in probability theory, and because no overflow occurs for large values of a. We have

$$P(a, z) + Q(a, z) = 1. \tag{5.10}$$

In Figure 5.2 we show the graphs of these functions, where we have used a parameter λ to scale the z-variable. In fact we give the graphs of the functions $P(a, \lambda a)$ and $Q(a, \lambda a)$ for $\lambda \in [0, 2]$ and $a = 10$ and $a = 100$. As a increases the graphs become steeper when λ passes the value $\lambda = 1$.

As is well known in the theory of cumulative distribution functions, many of these functions approach the normal or Gaussian probability functions when certain parameters become large. In probability theory the normal distribution functions are defined by

$$P(z) = \frac{1}{\sqrt{2\pi}} \int_{-\infty}^{z} e^{-\frac{1}{2}t^2}\, dt, \qquad Q(z) = \frac{1}{\sqrt{2\pi}} \int_{z}^{\infty} e^{-\frac{1}{2}t^2}\, dt, \qquad (5.11)$$

with the property $P(z) + Q(z) = 1$.

In our analysis we prefer the notation in terms of the error function and complementary error function, which are defined by

$$\operatorname{erf} z = \frac{2}{\sqrt{\pi}} \int_{0}^{z} e^{-t^2}\, dt, \qquad \operatorname{erfc} z = \frac{2}{\sqrt{\pi}} \int_{z}^{\infty} e^{-t^2}\, dt, \qquad (5.12)$$

with the symmetry relations $\operatorname{erf} z + \operatorname{erfc} z = 1$ and $\operatorname{erfc} z + \operatorname{erfc}(-z) = 2$. These are related to the normal distribution by

$$P(z) = \tfrac{1}{2}\operatorname{erfc}(-z/\sqrt{2}), \qquad Q(z) = \tfrac{1}{2}\operatorname{erfc}(z/\sqrt{2}). \qquad (5.13)$$

The uniform expansion of the incomplete gamma functions

In Temme (1996, pp. 283–286) we derived the uniform expansion by using saddle point methods for integrals. In that analysis the complementary error function appeared because a singularity (a pole) of the integrand approaches the saddle point, when $a \sim z$. For details on the role of the complementary error function in such situations, see Wong (2001, pp. 356–358).

We summarize the results by giving the following representations:

$$Q(a, z) = \tfrac{1}{2}\operatorname{erfc}(\eta\sqrt{a/2}) + R_a(\eta),$$

$$P(a, z) = \tfrac{1}{2}\operatorname{erfc}(-\eta\sqrt{a/2}) - R_a(\eta), \qquad (5.14)$$

where

$$\tfrac{1}{2}\eta^2 = \lambda - 1 - \ln \lambda, \qquad \lambda = \frac{z}{a}, \qquad (5.15)$$

and

$$R_a(\eta) = \frac{e^{-\frac{1}{2}a\eta^2}}{\sqrt{2\pi a}} S_a(\eta), \qquad S_a(\eta) \sim \sum_{n=0}^{\infty} \frac{C_n(\eta)}{a^n}, \qquad (5.16)$$

as $a \to \infty$.

The relation between η and λ in (5.15) becomes clear when we expand

$$\lambda - 1 - \ln \lambda = \tfrac{1}{2}(\lambda - 1)^2 - \tfrac{1}{3}(\lambda - 1)^3 + \tfrac{1}{4}(\lambda - 1)^4 + \cdots, \qquad (5.17)$$

and in fact the relation in (5.15) can also be written as

$$\eta = (\lambda - 1)\sqrt{\frac{2(\lambda - 1 - \ln \lambda)}{(\lambda - 1)^2}}, \qquad (5.18)$$

where the sign of the square root is positive for $\lambda > 0$. For complex values we use analytic continuation. An expansion for small values of $|\lambda - 1|$ reads

$$\eta = (\lambda - 1) - \tfrac{1}{3}(\lambda - 1)^2 + \tfrac{7}{36}(\lambda - 1)^3 + \cdots, \tag{5.19}$$

and, upon inverting this expansion,

$$\lambda = 1 + \eta + \tfrac{1}{3}\eta^2 + \tfrac{1}{36}\eta^3 + \cdots. \tag{5.20}$$

Note that the symmetry relation $P(a, z) + Q(a, z) = 1$ is preserved in the representations in (5.9) because erfc z + erfc$(-z) = 2$.

We give the steps on determining the coefficients $C_n(\eta)$ in (5.16). Differentiating the relation in (5.14) of $Q(a, z)$ with respect to η gives, on the one hand,

$$\frac{dQ(a, z)}{d\eta} = \frac{dQ(a, z)}{dz} \frac{dz}{d\eta} = -\frac{1}{\Gamma(a)} z^{a-1} e^{-z} \frac{dz}{d\eta}, \tag{5.21}$$

and on the other hand, by using (5.14) and (5.16),

$$\frac{dQ(a, z)}{d\eta} = \left[-\sqrt{\frac{a}{2\pi}} \frac{\lambda - 1}{\lambda} - \eta \sqrt{\frac{a}{2\pi}} S_a(\eta) + \frac{1}{\sqrt{2\pi a}} \frac{dS_a(\eta)}{d\eta} \right] e^{-\frac{1}{2}a\eta^2}, \tag{5.22}$$

where we have used

$$\frac{dz}{d\eta} = a\frac{d\lambda}{d\eta} = a\frac{\lambda\eta}{\lambda - 1}. \tag{5.23}$$

After straightforward manipulations we obtain

$$\frac{dS_a(\eta)}{d\eta} - a\eta S_a(\eta) = a\left[1 - \frac{\eta}{(\lambda - 1)\Gamma^*(a)} \right], \tag{5.24}$$

where $\Gamma^*(a)$ is defined by

$$\Gamma^*(a) = \sqrt{\frac{a}{2\pi}} e^a a^{-a} \Gamma(a), \quad a > 0. \tag{5.25}$$

We have the expansion

$$\frac{1}{\Gamma^*(a)} \sim \sum_{n=0}^{\infty} \gamma_n a^{-n}, \quad a \to \infty, \tag{5.26}$$

where the first few γ_n are

$$\gamma_0 = 1, \quad \gamma_1 = -\tfrac{1}{12}, \quad \gamma_2 = \tfrac{1}{288}, \quad \gamma_3 = \tfrac{139}{51840}. \tag{5.27}$$

The numbers γ_n also appear in the well-known asymptotic expansion of the Euler gamma function. That is,

$$\Gamma^*(a) \sim \sum_{n=0}^{\infty} (-1)^n \gamma_n a^{-n}, \quad a \to \infty. \tag{5.28}$$

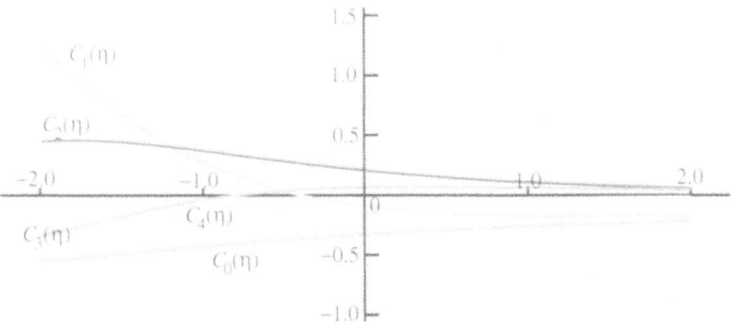

Figure 5.3. Graphs of the first five coefficients $C_n(\eta)$. Because of scaling we have drawn graphs of $\rho C_n(\eta)$, where $\rho = 1, 50, 50, 100, 100$ for $n = 0, 1, 2, 3, 4$, respectively.

We substitute the expansion of $S_a(\eta)$ given in (5.16) and the expansion (5.26) into the differential equation (5.24). Comparing, after these substitutions, equal powers of a in (5.24), we obtain

$$C_0(\eta) = \frac{1}{\lambda - 1} - \frac{1}{\eta}, \tag{5.29}$$

and the recurrence relation

$$\eta C_n(\eta) = \frac{\mathrm{d}}{\mathrm{d}\eta} C_{n-1}(\eta) + \frac{\eta}{\lambda - 1} \gamma_n, \quad n \geq 1. \tag{5.30}$$

For $C_1(\eta)$ we have

$$C_1(\eta) = \frac{1}{\eta^3} - \frac{1}{(\lambda - 1)^3} - \frac{1}{(\lambda - 1)^2} - \frac{1}{12(\lambda - 1)}. \tag{5.31}$$

The first two coefficients (and all higher coefficients) have a removable singularity at $\eta = 0$, that is, at $\lambda = 1$ or $z = a$. All $C_n(\eta)$ are analytic at the origin $\eta = 0$.

The expansion in (5.16) has no restrictions on the parameter λ. It holds uniformly with respect to $\lambda \geq 0$ (and for complex values of a and λ). So it is much more powerful than, for example, the asymptotic expansion (5.3). However, the computation of the coefficients $C_n(\eta)$ is not as easy as that of the coefficients in (5.3). In particular, near the transition point, that is, when $z \sim a$, the removable singularities in the representations of C_0 and C_1 as shown in (5.29) and (5.31) are difficult to handle in numerical computations. All higher coefficients show this type of cancellations, and the removable singularities in C_n are poles of order $2n + 1$.

In Figure 5.3 we show the graphs of $C_n(\eta)$ for $n = 0, 1, 2, 3, 4$, properly scaled in order to get them visible in one figure.

Expansions for the coefficients

We concentrate on the numerical aspects of the expansion of $S_a(\eta)$ in (5.16). We have already observed that the coefficients $C_k(\eta)$ in (5.16) are difficult to evaluate near the transition point $\eta = 0$, which corresponds to $\lambda = 1$, or $z = a$. We will give two methods for evaluating the coefficients and the expansion.

The coefficients $C_k(\eta)$ are analytic (Temme 1996, pp. 283–286) inside the disk $|\eta| < 2\sqrt{\pi} = 3.54\ldots$. So we can expand all coefficients in power series for $|\eta| < 2\sqrt{\pi}$. For numerical applications these expansions can be used for complex η, say, for $|\eta| \leq 1$. More efficiently, when the variables a and z are real and positive, we can expand the coefficients in terms of Chebyshev polynomials in intervals of the real η-axis.

We give the first terms in the Maclaurin expansions of the first coefficients:

$$C_0(\eta) = -\tfrac{1}{3} + \tfrac{1}{12}\eta - \tfrac{2}{135}\eta^2 + \tfrac{1}{864}\eta^3 + \tfrac{1}{2835}\eta^4 - \tfrac{139}{777600}\eta^5 + \cdots,$$

$$C_1(\eta) = -\tfrac{1}{540} - \tfrac{1}{288}\eta + \tfrac{1}{378}\eta^2 - \tfrac{77}{77760}\eta^3 + \tfrac{1}{4860}\eta^4 - \tfrac{1}{2488320}\eta^5 + \cdots,$$

$$C_2(\eta) = \tfrac{25}{6048} - \tfrac{139}{51840}\eta + \tfrac{1}{1296}\eta^2 + \tfrac{1}{497664}\eta^3 - \tfrac{6199}{57736800}\eta^4 + \cdots, \qquad (5.32)$$

$$C_3(\eta) = \tfrac{101}{155520} + \tfrac{571}{2488320}\eta - \tfrac{54179}{115473600}\eta^2 + \tfrac{41969}{156764160}\eta^3 - \tfrac{20639}{272937600}\eta^4 + \cdots,$$

$$C_4(\eta) = -\tfrac{3184811}{3695155200} + \tfrac{163879}{209018880}\eta - \tfrac{8707}{29113344}\eta^2 - \tfrac{47207}{32248627200}\eta^3 + \cdots.$$

In Section 5.3 we discuss alternative uniform expansions in which no coefficients occur that are difficult to compute. But first we give another numerical scheme for the expansion in (5.16).

Numerical algorithm for small values of η

Instead of expanding each coefficient $C_n(\eta)$ in powers of η, which needs the storage of many coefficients, we expand the function $S_a(\eta)$ of (5.16) in powers of η. The coefficients are functions of a, and we write

$$S_a(\eta) = \sum_{n=0}^{\infty} \alpha_n \eta^n, \qquad (5.33)$$

where the series again converges for $|\eta| < 2\sqrt{\pi}$.

To compute the coefficients α_n, we use the differential equation for $S_a(\eta)$ given in (5.24). Substituting the expansion (5.33) into (5.24), using the coefficients d_n in the expansion

$$\frac{\eta}{\lambda - 1} = \sum_{n=0}^{\infty} d_n \eta^n, \qquad (5.34)$$

we obtain for α_n the recurrence relation

$$\alpha_n = \frac{1}{a}(n+2)\alpha_{n+2} + \frac{d_{n+1}}{\Gamma^*(a)}, \quad n = 0,1,2,\ldots. \tag{5.35}$$

The coefficients d_n follow from the coefficients of $C_0(\eta)$, of which the first few are given in (5.32), because $\eta/(\lambda - 1) = 1 + \eta C_0(\eta)$. We have

$$d_0 = 1, \quad d_1 = -\tfrac{1}{3}, \quad d_2 = -\tfrac{1}{12}, \quad d_3 = -\tfrac{2}{135}, \quad d_4 = \tfrac{1}{864}, \quad d_5 = \tfrac{1}{2835}. \tag{5.36}$$

With these values we can compute the first terms with odd index:

$$\alpha_1 = \frac{a}{\Gamma^*(a)}\left[\Gamma^*(a) - 1\right],$$

$$\alpha_3 = \frac{a^2}{1 \cdot 3\,\Gamma^*(a)}\left[\Gamma^*(a) - 1 - \frac{1}{12a}\right], \tag{5.37}$$

$$\alpha_5 = \frac{a^2}{1 \cdot 3 \cdot 5\,\Gamma^*(a)}\left[\Gamma^*(a) - 1 - \frac{1}{12a} - \frac{1}{288a^2}\right].$$

We observe, see (5.27) and (5.28), that the computation of these α_n requires not only the value of $\Gamma^*(a)$, but also that of $\Gamma^*(a)$ with the first terms of the asymptotic expansion subtracted. The higher odd coefficients show the same pattern; more and more terms of the asymptotic expansions have to be subtracted. In fact we recur remainders of the asymptotic expansion of the gamma function. In particular, when a is large, this is a very unstable process. The same problems arise with the even coefficients. Note that $\alpha_0 = S_a(0)$, a quantity that can be computed from an asymptotic expansion, and the higher even terms follow from the recursion in (5.35), with more and more terms subtracted in this expansion of $S_a(0)$.

When we use (5.35) in the backward direction the recursion becomes stable. In addition, we do not need the computation of $S_a(0)$ and $\Gamma^*(a)$, because these values follow from the backward recursion process. We only need enough coefficients d_n of (5.34) for this recursion.

First we remove $\Gamma^*(a)$ from the recursion in (5.35) by writing

$$\alpha_n = \frac{\beta_n}{\Gamma^*(a)}, \quad n = 0,1,2,\ldots, \tag{5.38}$$

which gives for β_n the recursion

$$\beta_n = \frac{1}{a}(n+2)\beta_{n+2} + d_{n+1}, \quad n = 0,1,2,\ldots. \tag{5.39}$$

We choose a positive integer N, put $\beta_{N+2} = \beta_{N+1} = 0$ and compute the sequence

$$\beta_N, \beta_{N-1}, \ldots, \beta_1, \beta_0 \tag{5.40}$$

Table 5.1. Relative errors δ in the computation of $\Gamma(a)$ by using the backward recursion scheme (5.39) for several values of a and $N = 25$ and $N = 35$.

N	a	δ	a	δ	a	δ
$N = 25$	2	$0.39_{10}(-06)$	8	$0.11_{10}(-13)$	14	$0.79_{10}(-17)$
	3	$0.27_{10}(-08)$	9	$0.24_{10}(-14)$	15	$0.32_{10}(-17)$
	4	$0.75_{10}(-10)$	10	$0.61_{10}(-15)$	16	$0.14_{10}(-17)$
	5	$0.45_{10}(-11)$	11	$0.18_{10}(-15)$	17	$0.64_{10}(-18)$
	6	$0.44_{10}(-12)$	12	$0.58_{10}(-16)$	18	$0.30_{10}(-18)$
	7	$0.61_{10}(-13)$	13	$0.21_{10}(-16)$	19	$0.15_{10}(-18)$
$N = 35$	2	$0.80_{10}(-06)$	8	$0.76_{10}(-17)$	14	$0.19_{10}(-21)$
	3	$0.57_{10}(-09)$	9	$0.82_{10}(-18)$	15	$0.50_{10}(-22)$
	4	$0.30_{10}(-11)$	10	$0.11_{10}(-18)$	16	$0.15_{10}(-22)$
	5	$0.50_{10}(-13)$	11	$0.18_{10}(-19)$	17	$0.46_{10}(-23)$
	6	$0.17_{10}(-14)$	12	$0.35_{10}(-20)$	18	$0.15_{10}(-23)$
	7	$0.93_{10}(-16)$	13	$0.77_{10}(-21)$	19	$0.54_{10}(-24)$

from the recurrence relation (5.39). Because

$$\Gamma^*(a) = 1 + \frac{1}{a}\beta_1, \tag{5.41}$$

we have

$$S_a(\eta) \sim \frac{a}{a + \beta_1} \sum_{n=0}^{N} \beta_n \eta^n \tag{5.42}$$

as an approximation for $S_a(\eta)$.

We verify this algorithm by taking several values of a and $N = 25$ and $N = 35$. In Table 5.1 we give the relative errors of the approximations of $\Gamma(a)$, which are computed by using β_1 in (5.41), and by computing $\Gamma(a)$ from the relation in (5.25). For example, with $N = 25$ and $a = 5$ we obtain

$$\Gamma(a) = 23.999999999892\cdots, \quad \text{with relative error } 0.45_{10}(-11). \tag{5.43}$$

We observe that for the larger values of a the scheme gives better approximations, as is the case for the larger value of N.

We have used the approximation in (5.42) for computing the incomplete gamma functions in IEEE double precision for $a \geq 12$ and $|\eta| \leq 1$. We need the storage of 25 coefficients d_n, and in the series in (5.42) we need 25 terms or less.

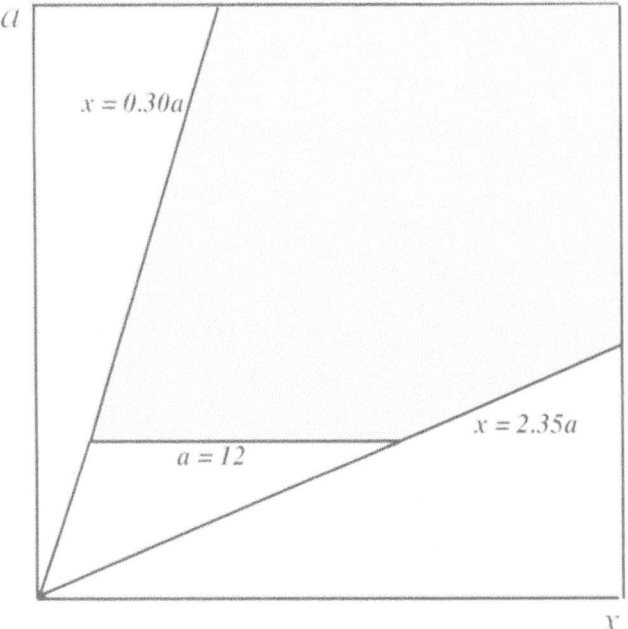

Figure 5.4. The domain of application (grey) where we can apply the backward recursion scheme (5.39) to obtain IEEE double precision values of $S_a(\eta)$. The grey domain is bounded by the lines $a = 12$, $x = 2.35a$ and $x = 0.30a$.

The value $\eta = -1$ corresponds to $\lambda = 0.30\ldots$, and the value $\eta = 1$ with $\lambda = 2.35\ldots$. In Figure 5.4 we show the area in the (x, a) quarter plane where we can apply the algorithm to obtain double precision. The domain is bounded by the lines $a = 12$, $x = 2.35a$ and $x = 0.30a$.

5.3. A simpler uniform expansion

The expansion considered in Section 5.2 can be modified to obtain an expansion with coefficients that can be evaluated much more easily than the coefficients $C_n(\eta)$. The modified expansion is again valid for large values of a; it is again valid in the transition area $a \sim z$. The restriction on $\lambda = z/a$, however, is $\lambda - 1 = o(a^{-\frac{1}{3}})$ as $a \to \infty$.

We start with the integral (see (5.1))

$$\Gamma(a + 1, z) = \int_z^\infty t^a e^{-t}\, dt, \qquad (5.44)$$

and we consider positive parameters a and z. We substitute $t = a(1 + s)$.

This gives

$$\Gamma(a+1,z) = a^{a+1}e^{-a} \int_\mu^\infty e^{-a[s-\ln(1+s)]}\,ds, \quad \mu = \lambda - 1 = \frac{z-a}{a}. \quad (5.45)$$

The exponential function has a maximum at the origin $s = 0$. We write

$$\Gamma(a+1,z) = a^{a+1}e^{-a} \int_\mu^\infty e^{-\frac{1}{2}as^2 - af(s)}\,ds, \quad (5.46)$$

where

$$f(s) = s - \tfrac{1}{2}s^2 - \ln(1+s) = \mathcal{O}(s^3), \quad s \to 0. \quad (5.47)$$

We expand

$$e^{-af(s)} = \sum_{n=0}^\infty D_n(a)s^n, \quad |s| < 1, \quad (5.48)$$

and upon substituting this expansion in (5.46), we obtain the expansion

$$\Gamma(a+1,z) \sim a^{a+1}e^{-a} \sum_{n=0}^\infty D_n(a)\Phi_n(a,z), \quad (5.49)$$

where

$$\Phi_n(a,z) = \int_\mu^\infty s^n e^{-\frac{1}{2}as^2}\,ds, \quad n = 0,1,2,\dots. \quad (5.50)$$

The first two Φ_n are

$$\Phi_0(a,z) = \sqrt{\frac{\pi}{2a}}\operatorname{erfc}(\mu\sqrt{a/2}), \qquad \Phi_1(a,z) = \frac{1}{a}e^{-\frac{1}{2}a\mu^2}. \quad (5.51)$$

By integrating by parts in (5.50) it easily follows that

$$a\Phi_{n+1}(a,z) = n\Phi_{n-1}(a,z) + \mu^n e^{-\frac{1}{2}a\mu^2}, \quad n = 1,2,3,\dots. \quad (5.52)$$

We can also derive a recurrence relation for the coefficients. We have the starting values $D_0(a) = 1$, $D_1(a) = 0$, and by differentiating (5.48), we obtain

$$(n+1)D_{n+1}(a) = aD_{n-2}(a) - nD_n(a), \quad n = 1,2,3,\dots. \quad (5.53)$$

The expansion in (5.50) is of interest because the coefficients $D_n(a)$ can be computed very easily by using the recursion in (5.53). Further, the recursion for the Φ_n is quite simple.

The expansion, with proofs for complex values of a and z, is derived in Ferreira, López and Pérez Sinusía (2005). In this reference several other expansions of the incomplete gamma function are considered. See also Paris (2002).

We observe that a similar expansion can be derived for the other incomplete gamma function $\gamma(a + 1, z)$. In that case the functions Φ_n should be replaced by functions Ψ_n defined by

$$\Psi_n(a, z) = \int_{-1}^{\mu} s^n e^{-\frac{1}{2}as^2}\, ds, \quad n = 0, 1, 2, \dots. \tag{5.54}$$

The function Ψ_0 can be expressed in terms of two error functions, and the other ones follow from integrating by parts.

From a numerical example we conclude that the asymptotic convergence of the expansion (5.49) is quite slow. When we take $z = 100$ and $a = 101$, and we sum the series with 12 terms, we obtain 4 significant digits. This result does not improve when we take more terms.

5.4. Airy-type expansions for Bessel functions

Airy functions are solutions of the differential equation

$$w'' - z\, w = 0. \tag{5.55}$$

Two linearly independent solutions that are real for real values of z are denoted by $\mathrm{Ai}(z)$ and $\mathrm{Bi}(z)$. Equation (5.55) is the simplest second-order linear differential equation that has a simple turning point (at $z = 0$). More general turning point equations have the standard form

$$\frac{\mathrm{d}^2 W}{\mathrm{d}\zeta^2} = \left[u^2 \zeta + \psi(\zeta)\right] W, \tag{5.56}$$

and the problem is to find an asymptotic approximation of $W(\zeta)$ for large values of u, that holds uniformly in a neighbourhood of $\zeta = 0$. A first approximation is obtained by neglecting $\psi(\zeta)$, which gives the solutions

$$\mathrm{Ai}\big(u^{\frac{2}{3}}\zeta\big), \qquad \mathrm{Bi}\big(u^{\frac{2}{3}}\zeta\big). \tag{5.57}$$

For a detailed discussion of this kind of problem we refer to Olver (1997, Chapter 11). Many solutions of physical problems and many special functions can be transformed into the standard form (5.56). Examples are Bessel functions, Whittaker functions, the classical orthogonal polynomials (in particular Hermite and Laguerre polynomials), and parabolic cylinder functions.

In all known cases the coefficients are difficult to compute in the neighbourhood of the turning point, and we saw a similar difficulty in the uniform expansion of the incomplete gamma functions in Section 5.2.

In this section we discuss two methods for computing the asymptotic series. One method is based on expanding the coefficients in the series into Maclaurin series. In the second method we consider the computation of auxiliary functions that can be computed more efficiently than the coefficients in the first method, and we do not need the tabulation of many coefficients. This method is similar to that described for the computation of incomplete gamma functions in Section 5.2.

The methods described in this section are quite general, but we only treat the case of Bessel functions, by using the differential equation of the Bessel functions, which has a turning point character when the order and argument of the Bessel functions are equal.

The Airy-type asymptotic expansions
The ordinary Bessel functions $J_\nu(z)$ and $Y_\nu(z)$, and all other Bessel functions, can be expanded in terms of Airy functions. We give the transformations of the Bessel differential equation

$$z^2 f'' + z f' + (z^2 - \nu^2) f = 0 \tag{5.58}$$

into the form (5.56). First we change the variable z into νz and apply a transformation to remove the first derivative term. We obtain the equation

$$F'' + \left(\nu^2 \frac{z^2 - 1}{z^2} + \frac{1}{4z^2} \right) F = 0, \tag{5.59}$$

with solutions $\sqrt{z} J_\nu(\nu z)$ and $\sqrt{z} Y_\nu(\nu z)$. The turning point character at $z = 1$ of this equation is visible now, and we transform this point to the origin by using the transformation

$$\zeta \left(\frac{d\zeta}{dz} \right)^2 = \frac{1 - z^2}{z^2}, \quad W = \sqrt{\zeta'} w, \tag{5.60}$$

This transformation gives the equation (5.56) with $\psi(\zeta)$ given by

$$\psi(\zeta) = \frac{5}{16\zeta^2} + \frac{\zeta z^2 (z^2 + 4)}{4(z^2 - 1)^3} \tag{5.61}$$

and solutions

$$\sqrt{z} \sqrt{\zeta'} J_\nu(\nu z), \qquad \sqrt{z} \sqrt{\zeta'} Y_\nu(\nu z). \tag{5.62}$$

The transformations used here are Liouville transformations; see also Olver (1997, p. 398).

Because the Airy functions given in (5.57) are solutions of (5.56) when $\psi(\zeta)$ is neglected, and because of asymptotic properties of the Bessel functions, the following representations are chosen:

$$J_\nu(\nu z) = \frac{\phi(\zeta)}{\nu^{\frac{1}{3}}} \left[\mathrm{Ai}(\nu^{\frac{2}{3}} \zeta) A_\nu(\zeta) + \nu^{-\frac{4}{3}} \mathrm{Ai}'(\nu^{\frac{2}{3}} \zeta) B_\nu(\zeta) \right],$$

$$Y_\nu(\nu z) = -\frac{\phi(\zeta)}{\nu^{\frac{1}{3}}} \left[\mathrm{Bi}(\nu^{\frac{2}{3}} \zeta) A_\nu(\zeta) + \nu^{-\frac{4}{3}} \mathrm{Bi}'(\nu^{\frac{2}{3}} \zeta) B_\nu(\zeta) \right], \tag{5.63}$$

where

$$\phi(\zeta) = \left(\frac{4\zeta}{1 - z^2}\right)^{\frac{1}{4}}, \qquad \phi(0) = 2^{\frac{1}{3}}. \tag{5.64}$$

The new variable ζ introduced in (5.60) can be written as

$$\frac{2}{3}\zeta^{3/2} = \ln\frac{1 + \sqrt{1 - z^2}}{z} - \sqrt{1 - z^2}, \qquad 0 \le z \le 1,$$

$$\frac{2}{3}(-\zeta)^{3/2} = \sqrt{z^2 - 1} - \arccos\frac{1}{z}, \qquad z \ge 1. \tag{5.65}$$

Next we introduce asymptotic expansions for the functions $A_\nu(\zeta)$ and $B_\nu(\zeta)$ of (5.63). It appears, after substituting the Bessel functions in (5.63) into (5.56) that we have the formal expansions

$$A_\nu(\zeta) \sim \sum_{s=0}^{\infty} \frac{a_s(\zeta)}{\nu^{2s}}, \qquad B_\nu(\zeta) \sim \sum_{s=0}^{\infty} \frac{b_s(\zeta)}{\nu^{2s}}. \tag{5.66}$$

The first coefficients

$$a_0(\zeta) = 1, \qquad b_0(\zeta) = -\frac{5}{48\zeta^2} + \frac{\phi^2(\zeta)}{48\zeta}\left[\frac{5}{1 - z^2} - 3\right], \tag{5.67}$$

where $\phi(\zeta)$ is given by (5.64). Higher coefficients follow certain recurrence relations. Further details on the coefficients are given later.

For the derivatives we have

$$J_\nu'(\nu z) = -\widehat{\phi}(\zeta)\left[\nu^{-\frac{4}{3}}\text{Ai}(\nu^{\frac{2}{3}}\zeta)\, C_\nu(\zeta) + \nu^{-\frac{2}{3}}\text{Ai}'(\nu^{\frac{2}{3}}\zeta)\, D_\nu(\zeta)\right],$$

$$Y_\nu'(\nu z) = \widehat{\phi}(\zeta)\left[\nu^{-\frac{4}{3}}\text{Bi}(\nu^{\frac{2}{3}}\zeta)\, C_\nu(\zeta) + \nu^{-\frac{2}{3}}\text{Bi}'(\nu^{\frac{2}{3}}\zeta)\, D_\nu(\zeta)\right], \tag{5.68}$$

where

$$\widehat{\phi}(\zeta) = -\frac{d\zeta}{dz}\phi(\zeta) = \frac{2}{z\phi(\zeta)}, \qquad \chi(\zeta) = \frac{\phi'(\zeta)}{\phi(\zeta)} = \frac{4 - z^2[\phi(\zeta)]^6}{16\zeta}, \tag{5.69}$$

and

$$C_\nu(\zeta) = \chi(\zeta)\, A_\nu(\zeta) + A_\nu'(\zeta) + \zeta\, B_\nu(\zeta),$$

$$D_\nu(\zeta) = A_\nu(\zeta) + \nu^{-2}\chi(\zeta)\, B_\nu(\zeta) + \nu^{-2}\, B_\nu'(\zeta). \tag{5.70}$$

Primes denote differentiation with respect to ζ.

The functions $C_\nu(\zeta), D_\nu(z)$ have the expansions

$$C_\nu(\zeta) \sim \sum_{s=0}^{\infty} \frac{c_s(\zeta)}{\nu^{2s}}, \qquad D_\nu(\zeta) \sim \sum_{s=0}^{\infty} \frac{d_s(\zeta)}{\nu^{2s}}, \tag{5.71}$$

where

$$c_s(\zeta) = \chi(\zeta)\, a_s(\zeta) + a_s'(\zeta) + \zeta\, b_s(\zeta),$$

$$d_s(\zeta) = a_s(\zeta) + \chi(\zeta)\, b_{s-1}(\zeta) + b_{s-1}'(\zeta). \tag{5.72}$$

The Airy-type asymptotic expansions of this section hold as $\nu \to \infty$, uniformly with respect to $z \in [0, \infty)$ and for complex values of ν and z (Olver 1997, Chapter 11).

The first coefficients c_s, d_s of (5.71) are

$$c_0(\zeta) = \frac{7}{48\zeta} + \frac{\phi^2(\zeta)}{48}\left[9 - \frac{7}{1 - z^2}\right], \quad d_0(\zeta) = 1. \tag{5.73}$$

A recursive scheme for evaluating a_s, b_s is given by

$$a_s''(\zeta) + 2\zeta b_s'(\zeta) + b_s(\zeta) - \psi(\zeta)\, a_s(\zeta) = 0,$$
$$2a_{s+1}'(\zeta) + b_s''(\zeta) - \psi(\zeta)\, b_s(\zeta) = 0, \tag{5.74}$$

where $a_0(\zeta) = 1$ and $\psi(\zeta)$ is given by (5.61).

Properties of the functions $A_\nu, B_\nu, C_\nu, D_\nu$
By using equation (5.56), with $\psi(\zeta)$ as given in (5.61), it is straightforward to derive the following system of differential equations for the functions $A_\nu(\zeta), B_\nu(\zeta)$:

$$A'' + 2\zeta B' + B - \psi(\zeta)A = 0,$$
$$B'' + 2\nu^2 A' - \psi(\zeta)B = 0, \tag{5.75}$$

where primes denote differentiation with respect to ζ.

A Wronskian for the system (5.75) follows by eliminating the terms with $\psi(\zeta)$. This gives

$$A''B - B''A + B^2 + 2\zeta B'B - 2\nu^2 A'A = 0, \tag{5.76}$$

which can be integrated as follows:

$$\nu^2 A_\nu^2(\zeta) + A_\nu(\zeta) B_\nu'(\zeta) - A_\nu'(\zeta) B_\nu(\zeta) - \zeta B_\nu^2(\zeta) = \nu^2. \tag{5.77}$$

The constant on the right-hand side follows by taking $\zeta = 0$ and from information given later in this section.

Expansions for $a_s(\zeta), b_s(\zeta), c_s(\zeta), d_s(\zeta)$
The singular points of the functions $z(\zeta), \psi(\zeta), \phi(\zeta), \widehat{\phi}(\zeta), \chi(\zeta)$ and those of the coefficients of the asymptotic expansions occur at

$$\zeta^{\pm} = \left(\tfrac{3}{2}\pi\right)^{\frac{2}{3}} e^{\pm i\pi/3}. \tag{5.78}$$

These points correspond to $z = e^{\mp \pi i}$. It follows that the radius of convergence of the Maclaurin series of these quantities equals $2.81\cdots$. In this section we give the expansions and mention the values of the early coefficients.

It is convenient to start with an expansion of z in powers of ζ. We substitute $z = 1 + z_1\zeta + \cdots$ into the first part of (5.60). This gives $z_1^3 = -1/2$.

Using the relations in (5.65) we obtain the correct branch: $z_1 = -2^{-\frac{1}{3}}$. We write

$$\zeta = 2^{\frac{1}{3}}\eta, \tag{5.79}$$

and we obtain in a straightforward way the following expansions:

$$z(\zeta) = \sum_{n=0}^{\infty} z_n \eta^n = \left[1 - \eta + \tfrac{3}{10}\eta^2 + \tfrac{1}{350}\eta^3 - \tfrac{479}{63000}\eta^4 + \cdots\right], \tag{5.80}$$

$$\psi(\zeta) = 2^{\frac{1}{3}}\sum_{n=0}^{\infty} \psi_n \eta^n = 2^{\frac{1}{3}}\left[\tfrac{1}{70} + \tfrac{2}{75}\eta + \tfrac{138}{13475}\eta^2 - \tfrac{296}{73125}\eta^3 - \tfrac{38464}{7074375}\eta^4 + \cdots\right],$$

$$\phi(\zeta) = 2^{\frac{1}{3}}\sum_{n=0}^{\infty} \phi_n \eta^n = 2^{\frac{1}{3}}\left[1 + \tfrac{1}{5}\eta + \tfrac{9}{350}\eta^2 - \tfrac{89}{15750}\eta^3 - \tfrac{4547}{1155000}\eta^4 + \cdots\right],$$

$$\widehat{\phi}(\zeta) = 2^{\frac{2}{3}}\sum_{n=0}^{\infty} \widehat{\phi}_n \eta^n = 2^{\frac{2}{3}}\left[1 + \tfrac{4}{5}\eta + \tfrac{18}{35}\eta^2 + \tfrac{88}{315}\eta^3 + \tfrac{79586}{606375}\eta^4 + \cdots\right],$$

$$\chi(\zeta) = 2^{-\frac{1}{3}}\sum_{n=0}^{\infty} \chi_n \eta^n = 2^{-\frac{1}{3}}\left[\tfrac{1}{5} + \tfrac{2}{175}\eta - \tfrac{64}{2625}\eta^2 - \tfrac{30424}{3031875}\eta^3 + \cdots\right].$$

Next we consider the coefficients a_s, b_s that are used in (5.66). We expand

$$a_s(\zeta) = \sum_{t=0}^{\infty} a_s^t \eta^t, \qquad b_s(\zeta) = 2^{\frac{1}{3}}\sum_{t=0}^{\infty} b_s^t \eta^t, \tag{5.81}$$

where η is given in (5.79). The coefficients a_s^t, b_s^t are rational numbers. We know that $a_0(\zeta) = 1$. Substituting the expansions in (5.74) we obtain recurrence relations for the coefficients a_s^t, b_s^t. We give the expansions of the first few terms:

$$a_0(\zeta) = 1, \tag{5.82}$$

$$a_1(\zeta) = -\tfrac{1}{225} - \tfrac{71}{38500}\eta + \tfrac{82}{73125}\eta^2 + \tfrac{5246}{3898125}\eta^3 + \tfrac{185728}{478603125}\eta^4 + \cdots,$$

$$a_2(\zeta) = \tfrac{151439}{218295000} + \tfrac{68401}{147262500}\eta - \tfrac{1796498167}{4193689500000}\eta^2 - \tfrac{583721053}{830718281250}\eta^3 + \cdots,$$

$$a_3(\zeta) = -\tfrac{887278009}{2504935125000} - \tfrac{3032321618951}{9708942993750000}\eta + \cdots,$$

$$b_0(\zeta) = 2^{\frac{1}{3}}\left[\tfrac{1}{70} + \tfrac{2}{225}\eta + \tfrac{138}{67375}\eta^2 - \tfrac{296}{511875}\eta^3 - \tfrac{38464}{63669375}\eta^4 + \cdots\right], \tag{5.83}$$

$$b_1(\zeta) = 2^{\frac{1}{3}}\left[-\tfrac{1213}{1023750} - \tfrac{3757}{2695000}\eta - \tfrac{3225661}{6700443750}\eta^2 + \tfrac{90454643}{336992906250}\eta^3 + \cdots\right],$$

$$b_2(\zeta) = 2^{\frac{1}{3}}\left[\tfrac{16542537833}{37743205500000} + \tfrac{115773498223}{162820783125000}\eta + \tfrac{548511920915149}{1721719224225000000}\eta^2 + \cdots\right],$$

$$b_3(\zeta) = 2^{\frac{1}{3}}\left[-\tfrac{9597171184603}{25476663712500000} - \tfrac{430990563936859253}{568167343994250000000}\eta + \cdots\right].$$

Expansions for the coefficients c_s, d_s are not really needed, because these quantities follow from the relations in (5.72), if expansions for the functions in the right-hand sides of (5.72) are available.

Evaluation of the functions $A_\nu(\zeta)$, $B_\nu(\zeta)$ by iteration
We now concentrate on solving the system of differential equations in (5.75) by using analytical techniques. Instead of expanding the coefficients a_s, b_s of the asymptotic series we expand the functions $A_\nu(\zeta)$, $B_\nu(\zeta)$ in Maclaurin series. As remarked earlier, the singular points of these functions occur at $\zeta^\pm = (3\pi/2)^{\frac{2}{3}} e^{\pm i\pi/3}$, and the radius of convergence of the series of $A_\nu(\zeta)$ and $B_\nu(\zeta)$ in powers of ζ equals $2.81\cdots$.

We expand

$$A_\nu(\zeta) = \sum_{n=0}^\infty f_n(\nu)\zeta^n, \qquad B_\nu(\zeta) = \sum_{n=0}^\infty g_n(\nu)\zeta^n, \qquad \psi(\zeta) = \sum_{n=0}^\infty h_n \zeta^n.$$

$$(5.84)$$

The coefficients $f_0, f_1, \ldots, g_0, g_1, \ldots$ are to be determined, while the coefficients h_n are known. The first few h_n follow from (5.79) and the second line in (5.80):

$$h_0 = \tfrac{1}{70} 2^{\frac{1}{3}}, \qquad h_1 = \tfrac{2}{75}, \qquad h_2 = \tfrac{69}{13475} 2^{\frac{2}{3}}, \qquad h_3 = \tfrac{148}{73125} 2^{\frac{1}{3}}. \qquad (5.85)$$

Upon substituting the expansions into (5.75), we obtain for $n = 0, 1, 2, \ldots$ the recurrence relations

$$(n+2)(n+1) f_{n+2} + (2n+1) g_n = \rho_n, \qquad \rho_n = \sum_{k=0}^n h_k f_{n-k},$$

$$(5.86)$$

$$(n+2)(n+1) g_{n+2} + 2\nu^2(n+1) f_{n+1} = \sigma_n, \qquad \sigma_n = \sum_{k=0}^n h_k g_{n-k}.$$

The solution $\{f_n, g_n\}$ of the set of recursions (5.86) cannot be obtained by forward recursion, because of instabilities. The same problems arise as in the case of the incomplete gamma functions described in Section 5.2. We give an iteration scheme to solve the recursions in the backward direction. The scheme runs as follows; for details we refer to Temme (1997).

We choose an appropriate pair of functions F_0, G_0, and define two sequences of functions $\{F_m\}$, $\{G_m\}$ by writing for $m = 1, 2, 3, \ldots$:

$$F_m'' + 2\zeta G_m' + G_m = \psi(\zeta) F_{m-1}, \qquad G_m'' + 2\nu^2 F_m' = \psi(\zeta) G_{m-1}. \quad (5.87)$$

We rewrite (5.86) in backward form:

$$f_n = \frac{1}{2\nu^2}\left[\frac{1}{n}\sigma_{n-1} - (n+1)g_{n+1} \right], \qquad (5.88)$$

$$g_{n-1} = \frac{1}{2n-1}\left[\rho_{n-1} - n(n+1)f_{n+1} \right], \qquad (5.89)$$

Table 5.2. Relative errors during five iterations (i) of f_0, g_0, f_5, g_5 compared with more accurate values f_0^a, etc. The final column shows the relative error in the Wronskian (5.77) at $\zeta = 1$.

i	$\lvert f_0 - f_0^a \rvert$	$\lvert g_0 - g_0^a \rvert$	$\lvert f_5 - f_5^a \rvert$	$\lvert g_5 - g_5^a \rvert$	Wronskian
			$\nu = 5$		
1	$6.11_{10}(-09)$	$1.76_{10}(-06)$	$1.12_{10}(-03)$	$6.14_{10}(-04)$	$4.36_{10}(-08)$
2	$4.54_{10}(-12)$	$1.03_{10}(-08)$	$6.14_{10}(-06)$	$8.33_{10}(-07)$	$2.05_{10}(-10)$
3	$2.56_{10}(-15)$	$1.60_{10}(-11)$	$1.52_{10}(-08)$	$8.48_{10}(-10)$	$3.29_{10}(-13)$
4	$1.21_{10}(-17)$	$1.83_{10}(-14)$	$6.40_{10}(-12)$	$5.25_{10}(-13)$	$6.72_{10}(-16)$
5	$0.00_{10}(-00)$	$4.19_{10}(-17)$	$4.64_{10}(-14)$	$4.04_{10}(-16)$	$2.23_{10}(-18)$
			$\nu = 10$		
1	$4.24_{10}(-10)$	$1.14_{10}(-07)$	$2.90_{10}(-04)$	$1.63_{10}(-04)$	$2.76_{10}(-09)$
2	$8.50_{10}(-14)$	$8.17_{10}(-10)$	$1.84_{10}(-06)$	$5.76_{10}(-08)$	$1.64_{10}(-11)$
3	$1.45_{10}(-17)$	$3.20_{10}(-13)$	$1.10_{10}(-09)$	$9.06_{10}(-11)$	$6.64_{10}(-15)$
4	$1.08_{10}(-19)$	$9.89_{10}(-17)$	$1.74_{10}(-12)$	$1.25_{10}(-15)$	$8.07_{10}(-18)$
5	$0.00_{10}(-00)$	$1.88_{10}(-19)$	$1.35_{10}(-15)$	$6.74_{10}(-17)$	$3.44_{10}(-19)$
			$\nu = 25$		
1	$1.12_{10}(-11)$	$2.94_{10}(-09)$	$4.70_{10}(-05)$	$2.66_{10}(-05)$	$7.10_{10}(-11)$
2	$3.66_{10}(-16)$	$2.22_{10}(-11)$	$3.09_{10}(-07)$	$1.52_{10}(-09)$	$4.47_{10}(-13)$
3	$0.00_{10}(-00)$	$1.40_{10}(-15)$	$2.95_{10}(-11)$	$2.66_{10}(-12)$	$2.91_{10}(-17)$
4	$0.00_{10}(-00)$	$0.00_{10}(-00)$	$5.63_{10}(-14)$	$7.25_{10}(-18)$	$1.75_{10}(-19)$
5	$0.00_{10}(-00)$	$0.00_{10}(-00)$	$6.45_{10}(-18)$	$3.40_{10}(-19)$	$1.76_{10}(-19)$

where $n \geq 1$. The coefficients are assumed to belong to the functions $F_m(\zeta), G_m(\zeta)$ of the iteration process described by (5.87), while the coefficients ρ_{n-1} and σ_{n-1} are assumed to be known, and contain Maclaurin coefficients of $F_{m-1}(\zeta), G_{m-1}(\zeta)$ and $\psi(\zeta)$. Observe that (5.88) does not define f_0. After having computed $f_1, f_2, \ldots$, and $g_0, g_1, g_2, \ldots$ by the backward recursion process, we compute f_0 from the Wronskian (5.77):

$$f_0 = \frac{-g_1 + \sqrt{g_1^2 + 4\nu^2(\nu^2 + f_1 g_0)}}{2\nu^2}, \qquad (5.90)$$

where the $+$sign of the square root is taken because of the known behaviour of $F_\nu(0)$ when ν is large.

Verifying the iterative scheme by numerical experiments

For numerical applications information is needed about the growth of the coefficients f_n, g_n. Since the Maclaurin series in (5.84) have a radius of convergence equal to $2.81\cdots$, for all values of ν, the size of the coefficients f_n, g_n is comparable with that of h_n. The number of coefficients f_n, g_n needed in (5.84) also depends on the size of ζ. When $|\zeta| = 1$ we need about 45 terms in the Maclaurin series in (5.84) in order to obtain an accuracy of about 20 decimal digits. The ζ-interval $[-1, 1]$ corresponds to the z-interval $[0.39, 1.98]$. When z is outside this interval many other efficient algorithms are available for the computation of $J_\nu(\nu z), Y_\nu(\nu z)$.

We have computed successive iterates of Maclaurin coefficients f_n, g_n defined in (5.84) for different values of ν.

During each iteration we start the backward recursions with $f_n = g_{n-1} = 0$, $n \geq 46$, and we compute $f_{45}, g_{44}, f_{44}, g_{43}, \ldots$ by using (5.88) and (5.89). We use $h_k, k = 0, 1, \ldots, 45$ and we recompute the coefficients $\rho_k, \sigma_k, k = 0, 1, 2, \ldots, 45$, using (5.86) with values f_k, g_k obtained in the previous iteration. In Table 5.2 we show the relative errors in the values f_0, g_0, f_5, g_5, when compared with more accurate values f_0^a, etc. Computations are done with Digits $= 20$. The accurate values are obtained by applying the backward recursion by using 10 iterations. We also give the relative error in the Wronskian relation (5.77) at $\zeta = 1$ during each iteration.

From Table 5.2 we conclude that for $\nu = 5$ we can already obtain an accuracy of 10^{-10} in the Wronskian after two iterations; further iterations improve the results. For larger values of ν the algorithm is very efficient.

5.5. Asymptotic expansions: Concluding remarks

The methods of this section can be used for Airy-type asymptotic expansions for other special functions. We mention as interesting cases parabolic cylinder functions, Coulomb wave functions, and other members of the class of Whittaker functions. To stay in the class of Bessel functions, we mention the modified Bessel function of the third kind $K_{i\nu}(z)$ of imaginary order, which plays an important role in the diffraction theory of pulses and in the study of certain hydrodynamical studies. Moreover, this function is the kernel of the Lebedev transform. The same functions $A_\nu, B_\nu, C_\nu, D_\nu$ can be used for this case.

We have described a method for handling the coefficients and auxiliary functions when the properties of these quantities can be derived from the differential equation of the functions to be approximated. The same or similar Airy-type expansions can be obtained when starting from integral representations in which two saddle points are close together (Wong 2001, Chapter 7). For the coefficients in the expansions the same difficulties may arise in numerical algorithms and the method for deriving Maclaurin expansions of the coefficients is quite different compared with the methods described earlier in this section. For details on the case of integrals we refer to Vidunas and Temme (2002).

6. The inversion of cumulative distribution functions

The inversion of cumulative distribution functions is an important topic in statistics, probability theory and econometrics, in particular for computing percentage points of chi-square, F, and Student's t-distributions. In the tails of these distributions the numerical inversion is not very easy, and for the standard distributions asymptotic formulas are available.

Here we use the uniform asymptotic expansions of the incomplete gamma functions (see Section 5.2) for inverting these functions for large values of a. We start with the relatively simple problem of inverting the complementary error function, again by using asymptotic methods. Finally we describe a different method, giving a high-order Newton-like iteration method.

6.1. Asymptotic inversion of the complementary error function

We recall the definition of the complementary error function (here we use real arguments)

$$\operatorname{erfc} x = \frac{2}{\sqrt{\pi}} \int_x^\infty e^{-t^2}\, dt, \tag{6.1}$$

and the asymptotic expansion

$$\operatorname{erfc} x \sim \frac{e^{-x^2}}{x\sqrt{\pi}} \sum_{k=0}^\infty \frac{(-1)^k (2k)!}{k!\, (2x)^{2k}} = \frac{e^{-x^2}}{x\sqrt{\pi}}\left(1 - \tfrac{1}{2}x^{-2} + \tfrac{3}{4}x^{-4} - \tfrac{15}{8}x^{-6} + \cdots\right), \tag{6.2}$$

as $x \to \infty$.

We derive an asymptotic expansion for the inverse $x(y)$ of the function $y(x) = \operatorname{erfc} x$ for small positive values of y.

Let t, α and β be defined by

$$t = \frac{2}{\pi y^2}, \qquad \alpha = \frac{1}{\ln t}, \qquad \beta = \ln(\ln t). \tag{6.3}$$

Then we have the expansion

$$x(y) \sim \frac{1}{\sqrt{2\alpha}}\left(1 + x_1\alpha + x_2\alpha^2 + x_3\alpha^3 + x_4\alpha^4 + \cdots\right). \tag{6.4}$$

The first coefficients x_k are given by

$$x_1 = -\tfrac{1}{2}\beta,$$

$$x_2 = -\tfrac{1}{8}\left(\beta^2 - 4\beta + 8\right), \tag{6.5}$$

$$x_3 = -\tfrac{1}{16}\left(\beta^3 - 8\beta^2 + 32\beta - 56\right),$$

$$x_4 = -\tfrac{1}{384}\left(15\beta^4 - 184\beta^3 + 1152\beta^2 - 4128\beta + 7040\right),$$

$$x_5 = -\tfrac{1}{768}\left(21\beta^5 - 352\beta^4 + 3056\beta^3 - 16752\beta^2 + 57536\beta - 97984\right).$$

We explain how these coefficients can be obtained. The equation $y = \operatorname{erfc} x$, with y small, will be solved for x by using the asymptotic expansion in (6.2). We square the equation, and write

$$\frac{1}{2}\pi y^2 = \frac{e^{-\xi}}{\xi}S^2(\xi), \tag{6.6}$$

where $\xi = 2x^2$ and $S(\xi)$ denotes the function that has the power series in (6.2) as its asymptotic expansion with $2x^2$ replaced by ξ.

We can rewrite (6.6) in the form

$$\xi\,e^{\xi} = t\,S^2(\xi), \tag{6.7}$$

where t is defined in (6.3). We solve this equation for ξ, with t large.

We observe that this equation has been discussed in detail in de Bruijn (1981, Section 2.4) for the case that $S(\xi) = 1$. We can apply the same method for constructing an asymptotic expansion of the equation in (6.7). We write

$$\xi = \ln t - \ln(\ln t) + \eta, \tag{6.8}$$

and find that η satisfies the relation

$$(\ln t - \ln(\ln t) + \eta)\,e^{\eta} = \ln t\, S^2(\ln t - \ln(\ln t) + \eta). \tag{6.9}$$

The quantity η can be expanded in the form

$$\eta = \eta_1\,\alpha + \eta_2\,\alpha^2 + \eta_3\,\alpha^3 + \cdots, \tag{6.10}$$

where α is defined in (6.3). By using a few terms in the expansion of $S(\xi)$ we find

$$\eta_1 = \beta - 2, \qquad \eta_2 = \frac{1}{2}\beta^2 - 3\beta + 7, \tag{6.11}$$

where β is defined in (6.3).

The expansion for η gives an expansion for ξ (see (6.8)), and by using $x = \sqrt{\xi/2}$ we obtain the expansion given in (6.4).

In Table 6.1 we give values of the approximation x_0 of the solution of the equation $y = \operatorname{erfc} x$ for several values of y. We have used the asymptotic expansion (6.4) with the terms up to and including x_4. The values x_1 are obtained by using one Newton step. We also give relative errors. The computations are done in Maple with Digits $= 10$.

Table 6.1. Solutions $x(y)$ of the equation $y = \operatorname{erfc} x$ by using (6.4).

y	x_0	$\lvert \operatorname{erfc}(x_0)/y - 1 \rvert$	x_1	$\lvert x_0/x_1 - 1 \rvert$
10^{-02}	1.820630554	$0.31_{10}(-02)$	1.821591563	$0.53_{10}(-03)$
10^{-03}	2.326648925	$0.51_{10}(-03)$	2.326782380	$0.57_{10}(-04)$
10^{-04}	2.751038248	$0.15_{10}(-03)$	2.751070914	$0.12_{10}(-04)$
10^{-05}	3.123404718	$0.56_{10}(-04)$	3.123415612	$0.35_{10}(-05)$
10^{-06}	3.458907270	$0.25_{10}(-04)$	3.458911685	$0.13_{10}(-05)$
10^{-07}	3.766560973	$0.13_{10}(-04)$	3.766563021	$0.54_{10}(-06)$
10^{-08}	4.052236421	$0.69_{10}(-05)$	4.052237469	$0.26_{10}(-06)$
10^{-09}	4.320004932	$0.41_{10}(-05)$	4.320005509	$0.14_{10}(-06)$
10^{-10}	4.572824704	$0.25_{10}(-05)$	4.572825039	$0.74_{10}(-07)$

For other methods of the inversion of the error functions we refer to Strecok (1968), where coefficients of the Maclaurin expansion of $x(y)$, the inverse of $y = \operatorname{erf} x$, are given, with Chebyshev coefficients for an expansion on the y-interval $[-0.8, 0.8]$. For small values of y (not smaller than 10^{-300}) high-precision coefficients of Chebyshev expansions are given for the numerical evaluation of the inverse of $y = \operatorname{erfc} x$. For rational Chebyshev (near-minimax) approximations for the inverse of the complementary error function $y = \operatorname{erfc} x$ we refer to Blair, Edwards and Johnson (1976), where y-values are considered in the y-interval $[10^{-10000}, 1]$, with relative errors ranging down to 10^{-23}. An asymptotic formula for the region $y \to 0$ is also given.

Remark 6.1. We can use similar asymptotic inversion methods for finding complex zeros of the complementary error function; see Gil *et al.* (2007b). The present case is simpler because we only want to find one real solution with the inversion method.

Remark 6.2. The solution of the equation $\xi e^{\xi} = t$ (see (6.7)) can be expressed in terms of Lambert's W-function: $\xi = W(t)$.

6.2. Asymptotic inversion of incomplete gamma functions

We solve the equations

$$P(a, x) = p, \qquad Q(a, x) = q, \qquad 0 \le p \le 1, \quad 0 \le q \le 1, \qquad (6.12)$$

where $P(a, x)$ and $Q(a, x)$ are the incomplete gamma functions introduced in Section 5.1. We invert the equations for x, with a as a large positive parameter. This problem is of importance in probability theory and mathematical statistics. Several approaches are available in the (statistical) literature, where often a first approximation of x is constructed, based on asymptotic expansions, but this first approximation is not always reliable. Higher approximations may be obtained by numerical inversion techniques, which require evaluation of the incomplete gamma functions. This may be rather time-consuming, especially when a is large.

In the present method we also use an asymptotic result. The approximation is quite accurate, especially when a is large. It follows from numerical results, however, that a three-term asymptotic expansion already gives an accuracy of 4 significant digits for $a = 2$, uniformly with respect to $p, q \in [0, 1]$.

The method is rather general. In a later section we mention application of the same method on a wider class of cumulative distribution functions.

The approximations are obtained by using the uniform asymptotic expansions of the incomplete gamma functions given in Section 5.1, in which the complementary error function is the dominant term. The inversion problem is started by inverting this error function term. For more details we refer to Temme (1992a); for the asymptotic inversion of the incomplete beta function, see Temme (1992b).

6.3. The asymptotic inversion method

We perform the inversion of the equations (6.12) with respect to the parameter η, by using the representations (5.14), with z replaced by x throughout, and large positive values of a. Afterwards we have to compute λ and x from the relation for η in (5.15) and $\lambda = x/a$. We concentrate on the second equation in (6.12). Let us rewrite the inversion problem in the form

$$\tfrac{1}{2}\operatorname{erfc}\left(\eta\sqrt{a/2}\right) + R_a(\eta) = q, \quad q \in [0, 1], \tag{6.13}$$

which is equivalent to the second equation in (6.12), and we denote the solution of the above equation by $\eta(q, a)$.

To start the procedure we consider $R_a(\eta)$ in (6.13) as a perturbation, and we define the number $\eta_0 = \eta_0(q, a)$ as the real number that satisfies the equation

$$\tfrac{1}{2}\operatorname{erfc}\left(\eta_0\sqrt{a/2}\right) = q. \tag{6.14}$$

Known values are

$$\eta_0(0, a) = +\infty, \qquad \eta_0(\tfrac{1}{2}, a) = 0, \qquad \eta_0(1, a) = -\infty. \tag{6.15}$$

We note the symmetry $\eta_0(q, a) = -\eta_0(p, a)$. Computation of η_0 requires an inversion of the error function, but this problem has been satisfactorily solved in the literature; see Blair et al. (1976) and Strecok (1968). The value η defined by (6.13) is, for large values of a, approximated by the value η_0. We write

$$\eta(q, a) = \eta_0(q, a) + \varepsilon(\eta_0, a), \tag{6.16}$$

and we try to determine the function ε. It appears that we can expand this quantity in the form

$$\varepsilon(\eta_0, a) \sim \frac{\varepsilon_1}{a} + \frac{\varepsilon_2}{a^2} + \frac{\varepsilon_3}{a^3} + \cdots ,$$ (6.17)

as $a \to \infty$. The coefficients ε_i will be written explicitly as functions of η_0.

We first remark that (6.13) yields the relation

$$\frac{dq}{d\eta} = \frac{d}{d\eta} Q(a, x) = \frac{d}{dx} Q(a, x) \frac{dx}{d\eta}.$$ (6.18)

Using the definition of $Q(a, x)$ and the relation for η in (5.15), we obtain, after straightforward calculations,

$$\frac{dq}{d\eta} = -\frac{1}{\Gamma^*(a)} \sqrt{\frac{a}{2\pi}} f(\eta) e^{-\frac{1}{2} a \eta^2},$$ (6.19)

where $\Gamma^*(a)$ is defined in (5.25), and

$$f(\eta) = \frac{\eta}{\lambda - 1},$$ (6.20)

the relation between η and λ being given in (5.15). For small values of η we can expand

$$f(\eta) = 1 - \tfrac{1}{3}\eta + \tfrac{1}{12}\eta^2 + \cdots .$$ (6.21)

From (6.14) we obtain

$$\frac{dq}{d\eta_0} = -\sqrt{\frac{a}{2\pi}} e^{-\frac{1}{2} a \eta_0^2}.$$ (6.22)

Upon dividing (6.19) and (6.22) we eliminate q, although it is still present in η_0. So we obtain

$$\frac{d\eta}{d\eta_0} = \frac{\Gamma^*(a)}{f(\eta)} e^{\frac{1}{2} a(\eta^2 - \eta_0^2)}, \qquad -\infty < \eta_0 < \infty.$$ (6.23)

Substitution of (6.16) gives the differential equation

$$f(\eta_0 + \varepsilon)\left[1 + \frac{d\varepsilon}{d\eta_0}\right] = \Gamma^*(a) e^{a\varepsilon(\eta_0 + \frac{1}{2}\varepsilon)},$$ (6.24)

a relation between ε and η_0, with a as (large) parameter.

It is convenient to write η in place of η_0. That is, we try to find the function $\varepsilon = \varepsilon(\eta, a)$ that satisfies the equation

$$f(\eta + \varepsilon)\left[1 + \frac{d\varepsilon}{d\eta}\right] = \Gamma^*(a) e^{a\varepsilon(\eta + \frac{1}{2}\varepsilon)}.$$ (6.25)

When we have obtained the solution $\varepsilon(\eta, a)$ (in fact we find an approximation of the form (6.17)), we write it as $\varepsilon(\eta_0, a)$ and the final value of η

follows from (6.16). The parameters λ and x of the incomplete gamma function then follow from inversion of the relation for η in the first relation of (5.15).

6.4. Determination of the coefficients ε_i

For large values of a we have $\Gamma^*(a) = 1 + \mathcal{O}(a^{-1})$; see (5.26). Comparing dominant terms in (6.25), we infer that the first coefficient ε_1 in (6.17) is defined by

$$f(\eta) = e^{\eta \varepsilon_1}, \tag{6.26}$$

giving

$$\varepsilon_1 = \frac{1}{\eta} \ln f(\eta). \tag{6.27}$$

It is not difficult to verify that f is positive on $\mathbb{R}$, $f(0) = 1$, and that f is analytic in a neighbourhood of $\eta = 0$. It follows that $\varepsilon_1 = \varepsilon_1(\eta)$ is an analytic function on $\mathbb{R}$. For small values of η we have, using (6.21),

$$\varepsilon_1 = -\tfrac{1}{3} + \tfrac{1}{36}\eta + \tfrac{1}{1620}\eta^2 + \cdots. \tag{6.28}$$

The function $\varepsilon_1(\eta)$ is non-vanishing on $\mathbb{R}$ (and hence negative). To show this, consider the equation $f^2(\eta) = 1$. From (6.20) and the relation for η in (5.15), it follows that the corresponding λ-value should satisfy

$$-\ln \lambda = (\lambda - 1)(2\lambda - 3). \tag{6.29}$$

This equation has only one real solution $\lambda = 1$, which gives $\eta = 0$. However, for this value ε_1 equals $-\tfrac{1}{3}$.

Further coefficients in (6.17) are obtained by using standard perturbation methods. We need the expansion of $\Gamma^*(a)$ given in (5.26), and

$$f(\eta + \varepsilon) = f(\eta) + \varepsilon f'(\eta) + \tfrac{1}{2}\varepsilon^2 f''(\eta) + \cdots, \tag{6.30}$$

in which (6.17) is substituted to obtain an expansion in powers of a^{-1}. Putting all this into (6.25), we find by comparing terms with equal powers of a^{-1}

$$\varepsilon_2 = \frac{1}{12\eta f}(12 f \varepsilon_1' + 12 f' \varepsilon_1 - f - 6 f \varepsilon_1^2), \tag{6.31}$$

$$\varepsilon_3 = \frac{1}{288\eta f}(288 f \varepsilon_2' + 288 f' \varepsilon_1 \varepsilon_1' - 24 f \varepsilon_1' + 288 f' \varepsilon_2 + 144 f'' \varepsilon_1^2 \tag{6.32}$$

$$+ f - 288 f \varepsilon_1 \varepsilon_2 - 144 f \varepsilon_2^2 \eta^2 - 144 f \varepsilon_2 \eta \varepsilon_1^2 - 36 f \varepsilon_1^4 - 24 f' \varepsilon_1).$$

The derivatives f' and ε_j' are with respect to η, and evaluated at η. It will be obvious that the complexity for obtaining higher-order terms is considerable. The terms shown so far have been obtained by symbolic manipulation.

For numerical evaluations it is very convenient to have representations free of derivatives.

The derivatives of f can be eliminated by using

$$f' = f(1 - f^2 - f\eta)/\eta, \tag{6.33}$$

$$f'' = f^2(-3\eta - 3f + 3f^3 + 5f^2\eta + 2\eta^2 f)/\eta^2. \tag{6.34}$$

The first relation easily follows from (6.20) and the relation between η and λ. Using these relations in ε_i, and eliminating the derivatives of previous ε_j, it follows that we can write $\eta^{2j-1}\varepsilon_j$ as a polynomial in η, f, ε_1. We have

$$12\eta^3\varepsilon_2 = +12 - 12f^2 - 12f\eta - 12f^2\eta\varepsilon_1 - 12f\eta^2\varepsilon_1 - \eta^2 - 6\eta^2\varepsilon_1^2, \tag{6.35}$$

$$12\eta^5\varepsilon_3 = -30 + 12f^2\eta\varepsilon_1 + 12f\eta^2\varepsilon_1 + 24f^2\eta^3\varepsilon_1 + 6\varepsilon_1^3\eta^3 - 12f^2 \tag{6.36}$$

$$+ 31f^2\eta^2 + 72f^3\eta + 42f^4 + 18f^3\eta^3\varepsilon_1^2 + 6f^2\eta^4\varepsilon_1^2 + 36f^4\eta\varepsilon_1 + 60f^3\eta^2\varepsilon_1$$

$$+ 12\varepsilon_1^2\eta^3 f + 12\varepsilon_1^2\eta^2 f^2 - 12\eta\varepsilon_1 + \eta^3\varepsilon_1 + f\eta^3 - 12f\eta + 12\varepsilon_1^2\eta^2 f^4,$$

The coefficients $\varepsilon_1, \ldots, \varepsilon_3$ are bounded on $\mathbb{R}$. To show this one needs

$$f(\eta) \sim -\eta, \quad \eta \to -\infty, \qquad f(\eta) \sim 2\eta^{-1}, \quad \eta \to +\infty, \tag{6.37}$$

and the above representations of ε_i. We find

$$\varepsilon_1 \sim \mp\frac{\ln|\eta|}{\eta}, \qquad \varepsilon_2 \sim -\frac{1}{12\eta}, \qquad \varepsilon_3 \sim \frac{\varepsilon_1}{12\eta^2}, \tag{6.38}$$

as $\eta \to \pm\infty$. In deriving the behaviour at $-\infty$ one should take into account that (see (6.20) and the relation between η and λ)

$$f(\eta) + \eta = \frac{\lambda\eta}{\lambda - 1} \sim -\eta e^{-\frac{1}{2}\eta^2}, \quad \eta \to -\infty. \tag{6.39}$$

6.5. Expansions of the coefficients ε_i

As explained in Section 5.2, the function f is analytic in a strip $|\text{Im}\,\eta| < \sqrt{2\pi}$, and it can be expanded in a Taylor series around the origin with radius of convergence $2\sqrt{\pi}$. All ε_i have similar analytic properties. That is, the coefficients ε_i can be expanded in series as follows:

$$\varepsilon_i = \sum_{n=0}^{\infty} c_{i,n}\eta^n, \quad |\eta| < 2\sqrt{\pi}, \quad i = 1, 2, 3, \ldots. \tag{6.40}$$

The representations of ε_i given in the previous section are not suitable for numerical computation, first because of the appearance of derivatives of f and ε_i, second because of the complexity of the expressions. Therefore, to facilitate numerical evaluations of $\varepsilon_1, \ldots, \varepsilon_3$ we provide the following Taylor expansions:

$$\varepsilon_1 = -\frac{1}{3} + \frac{1}{36}\eta + \frac{1}{1620}\eta^2 - \frac{7}{6480}\eta^3 + \frac{5}{18144}\eta^4 - \frac{11}{382725}\eta^5 - \frac{101}{16329600}\eta^6 \quad (6.41)$$

$$+ \frac{37}{9797760}\eta^7 - \frac{454973}{498845952000}\eta^8 + \frac{1231}{15913705500}\eta^9 + \frac{2745493}{84737299046400}\eta^{10}$$

$$- \frac{2152217}{127673385840000}\eta^{11} + \frac{119937661}{30505427656704000}\eta^{12}$$

$$- \frac{449}{1595917323000}\eta^{13} - \frac{756882301459}{4455179048226816000000}\eta^{14}$$

$$+ \frac{12699400547}{153146779782796800000}\eta^{15} - \frac{3224618478943}{17026421414023397360000}\eta^{16} + \cdots,$$

$$\varepsilon_2 = -\frac{7}{405} - \frac{7}{2592}\eta + \frac{533}{204120}\eta^2 - \frac{1579}{2099520}\eta^3 + \frac{109}{1749600}\eta^4 + \frac{10217}{251942400}\eta^5 \quad (6.42)$$

$$- \frac{9281803}{436490208000}\eta^6 + \frac{919081}{185177664000}\eta^7 - \frac{100824673}{5719767685632000}\eta^8$$

$$- \frac{311266223}{899963447040000}\eta^9 + \frac{52310527831}{343186061137920000}\eta^{10} + \cdots,$$

$$\varepsilon_3 = \frac{449}{102060} - \frac{63149}{20995200}\eta + \frac{29233}{36741600}\eta^2 + \frac{346793}{5290790400}\eta^3 \quad (6.43)$$

$$- \frac{18442139}{130947062400}\eta^4 + \frac{14408797}{246903552000}\eta^5 - \frac{1359578327}{129994720128000}\eta^6$$

$$- \frac{69980826653}{39598391669760000}\eta^7 + \frac{987512909021}{514779091706880000}\eta^8 + \cdots,$$

6.6. Numerical examples

When $p = q = \frac{1}{2}$, the asymptotics are quite simple. Then η_0 of (6.14) equals zero, and from (6.28) and the expansions in the previous section we obtain (6.16) and (6.17) in the form

$$\eta \sim -\frac{1}{3}a^{-1} - \frac{7}{405}a^{-2} + \frac{449}{102060}a^{-3} + \cdots. \quad (6.44)$$

In this case we give an expansion of the requested value x. Recall that $x = a\lambda$ and that λ can be obtained from the relation between η and λ in (5.15) with η given by (6.44). Inverting

$$\frac{1}{2}\eta^2 = \frac{1}{2}(\lambda - 1)^2 - \frac{1}{3}(\lambda - 1)^3 + \frac{1}{4}(\lambda - 1)^4 + \cdots, \quad (6.45)$$

we obtain

$$\lambda = 1 + \eta + \frac{1}{3}\eta^2 + \frac{1}{36}\eta^3 - \frac{1}{270}\eta^4 + \frac{1}{4320}\eta^5 + \cdots. \quad (6.46)$$

Substituting (6.44), we have

$$x \sim a\left(1 - \frac{1}{3}a^{-1} + \frac{8}{405}a^{-2} + \frac{184}{25515}a^{-3} + \cdots\right). \quad (6.47)$$

Table 6.2. Relative errors $|x_a - x|/x$ and $|Q(a, x_a) - q|/q$ for several values of q and a; x_a is obtained by asymptotic expansion (6.17), x is a more accurate value.

q	$a = 1$		$a = 5$		$a = 10$	
0.0001	$2.3_{10}(-4)$	$2.1_{10}(-3)$	$1.1_{10}(-6)$	$1.6_{10}(-5)$	$9.4_{10}(-8)$	$1.7_{10}(-6)$
0.1	$6.6_{10}(-4)$	$1.5_{10}(-3)$	$2.0_{10}(-6)$	$9.3_{10}(-6)$	$1.4_{10}(-7)$	$8.8_{10}(-7)$
0.3	$8.7_{10}(-4)$	$1.0_{10}(-3)$	$2.3_{10}(-6)$	$6.4_{10}(-6)$	$1.6_{10}(-7)$	$6.0_{10}(-7)$
0.5	$7.0_{10}(-4)$	$4.8_{10}(-4)$	$6.7_{10}(-7)$	$1.2_{10}(-6)$	$5.4_{10}(-8)$	$1.4_{10}(-7)$
0.7	$4.9_{10}(-4)$	$1.7_{10}(-4)$	$2.7_{10}(-6)$	$2.6_{10}(-6)$	$1.7_{10}(-7)$	$2.6_{10}(-7)$
0.9	$1.9_{10}(-3)$	$2.0_{10}(-4)$	$2.5_{10}(-6)$	$8.8_{10}(-7)$	$1.8_{10}(-7)$	$9.3_{10}(-8)$
0.9999	$5.1_{10}(-3)$	$5.1_{10}(-7)$	$3.9_{10}(-6)$	$1.8_{10}(-9)$	$6.0_{10}(-8)$	$4.8_{10}(-11)$

When $a = 1$, $q = 1/2$, the equations in (6.12) reduce to $e^{-x} = 1/2$, with solution $x = \ln 2 \doteq 0.693147$, while expansion (6.47) gives $x \sim 0.693631$, an accuracy of 3 digits. When $a = 2$, $q = 1/2$, the equations in (6.12) become $(1 + x)e^{-x} = 1/2$, with solution $x \doteq 1.6783469$; in this case our expansion (6.47) gives $x \sim 1.6783461$, an accuracy of 6 significant digits. This shows that (6.47) is quite accurate for small values of the (large) parameter a. Computer experiments show that for other q-values the results are of the same kind. See Table 6.2.

In Table 6.2 we give more results of numerical experiments. We have used (6.17) with three terms. The first column under each a-value gives the relative accuracy $|x_a - x|/x$, where x_a is the result of the asymptotic method, and x is a more accurate value obtained by Newton's method. The second column under each a-value gives the relative errors $|Q(a, x_a) - q|/q$.

6.7. Generalizations

The method described in the previous sections can be applied to other cumulative distribution functions. Consider the function

$$F_a(\eta) = \sqrt{\frac{a}{2\pi}} \int_{-\infty}^{\eta} e^{-\frac{1}{2}a\zeta^2} f(\zeta) \, d\zeta, \qquad (6.48)$$

where $a > 0$ and $\eta \in \mathbb{R}$. We assume that f is an analytic function in a domain containing the real axis, and that f is positive on $\mathbb{R}$ with the normalization $f(0) = 1$. In Temme (1982) it is shown that several well-known distribution functions can be written in this form, including the incomplete gamma and beta functions. It is also shown that the following representation holds:

$$F_a(\eta) = \tfrac{1}{2} \operatorname{erfc}\left(-\eta\sqrt{a/2}\right) F_a(\infty) + R_a(\eta), \qquad (6.49)$$

where $R_a(\eta)$ can be expanded as in (5.16). $F_a(\infty)$ is the complete integral, and can be expanded in the form

$$F_a(\infty) \sim \sum_{n=0}^{\infty} \frac{A_n}{a^n}, \quad \text{as } a \to \infty, \quad A_0 = 1. \tag{6.50}$$

By dividing both sides of (6.48) by $F_a(\infty)$, we obtain a further normalization, which is typical for distribution functions.

The inversion of the equation $F_a(\eta)/F_a(\infty) = q$, with $q \in [0,1]$ and a a given (large) number, can be performed as in the case of the incomplete gamma functions. As in (6.14), let η_0 be the real number satisfying the equation

$$\tfrac{1}{2}\operatorname{erfc}\left(-\eta_0 \sqrt{a/2}\right) = q. \tag{6.51}$$

Then the requested value η is written as in (6.16), and an expansion like (6.17) can be obtained by deriving the differential equation (6.24), with f of (6.48) and $\Gamma^*(a)$ replaced with $F_a(\infty)$.

6.8. High-order Newton-like methods

Special functions usually satisfy a simple ordinary differential equation, and this equation can be used to construct Newton-like methods of high order.

Let $f(z)$ be the function, the zero ζ of which has to be computed. We put $\zeta = \zeta_0 + h$, where ζ_0 is an approximation of this zero and we assume that we can expand in a neighbourhood of this point

$$f(\zeta) = f(\zeta_0 + h) = f(\zeta_0) + h f_1 + \frac{1}{2!}h^2 f_2 + \frac{1}{3!}f_3 + \cdots, \tag{6.52}$$

where f_k denotes the kth derivative of f at ζ_0. We assume that $f(\zeta_0)$ is small and we expand

$$h = c_1 f(\zeta_0) + c_2 f^2(\zeta_0) + c_3 f^3(\zeta_0) + \cdots. \tag{6.53}$$

Substituting this expansion into (6.52), using that $f(\zeta) = 0$, and comparing equal powers of $f(\zeta_0)$, we find, when $f_1 \neq 0$,

$$c_1 = -\frac{1}{f_1}, \qquad c_2 = -\frac{f_2}{2f_1^3},$$

$$c_3 = \frac{-3f_2^2 + f_3 f_1}{6f_1^5}, \qquad c_4 = -\frac{f_4 f_1^2 + 15f_2^3 - 10f_2 f_3 f_1}{24 f_1^7}. \tag{6.54}$$

When we neglect in (6.53) the coefficients c_k with $k \geq 2$ we obtain Newton's method, with $\zeta \doteq \zeta_0 - f(\zeta_0)/f'(\zeta_0)$.

When $f(z)$ satisfies a simple ordinary differential equation the higher derivatives can be replaced by combinations of lower derivatives.

Example 6.3. (The inversion of the incomplete gamma function)
In Section 6.2 we consider the inversion of the equations $P(a, x) = p$, $Q(a, x) = q$, where $0 < p < 1$, $0 < q < 1$, for large positive values of a. When a is small the asymptotic methods cannot be applied, although for $a = 1$ the results can be used as a first approximation. We take $a \in (0, 1]$, $f(x) = P(a, x) - p$ and an initial value $x_0 > 0$. We derive from $f_1 = f'(x) = x^{a-1}e^{-x}/\Gamma(a)$ and (6.54) the values

$$c_1 = -x_0^{1-a}e^{x_0}\Gamma(a), \tag{6.55}$$

$$c_2 = \frac{x_0 + 1 - a}{2x_0}c_1^2,$$

$$c_3 = \frac{2x_0^2 + 4x_0(1 - a) + 2a^2 - 3a + 1}{6x_0^2}c_1^3,$$

$$c_4 = \frac{6x_0^3 + 18x_0^2(1 - a) + x_0(18a^2 - 29a + 11) - 6a^3 + 11a^2 - 6a + 1}{24x_0^3}c_1^4.$$

For $a = 1$ the equation $f(x) = 0$ is simple, because $P(1, x) = 1 - e^{-x}$ and the solution of $f(x) = 0$ is $x = -\ln(1 - p)$. The values c_k are in this case $c_k = (-1)^k e^{kx_0}/k$, $k = 1, 2, 3, \ldots$, and h of (6.53) becomes

$$h = \sum_{k=1}^{\infty} c_k f^k = -\ln(1 + e^{x_0} f(x_0)).$$

Using this value of h we obtain

$$x = x_0 + h = x_0 - \ln(1 + e^{x_0} f(x_0)) = -\ln(1 - p), \tag{6.56}$$

which gives the exact solution of $f(x) = 0$, for any $x_0 > 0$.

For general $a \in (0, 1]$ we derive a convenient starting value x_0. We observe that

$$P(a, x) = \frac{1}{\Gamma(a)} \int_0^x t^{a-1}e^{-t}\, dt < \frac{1}{\Gamma(a)} \int_0^x t^{a-1}\, dt = \frac{x^a}{\Gamma(a + 1)}. \tag{6.57}$$

Hence, the solution x_0 of the equation $x^a = p\Gamma(a + 1)$ satisfies $0 < x_0 < x$, where x is the exact solution of $f(x) = 0$.

The case $a = 1/2$ is of special interest, because $P(1/2, x) = \text{erf}\sqrt{x}$, the error function; see also Section 6.1. For a numerical example we take $p = 0.5$. We have $x_0 = \pi/16 = 0.196349540849362$ and $f(x_0) = \text{erf}\sqrt{x_0} - 1/2 = -0.030084051069941$. Using the values values c_1, c_2, c_3, c_4 from (6.57), we have $h = 0.0311185517296367$. This gives the new approximation $x \doteq x_0 + h = 0.227468092579000$ and with this value we have $f(x) = -1.12\cdots 10^{-7}$. It is easy to iterate and to obtain much higher accuracy.

7. How to handle series with special functions

When we have solved the problem of writing good software for a class of special functions for a wide range of the real or complex parameters, some application problems remain unsolved. We consider two examples in which special functions arise in representations of solutions of certain problems from probability theory and mathematical physics. We consider one series with incomplete gamma functions and one with modified Bessel functions.

7.1. The non-central chi-squared distribution functions

We start from Abramowitz and Stegun (1964, equation 26.4.25) and consider for positive x, y, μ the non-central chi-squared distribution functions (which are also called non-central gamma distributions)

$$P_\mu(x, y) = e^{-x} \sum_{n=0}^{\infty} \frac{x^n}{n!} P(\mu + n, y), \qquad Q_\mu(x, y) = e^{-x} \sum_{n=0}^{\infty} \frac{x^n}{n!} Q(\mu + n, y),$$

(7.1)

in terms of the incomplete gamma functions, which are related to the standard chi-squared probability functions,

$$P(a, x) = \frac{1}{\Gamma(a)} \int_0^x t^{a-1} e^{-t} \, dt, \qquad Q(a, x) = \frac{1}{\Gamma(a)} \int_x^\infty t^{a-1} e^{-t} \, dt. \quad (7.2)$$

Because $P(a, x) + Q(a, x) = 1$ we also have

$$P_\mu(x, y) + Q_\mu(x, y) = 1. \tag{7.3}$$

Integral representations in terms of modified Bessel functions follow from replacing the incomplete gamma functions in (7.1) by their integral representations, and by using

$$I_\mu(z) = (\tfrac{1}{2}z)^\mu \sum_{n=0}^{\infty} \frac{(\tfrac{1}{2}z)^{2n}}{\Gamma(\mu + n + 1)\, n!}, \tag{7.4}$$

which gives

$$P_\mu(x, y) = e^{-x} \int_0^y \left(\frac{t}{x}\right)^{\frac{1}{2}(\mu-1)} e^{-t} I_{\mu-1}(2\sqrt{xt}) \, dt, \tag{7.5}$$

$$Q_\mu(x, y) = e^{-x} \int_y^\infty \left(\frac{t}{x}\right)^{\frac{1}{2}(\mu-1)} e^{-t} I_{\mu-1}(2\sqrt{xt}) \, dt. \tag{7.6}$$

The function $Q_\mu(x, y)$ plays a role in physics and engineering, for instance in problems on radar communications, where it is called the generalized Marcum Q-function. In Marcum (1960) the function is considered with $\mu = 1$. The parameter μ is related to the degrees of freedom and y to the non-centrality. The recurrence relations of the incomplete gamma functions

$$P(a+1, x) = P(a, x) - \frac{x^a e^{-x}}{\Gamma(a+1)}, \qquad Q(a+1, x) = Q(a, x) + \frac{x^a e^{-x}}{\Gamma(a+1)}, \quad (7.7)$$

give the recursions

$$P_{\mu+1}(x, y) = P_\mu(x, y) - \left(\frac{y}{x}\right)^{\frac{1}{2}\mu} e^{-x} I_\mu(2\sqrt{xy}), \qquad (7.8)$$

$$Q_{\mu+1}(x, y) = Q_\mu(x, y) + \left(\frac{y}{x}\right)^{\frac{1}{2}\mu} e^{-x} I_\mu(2\sqrt{xy}). \qquad (7.9)$$

We can eliminate the Bessel function in (7.9) using $I_{\mu-1}(z) = I_{\mu+1}(z) + (2\mu/z) I_\mu(z)$. This gives the homogeneous third-order recurrence relation:

$$xQ_{\mu+2}(x, y) = (x - \mu)Q_{\mu+1}(x, y) + (y + \mu)Q_\mu(x, y) - yQ_{\mu-1}(x, y). \quad (7.10)$$

Because a constant satisfies this equation, it also holds for $P_\mu(x, y)$.

By using the integral representation of the modified Bessel function (see Abramowitz and Stegun (1964, 29.3.81))

$$\left(\frac{z}{x}\right)^{\frac{1}{2}(\mu-1)} I_{\mu-1}(2\sqrt{xz}) = \frac{1}{2\pi i} \int e^{sz + x/s} s^{-\mu} ds, \quad \mu > 0, \qquad (7.11)$$

where the path of integration may be any vertical line in the half-plane $\operatorname{Re} s > 0$. The path may be deformed into a Hankel contour $\mathcal{L}$ shown in Figure 3.1. Substituting the loop integral into the integral in (7.6), we obtain

$$Q_\mu(x, y) = \frac{e^{-x-y}}{2\pi i} \int_{c-i\infty}^{c+i\infty} \frac{e^{x/s+ys}}{(1-s)s^\mu} ds, \quad 0 < c < 1. \qquad (7.12)$$

When we move the vertical line to the right, across the pole at $s = 1$, and take into account the residue, we obtain

$$P_\mu(x, y) = \frac{e^{-x-y}}{2\pi i} \int_{c-i\infty}^{c+i\infty} \frac{e^{x/s+ys}}{(1-s)s^\mu} ds, \quad c > 1. \qquad (7.13)$$

These representations are useful for deriving asymptotic expansions. In problems in radar communications very large values of μ, x, y are used, say, about 10,000. Asymptotic analysis shows a transition when y passes the value $x + \mu$. There is a fast transition from 0 to 1. In fact we have

$$Q_\mu(x, y) \sim \begin{cases} 1 & \text{if } x + \mu > y, \\ \frac{1}{2} & \text{if } x + \mu = y, \\ 0 & \text{if } x + \mu < y, \end{cases} \qquad (7.14)$$

and complementary behaviour for $P_\mu(x, y) = 1 - Q_\mu(x, y)$.

In Temme (1993, 1996) details can be found on two types of asymptotic approximations, one for x and y large, with μ fixed, and one for all parameters large. In both approximations the complementary error function is used that describes the transition from 0 to 1, as shown for $Q_\mu(x, y)$ in (7.14).

Numerical aspects

In applications it is of interest to have available algorithms for $Q_\mu(x, y)$ when $0 < Q_\mu(x, y) \le \frac{1}{2}$ (see (7.14)) and for $P_\mu(x, y)$ otherwise.

The recurrence relation (7.9) is useful for computing $Q_\mu(x, y)$. It is numerically stable in the forward direction, since the right-hand side of (7.9) has positive terms. An algorithm for the modified Bessel function is needed. A point of warning: the recurrence relation for the modified Bessel function should not be used in the forward direction: see Example 4.2.

Observe that the function $P_\mu(x, y)$ satisfies the recursion

$$P_\mu(x, y) = P_{\mu+1}(x, y) + \left(\frac{y}{x}\right)^{\mu/2} e^{-x} I_\mu(2\sqrt{xy}), \qquad (7.15)$$

which is stable in the backward direction.

In the homogeneous recurrence relation (7.10) Bessel functions do not occur. It is attractive to use this equation for $P_\mu(x, y)$ and $Q_\mu(x, y)$ in order to avoid the recursion of the Bessel functions. However, one needs to investigate the stability of (7.10) in more detail, and for several combinations of the parameters. We know three linearly independent solutions: $P_\mu(x, y)$, $Q_\mu(x, y)$ and the constant function (with respect to μ). This indicates that it may be stable for $Q_\mu(x, y)$ in the forward direction and for $P_\mu(x, y)$ in the backward direction.

For small and moderate values of x, y, μ the expansions in (7.1) can be used. Both series have positive terms and both series require the evaluation of one incomplete gamma function. The series for $Q_\mu(x, y)$ requires the value $Q(\mu, y)$, and the remaining terms follow from the stable recursion in (7.7) for $Q(a, x)$.

The series in (7.1) for $P_\mu(x, y)$ requires an initial value of $P(a, x)$. The recursion in (7.7) should be used in the backward direction. Let n_0 be the (smallest) number such that

$$P_\mu(x, y) \doteq e^{-x} \sum_{n=0}^{n_0} \frac{x^n}{n!} P(\mu + n, y), \qquad (7.16)$$

within the required relative accuracy. Then as starting value we need to compute $P(\mu + n_0, y)$, and the remaining values follow from the recursion in (7.7) for $P(a, x)$. To estimate n_0 we use

$$P(\mu + n, y) \sim \frac{y^{n+\mu} e^{-y}}{\Gamma(\mu + n + 1)}, \qquad \mu + n \to \infty. \qquad (7.17)$$

Table 7.1. The number n_0 is as in (7.16), or a similar series for $Q_\mu(x, y)$; $\mu = 8192$, $y = 1.05\mu$; the relative accuracy is 10^{-10}.

x/μ	n_0	$Q_\mu(x, y)$	$P_\mu(x, y)$
0.01	150	$1.984527803_{10}(-04)$	$9.998015472_{10}(-01)$
0.03	355	$4.000364970_{10}(-02)$	$9.599963503_{10}(-01)$
0.05	543	$4.985354536_{10}(-01)$	$5.014645464_{10}(-01)$
0.07	727	$9.556573418_{10}(-01)$	$4.434265825_{10}(-02)$
0.09	894	$9.996249724_{10}(-01)$	$3.750276164_{10}(-04)$
0.11	1054	$9.999997188_{10}(-01)$	$2.811864384_{10}(-07)$
0.13	1207	$1.000000000_{10}(-00)$	$1.999694515_{10}(-11)$

For obtaining relative accuracy, we need an estimate of $P_\mu(x, y)$. One can use the value of the integrand of (7.5) at $t = y$, that is,

$$P_\mu(x, y) \sim \left(\frac{y}{x}\right)^{\frac{1}{2}\mu} e^{-x-y} I_\mu(\xi), \quad \xi = 2\sqrt{xy}, \tag{7.18}$$

and we replace $I_\mu(\xi)$ by the dominant part of the uniform asymptotic approximation (Abramowitz and Stegun 1964, equation 9.7.7). That is, we replace $I_\mu(\xi)$ by $e^{\mu\eta}$, where

$$\eta = \coth\gamma - \gamma, \quad \gamma = \operatorname{arcsinh}\frac{\mu}{\xi}. \tag{7.19}$$

Combining these estimates, and using the dominant factor in Stirling's approximation of the gamma function, we infer that the equation

$$\frac{e^{-x}x^n P(\mu + n, y)}{n! P_\mu(x, y)} = \varepsilon \tag{7.20}$$

can be replaced by the equation

$$n\log\frac{n}{ex} + (\mu + n)\log\frac{\mu + n}{ey} + \mu\left(\tfrac{1}{2}\log(y/x) + \coth\gamma - \gamma\right) + \log\varepsilon = 0. \tag{7.21}$$

The left-hand side assumes a minimal value at $n = \frac{1}{2}\xi\exp(-\gamma)$. A Newton process (a safe starting value is $\xi\exp(-\gamma)$) gives the desired values of n, which is taken as the number n_0 in (7.16).

Table 7.1 shows the number of terms n_0 used in the series in (7.16), or a similar series for $Q_\mu(x, y)$, for several values of x. In all cases $\mu = 8192$, $y = 1.05\mu$. These numbers are as in Robertson (1969, Table I). For larger values of the parameters the computation can be based on asymptotic expansions, in particular when $y \sim x + \mu$. For these expansions we refer to Temme (1993, 1996).

7.2. A series of modified Bessel function

In the well-known Fourier series (see Abramowitz and Stegun (1964, equation 9.6.34))

$$e^{z \cos \theta} = I_0(z) + 2 \sum_{n=1}^{\infty} I_n(z) \cos n\theta \qquad (7.22)$$

the Bessel coefficients are large when $\mathrm{Re}\, z$ is large and positive. We have

$$I_\nu(z) \sim \frac{e^z}{\sqrt{2\pi z}}, \qquad \mathrm{Re}\, z \to +\infty. \qquad (7.23)$$

When $\cos \theta < 0$, the left-hand side in (7.22) is very small when $\mathrm{Re}\, z \to +\infty$. Hence a lot of cancellations occur in the series, and from a numerical point of view the summation of the series is a very unstable process. The same happens in the simpler expansion $e^x = \sum_{n=0}^{n} x^n/n!$ when x is a large negative number.

In a more general Fourier series

$$a_0 I_0(x) + 2 \sum_{n=1}^{\infty} a_n I_n(x) \cos n\theta, \qquad (7.24)$$

where a_n is slowly varying when n is large, the same instabilities may occur, and in the numerical evaluation for large positive x and $\cos \theta < 0$ of such a series serious problems arise.

We have met such a series when solving a singular perturbation problem inside a circle. The equation is a second-order elliptic equation with a small parameter multiplying the Laplace operator and reads

$$\varepsilon \left(\frac{\partial^2 \Phi}{\partial x^2} + \frac{\partial^2 \Phi}{\partial y^2} \right) - \frac{\partial \Phi}{\partial y} = 1, \quad x^2 + y^2 < 1, \qquad (7.25)$$

with boundary condition

$$\Phi(\cos \theta, \sin \theta) = 0 \qquad (7.26)$$

on the boundary of the circle $r = 1$. We use polar coordinates

$$x = r \cos \theta, \qquad y = r \sin \theta. \qquad (7.27)$$

By separating the variables, using the differential equation of the modified Bessel function, and the series in (7.22) for fitting the boundary value, we obtain

$$\Phi(x, y) = -y - e^{\omega r \sin \theta} \sum_{n=-\infty}^{\infty} \frac{I_n'(\omega)}{I_n(\omega)} I_n(\omega r) \cos n(\theta + \pi/2), \qquad (7.28)$$

where $\omega = \frac{1}{2\varepsilon}$.

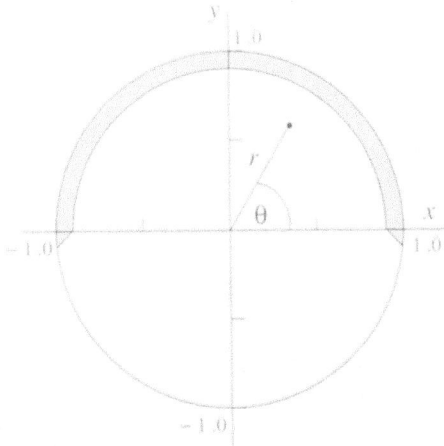

Figure 7.1. Boundary layer inside the circle along the
upper boundary $r = 1, y > 0$ and near the points $(\pm 1, 0)$.

When $\varepsilon \to 0$ the second-order elliptic operator in (7.25) reduces (in the
limit $\varepsilon = 0$) to a first-order operator. The solution of the reduced equation
cannot satisfy the boundary condition on the whole circle. For small values
of ε the solution behaves like (the first-order operator in (7.25) has the
upper hand)

$$\Phi(x, y) \sim w_0(x, y), \qquad w_0(x, y) = -y - \sqrt{1 - x^2}, \qquad (7.29)$$

which indeed solves the first-order part of the equation (7.25), and fits the
boundary condition at the lower side of the circle, but not at the upper
side. There is a sudden change from regular behaviour inside the circle to
steep behaviour at the upper side of the circle. At this part of the circle a
boundary layer occurs: see Figure 7.1. The boundary layer occurs at the
upper part of the circle, and not at the lower part of it. This is because the
linear operator in (7.25) has a minus sign.

It is quite easy to solve the singular perturbation problem in the lower
part of the disk by using the singular perturbation method (*cf.* for instance
Eckhaus and de Jager (1966)) When we substitute the formal series

$$\Phi(x, y) \sim \sum_{n=0}^{\infty} \varepsilon^n w_n(x, y) \qquad (7.30)$$

into (7.25) and equate equal powers of ε, we find

$$\frac{\partial w_0(x, y)}{\partial y} = -1, \qquad \frac{\partial w_n(x, y)}{\partial y} = \Delta w_{n-1}(x, y), \quad n = 1, 2, \ldots, \qquad (7.31)$$

and all w_n should vanish on the lower part of the unit circle. This gives w_0

as in (7.29) and

$$w_n(x, y) = \int_{-\sqrt{1-x^2}}^{y} \Delta w_{n-1}(x, \eta) \, d\eta, \quad n = 1, 2, \ldots. \quad (7.32)$$

It is easily verified that

$$w_1(x, y) = \frac{y + R}{R^3}, \quad w_2(x, y) = \frac{y + R}{2R^7}(3y + 12yx^2 + R), \quad (7.33)$$

where $R = \sqrt{1 - x^2}$. We observe that these w_n become singular at the points $(\pm 1, 0)$ and that they do not satisfy the boundary condition $w_n = 0$ on the upper part of the unit circle. An expansion like (7.30) is called an outer expansion, because it is valid outside the boundary layer.

To satisfy the boundary conditions along the upper part of the unit circle the inner expansion is constructed with so-called boundary layer terms. These functions have the property of being of order $\mathcal{O}(\varepsilon^n)$ for all n everywhere inside the unit circle, except for a small neighbourhood of the upper part of the circle. Following the construction of the boundary layer term given in Eckhaus and de Jager (1966), we can write in first approximation

$$\Phi(x, y) = -y - \sqrt{1 - x^2} + 2\sin\theta \, e^{-\frac{1}{\varepsilon}(1-r)\sin\theta} \, \psi(x, y) + z_0(x, y, \varepsilon), \quad (7.34)$$

where $z_0(x, y, \varepsilon) = \mathcal{O}(\varepsilon)$, uniformly inside the unit disk, with the exception of small neighbourhoods of the points $(\pm 1, 0)$. The function ψ is a C^∞-function, a smoothing factor, on the disk, which equals unity on a neighbourhood of the upper part of the circle, say the domain given by $\frac{2}{3} < r \leq 1, y > y_0$, where y_0 is a fixed positive small number, and ψ vanishes in the lower part of the disk.

It is not possible to describe with simple expansions the behaviour of the solution $\Phi(x, y)$ near the points $(\pm 1, 0)$. At these points the characteristics of the linear operator touch the boundary of the domain, and it is known that the boundary layers near these points are rather complicated; see Grasman (1971) and Eckhaus and de Jager (1966).

Our interest in this type of simple equations is the following.

(1) Can we find the outer expansion given in (7.30) directly from the exact solution given in (7.28)? Observe that it is quite simple to construct this expansion by using perturbation analysis.

(2) Can we find the inner expansion directly from the exact solution that holds near the upper part of the unit circle? Although it is more complicated than for the outer expansion, the construction of the inner expansion is quite straightforward by using singular perturbation analysis.

(3) Even more challenging, can we find approximations that are valid near the points $(\pm 1, 0)$?

(4) Can we use the exact solution given in (7.28) to compute the solution in all parts of the unit disk when $\varepsilon \downarrow 0$ with high accuracy? This is of interest for research in numerical methods for singular perturbation problems when numerical algorithms for more general problems need test problems with known exact solution, and where the test problems, although being quite simple, contain essential difficult elements for numerical and asymptotic analysis.

The first two points are considered in a recent paper, Temme (2007). It turns out that the asymptotic analysis can be based on transforming the series by means of the Poisson summation formula (3.50) of which the first term is an integral with respect to the order of the modified Bessel function. By using certain asymptotic approximations of the Bessel functions we can obtain the outer expansion from this first term.

8. Software for computing special functions

- For web links to software packages for evaluating special functions we refer to the repository GAMS: Guide to Available Mathematical Software, http://gams.nist.gov/.

- In 1994 a complete survey of the available software was published: Lozier and Olver (1994). The latest update of this project appeared in December 2000.
  See http://math.nist.gov/mcsd/Reports/2001/nesf/paper.pdf.

- Many interactive systems based on computer algebra, such as MATLAB, Maple, and Mathematica, have a vast collection of special functions, for symbolic and numerical purposes.

- Mathematical libraries: NETLIB, SPECFUNC, CALGO, SLATEC, CERN, IMSL, NAG. Some are available on a commercial basis, whilst others are available free (consult the GAMS repository). See also the software published in the journals *Computer Physics Communications* and *Applied Statistics*.

- Books, usually with software: Baker (1992), Moshier (1989), Press, Teukolsky, Vetterling and Flannery (2002), Thompson (1997), Wang and Guo (1989) and Zhang and Jin (1996).

9. Concluding remarks

This paper deals with special functions that can be represented by an integral, a differential or difference equation, or by other standard ways. We focus on the functions that are useful in applications and we forget about the more exotic functions such as the Fox H-function, the Meijer G-function, and so on, which may be useful in certain analytic descriptions, but are too general to handle from a numerical point of view.

Also, my choice of the numerical aspects is rather personal, and frequently based on my experience in complex analysis and (uniform) asymptotic approximations. Many other topics are left out, such as continued fractions, best rational approximations, Padé-type approximations, sequence transformations, to name a few, not because they are not of mathematical or computational interest, but simply as a matter of selection for the present paper.

Acknowledgements

The author acknowledges financial support from the Spanish *Ministry of Education and Science* (Project MTM2004–01367).

The author thanks SIAM, the *Society for Industrial and Applied Mathematics*, for allowing this paper to share several topics, parts, and examples in common with the book entitled *Numerical Methods for Special Functions* written by Amparo Gil, Javier Segura, and the present author, to be published by SIAM in 2007.

REFERENCES

M. Abramowitz and I. A. Stegun (1964), *Handbook of Mathematical Functions with Formulas, Graphs, and Mathematical Tables*, Vol. 55 of *National Bureau of Standards Applied Mathematics* series, US Printing Office.

G. E. Andrews, R. Askey and R. Roy (1999), *Special Functions*, Vol. 71 of *Encyclopedia of Mathematics and its Applications*, Cambridge University Press, Cambridge.

L. Baker (1992), *C Mathematical Function Handbook: Programming Tools For Engineers and Scientists*, McGraw-Hill, New York.

W. G. Bickley, L. J. Comrie, J. C. P. Miller, D. H. Sadler and A. J. Thompson (1952), *Bessel Functions, Part II: Functions of Positive Integer Order*, Vol. X of British Association for the Advancement of Science, Mathematical Tables, Cambridge University Press, Cambridge.

J. M. Blair, C. A. Edwards and J. H. Johnson (1976), 'Rational Chebyshev approximations for the inverse of the error function', *Math. Comp.* **30**, 7–68.

M. Blakemore, G. A. Evans and J. Hyslop (1976), 'Comparison of some methods for evaluating infinite range oscillatory integrals', *J. Comput. Phys.* **22**, 352–376.

N. G. de Bruijn (1981), *Asymptotic Methods in Analysis*, 3rd edn, Dover, New York.

W. Bühring (1987), 'An analytic continuation of the hypergeometric series', *SIAM J. Math. Anal.* **18**, 884–889.

W. W. Clendenin (1966), 'A method for numerical calculation of Fourier integrals', *Numer. Math.* **8**, 422–436.

C. W. Clenshaw (1957), 'The numerical solution of linear differential equations in Chebyshev series', *Proc. Cambridge Philos. Soc.* **53**, 134–149.

P. J. Davis and P. Rabinowitz (1984), *Methods of Numerical Integration*, Academic Press, Orlando, FL.

A. Deaño and J. Segura (2007), 'Transitory minimal solutions of hypergeometric recursions and pseudoconvergence of associated continued fractions', *Math. Comp.* **76**, 879–901.

W. Eckhaus and E. M. de Jager (1966), 'Asymptotic solutions of singular perturbation problems for linear differential equations of elliptic type', *Arch. Rational Mech. Anal.* **23**, 26–86.

N. Eggert and J. Lund (1989), 'The trapezoidal rule for analytic functions of rapid decrease', *J. Comput. Appl. Math.* **27**, 389–406.

S. Elaydi (2005), *An Introduction to Difference Equations*, Undergraduate Texts in Mathematics, 3rd edn, Springer, New York.

D. Elliott (1998/99), 'Sigmoidal transformations and the trapezoidal rule', *J. Austral. Math. Soc. Ser. B* **40**, E77–E137 (electronic).

C. Ferreira, J. L. López and E. Pérez Sinusía (2005), 'Incomplete gamma functions for large values of their variables', *Adv. in Appl. Math.* **34**, 467–485.

L. N. G. Filon (1928), 'On a quadrature formula for trigonometric integrals', *Proc. Roy. Soc. Edinburgh* **49**, 38–47.

R. C. Forrey (1997), 'Computing the hypergeometric function', *J. Comput. Phys.* **137**, 79–100.

W. Gautschi (1967), 'Computational aspects of three-term recurrence relations', *SIAM Rev.* **9**, 24–82.

W. Gautschi (1977), 'Anomalous convergence of a continued fraction for ratios of Kummer functions', *Math. Comp.* **31**, 994–999.

W. Gautschi (2004), *Orthogonal Polynomials: Computation and Approximation*, Numerical Mathematics and Scientific Computation series, Oxford University Press, New York.

A. Gil and J. Segura (1997), 'Evaluation of Legendre functions of argument greater than one', *Comput. Phys. Comm.* **105**, 273–283.

A. Gil and J. Segura (2000), 'Evaluation of toroidal harmonics', *Comput. Phys. Comm.* **124**, 104–122.

A. Gil, J. Segura and N. M. Temme (2006), 'The ABC of hyper recursions', *J. Comput. Appl. Math.* **190**, 270–286.

A. Gil, J. Segura and N. M. Temme (2007*a*), 'Numerically satisfactory solutions of hypergeometric recursions', *Math. Comp.*, to appear.

A. Gil, J. Segura and N. M. Temme (2007*b*), *Numerical Methods for Special Functions*, SIAM, Philadelphia, PA. To appear.

E. T. Goodwin (1949), 'The evaluation of integrals of the form $\int_{-\infty}^{\infty} f(x)e^{-x^2}\,dx$', *Proc. Cambridge Philos. Soc.* **45**, 241–245.

J. Grasman (1971), *On the Birth of Boundary Layers*, No. 36 of *Mathematical Centre Tracts*, Mathematisch Centrum, Amsterdam.

S. Haber (1977), 'The tanh rule for numerical integration', *SIAM J. Numer. Anal.* **14**, 668–685.

A. Iserles (2004), 'On the numerical quadrature of highly-oscillating integrals I: Fourier transforms', *IMA J. Numer. Anal.* **24**, 365–391.

A. Iserles (2005), 'On the numerical quadrature of highly-oscillating integrals II: Irregular oscillators', *IMA J. Numer. Anal.* **25**, 25–44.

A. Iserles and S. P. Nørsett (2005), 'Efficient quadrature of highly oscillatory integrals using derivatives', *Proc. R. Soc. Lond. Ser. A, Math. Phys. Eng. Sci.* **461**, 1383–1399.

R. Kreß (1972), 'On the general Hermite cardinal interpolation', *Math. Comp.* **26**, 925–933.

R. Kress and E. Martensen (1970), 'Anwendung der Rechteckregel auf die reelle Hilbertransformation mit unendlichem Intervall', *Z. Angew. Math. Mech.* **50**, T61–T64.

H. Krumhaar (1965), 'Error estimates for Luke's approximation formulas for Bessel and Hankel functions', *Z. Angew. Math. Mech.* **45**, 245–255.

I. M. Longman (1956), 'Note on a method for computing infinite integrals of oscillatory functions', *Proc. Cambridge Philos. Soc.* **52**, 764–768.

D. W. Lozier and F. W. J. Olver (1994), Numerical evaluation of special functions, in *Mathematics of Computation 1943–1993: A Half-Century of Computational Mathematics* (Vancouver, BC, 1993), Vol. 48 of *Proc. Sympos. Appl. Math.*, AMS, Providence, RI, pp. 79–125. Updates are available at http://math.nist.gov/mcsd/Reports/2001/nesf/.

S. K. Lucas and H. A. Stone (1995), 'Evaluating infinite integrals involving Bessel functions of arbitrary order', *J. Comput. Appl. Math.* **64**, 217–231.

Y. L. Luke (1969a), *The Special Functions and their Approximations I*, Vol. 53 of *Mathematics in Science and Engineering*, Academic Press, New York.

Y. L. Luke (1969b), *The Special Functions and their Approximations II*, Vol. 53 of *Mathematics in Science and Engineering*, Academic Press, New York.

J. N. Lyness (1985), Integrating some infinite oscillating tails, in *Proc. International Conference on Computational and Applied Mathematics* (Leuven, 1984), Vol. 12/13, pp. 109–117.

J. I. Marcum (1960), 'A statistical theory of target detection by pulsed radar', *Trans. IRE* **IT-6**, 59–267.

F. Matta and A. Reichel (1971), 'Uniform computation of the error function and other related functions', *Math. Comp.* **25**, 339–344.

M. Mori (1974), 'On the superiority of the trapezoidal rule for the integration of periodic analytic functions', *Mem. Numer. Math.* **1**, 11–19.

S. L. B. Moshier (1989), *Methods and Programs for Mathematical Functions*, Ellis Horwood series: Mathematics and its Applications, Ellis Horwood, Chichester.

A. Murli and M. Rizzardi (1990), 'Algorithm 682: Talbot's method for the Laplace inversion problem', *ACM Trans. Math. Software* **16**, 158–168.

F. W. J. Olver (1967), 'Numerical solution of second-order linear difference equations', *J. Res. Nat. Bur. Standards Sect. B* **71B**, 111–129.

F. W. J. Olver (1997), *Asymptotics and Special Functions*, AKP Classics, A. K. Peters, Wellesley, MA. Reprint of the 1974 original (Academic Press, New York).

R. B. Paris (2002), 'A uniform asymptotic expansion for the incomplete gamma function', *J. Comput. Appl. Math.* **148**, 323–339.

T. N. L. Patterson (1976/77), 'On high precision methods for the evaluation of Fourier integrals with finite and infinite limits', *Numer. Math.* **27**, 41–52.

W. H. Press, S. A. Teukolsky, W. T. Vetterling and B. P. Flannery (2002), *Numerical Recipes in C++: The Art of Scientific Computing*, 2nd edn, Cambridge University Press, Cambridge.

S. O. Rice (1973), 'Efficient evaluation of integrals of analytic functions by the trapezoidal rule', *Bell System Tech. J.* **52**, 707–722.

M. Rizzardi (1995), 'A modification of Talbot's method for the simultaneous approximation of several values of the inverse Laplace transform', *ACM Trans. Math. Software* **21**, 347–371.

G. Robertson (1969), 'Computation of the noncentral chi-square distribution', *Bell. System Tech. J.* **48**, 201–207.

W. Squire (1976a), 'An efficient iterative method for numerical evaluation of integrals over a semi-infinite range', *Internat. J. Numer. Methods Engrg.* **10**, 478–484.

W. Squire (1976b), 'A quadrature method for finite intervals', *Internat. J. Numer. Methods Engrg.* **11**, 708–712.

A. Strecok (1968), 'On the calculation of the inverse of the error function', *Math. Comp.* **22**, 144–158.

H. Takahasi and M. Mori (1970), 'Error estimation in the numerical integration of analytic functions', *Rep. Comput. Centre Univ. Tokyo* **3**, 41–108.

H. Takahasi and M. Mori (1971), 'Estimation of errors in the numerical quadrature of analytic functions.', *Applicable Anal.* **1**, 201–229.

H. Takahasi and M. Mori (1973/74), 'Quadrature formulas obtained by variable transformation', *Numer. Math.* **21**, 206–219.

A. Talbot (1979), 'The accurate numerical inversion of Laplace transforms', *J. Inst. Math. Appl.* **23**, 97–120.

N. M. Temme (1982), 'The uniform asymptotic expansion of a class of integrals related to cumulative distribution functions', *SIAM J. Math. Anal.* **13**, 239–253.

N. M. Temme (1992a), 'Asymptotic inversion of incomplete gamma functions', *Math. Comp.* **58**, 755–764.

N. M. Temme (1992b), 'Asymptotic inversion of the incomplete beta function', *J. Comput. Appl. Math.* **41**, 145–157.

N. M. Temme (1993), 'Asymptotic and numerical aspects of the noncentral chi-square distribution', *Comput. Math. Appl.* **25**, 55–63.

N. M. Temme (1996), *Special Functions: An Introduction to the Classical Functions of Mathematical Physics*, Wiley-Interscience, New York.

N. M. Temme (1997), 'Numerical algorithms for uniform Airy-type asymptotic expansions', *Numer. Algorithms* **15**, 207–225.

N. M. Temme (2007), 'Analytical methods for an elliptic singular perturbation problem in a circle', *J. Comput. Appl. Math.*, to appear.

W. Thompson (1997), *An Atlas for Computing Mathematical Functions: An Illustrated Guide for Practitioners, with program in Fortran 90 and Mathematica*, Wiley-Interscience, New York.

L. N. Trefethen, J. Weideman and T. Schmelzer (2005), Talbot quadrature and rational approximations. Technical report, Oxford University Computing Laboratory, Numerical Analysis Group.

R. Vidunas and N. M. Temme (2002), 'Symbolic evaluation of coefficients in Airy-type asymptotic expansions', *J. Math. Anal. Appl.* **269**, 317–331.

Z. X. Wang and D. R. Guo (1989), *Special Functions*, World Scientific, Teaneck, NJ. Translated from the Chinese by Guo and X. J. Xia.

J. A. C. Weideman (2002), 'Numerical integration of periodic functions: A few examples', *Amer. Math. Monthly* **109**, 21–36.

J. A. C. Weideman (2005), Optimizing Talbot's contours for the inversion of the Laplace transform. Technical Report NA 05/05, Oxford University Computing Laboratory.

E. J. Weniger (1989), 'Nonlinear sequence transformations for the acceleration of convergence and the summation of divergent series', *Computer Physics Reports* **10**, 189–371.

J. Wimp (1984), *Computation with Recurrence Relations*, Applicable Mathematics series, Pitman (Advanced Publishing Program), Boston, MA.

R. Wong (1982), 'Quadrature formulas for oscillatory integral transforms', *Numer. Math.* **39**, 351–360.

R. Wong (2001), *Asymptotic Approximations of Integrals*, Vol. 34 of *Classics in Applied Mathematics*, SIAM, Philadelphia, PA. Corrected reprint of the 1989 original.

R. Wong and H. Li (1992a), 'Asymptotic expansions for second-order linear difference equations', *J. Comput. Appl. Math.* **41**, 65–94.

R. Wong and H. Li (1992b), 'Asymptotic expansions for second-order linear difference equations II', *Stud. Appl. Math.* **87**, 289–324.

S. Zhang and J. Jin (1996), *Computation of Special Functions*, Wiley-Interscience, New York.